COMPLEXITY IN NUMERICAL OPTIMIZATION

COMPLEXITY IN NUMERICAL OPTIMIZATION

Editor

Panos M. Pardalos

Department of Industrial and Systems Engineering
University of Florida and
Technical University of Crete

World Scientific
Singapore • New Jersey • London • Hong Kong

Published by

World Scientific Publishing Co. Pte. Ltd.
P O Box 128, Farrer Road, Singapore 9128
USA office: Suite 1B, 1060 Main Street, River Edge, NJ 07661
UK office: 73 Lynton Mead, Totteridge, London N20 8DH

ISBN 981-02-1415-4

Printed in Singapore by Utopia Press.

Preface

Computational complexity, originated from the interactions between computer science and numerical optimization, is one of the major theories that have revolutionized the approach to solving optimization problems and to analyzing their intrinsic difficulty. The main focus of complexity is the study of whether existing algorithms are efficient for the solution of problems, and which problems are likely to be tractable. The quest for developing efficient algorithms leads also to elegant general approaches for solving optimization problems, and reveals surprising connections among problems and their solutions.

This book is a collection of articles on recent complexity developments in numerical optimization. The topics covered include complexity of approximation algorithms, new polynomial time algorithms for convex quadratic minimization, interior point algorithms, complexity issues regarding test generation of NP-hard problems, complexity of scheduling problems, min-max, fractional combinatorial optimization, fixed point computations, and network flow problems.

The collection of articles provide a broad spectrum of the direction in which research is going and help to elucidate the nature of computational complexity in optimization. The book will be a valuable source of information to faculty, students and researchers in numerical optimization and related areas.

I would like to take the opportunity to thank the authors of the papers, the anonymous referees, and the publisher for helping me to produce this beautiful book.

Panos M. Pardalos
University of Florida
April 1993

" He who does not expect the unexpected will not detect it" - Heraclitus

Contents

A Note on the Complexity of Fixed-Point Computation for Noncontractive Maps **448**
C. W. Tsay and K. Sikorski

A Technique for Bounding the Number of Iterations in Path Following Algorithms **462**
P. M. Vaidya and D. S. Atkinson

Polynomial Time Weak Approximation Algorithms for Quadratic Programming **490**
S. A. Vavasis

Complexity Results for a Class of Min-Max Problems with Robust Optimization Applications **501**
G. Yu and P. Kouvelis

Complexity in Numerical Optimization, pp. 1-15
P.M. Pardalos, Editor
©1993 World Scientific Publishing Co.

Average Performance of a Self–Dual Interior Point Algorithm for Linear Programming

Kurt M. Anstreicher[1]
*Department of Management Sciences, University of Iowa,
Iowa City, IA 52242 USA*

Jun Ji[1]
*Department of Mathematics, University of Iowa,
Iowa City, IA 52242 USA*

Florian A. Potra[1]
*Department of Mathematics, University of Iowa,
Iowa City, IA 52242 USA*

Yinyu Ye[2]
*Department of Management Sciences, University of Iowa,
Iowa City, IA 52242 USA*

Abstract

We consider the self–dual algorithm of Ye, Todd and Mizuno, with a modified finite termination scheme, applied to random linear programs generated according to a model of Todd. Such problems have degenerate optimal solutions, and possess no feasible starting point. Our main result is that the expected number of iterations before termination with an exact optimal solution is bounded above by $O(\sqrt{n}\ln n)$. In terms of generality of the probabilistic model, computational effort per iteration, and overall bound on the expected number of iterations, our analysis provides the best probabilistic result to date for an interior point algorithm.

Keywords: Linear Programming, Average Case Behavior, Interior Point Algorithm, Self–Dual Algorithm.

---

[1]Supported by an Interdisciplinary Research Grant from the Center for Advanced Studies, University of Iowa.
[2]Supported by NSF Grant DDM-8922636.

1 Introduction

This paper continues a series of recent works regarding the average complexity of interior point algorithms for linear programming (LP). The idea of average case analysis is to obtain rigorous probabilistic bounds on the number of iterations required by an algorithm to reach some termination criterion, when the algorithm is applied to a random instance of a problem drawn from some probability distribution. In the case of the simplex method for LP, average case analysis (see for example [1], [4], [6], [16], and [17]) has provided some theoretical justification for the observed practical efficiency of the method, despite its exponential worst case bound. Although many interior point algorithms devised in the last several years are polynomial time methods, in practice they too generally perform much better than worst case bounds would indicate, a "gap" between theory and practice that average case analysis might (at least partially) close.

The first rigorous average case analysis for an interior point method was provided by Ye [20]. Earlier efforts (see for example [12] and [13]) considered the "average" performance on one step of an algorithm, but do not provide a rigorous bound because the steps cannot be assumed to be independent of one another. Ye [20] proves a "high probability" bound of $O(\sqrt{n}\ln n)$ iterations for termination, using a variety of algorithms applied to several different probabilistic models. Here n is the number of variables in a standard form problem, and "high probability" means that the probability that the bound holds goes to one, as $n \to \infty$. The analysis of [20] uses the finite termination scheme of [19] and [10] to produce an exact optimal solution for the problem being solved. Subsequently [7] obtained an $O(\sqrt{n}\ln n)$ bound for the expected number of iterations before termination, using the same finite termination criterion as [20], and a particular random LP model (Model I) from Todd [18]. Model I produces linear programs with initial feasible primal and dual interior solutions, and is nondegenerate with probability one. Next, [3] obtained expected value, and high probability, bounds of $O(\sqrt{n}\ln n)$ iterations for termination with an exact solution, using Model I and a different ellipsoid–based termination criterion. Expected value and high probability bounds of $O(\sqrt{n}\ln n) + \log_2(|\ln \epsilon|)$ iterations to obtain an ϵ-optimal solution, using the predictor–corrector algorithm of [12] applied to Model I, were obtained in [8].

A more general model, from Section 4 of [18], allows for degenerate optimal solutions, and does not provide a feasible starting point. We refer to this latter model as "Todd's degenerate model." The lack of an initial solution in the degenerate model is problematic for many interior point algorithms, which require an interior solution to start from. The high probability analysis of [20] successfully handles the degenerate model by forming a combined primal–dual feasibility problem, which unfortunately increases the computational effort on each iteration substantially. An expected value result for the degenerate model, which also avoids working with the larger primal–dual system, was obtained in [2]. The analysis of [2], based on Zhang's [22] complexity analysis of an infeasible primal–dual algorithm proposed in [9], obtains a bound of $O(n^2 \ln n)$ iterations for the expected number of iterations before termination with an exact optimal solution. Termination in [2] is based on an infeasible extension of the projection technique from [19]. The results of [9] and [22] stimulated a number of recent papers which obtain complexity results for various infeasible primal–dual algorithms; see for example [11], [14], [15], and [21]. The result of [21], which is based on a cleverly constructed homogenous "self–dual" feasibility problem, has many attractive features. In particular, the self–dual property allows for the use of a primal–dual system with no increase in the work per iteration. In addition, a variety of algorithms can be applied to the self–dual problem ([21] uses the predictor–corrector algorithm of [12]) to obtain $O(\sqrt{n}L)$ methods which apply to any linear program, with no assumed feasible point and no regularity assumptions whatsoever. Here L is the bit length for a standard form problem with all integer data.

In this paper we obtain a bound of $O(\sqrt{n} \ln n)$ iterations for the expected number of iterations before termination with an exact optimal solution, using the self–dual algorithm of [21], applied to the degenerate model from Section 4 of [18]. Our termination scheme, which is again based on [19], improves on the scheme suggested in [21] by avoiding computations that use the combined primal–dual system. We make considerable use of the analysis of the degenerate model developed in [2]. In terms of generality of the probabilistic model, computational effort per iteration, and overall bound on the expected number of iterations, our analysis provides the best probabilistic result to date for an interior point algorithm.

2 The HLP problem and the self–dual algorithm

In this section, we briefly describe a homogeneous, self–dual artificial linear program and the self–dual linear programming algorithm of Ye-Todd-Mizuno [21]. Consider the primal and dual linear programs:

$$\text{(LP)} \qquad \min \quad \{c^T x : Ax = b,\ x \geq 0\},$$
$$\text{(LD)} \qquad \max \quad \{b^T y : A^T y + s = c,\ s \geq 0\},$$

where A is an $m \times n$ real matrix with independent rows, $c \in \mathbf{R}^n$, and $b \in \mathbf{R}^m$. Instead of solving (LP) and (LD) directly, [21] introduces a homogeneous and self–dual artificial linear program (HLP):

$$
\begin{aligned}
\text{(HLP)} \qquad &\min((x^0)^T s^0 + 1)\theta \\
\text{s.t.} \qquad & Ax - b\tau + \bar{b}\theta && = 0, &&(1)\\
& -A^T y && + c\tau - \bar{c}\theta - s && = 0, &&(2)\\
& b^T y - c^T x && + \bar{z}\theta - \kappa && = 0, &&(3)\\
& -\bar{b}^T y + \bar{c}^T x - \bar{z}\tau && && = -(x^0)^T s^0 - 1, &&(4)\\
& x \geq 0,\ \tau \geq 0,\ s \geq 0,\ \kappa \geq 0,
\end{aligned}
$$

where $x^0 > 0$, $s^0 > 0$, y^0 are given, and

$$
\bar{b} = b - Ax^0, \quad \bar{c} = c - A^T y^0 - s^0, \quad \bar{z} = c^T x^0 + 1 - b^T y^0.
$$

Here $\bar{b}$, $\bar{c}$ and $\bar{z}$ represent the "infeasibility" of the initial primal and dual points, and the initial primal-dual gap, respectively. Denote by F_h the set of all points that are feasible for (HLP). Denote by F_h^0 the set of strictly feasible (interior) points in F_h, with $(x, \tau, s, \kappa) > 0$. It is easily seen that (HLP) has a strictly feasible point: $y = y^0$, $x = x^0 > 0$, $\tau = 1$, $\theta = 1$, $s = s^0 > 0$, $\kappa = 1$. Premultiplying (1) by y^T, (2) by x^T, (3) by τ and (4) by θ, respectively, and adding them up leads to

$$
((x^0)^T s^0 + 1)\theta = x^T s + \tau\kappa. \tag{5}
$$

The problem (HLP) is formulated so that the dual of (HLP) has exactly the same form as (HLP), and the optimal value of (HLP) is zero. A major component in [21] is the following result (see Theorems 1 and 4 of [21]).

Proposition 2.1 *There is a strictly self-complementary solution to (HLP), i.e., there is an optimal solution to (HLP) such that $\theta^* = 0$, $x^* + s^* > 0$ and $\tau^* + \kappa^* > 0$. Let $(y^*, x^*, \tau^*, \theta^*, s^*, \kappa^*)$ be a strictly self-complementary solution to (HLP). Then*

(1) *(LP) has a solution if and only if $\tau^* > 0$. In this case, x^*/τ^* is an optimal solution for (LP), and $(y^*/\tau^*, s^*/\tau^*)$ is an optimal solution for (LD);*

(2) *If $\tau^* = 0$ then $\kappa^* > 0$, which implies that $c^T x^* - b^T y^* < 0$. If $c^T x^* < 0$ then (LD) is infeasible; if $-b^T y^* < 0$ then (LP) is infeasible; and if both $c^T x^* < 0$ and $-b^T y^* < 0$ then (LP) and (LD) are both infeasible.*

From Proposition 2.1, it is clear that the key to solving (LP), or alternatively detecting its infeasibility or unboundedness, is to find a strictly self-complementary solution to (HLP). According to Güler and Ye [5], many interior point algorithms might be used to solve (HLP), since they generate a sequence or subsequence of feasible pairs which converge to a strictly complementary optimal solution of the problem being solved. In fact, in [21] the Mizuno-Todd-Ye predictor-corrector algorithm [12] is applied to solving (HLP) by following the central path of (HLP) (as defined by Theorem 2 of [21]). In what follows, we simply choose

$$y^0 = 0, \ x^0 = e, \ \tau^0 = 1, \ \theta^0 = 1, \ s^0 = e, \ \kappa^0 = 1, \tag{6}$$

where $e \in \mathbf{R}^n$ is the vector of ones, and then have

$$\bar{b} = b - Ae, \ \bar{c} = c - e, \ \bar{z} = c^T e + 1, \ \text{and} \ (x^0)^T s^0 + 1 = n + 1.$$

We can define a neighborhood of the central path as

$$N(\beta) = \left\{ (y, x, \tau, \theta, s, \kappa) \in F_h^0 : \left\| \begin{pmatrix} Xs \\ \tau\kappa \end{pmatrix} - \mu e \right\| \leq \beta\mu, \ \text{where} \ \mu = \frac{x^T s + \tau\kappa}{n+1} \right\},$$

for $\beta \in (0, 1)$. Note that from (5) we have $\theta = \mu$ for any feasible point.

Predictor-Corrector Algorithm

Given an interior feasible point $(y, x, \tau, \theta, s, \kappa) \in F_h^0$, the predictor-corrector algorithm solves the system of linear equations in $(d_y, d_x, d_\tau, d_\theta, d_s, d_\kappa)$:

$$\begin{pmatrix} Xd_s + Sd_x \\ \tau d_\kappa + \kappa d_\tau \end{pmatrix} = \gamma\mu e - \begin{pmatrix} Xs \\ \tau\kappa \end{pmatrix}, \tag{7}$$

$$(d_y, d_x, d_\tau, d_\theta, d_s, d_\kappa) \in Q, \tag{8}$$

where $\gamma \in \mathbf{R}$, and Q denotes the null space of the coefficient matrix in (HLP).

Predictor Step. Given $(y^k, x^k, \tau^k, \theta^k, s^k, \kappa^k) \in N(\beta)$ with $\beta < 1/2$, solve the system (7)–(8) with $(y, x, \tau, \theta, s, \kappa) = (y^k, x^k, \tau^k, \theta^k, s^k, \kappa^k)$ and $\gamma = 0$. Denote the solution $(d_y^p, d_x^p, d_\tau^p, d_\theta^p, d_s^p, d_\kappa^p)$, and set:

$$y(\alpha) = y^k + \alpha d_y^p, \quad x(\alpha) = x^k + \alpha d_x^p, \quad \tau(\alpha) = \tau^k + \alpha d_\tau^p,$$
$$\theta(\alpha) = \theta^k + \alpha d_\theta^p, \quad s(\alpha) = s^k + \alpha d_s^p, \quad \kappa(\alpha) = \kappa^k + \alpha d_\kappa^p,$$

where the stepsize is determined by

$$\bar{\alpha} = \max\left\{ \alpha : (y(\alpha), x(\alpha), \tau(\alpha), \theta(\alpha), s(\alpha), \kappa(\alpha)) \in N(2\beta) \right\}.$$

Corrector Step. Solve the system (7)–(8) again with $\gamma = 1$, and $(y, x, \tau, \theta, s, \kappa) = (y(\bar{\alpha}), x(\bar{\alpha}), \tau(\bar{\alpha}), \theta(\bar{\alpha}), s(\bar{\alpha}), \kappa(\bar{\alpha}))$. Let $(d_y^c, d_x^c, d_\tau^c, d_\theta^c, d_s^c, d_\kappa^c)$ be the solution, and set

$$y^{k+1} = y(\bar{\alpha}) + d_y^c, \quad x^{k+1} = x(\bar{\alpha}) + d_x^c, \quad \tau^{k+1} = \tau(\bar{\alpha}) + d_\tau^c,$$
$$\theta^{k+1} = \theta(\bar{\alpha}) + d_\theta^c, \quad s^{k+1} = s(\bar{\alpha}) + d_s^c, \quad \kappa^{k+1} = \kappa(\bar{\alpha}) + d_\kappa^c.$$

By using the analysis employed in [12], with $\beta = 1/4$, Ye et al. [21] proved that $(y^{k+1}, x^{k+1}, \tau^{k+1}, \theta^{k+1}, s^{k+1}, \kappa^{k+1}) \in N(\beta)$, so the algorithm is well defined, and

$$\frac{\theta^{k+1}}{\theta^k} = \frac{(x^{k+1})^T s^{k+1} + \tau^{k+1}\kappa^{k+1}}{(x^k)^T s^k + \tau^k\kappa^k} \leq (1 - 8^{-.25}/\sqrt{n+1}). \tag{9}$$

Coupled with a version of the termination technique described in [19] and [10], the algorithm will then generate an (exact) strictly self-complementary solution for (HLP) in at most $O(\sqrt{n}L)$ iterations.

3   A new projection scheme for finite termination

In this section we consider the problem of generating an exact optimal solution to (HLP). For simplicity, we denote

$$u = \left(\begin{array}{c} x \\ \tau \end{array} \right), \quad v = \left(\begin{array}{c} s \\ \kappa \end{array} \right).$$

To begin, let $(y^*, u^*, \theta^* = 0, v^*)$ be any strictly self-complementary solution (from Proposition 2.1 such a solution exists). Define

$$\sigma_h^* = \{i : 0 \le i \le n+1, \ u_i^* > 0\}, \quad \text{and} \quad \xi_h^* = \min_i(u_i^* + v_i^*).$$

We refer to σ_h^* as the "optimal partition" of (HLP), and clearly $\xi_h^* > 0$. Our goal here is to use the iterates (u^k, v^k) of the algorithm to eventually identify the optimal partition, and to generate an exact optimal solution of (HLP). Using the techniques developed in [20] we can prove the following result.

Lemma 3.1 *Let $\zeta = (1-\beta)\xi_h^*/(n+1)$. Then in order to obtain $v_j^k < \zeta \le u_j^k$, $j \in \sigma_h^*$, and $u_j^k < \zeta \le v_j^k$, $j \notin \sigma_h^*$, it suffices to have*

$$\theta^k < \frac{1-\beta}{(n+1)^2}(\xi_h^*)^2. \tag{10}$$

Proof. Note that $(y^k - y^*, u^k - u^*, \theta^k - \theta^*, v^k - v^*)$ is in the null space of the coefficient matrix in (HLP), which implies (see Theorem 2 of [21])

$$(u^k - u^*)^T(v^k - v^*) = 0,$$

i.e.,

$$\sum_{i \in \sigma_h^*} u_i^* v_i^k + \sum_{i \notin \sigma_h^*} u_i^k v_i^* = (u^k)^T v^k. \tag{11}$$

For $j \in \sigma_h^*$, we deduce from (10)–(11) that

$$\xi_h^* v_j^k \le (u^k)^T v^k = (n+1)\theta^k < \frac{1-\beta}{n+1}(\xi_h^*)^2,$$

and therefore

$$v_j^k < \frac{1-\beta}{n+1}\xi_h^*. \tag{12}$$

On the other hand, (11) implies

$$\frac{u_j^*}{u_j^k} u_j^k v_j^k \leq (u^k)^T v^k,$$

which further implies

$$u_j^k \geq \frac{u_j^k v_j^k}{(u^k)^T v^k} u_j^* \geq \frac{1-\beta}{n+1} \xi_h^*. \tag{13}$$

The last inequality is deduced from the fact that $(y^k, x^k, \tau^k, \theta^k, \kappa^k) \in N(\beta)$ for all $k \geq 0$. Combining (12) with (13), we get $v_j^k < \zeta \leq u_j^k$ for all $j \in \sigma_h^*$. The argument for $j \notin \sigma_h^*$ is similar. $\qquad\square$

Given an iterate (u^k, v^k), let $B = B^k$ denote the columns of A having $u_j^k > v_j^k$, and let u_B denote the corresponding components of u. Similarly let N and v_N denote the remaining columns of A, and the corresponding components of v. Also, we use $\mathcal{B}$ and $\mathcal{N}$ to denote the corresponding index sets. Note that (9) implies that $\theta^k \to 0$, so Lemma (3.1) implies that eventually we always have $\mathcal{B} = \sigma_h^* \setminus \{n+1\}$. In what follows, we assume that k is in fact large enough so that $\mathcal{B} = \sigma_h^* \setminus \{n+1\}$. Instead of using the projection scheme proposed in [21], we employ the following projections to generate an exact optimal solution. We distinguish two cases:

Case 1. $(u_{n+1}^k > v_{n+1}^k)$. Find the solution $\hat{u}_B^k$ of

$$\text{(PP1)} \qquad \begin{aligned} \min \quad & \|u_B - u_B^k\| \\ \text{s.t.} \quad & Bu_B = bu_{n+1}^k. \end{aligned}$$

If $\hat{u}_B^k \geq 0$, then compute the solution $\hat{y}^k$ of

$$\text{(DP1)} \qquad \begin{aligned} \min \quad & \|y - y^k\| \\ \text{s.t.} \quad & B^T y = c_B u_{n+1}^k, \end{aligned}$$

and set $\hat{v}_N^k = c_N u_{n+1}^k - N^T \hat{y}^k$.

Case 2. $(u_{n+1}^k \leq v_{n+1}^k)$. Find the solution $\hat{u}_B^k$ of

$$\text{(PP2)} \qquad \begin{aligned} \min \quad & \|u_B - u_B^k\| \\ \text{s.t.} \quad & Bu_B = 0. \end{aligned}$$

If $\hat{u}_B^k \geq 0$, then compute the solution $\hat{y}^k$ of

$$\text{(DP2)} \qquad \begin{aligned} \min \quad & \|y - y^k\| \\ \text{s.t.} \quad & B^T y = 0, \end{aligned}$$

and set $\hat{v}_N^k = -N^T \hat{y}^k$, and $\hat{v}_{n+1}^k = b^T \hat{y}^k - c_B^T \hat{u}_B^k$.

According to Lemma 3.1, exactly one of the above cases occurs for all sufficiently large k. Also, Case 1 eventually occurs exactly when (LP) has an optimal solution, in which case (PP1) and (DP1) are both clearly feasible. (Note that (PP2) and (DP2) are trivially always feasible.) We remark that the projections used here are more efficient than those used in [21] because we avoid doubling the dimension and, in both cases, if (PPi) fails then (DPi) is not performed ($i = 1, 2$). In addition, our projection scheme is better suited for the probabilistic analysis below. It is easily seen from the definition of (HLP) that:
(PP1) is equivalent to

$$
\begin{aligned}
\min \quad & \|u_B - u_B^k\| \\
\text{s.t.} \quad & B(u_B - u_B^k) = Nu_N^k + \bar{b}\theta^k;
\end{aligned}
\tag{14}
$$

(DP1) is equivalent to

$$
\begin{aligned}
\min \quad & \|y - y^k\| \\
\text{s.t.} \quad & B^T(y - y^k) = \bar{c}_B\theta^k + v_B^k;
\end{aligned}
\tag{15}
$$

(PP2) is equivalent to

$$
\begin{aligned}
\min \quad & \|u_B - u_B^k\| \\
\text{s.t.} \quad & B(u_B - u_B^k) = Nu_N^k - bu_{n+1}^k + \bar{b}\theta^k;
\end{aligned}
\tag{16}
$$

(DP2) is equivalent to

$$
\begin{aligned}
\min \quad & \|y - y^k\| \\
\text{s.t.} \quad & B^T(y - y^k) = -c_Bu_{n+1}^k + \bar{c}_B\theta^k + v_B^k.
\end{aligned}
\tag{17}
$$

Also, in Case 1 we can write

$$
\hat{v}_N^k - v_N^k = N^T(y^k - \hat{y}^k) + \bar{c}_N\theta^k,
\tag{18}
$$

while in Case 2 we have

$$
\begin{aligned}
\hat{v}_N^k - v_N^k &= N^T(y^k - \hat{y}^k) - c_Nu_{n+1}^k + \bar{c}_N\theta^k, \tag{19} \\
\hat{v}_{n+1}^k - v_{n+1}^k &= b^T(\hat{y}^k - y^k) - c_B^T(\hat{u}_B^k - u_B^k) + c_N^Tu_N^k - \bar{z}\theta^k. \tag{20}
\end{aligned}
$$

From (11), and $\theta^k \to 0$, we conclude that $(u_N^k, v_B^k) \to 0$ as $k \to \infty$, and also $u_{n+1}^k \to 0$ if $\{n+1\} \notin \sigma_h^*$. Using these facts and (14)–(20) we can easily deduce that

$$
(\hat{u}_B^k - u_B^k) \to 0 \text{ and } (\hat{v}_N^k - v_N^k) \to 0 \quad \text{as} \quad k \to \infty,
\tag{21}
$$

and in addition, in Case 2,

$$
(\hat{v}_{n+1}^k - v_{n+1}^k) \to 0 \quad \text{as} \quad k \to \infty.
\tag{22}
$$

From (21)–(22) and Lemma 3.1 it follows that we have, in both cases, $(\hat{u}_B^k, \hat{v}_N^k) > 0$, and, in Case 2, $\hat{v}_{n+1}^k > 0$, if k is large enough. Then, by proper scaling, a strictly complementary solution to (HLP) can be recovered.

The above discussion shows that our projection scheme works provided k is large enough. Below we give a more precise characterization of this fact, for Case 1. For our probabilistic model, to be described in Section 4, only Case 1 occurs provided k is large enough. Therefore in what follows we assume that k is large enough and that we are always in Case 1. Let $\hat{B}$ be any set of rows of B having maximal rank ($\hat{B} = B$ if the rows of B are independent). Let $\hat{N}$, $\hat{A}$, $\hat{b}$, and $\check{b}$ denote the corresponding rows of N and A, and components of b and $\bar{b}$, respectively. Let B_1 denote any square, nonsingular submatrix of $\hat{B}$, and let $\sigma_1^* \subset \sigma_h^*$ be the corresponding index set. Let B_2 denote any remaining columns of $\hat{B}$, and let $\sigma_2^* \subset \sigma_h^*$ be the corresponding index set. Finally let u_1^* denote the components of u^* corresponding to σ_1^*, and u_2^* denote the components of u^* corresponding to σ_2^*. We denote $\{1, 2, \cdots, n\} \setminus \sigma_1^*$ by $\bar{\sigma}^*$.

Lemma 3.2 *Assume that $u_{n+1}^* > 0$. Then in Case 1 the solution of (PP1) generates $\hat{u}_B^k > 0$ whenever*

$$\theta^k \leq \frac{(1 - \beta)(\xi_h^*)^2}{(n + 1)^2 (1 + \rho_h^*)^2 (1 + \sum_{j \in \bar{\sigma}^*} \|B_1^{-1} \hat{a}_j\|)}, \tag{23}$$

where $\rho_h^* = \|u^* + v^*\|$, and $\hat{a}_j$ is the j'th column of $\hat{A}$.

Proof. Assume that (23) holds. Since (23) implies (10), $\mathcal{B} \cup \{n + 1\}$ must be the "optimal partition" σ_h^*, by Lemma 3.1. From (14), (PP1) is clearly equivalent to

$$\begin{aligned} \min \quad & \|u_B - u_B^k\| \\ \text{s.t.} \quad & \hat{B}(u_B - u_B^k) = \hat{N} u_N^k + \check{b}\theta^k. \end{aligned} \tag{24}$$

The solution $\hat{u}_B^k$ of (24) satisfies

$$\hat{u}_B^k - u_B^k = \hat{B}^T (\hat{B}\hat{B}^T)^{-1} \left(\hat{N} u_N^k + \check{b}\theta^k\right),$$

and $\xi^T (\hat{B}\hat{B}^T)^{-1}\xi \leq \xi^T (B_1 B_1^T)^{-1}\xi = \|B_1^{-1}\xi\|^2$ for all ξ then implies

$$\|\hat{u}_B^k - u_B^k\| \leq \|B_1^{-1}\hat{N} u_N^k\| + \|B_1^{-1}\check{b}\|\theta^k. \tag{25}$$

For the first term in (25), we have

$$\begin{aligned} \|B_1^{-1}\hat{N} u_N^k\| &\leq \max_{j \in \mathcal{N}} u_j^k \sum_{j \in \mathcal{N}} \|B_1^{-1}\hat{a}_j\| \\ &\leq \frac{(u^k)^T v^k}{\xi_h^*} \left(\sum_{j \in \mathcal{N}} \|B_1^{-1}\hat{a}_j\|\right), \end{aligned} \tag{26}$$

where the last inequality is from (11). To bound the second term of (25), we use $Bu_B^* = bu_{n+1}^*$. Since $u_{n+1}^* > 0$, by assumption, we may write

$$
\begin{aligned}
\|B_1^{-1}\breve{b}\|\theta^k \;=\;& \|B_1^{-1}(\hat{b} - \hat{A}e)\|\theta^k \\[2mm]
\leq\;& \left(\frac{1}{u_{n+1}^*}\|B_1^{-1}\hat{b}u_{n+1}^*\| + \|B_1^{-1}\hat{A}e\| \right)\theta^k \\[2mm]
\leq\;& \left(\frac{1}{u_{n+1}^*}\|B_1^{-1}(B_1 u_1^* + B_2 u_2^*)\| + \|[I|B_1^{-1}B_2]e\| + \|B_1^{-1}\hat{N}e\| \right)\theta^k \\[2mm]
\leq\;& \left(\frac{1}{\xi_h^*}(\|u_1^*\| + \|u_2^*\|\sum_{j\in\sigma_2^*}\|B_1^{-1}\hat{a}_j\|) + \sqrt{n+1} + \sum_{j\in\bar{\sigma}^*}\|B_1^{-1}\hat{a}_j\| \right)\theta^k \\[2mm]
\leq\;& \frac{1}{\xi_h^*}\left(\rho_h^*(1 + \sum_{j\in\sigma_2^*}\|B_1^{-1}\hat{a}_j\|) + \rho_h^*(1 + \frac{1}{\sqrt{n+1}}\sum_{j\in\bar{\sigma}^*}\|B_1^{-1}\hat{a}_j\|) \right)\theta^k.
\end{aligned}
$$

Substituting the above inequality, and (26), into (25) results in

$$
\begin{aligned}
\|\hat{u}_B^k - u_B^k\| \;\leq\;& \frac{2\rho_h^* + (n+1)}{\xi_h^*}(1 + \sum_{j\in\bar{\sigma}^*}\|B_1^{-1}\hat{a}_j\|)\theta^k \\[2mm]
\leq\;& \frac{n+1}{\xi_h^*}(1 + \rho_h^*)^2(1 + \sum_{j\in\bar{\sigma}^*}\|B_1^{-1}\hat{a}_j\|)\theta^k,
\end{aligned}
\tag{27}
$$

so (23), (27) and (13) imply that $\hat{u}_B^k > 0$. This completes the proof. $\qquad\square$

4   Random model and probabilistic analysis

In this section we describe a random LP model proposed in Section 4 of Todd [18], and perform a probabilistic analysis of the behavior of the algorithm given in Section 2, using the finite termination scheme of Section 3, when applied to this model. As in the Introduction, we will refer to the model under consideration as "Todd's degenerate model."

Todd's Degenerate Model. Let $A = (\hat{A}_1, \hat{A}_2, \hat{A}_3)$, where $\hat{A}_i$ is $m \times n_i$, $n_1 < m, n_1 + n_2 > m$, $n_1 + n_2 + n_3 = n$, and each component of A is i.i.d. from the $N(0,1)$ distribution. Let

$$
\hat{x} = \begin{pmatrix} \hat{x}_1 \\ 0 \\ 0 \end{pmatrix}, \qquad \hat{s} = \begin{pmatrix} 0 \\ 0 \\ \hat{s}_3 \end{pmatrix},
$$

where the components of $\hat{x}_1$ and $\hat{s}_3$ are i.i.d. from the $|N(0,1)|$ distribution. Finally let $b = A\hat{x} = \hat{A}_1\hat{x}_1, c = \hat{s} + A^T\hat{\pi}$. We assume that either $\hat{\pi} = 0$, or the components of $\hat{\pi}$ are i.i.d. from any distribution with $O(1)$ mean and variance.

Clearly this model allows for degenerate solutions, and produces instances of (LP) having no feasible starting point. This presents an obstacle for most interior point methods, which require interior feasible points for initialization. A very important feature of the self–dual algorithm is that it can start at any positive primal–dual pair, feasible or infeasible, near the central ray of the positive orthant. Since an instance of Todd's degenerate model always has an optimal solution, it follows from Proposition 2.1 that $n + 1 \in \sigma_h^*$. Therefore, if the self–dual algorithm described in Section 2 is applied to an instance of Todd's degenerate model, we are eventually always in Case 1.

Now, we begin a probabilistic analysis of the self–dual algorithm equipped with the new projection scheme described in Section 3. Since our finite termination criterion in Lemma 3.2 depends on ξ_h^*, from a strictly complementary solution (x^*, s^*) to (HLP), we must first infer a valid value of ξ_h^* from the given, but *not* strictly complementary, solution $(\hat{x}, \hat{s})$. This can be done using the following result of [20] and [2]. We use the notation $\min(\cdot)$ and $\max(\cdot)$ to denote the minimum and maximum components of a nonnegative vector, respectively. Let

$$\hat{\xi} = \min \begin{pmatrix} \hat{x}_1 \\ \hat{s}_3 \end{pmatrix}, \; \hat{\rho} = \left\| \begin{pmatrix} \hat{x}_1 \\ \hat{s}_3 \end{pmatrix} \right\|, \; \xi^* = \min(x^* + s^*), \; \rho^* = \left\| \begin{pmatrix} x^* \\ s^* \end{pmatrix} \right\|.$$

Proposition 4.1 *Consider Todd's degenerate model, with optimal solution $(\hat{x}, \hat{s})$. Then there is a strictly complementary solution (x^*, s^*) such that $\rho^* \leq (3/2)\sqrt{n}\hat{\rho}$, and $\xi^* \geq \hat{\xi} \min(\hat{\lambda})/[2\sqrt{n}\max(\hat{\lambda})]$, where the components of $\hat{\lambda} \in \mathbf{R}^n$ are i.i.d. $|N(0,1)|$, and $\hat{\lambda}$ is independent of $(\hat{x}, \hat{s})$.*

Lemma 4.2 *Consider an instance of Todd's degenerate model, and let $\hat{\lambda}$ be as in Proposition 4.1. Suppose that k is large enough so that the following inequality is satisfied:*

$$\theta^k \leq \frac{(1 - \beta)\hat{\xi}^2[\min(\hat{\lambda})]^2}{16(n + 1)^4(1 + 3\sqrt{n}\hat{\rho}/2)^6[\max(\hat{\lambda})]^2(1 + \sum_{j \in \bar{\sigma}^*} \|B_1^{-1}\hat{a}_j\|)}. \tag{28}$$

Then (PP1) generates a solution $\hat{u}_B^k > 0$, so that $\hat{x} = (\hat{x}_B, \hat{x}_N)$ solves (LP), where $\hat{x}_B = \hat{u}_B^k/u_{n+1}^k$, and $\hat{x}_N = 0$.

Proof. Let (x^*, s^*) be the strictly complementary solution constructed in the previous proposition. Then it is easy to check (see [20]) that

$$u^* = \alpha \begin{pmatrix} x^* \\ 1 \end{pmatrix}, \; v^* = \alpha \begin{pmatrix} s^* \\ 0 \end{pmatrix}, \; \text{with } \alpha = \frac{n + 1}{e^T x^* + e^T s^* + 1},$$

is a strictly complementary solution to (HLP). Note that

$$\alpha \geq \frac{n + 1}{1 + \sqrt{n}\rho^*} \geq \frac{\sqrt{n + 1}}{1 + \rho^*},$$

and

$$u^* + v^* = \alpha \begin{pmatrix} x^* + s^* \\ 1 \end{pmatrix} \geq \alpha \begin{pmatrix} \xi^* e \\ 1 \end{pmatrix} \geq \frac{\alpha \xi^*}{1 + \rho^*} \begin{pmatrix} e \\ 1 \end{pmatrix} \geq \frac{\sqrt{n+1}\,\xi^*}{(1 + \rho^*)^2} e.$$

Therefore

$$\xi_h^* \geq \frac{\sqrt{n+1}}{(1 + \rho^*)^2} \xi^* \geq \frac{\hat{\xi}\min(\hat{\lambda})}{2(1 + 3\sqrt{n}\hat{\rho}/2)^2 \max(\hat{\lambda})}, \tag{29}$$

using Proposition 4.1. Finally we have

$$\begin{aligned}
1 + \rho_h^* &= 1 + \alpha \left\| \begin{pmatrix} x^* + s^* \\ 1 \end{pmatrix} \right\| = 1 + \frac{n+1}{e^T(x^* + s^*) + 1}\sqrt{1 + (\rho^*)^2} \\
&\leq 1 + (n+1)(1 + \rho^*) \quad \leq \quad 2(n+1)(1 + 3\sqrt{n}\hat{\rho}/2),
\end{aligned} \tag{30}$$

again using Proposition 4.1. Combining (29) and (30) with (23), we obtain the desired result. $\square$

The bound in (28) is not the simplest, or tightest, possible, but is chosen for applicability in our probabilistic analysis. Using the criterion in the previous lemma, we can terminate the algorithm once (28) holds. From (6), (9), and (28), this definitely happens if

$$(1 - 8^{-0.25}/\sqrt{n+1})^k \leq \frac{(1 - \beta)(\hat{\xi})^2[\min(\hat{\lambda})]^2}{16(n+1)^4(1 + 3\sqrt{n}\hat{\rho}/2)^6[\max(\hat{\lambda})]^2(1 + \sum_{j \in \bar{\sigma}^*} \|B_1^{-1}\hat{a}_j\|)},$$

which requires

$$k = O(\sqrt{n})[\ln n - \ln \hat{\xi} - \ln \min(\hat{\lambda}) + \ln(1 + 3\sqrt{n}\hat{\rho}/2) + \ln \max(\hat{\lambda}) + \ln(1 + \sum_{j \in \bar{\sigma}^*} \|B_1^{-1}\hat{a}_j\|)].$$

Since $\hat{\rho}^2$ is distributed as a chi-square random variable, with $n_1 + n_3 \leq n$ degrees of freedom, we have

$$E[\ln(1 + 3\sqrt{n}\hat{\rho}/2)] \leq \ln(1 + 3\sqrt{n}E[\hat{\rho}]/2) \leq \ln(1 + 3\sqrt{n}\sqrt{E[\hat{\rho}^2]}/2) = O(\ln n).$$

It is also straightforward to show that

$$E[\ln \min(\hat{\lambda})] \geq -O(\ln n), \quad E[\ln \max(\hat{\lambda})] \leq O(\ln n), \quad \text{and} \quad E[\ln \hat{\xi}] \geq -O(\ln n).$$

Moreover, it is known (see Lemma 5.4 of [2]) that

$$E[\ln(1 + \sum_{j \in \bar{\sigma}^*} \|B_1^{-1}\hat{a}_j\|)] \leq O(\ln n).$$

Therefore, termination occurs on an iteration k, whose expected value is bounded as

$$
\begin{aligned}
E[k] \;\leq\;& O(\sqrt{n})(\ln n - E[\ln \min(\hat{\lambda})] + E[\ln \hat{\xi}] + E[\ln(1 + 3\sqrt{n}\hat{\rho}/2)] \\
& + E[\ln \max(\hat{\lambda})] + E[\ln(1 + \sum_{j \in \bar{\sigma}^*} \|B_1^{-1}\hat{a}_j\|)]) \\
\leq\;& O(\sqrt{n}\ln n).
\end{aligned}
$$

Thus we have proved the main result of our paper:

Theorem 4.3 *Assume that the self–dual algorithm, using the projection scheme described in Section 3, is applied to an instance of Todd's degenerate model. Then the expected number of iterations before termination with an exact optimal solution of (LP) is bounded above by* $O(\sqrt{n}\ln n)$.

References

[1] Adler, I. and Megiddo, N. (1985), "A simplex algorithm whose average number of steps is bounded between two quadratic functions of the smaller dimension," *Journal of the Association of Computing Machinery* 32, 891-895.

[2] Anstreicher, K.M., Ji, J., Potra, P.A. and Ye, Y. (1992), "Probabilistic analysis of an infeasible primal-dual algorithm for linear programming," Reports on Computational Mathematics No. 27, Dept. of Mathematics, University of Iowa (Iowa City, IA).

[3] Anstreicher, K.M., Ji, J. and Ye, Y. (1992), "Average performance of an ellipsoid termination criterion for linear programming interior point programming," Working paper No. 92-1, Dept. of Management Sciences, University of Iowa (Iowa City, IA).

[4] Borgwardt, K.H. (1987), *The Simplex Method – A Probabilistic Approach* (Springer-Verlag, New York).

[5] Güler, O. and Ye, Y. (1991), "Convergence behavior of some interior-point algorithms," Working paper No. 91-4, Dept. of Management Sciences, University of Iowa (Iowa City, IA).

[6] Haimovich, M. (1983), "The simplex method is very good! – On the expected number of pivot steps and related properties of random linear programs," Working paper, Dept. of IE and OR, Columbia University (New York, NY).

[7] Huang, S. and Ye, Y. (1991), "On the average number of iterations of the polynomial interior-point algorithms for linear programming," Working paper No. 91-16, Dept. of Management Sciences, University of Iowa (Iowa City, Iowa).

[8] Ji, J. and Potra, F.A. (1992), "On the average complexity of finding an ϵ-optimal solution for linear programming," Reports on Computational Mathematics No. 25, Dept. of Mathematics, University of Iowa (Iowa City, IA).

[9] Kojima, M., Megiddo, N. and Mizuno, S. (1991), "A primal-dual infeasible-interior-point algorithm for linear programming," Report RJ 8500, IBM Almaden Research Center (San Jose, CA).

[10] Mehrotra, S. and Ye, Y. (1991), "Finding an interior point in the optimal face of linear programs". Technical Report, Dept. of IE and MS, Northwestern University (Evanston, IL).

[11] Mizuno, S. (1992), "Polynomiality of the Kojima-Megiddo-Mizuno infeasible interior point algorithm for linear programming," Technical Report No. 1006, School of OR/IE, Cornell University (Ithaca, NY).

[12] Mizuno, S., Todd, M. and Ye, Y. (1990), "On adaptive-step primal-dual interior-point algorithms for linear programming," Technical Report No. 944, School of OR/IE, Cornell University (Ithaca, NY).

[13] Nemirovsky, A.S. (1987), "An algorithm of the Karmarkar type," *Tekhnicheskaya Kibernetika* 1, 105-118.

[14] Potra, F.A. (1992), "An infeasible interior-point predictor-corrector algorithm for linear programming," Reports on Computational Mathematics No. 26, Dept. of Mathematics, University of Iowa (Iowa City, IA).

[15] Potra, F.A. (1992), "A quadratically convergent infeasible interior-point algorithm for linear programming," Reports on Computational Mathematics No. 28, Dept. of Mathematics, University of Iowa (Iowa City, IA).

[16] Smale, S. (1983), "On the average number of steps of the simplex method of linear programming," *Mathematical Programming* 27, 241-262.

[17] Todd, M.J. (1986), "Polynomial expected behavior of a pivoting algorithm for linear complementarity and linear programming problems," *Mathematical Programming* 35, 173-192.

[18] Todd, M.J. (1991), "Probabilistic models for linear programming," *Mathematics of Operations Research* 16, 671-693.

[19] Ye, Y. (1991) "On the finite convergence of interior-point algorithms for linear programming," Working paper No. 91-5, Dept. of Management Sciences, University of Iowa (Iowa City, IA).

[20] Ye, Y. (1991), "Towards probabilistic analysis of interior-point algorithms for linear programming," Working paper No. 91-13, Dept. of Management Sciences, University of Iowa (Iowa City, IA).

[21] Ye, Y., Todd, M.J. and Mizuno, S. (1992), "An $O(\sqrt{n}L)$–iteration homogeneous and self–dual linear programming algorithm," Working paper No. 92-11, Dept. of Management Sciences, University of Iowa (Iowa City, IA).

[22] Zhang, Y. (1992), "On the convergence of an infeasible interior-point algorithm for linear programming and other problems," Dept. of Mathematics and Statistics, University of Maryland, Baltimore County (Catonsville, MD).

Complexity in Numerical Optimization, pp. 16-32
P.M. Pardalos, Editor
©1993 World Scientific Publishing Co.

The Complexity of Approximating a Nonlinear Program

Mihir Bellare
High Performance Computing and Communications, IBM T.J. Watson Research Center, P.O. Box 704, Yorktown Heights, NY 10598 USA

Phillip Rogaway
LAN System Design, IBM Personal Software Products, 11400 Burnet Road, Austin, TX 78758 USA

Abstract

We consider the problem of finding the maximum of a multivariate polynomial inside a convex polytope. We show that there is no polynomial time approximation algorithm for this problem, even one with a very poor guarantee, unless P = NP. We show that even when the polynomial is quadratic (i.e. quadratic programming) there is no polynomial time approximation unless NP is contained in quasi-polynomial time.

Our results rely on recent advances in the theory of interactive proof systems. They exemplify an interesting interplay of discrete and continuous mathematics—using a combinatorial argument to get a hardness result for a continuous optimization problem.

Keywords: approximation, complexity theory, interactive proofs, optimization, probabilistically checkable proofs, quadratic programming.

1　Introduction

Many nonlinear optimization problems are not known to admit polynomial time algorithms. In fact, most are NP-hard, so that finding a polynomial time solution is unlikely. Despite this, we often need to solve these "intractable" computational problems. As with NP-hard problems in combinatorial optimization, interest is turning to the development of (efficient) approximation algorithms—algorithms which run in polynomial time and find a solution not too far from an optimal one.

Will approximation succeed? While approximation algorithms for some nonlinear optimization problems do exist, we have no indication of the complexity of approximation in many important cases. Yet in this fledgling field, this seems an important thing to gauge. In particular, for those problems of significant practical interest, it is highly desirable to find "hardness of approximation" results which indicate when it is not worthwhile to seek an approximation algorithm—just as NP-hardness results indicate when it is not worthwhile to seek a polynomial time algorithm.

In general, obtaining "hardness of approximation" results has not been easy. In combinatorial optimization, many important problems defied such efforts for years. Recently, however, powerful techniques to indicate hardness of approximation have emerged; using interactive proofs, this exciting work has been able to settle the approximation complexity of a host of combinatorial optimization problems about which little was known before. Typically, these results indicate hardness of approximation by showing that the existence of an approximation algorithm would imply an unbelievable efficient deterministic algorithm for NP.

Here we apply these techniques to nonlinear optimization, and derive similar results about the complexity of approximating nonlinear optimization problems. We will show that several important nonlinear optimization problems don't possess (efficient) approximation algorithms unless NP has efficient deterministic solutions. The results are strong in that the conclusions hold even for approximation algorithms with very poor guarantees. Yet our constructions and proofs are simple; the strength of our conclusions derives from the powerful results about interactive proofs that underlie this work. We will begin by looking at polynomial programming, and then turn to the special case of most interest: quadratic programming.

1.1　The Complexity of Polynomial Programming

POLYNOMIAL PROGRAMMING is the problem of finding the maximum of a multivariate polynomial $f(x_1, \ldots, x_n)$ inside a "feasible region" $S \subseteq \mathbb{R}^n$. The latter is specified by a set of linear constraints: $S = \{ x \in [0,1]^n : Ax \leq b \}$. POLYNOMIAL PROGRAMMING is known to be solvable in polynomial space [10] but is not known to be in NP.

Denote by f^* and f_* the maximum and minimum of f inside S, respectively. Following [21, 3], we say an algorithm is a μ-approximation, for $\mu\colon \mathbb{N} \to [0,1]$, if it computes $\tilde{f}$ satisfying $|\tilde{f} - f^*| \leq \mu(n)[f^* - f_*]$. As pointed out by Vavasis [25, 24] it is important to use this definition, as opposed to ones (more frequently used in combinatorial optimization) which measure the quality of an approximation compared only to f^*; see Section 2.1 and [25, 24] for more information. Notice that a 0-approximation is optimal, while the value of f at *any* feasible point is a 1-approximation. A 1-approximation is therefore easy to find. Our result says that efficiently computing an approximation which is even marginally better is as hard as deciding NP in polynomial time.

Theorem 1.1 *There is a constant $\delta > 0$ such that the following is true. Suppose* POLYNOMIAL PROGRAMMING *has a polynomial time, μ-approximation, where $\mu(n) = 1 - n^{-\delta}$. Then* P = NP.

We prove this theorem by reducing the problem of computing the size of a maximal independent set in a graph to POLYNOMIAL PROGRAMMING in an approximation-preserving way. This reduction underlies a simple special case of a theorem of Ebenegger, Hammer and de Werra [12], who show that the maximal size of an independent set in a graph is the maximum of some multivariate polynomial associated to it. One concludes by using the result of Arora, Lund, Motwani, Sudan and Szegedy [2], which shows that if P $\neq$ NP then the size of a maximum independent set in a graph cannot be estimated, even quite poorly.

Problem instances in our results include only integers for their numbers, and the results hold even when these integers are encoded in unary. Since the results are negative, this makes them stronger.

1.2   The Complexity of Quadratic Programming

QUADRATIC PROGRAMMING is the special case of POLYNOMIAL PROGRAMMING in which the polynomial f is of total degree 2; that is, maximize $f(x_1, \ldots, x_n) = \sum_{i \geq j} c_{ij} x_i x_j$ inside $S = \{\, x \in [0,1]^n : Ax \leq b \,\}$. It is probably the most important of the nonlinear optimization problems, with applications including economics, planning and genetics.

On the positive side, QUADRATIC PROGRAMMING is known to be in NP [23]. The convex case admits a polynomial time solution [17]. The concave and indefinite cases admit μ-approximation algorithms which, for any constant $\mu \in (0,1)$, are polynomial time under certain conditions on the objective function [24, 25]. The general case admits a weak polynomial time approximation algorithm; specifically, one which achieves a $(1 - \Theta(n^{-2}))$-approximation [26].

On the negative side, QUADRATIC PROGRAMMING is NP-hard [22]. In fact, the existence of a polynomial time $.75n^{-1}$-approximation algorithm for this problem already implies P = NP [25]. In other words, finding an excellent approximation algorithm is unlikely. We improve this result to show that even finding a terrible approximation algorithm is unlikely. We say that a function of n is *quasi-polynomial* if it is bounded above by $n^{\log^c n}$ for some constant $c > 0$.

Theorem 1.2 *There is a constant $\delta > 0$ such that the following is true. Suppose* QUADRATIC PROGRAMMING *has a polynomial time, μ-approximation, where $\mu(n) = 1 - 2^{-\log^\delta n}$. Then any problem in* NP *can be solved in quasi-polynomial time.*

The conclusion can be strengthened to P = NP at the cost of raising the quality of approximation shown hard.

Theorem 1.3 *There is a constant $\mu \in (0,1)$ such that the following is true. Suppose* QUADRATIC PROGRAMMING *has a polynomial time μ-approximation. Then* P = NP.

The value of the constant μ in Theorem 1.2 is a function of the number of "queries" in a "probabilistically checkable proof system" for NP; a rough estimate of this value, based on the result of [2], is $\mu = 10^{-4}$. Recent constructions of more efficient proof systems lead to significant improvements: based on the results of [7] we can easily get $\mu = 1/143$, and a better analysis will probably yield an even larger constant.

We conjecture that these theorems can be improved to show that there is a constant $\delta > 0$ such that the following is true: if QUADRATIC PROGRAMMING has a polynomial time, μ-approximation, where $\mu(n) = 1 - n^{-\delta}$, then P = NP. This would mean that the result of [26] mentioned above is essentially optimal. To improve our theorems in this manner we would need new and more efficient two prover, one round proof systems for NP (cf. Section 4).

1.3   Techniques from Interactive Proofs

Our results rely on recent advances in the theory of interactive proof systems and the connection of these to approximation problems. We give a brief summary of relevant work in this area.

Interactive proofs were introduced by Goldwasser, Micali and Rackoff [16] and Babai [4]. Ben-Or, Goldwasser, Kilian and Wigderson [9] extended these ideas to define a notion of multi-prover interactive proofs. Applications of interactive proof based ideas to the derivation of hardness of approximation results emerged in the work of Condon [11] and Feige, Goldwasser, Lovász, Safra and Szegedy [14]. The latter showed that the size of a maximum independent set in a graph is hard to approximate. Their proof exploited a powerful result of Babai, Fortnow and Lund [5] which equates the class MIP of languages possessing multi-prover interactive proofs of membership with the class NEXP of languages recognizable in non-deterministic exponential time. Subsequent constructions of proof systems of lower complexity has lead to better results on the hardness of approximation [1, 2]. We exploit the most recent result on approximating the size of a maximum independent set to prove Theorem 1.1.

Two prover, one round proofs are multi-prover proofs in which there are only two provers and the interaction is restricted to one round. Using techniques of Lapidot and Shamir [18], it was shown by Feige [13] that two provers and one round of interaction suffice to recognize any NEXP language with exponentially small error probability. This result, "scaled down" to NP, is the basis for our proof of Theorem 1.2. A different result about two prover, one round proofs (cf. [2]) is the basis of the proof of Theorem 1.3.

1.4   Related Work

The particular association of a quadratic program to a two-prover, one-round interactive proof that we use was independently discovered by Feige and Lovász [15]. Meanwhile the original work of Feige [13] on which some of our results were based has also been incorporated into this same joint paper with Lovász [15].

Other works using two prover, one round proofs to show hardness of approximation results include [6] and [19].

QUARTIC PROGRAMMING is the special case of POLYNOMIAL PROGRAMMING in which the objective function is a polynomial of degree four. Since QUADRATIC PROGRAMMING is a special case of QUARTIC PROGRAMMING, the results of Theorems 1.2 and 1.3 apply. For the quartic case, however, a stronger result than Theorem 1.3 was recently obtained in [7]; the authors show that for *any* constant $\mu \in (0, 1)$, if QUARTIC PROGRAMMING has a polynomial time, μ-approximation, then P = NP.

A preliminary version of this paper appeared as [8].

2   Preliminaries

We summarize the needed definitions and results for dealing with optimization, approximation, and interactive proofs. The notation $|\cdot|$ will be used to denote the absolute value of a number, the length of a string, or the size of a set; the context will disambiguate.

2.1   Optimization and Approximation

An optimization problem specifies for each instance $w \in \{0, 1\}^*$ a "solution space" $S(w)$, and for each solution $y \in S(w)$ a "utility" $g(w, y)$. The problem is to maximize the utility over the feasible region. (Minimization problems can be accommodated by modifying these definitions in the obvious ways.) Formally:

Definition 2.1 *An* optimization problem *is a pair* (S, g); S *is a map assigning to each* instance w *a set* $S(w)$ *called the* solution space, *and* $g(w, y)$ *is, for each solution* $y \in S(w)$, *a real number called the* utility *of* y *with respect to* w. *An instance* w *is* degenerate *if* $S(w) = \emptyset$.

For programming problems the solution space is also called the "feasible region." When an instance is degenerate the optimization problem itself is not well defined, and hence we will restrict our attention in what follows to non-degenerate instances. In order to define what is a good approximation, we need to look at both the sup and the inf of the value of $g(w, \cdot)$ over $S(w)$. For technical reasons we need to allow these to take on values in the extended real numbers $\overline{R} = R \cup \{\infty, -\infty\}$.

Definition 2.2 *Let* (S, g) *be an optimization problem. For each non-degenerate instance* w *we define* $g^*(w), g_*(w) \in \overline{R}$ *by* $g^*(w) = \sup_{y \in S(w)} g(w, y)$ *and* $g_*(x) = \inf_{y \in S(w)} g(w, y)$. *A non-degenerate instance* w *is* bounded *if* $g^*(w)$ *and* $g_*(w)$ *are finite.*

Following [21, 3] we measure the quality of an approximation $\tilde{g}$ by seeing how much it differs from g^*, as measured in units of $|g^* - g_*|$. That is, we will speak of an approximation as being multiplicatively within some factor μ of $|g^* - g_*|$. For simplicity we will only talk of approximation when the instance is (non-degenerate and) bounded, so that the unit of measurement $|g^* - g_*|$ is finite.

We allow μ to be a function of the input. It is convenient, for different problems, to express μ as a function of different *aspects* of the input; for example, for graph problems we will express μ as a function of the number of nodes, and for programming problems as a function of the number of variables in the program. To capture this we associate to an optimization problem a norm $\| \cdot \|$ which equals that aspect of the input in terms of which μ is expressed.

Definition 2.3 *Let* (S, g) *be an optimization problem,* $\| \cdot \|$ *a map from strings to* N, *and* μ *a map from* N *to* $[0, 1]$. *A* μ-approximation *for* (S, g) *and norm* $\| \cdot \|$ *is a function* $\tilde{g}$ *which, on any non-degenerate, bounded instance* w *gives a number* $\tilde{g}(w) \in R$ *for which*

$$|g^*(w) - \tilde{g}(w)| \leq \mu(\|w\|) \cdot |g^*(w) - g_*(w)| \ .$$

A constant factor approximation *is a* μ-approximation *in which* μ *is a* constant less than 1.

An attribute of this definition which renders it preferable, in this context, to other definitions more commonly used in combinatorial optimization is the invariance under scaling: if $\tilde{g}$ is a μ-approximation to (S, g), then $a\tilde{g} + b$ is a μ-approximation to $(S, ag + b)$, for any constants a and b; this corresponds, for example, to the fact that measuring utility in different units should not affect the quality of an approximation. Another such attribute is the invariance under affine linear transformations of the feasible region and the objective function. For more information on the appropriateness of this definition in this context we refer the reader to [25, 24].

Note that it is not required that an approximation algorithm "find" the point with the specified utility, and it is not even required that such a point exist. That is, when the algorithm reports a value $\tilde{g}$ for an instance w, there need not even be a $\tilde{y} \in S(w)$ such that $\tilde{g} = g(w, \tilde{y})$. This is therefore a weak notion of approximation. Since our results are negative, this serves only to strengthen them.

We will be interested in polynomial time approximation algorithms. To avoid confusion, we emphasize that while the quality of the approximation is measured in terms of the norm, the running time of the approximation algorithm is measured, as usual, as a function of the length of the encoding of the instance. Approximation algorithms will return rationals, encoded as pairs of integers, each integer itself encoded as usual in binary.

2.2   List of Optimization Problems

As usual, SAT denotes the decision problem for satisfiability of Boolean formulas. By a "graph" we mean a simple, finite, undirected graph. For the graph-theoretic problems, the norm will be the number of nodes, while for the polynomial and quadratic programming problems, the norm will be the number of variables. All numbers in problem instances are integers; this eliminate issues concerning computational complexity over the reals. Furthermore, the integers in problem instances are specified in *unary*; since our results are negative, this makes them stronger. In all the programming problems the feasible region is restricted to a subset of $[0, 1]^n$ and the utility functions are continuous, so all (non-degenerate) instances are bounded.

INDEPENDENT SET
Instance: A graph $G = (V, E)$.
Solutions: $W \subseteq V$ is a solution if it is an independent set: for each $u, v \in W$, $\{u, v\} \notin E$.
Utility of Solutions: A solution W for the instance G has utility $|W|$.

POLYNOMIAL PROGRAMMING
Instance: Number n and t, and for each $k \in [1..t]$, an integer c_k and a subset A_k of $\{1, \ldots, n\}$. Together this encodes a polynomial $f(x_1, \ldots, x_n) = \sum_{k=1}^{t} c_k \left[\left(\prod_{i \in A_k} x_i \right) \right]$. Also an $m \times n$ integer matrix A and an integer m-vector b.
Solutions: A vector $x \in [0, 1]^n$ is a solution if $Ax \leq b$.
Utility of Solutions: A solution x has utility $f(x)$.

POLYNOMIAL PROGRAMMING–RESTRICTED CASE

Instance: Numbers n and t, and for each $k \in [1..t]$, a pair A_k, B_k of disjoint subsets of $\{1, \ldots, n\}$. Together, this encodes the polynomial $f(x_1, \ldots, x_n) = \sum_{k=1}^{t} \left[(\prod_{i \in A_k} x_i) \cdot (\prod_{j \in B_k} (1 - x_j)) \right]$.
Solutions: Any vector $x \in [0, 1]^n$ is a solution.
Utility of Solutions: A solution x has utility $f(x)$.

QUADRATIC PROGRAMMING
Instance: A number n and, for each $i, j \in \{1, \ldots, n\}$ with $i \leq j$, an integer c_{ij}. Together this encodes the quadratic polynomial $f(x) = \sum_{i \leq j} c_{ij} x_i x_j$. Also an $m \times n$ integer matrix A and an integer m-vector b.
Solutions: A vector $x \in [0, 1]^n$ is a solution if $Ax \leq b$.
Utility of Solutions: A solution x has utility $f(x)$.

2.3 Independent Set Approximation

INDEPENDENT SET was shown hard to approximate by [14, 1, 2]. Stating the last of these results in terms of our definition we get the result we will use.

Theorem 2.4 *There is a constant $\delta > 0$ such that the following is true. Suppose* INDEPENDENT SET *has a polynomial time, μ-approximation, where $\mu(n) = 1 - n^{-\delta}$. Then* P = NP.

2.4 Two-Prover, One-Round Proofs

A two-prover, one-round interactive proof system involves a probabilistic, polynomial time *verifier*, V, and a pair of (computationally unbounded, deterministic) *provers*, A and B. Formally:

Definition 2.5 *A verifier is a pair of functions (π, ρ), each computable in time polynomial in the length of its first argument; π takes two string arguments and returns a string, and ρ takes five string arguments and returns a bit. A prover is a function which takes two string arguments and returns a string.*

Each prover can communicate with the verifier, but they can neither talk to one another once the protocol begins, nor can either prover see the communication between the verifier and the other prover. The parties share a common input w, and it is the provers (joint) goal to convince V to accept this string. To this end, the parties engage in a simple interaction, which is begun by the verifier. The latter applies π to the common input w and a string R (the verifier's random tape) to get a pair of "questions" p, q. He then sends p to A and q to B. The provers then provide answers, A

sending the answer $a = A(w, p)$, and B sending $b = B(w, q)$. After the verifier receives his answers, he computes $\rho(w, p, q, a, b)$. If this value is 1 he is considered to "accept" else to "reject."

The number of coins flipped by the verifier and the size of answers sufficient to convince him are the attributes of the verifier which are important in our construction. Specifically, let l be a function from N to N. We say that a verifier V has complexity l if, when the common input has length n, a random tape of length $l(n)$ suffices to produce the questions p, q, and $\rho(w, p, q, a, b) = 0$ if either a or b have length different from $l(n)$. It is convenient, although not necessary, to also assume that the lengths of the questions p, q are equal to $l(n)$, and unless otherwise stated it is to be assumed that this is the case. We denote by $\pi^i(w, R)$ the question to the i-th prover, $i = 1, 2$.

Definition 2.6 *Let $V = (\pi, \rho)$ be a verifier of complexity l. Let (A, B) be a pair of provers. For each w, let $\mathrm{ACC}_{V,(A,B)}(w)$ denote the probability that*

$$\rho(w, \pi^1(w, R), \pi^2(w, R), A(w, \pi^1(w, R)), B(w, \pi^2(w, R))) = 1$$

when R is chosen at random from $\{0, 1\}^{l(|w|)}$. The **accepting probability of the verifier** *V at w is the maximum of $\mathrm{ACC}_{V,(A,B)}(w)$ over all possible pairs (A, B) of provers. We denote it by $\mathrm{ACC}_V(w)$. If L is a language and ϵ a function of N to $[0, 1]$, we say that V has error probability ϵ with respect to L if the following two conditions hold: first, $w \in L$ implies $\mathrm{ACC}_V(w) = 1$; second, $w \notin L$ implies $\mathrm{ACC}_V(w) < \epsilon(|w|)$.*

We say that a language L has a two-prover, one-round proof with complexity l and error probability ϵ if there exists a verifier having complexity l and error probability ϵ with respect to L. Usually we say that L has a two prover, one round proof if it has a two prover one round proof with complexity poly(n) and error probability $1/2$. Important to our results is the fact that NP-complete languages have two prover one round proofs of very low complexity.

Theorem 2.7 [13] *There is a constant $c > 0$ such that* SAT *has a two prover, one round proof with complexity $O(\log^c n)$ and error probability $1/n$.*

If constant error probability suffices, the complexity can be reduced to logarithmic.

Theorem 2.8 [2] *There is a constant $\epsilon < 1$ such that* SAT *has a two prover, one round proof with complexity $O(\log n)$ and error probability ϵ.*

The current best values of the constants in these two theorems are $c = 3$ in the first [13] and $\epsilon = 71/72$ in the second [7].

3 The Complexity of Polynomial Programming

We restate Theorem 1.1 and provide a proof.

Theorem 3.1 *There is a constant $\epsilon > 0$ such that the following is true. Suppose* POLYNOMIAL PROGRAMMING *has a polynomial time, μ-approximation, where $\mu(n) = 1 - n^{-\epsilon}$. Then* P = NP.

Proof: Let $G = (V, E)$ be an instance of INDEPENDENT SET. We will construct from G an instance f of POLYNOMIAL PROGRAMMING–RESTRICTED CASE where $f^* = G^*$ and $\|f\|$ is polynomially bounded in $\|G\|$. We will then explain why our reduction is enough to establish the theorem.

Without loss of generality, assume G has no isolated nodes. The program f is constructed as follows. Introduce a formal variable x_e for each edge $e \in E$. To each edge $e = \{u, v\}$, arbitrarily order its endpoints (u, v) and associate the polynomial x_e with endpoint u and the polynomial $1 - x_e$ with endpoint v, defining $x_{uv} = x_e$ and $x_{vu} = 1 - x_e$. The polynomial f is defined as the sum, over all vertices, of the product of the polynomials associated to that vertex: $f(x) = \sum_{u \in V} \prod_{v \in N(u)} x_{uv}$, where $N(u)$ is the set of all vertices adjacent to u. This is a degree $\Delta = \max_u deg(u)$ polynomial in $m = |E|$ variables.

An algorithm which reduces INDEPENDENT SET to POLYNOMIAL PROGRAMMING–RESTRICTED CASE constructs from graph G the polynomial f described above, obtains an estimate of its maximum in $[0, 1]^m$, and then returns this as its own estimate for the size of the maximal independent set in G. Note that f is easily constructed from G in polynomial time, and f has norm (number of variables) which is at most the square of the norm of G (the number of nodes).

Let $f^* = \max_{0 \leq x_e \leq 1} f(x)$ denote f's maximum inside the m-dimensional unit hypercube, and let G^* denote the size of a maximum independent set of G. We show that $G^* = f^*$.

First, we claim that $f^* \geq G^*$. For given an independent set W of cardinality ω, one constructs an assignment $x = \{x_e\}$ of utility at least ω by setting $x_e = 1$ if $u \in W$ and e is ordered (u, v); by setting $x_e = 0$ if $u \in W$ and e is ordered (v, u); and by setting x_e arbitrarily otherwise. The assignment is well defined by W being an independent set. We have $f^* \geq G^*$ because $f(x) \geq \sum_{u \in W} \prod_{v \in N(u)} x_{uv} = \omega$.

Conversely, $G^* \geq f^*$. For given an assignment $x = \{x_e\}$, construct an independent set W of cardinality at least $\lceil f(x) \rceil$ as follows: Choose an edge $e = \{u, v\}$ and set $\pi_u = \prod_{r \in N(u) - \{v\}} x_{ur}$, and $\pi_v = \prod_{r \in N(v) - \{u\}} x_{vr}$. Now if $\pi_u \leq \pi_v$, then adjust the assignment by "pushing" all x_{uv} units from u to v and obtain an assignment x' of at least as great a value as that of x; that is, letting $x'_e = x_e$ apart from setting $x_e = 0$ if $e = (u, v)$ and $x_e = 1$ if $e = (v, u)$, we have $g(x') \geq g(x)$, as $g(x') - g(x) =$

$x_{uv}(\pi_v - \pi_u) \geq 0$. If, instead, $\pi_u > \pi_v$, then let $x' = x$ except $x_e = 1$ if $e = (u, v)$ and $x_e = 0$ if $e = (v, u)$; this again ensures that $g(x') \geq g(x)$. Repeating this process for each edge of G gives an assignment x'' with $g(x'') \geq g(x)$ and each $x_e'' \in \{0, 1\}$. Consider the set of vertices $W = \{u \in V : x_{uv}'' = 1 \text{ for all } v \in N(u)\}$. Then W is an independent set of vertices and $|W| = g(x'') \geq g(x)$.

We have shown how to efficiently map G to f and $\tilde{f}$ to $\tilde{G}$. Now suppose we had a $(1 - \|f\|^{-2\delta})$-approximation for POLYNOMIAL PROGRAMMING–RESTRICTED CASE. Since $f^* = G^*$, $f_* = G_*$ (both are 0), and $\|f\| \leq \|G\|^2$, we immediately get a $(1 - n^{-4\delta})$-approximation for INDEPENDENT SET. Likewise, the straightforward reduction from POLYNOMIAL PROGRAMMING–RESTRICTED CASE to POLYNOMIAL PROGRAMMING, in which each expression $1 - x_j$ is replaced by a formal variable x_j' and constraints are added to enforce that $x_j' = 1 - x_j$, also preserves the optimal value, the worst value, is efficient, and at most doubles the norm. Thus a $(1 - n^{-\delta})$-approximation for POLYNOMIAL PROGRAMMING easily gives a $(1 - n^{-2\delta})$-approximation for POLYNOMIAL PROGRAMMING–RESTRICTED CASE. Putting this all together and using Theorem 2.4, with the constant δ of the present theorem being 4 times the constant of Theorem 2.4, we conclude our result. ∎

Note the reduction above is adequate for positive results, too: independent sets and assignments map efficiently between one another by the construction above.

4 The Complexity of Quadratic Programming

Our hardness results for quadratic programming will be obtained by associating to any V (verifier of a two prover, one round proof) and w (common input) an instance of quadratic programming with the property that solutions to this instance correspond to provers, and the maximum corresponds to the value of the interaction. Let us begin by describing the association and developing its basic properties. We will then apply the results of Section 2.4 to obtain our results.

Definition 4.1 *Let $V = (\pi, \rho)$ be a verifier of complexity l, and let w be an N bit string. Write l for $l(N)$. If $\rho(w, p, q, a, b) = 1$ then let $\hat{\rho}(p, q, a, b)$ be the number of strings $R \in \{0, 1\}^l$ satisfying $\pi(w, R) = (p, q)$. Otherwise, let $\hat{\rho}(p, q, a, b) = 0$. For each $p, a \in \{0, 1\}^l$ we introduce a variable $x_{p,a}$, and for each $q, b \in \{0, 1\}^l$ we introduce a variable $y_{q,b}$. We then introduce the 2^{2l+1} variable quadratic program defined as follows:*

$$\text{maximize:} \quad f(xy) = \textstyle\sum_{p,a,q,b \in \{0,1\}^l} \hat{\rho}(p, q, a, b) \cdot x_{p,a} \cdot y_{q,b}$$

$$\text{subject to:} \begin{cases} \sum_{a \in \{0,1\}^l} x_{p,a} = 1 & \text{\textit{for each} } p \in \{0,1\}^l, \\ \sum_{b \in \{0,1\}^l} y_{q,b} = 1 & \text{\textit{for each} } q \in \{0,1\}^l, \text{\textit{and}} \\ 0 \leq x_{p,a}, y_{q,b} \leq 1 & \text{\textit{for all} } p, q, a, b \in \{0,1\}^l. \end{cases}$$

We refer to (f, A, b) as the the quadratic program *associated to V and w, where A and b are defined so that $A(xy) \leq b$ captures the above set of constraints.*

We now develop the basic properties of this association. A boolean point is one all of whose coordinates are 0 or 1, and $f^* = \max_{xy:A(xy)\leq b} f(xy)$ denotes the maximum value of f over its feasible region. The first lemma says that this maximum of f occurs at a boolean point.

Lemma 4.2 *Let (f, A, b) be the quadratic program associated to a verifier $V = (\pi, \rho)$ and string w. Then there exists a boolean point x^*y^* in the feasible region of this program such that $f^* = f(x^*y^*)$.*

Proof: Let xy be any point in the feasible region. For each p fix a string $\bar{a}(p)$ such that $\sum_{q,b} \hat{\rho}(p, q, \bar{a}(p), b) \cdot y_{q,b} = \max_a \sum_{q,b} \hat{\rho}(p, q, a, b) \cdot y_{q,b}$. Then for each a set $x^*_{p,a} = 1$ if $a = \bar{a}(p)$ and 0 otherwise. The vector x^* is boolean, and x^*y lies in the feasible region. Moreover, for each p

$$\sum_a \left(\sum_{q,b} \hat{\rho}(p, q, a, b) \cdot y_{q,b} \right) x_{p,a} \leq \sum_a \left(\sum_{q,b} \hat{\rho}(p, q, \bar{a}(p), b) \cdot y_{q,b} \right) x_{p,a}$$

$$= \sum_{q,b} \hat{\rho}(p, q, \bar{a}(p), b) \cdot y_{q,b}$$

$$= \sum_a \left(\sum_{q,b} \hat{\rho}(p, q, a, b) \cdot y_{q,b} \right) x^*_{p,a} .$$

So $f(x^*y) \geq f(xy)$. We now apply the same argument to the point x^*y, this time with respect to y, and thereby transform y into a boolean point y^* such that x^*y^* is in the feasible region and $f(x^*y^*) \geq f(x^*y)$. Putting this together we get $f(x^*y^*) \geq f(xy)$. The lemma follows. ∎

The next lemma says that the maximum value of f equals, apart from a scaling factor, the value of the interaction defined by V.

Lemma 4.3 *Let (f, A, b) be the quadratic program associated to a verifier V and string w. Then $\mathrm{ACC}_V(w) = 2^{-l} \cdot f^*$, where l is the complexity of V.*

Proof: Let (A, B) be a pair of provers. Set $x_{p,a} = 1$ if $a = A(w, p)$ and 0 otherwise, and $y_{q,b} = 1$ if $b = B(w, q)$ and 0 otherwise. Then xy is a (boolean) point in the feasible region, and $2^{-l} \cdot f(xy) = \text{ACC}_{V,(A,B)}(w)$. Since (A, B) was arbitrary it follows that $2^{-l} \cdot f^* \geq \text{ACC}_V(w)$.

Conversely, by Lemma 4.2 we know that there is a boolean point xy in the feasible region such that $f^* = f(xy)$. Since xy is both boolean and feasible it must be that for each p there is a unique $\bar{a}$ such that $x_{p,a} = 1$ if $a = \bar{a}$ and $x_{p,a} = 0$ otherwise. Set $A(w, p) = \bar{a}$. Construct B correspondingly from y. Then note that $\text{ACC}_{V,(A,B)}(w) = 2^{-l} \cdot f(xy)$. So $\text{ACC}_V(w) \geq 2^{-l} \cdot f^*$. ∎

Finally, we observe that the quadratic program associated to V and w is easily constructible.

Lemma 4.4 *Let V be a verifier of complexity l. Then there is a polynomial in $2^{l(N)}$ time algorithm which on input $w \in \{0,1\}^N$ outputs (an encoding of) the quadratic program associated to V and w.* ∎

Let us put the above and Theorem 2.7 together to derive the hardness results on approximating quadratic programming.

Theorem 4.5 *There is a constant $\delta > 0$ such that the following is true. Suppose* QUADRATIC PROGRAMMING *has a polynomial time, μ-approximation, where $\mu(n) = 1 - 2^{-\lg^\delta n}$. Then* NP *can be solved in quasi-polynomial time.*

Proof: It suffices to show that the conditions of the theorem imply that SAT is solvable in quasi-polynomial time. Let $V = (\pi, \rho)$ be the verifier and c the constant of the two prover one round proof of Theorem 2.7, and let l be the complexity of this verifier as specified by the theorem. Let $\tilde{g}$ be a quasi-polynomial time, μ-approximation algorithm for QUADRATIC PROGRAMMING. We specify a quasi-polynomial time decision procedure M for SAT.

We let w denote the input to M, with N denoting its length. We write l for $l(N)$. On input w, algorithm M constructs the quadratic program (f, A, b) associated to V and w. We let n denote the number of variables in this quadratic program, so that $n = 2^{\Theta(l)}$ (cf. Definition 4.1). M now applies the approximation algorithm $\tilde{g}$ to this program, and lets $\tilde{f}$ denote the output. If $\tilde{f} \geq 2^l(1 - \mu(n))$ then M accepts; otherwise, it rejects.

First, note that the length of an encoding of (f, A, b) is $2^{\lg^{O(1)} N}$, even with the encoding in unary as we assume. And this program can easily be computed in $2^{\lg^{O(1)} N}$ time (cf. Lemma 4.4). M will run the quasi-polynomial time algorithm $\tilde{g}$ on an input of quasi-polynomial size in N, so the total running time is quasi-polynomial in N. All together, then, M is quasi-polynomial time.

We note that the minimum f_* of f over the feasible region is nonnegative. The approximation thus guarantees $|f^* - \tilde{f}| \leq \mu(n) f^*$. Now if $w \in$ SAT, then the program constructed has maximum $f^* = 2^l$, while if $w \notin$ SAT, then it has maximum $f^* < \epsilon 2^l$ where $\epsilon = 1/N$ is the error probability of V with respect to SAT. (cf. Theorem 2.7 and Lemma 4.3). So

(1) if $w \in$ SAT then $\tilde{f} \geq 2^l(1 - \mu(n))$, and

(2) if $w \notin$ SAT then $\tilde{f} < \epsilon 2^l(1 + \mu(n))$.

Thus our decision procedure for SAT is correct as long as $\epsilon 2^l(1 + \mu(n)) \leq 2^l(1 - \mu(n))$. Simplifying this expression yields that $\mu(n)$ must be at most $(1 - \epsilon)/(1 + \epsilon) = (N - 1)/(N + 1)$. Since $l = \lg^c N$ and $n = 2^{\Theta(l)}$ we may certainly find a positive δ so that defining μ as in the theorem statement does indeed guarantee $\mu(n) \leq (N - 1)/(N + 1)$. ∎

An analogous proof with Theorem 2.8 substituted for Theorem 2.7 yields the following. Here the quality of approximation is worse, but the conclusion P = NP is better.

Theorem 4.6 *There is a constant $\mu \in (0, 1)$ such that the following is true. Suppose* QUADRATIC PROGRAMMING *has a polynomial time, μ-approximation. Then* P = NP.

We believe these results could be improved to show that there is a constant $\delta > 0$ such that the following is true: if QUADRATIC PROGRAMMING has a polynomial time, μ-approximation, where $\mu(n) = 1 - n^{-\delta}$, then P = NP. One way to do this would be to construct two prover, one round proof systems for SAT which have appropriate complexity and error probability. Specifically, it suffices that the verifier use logarithmic randomness and answer sizes to achieve error probability $1/n$. (The length of the questions p, q is not important; our construction is easily modified so that the size of the quadratic program associated to the verifier and a string w depends only on the randomness and answer sizes.) This result would be interesting in its own right, and would have many other applications.

It is natural to ask whether the hardness of approximation of QUADRATIC PROGRAMMING, like the hardness of approximation of POLYNOMIAL PROGRAMMING, could be derived by a reduction from INDEPENDENT SET rather than via two-prover interactive proofs. While any reduction from INDEPENDENT SET to QUADRATIC PROGRAMMING does yield some conclusion about the hardness of approximating the latter problem, the quality of results obtained by this method depends very much on the nature of the particular reduction used, and we know of no reduction leading to results as strong as those we obtain via two-prover, one-round proofs. In particular, from the reduction of Motzkin and Strauss [20], in conjunction with the result of [2], say, the best one may (directly) conclude is that there is a constant $c > 0$ such that QUADRATIC PROGRAMMING has no polynomial time, n^{-c}-approximation (as long P $\neq$ NP). Intuitively, the problem is with the particular functional relationship that [20] establish between the size of a maximum size independent set in a graph and the maximum of its associated program: for this function, even a big change in maximum independent set size translates into a small change in the maximum of the associated program.

Acknowledgments

We are grateful to Peter Hammer, who, during a visit to Dartmouth College, described the work in [12] which inspired our initial results. We thank Stephen Vavasis for much helpful information on the subject of nonlinear optimization, and especially for explaining to us the importance of using the right definition of a μ-approximation. We thank Rajeev Motwani for drawing our attention to [3]. We thank an anonymous referee for various helpful comments.

References

[1] Arora, S. and Safra, S. (1992), "Probabilistic checking of proofs; a new characterization of NP," *Proceedings of the 33rd Annual IEEE Symposium on the Foundations of Computer Science*, IEEE.

[2] Arora, S., Lund, C., Motwani, M., Sudan, M. and Szegedy, M. (1992), "Proof verification and hardness of approximation problems," *Proceedings of the 33rd Annual IEEE Symposium on the Foundations of Computer Science*, IEEE.

[3] Ausiello, G., D'Atri, A., and Protasi, M. (1980), "Structure preserving reductions among convex optimization problems," *Journal of Computer and System Sciences* **21**, 136–153 (1980).

[4] Babai, L. (1985), "Trading Group Theory for Randomness," *Proceedings of the 17th Annual ACM Symposium on the Theory of Computing*, ACM.

[5] Babai, L., Fortnow, L. and Lund, C. (1991) , "Non-deterministic exponential time has two-prover interactive protocols," *Computational Complexity* **1**, 3–40.

[6] Bellare, M. (1992), "Interactive proofs and approximation," *IBM Research Report* RC 17969. Also *Proceedings of the second Israel Symposium on theory of computing and systems, 1993*.

[7] Bellare, M., Goldwasser, S., Lund, C., and Russell, A. (1993), "Efficient probabilistically checkable proofs and applications to approximation," *Proceedings of the 25th Annual ACM Symposium on the Theory of Computing*, ACM.

[8] Bellare, M. and Rogaway, P. (1992), "The complexity of approximating a non-linear program," *IBM Research Report* RC 17831.

[9] Ben-Or, M., Goldwasser, S., Kilian, J. and Wigderson, A. (1988), "Multi-prover interactive proofs: how to remove intractability assumptions," *Proceedings of the 20th Annual ACM Symposium on the Theory of Computing*, ACM.

[10] Canny, J. (1988), "Some algebraic and geometric computations in PSPACE," *Proceedings of the 20th Annual ACM Symposium on the Theory of Computing*, ACM.

[11] Condon, A. (1991), "The complexity of the max word problem and the power of one-way interactive proof systems," *Proceedings of the 8th Annual Symposium on Theoretical Aspects of Computer Science*, Lecture Notes in Computer Science Vol. 480, Springer-Verlag.

[12] Ebenegger, C., Hammer, P. and de Werra, D. (1984), "Pseudo-boolean functions and stability of graphs," in *Algebraic and Combinatorial Methods in Operations Research*, Annals of Discrete Mathematics, vol. 19, 83–97. (North Holland Mathematics Studies, 95.)

[13] Feige, U. (1991), "NEXPTIME has two-provers one-round proof systems with exponentially small error probability," Manuscript.

[14] Feige, U., Goldwasser, S., Lovász, L., Safra, S. and Szegedy, M. (1991), "Approximating clique is almost NP-complete," *Proceedings of the 32nd Annual IEEE Symposium on the Foundations of Computer Science*, IEEE.

[15] Feige, U. and Lovász, L. (1992), "Two-prover one round proof systems: their power and their problems," *Proceedings of the 24th Annual ACM Symposium on the Theory of Computing*, ACM.

[16] Goldwasser, S., Micali, S. and Rackoff, C. (1989), "The knowledge complexity of interactive proofs," *SIAM J. Computing* **18**(1), 186–208.

[17] Kozlov, M., Tarasov, S. and Hačijan, L. (1979), "Polynomial solvability of convex quadratic programming," *Dokl. Akad. Nauk SSSR* **248**, 1049–1051. Translated in *Soviet Math Dokl.* **20**, 1108-1111.

[18] Lapidot, D. and Shamir, A. (1991), "Fully parallelized multi-prover protocols for NEXP-time," *Proceedings of the 32nd Annual IEEE Symposium on the Foundations of Computer Science*, IEEE.

[19] Lund, C., and Yannakakis, M. (1993), "On the hardness of approximating minimization problems," *Proceedings of the 25th Annual ACM Symposium on the Theory of Computing*, ACM.

[20] Motzkin, T. and Straus, E. (1964), "Maxima for graphs and a new proof of a theorem by Tuán," *Notices of the American Mathematical Society* **11**, 533–540.

[21] Nemirovsky, A. and Yudin, D. (1979), *Slozhnost' Zadach i Effektivnost' Metodov Optimizatsii.* Translated by E. Dawson as *Problem Complexity and Method Efficiency in Optimization,* (John Wiley and Sons), 1983.

[22] Sahni, S. (1974), "Computationally related problems," *SIAM J. of Computing* **3**, No. 4, 262–279.

[23] Vavasis, S. (1990), "Quadratic programming is in NP," *Info. Proc. Letters* **36**, 73–77.

[24] Vavasis, S. (1991), "Approximation algorithms for indefinite quadratic programming," TR 91-1228, Dept. of Computer Science, Cornell University, August 1991. *Math. Prog.*, to appear.

[25] Vavasis, S. (1992), "On approximation algorithms for concave programming," *Recent Advances in Global Optimization,* C. A. Floudas and P.M. Pardalos, pp. 3-18, Princeton University Press.

[26] Vavasis, S. (1993), "Polynomial time weak approximation algorithms for quadratic programming," in *Complexity in Numerical Optimization,* Panos Pardalos, editor.

Complexity in Numerical Optimization, pp. 33-56
P.M. Pardalos, Editor
©1993 World Scientific Publishing Co.

Algorithms for the Least Distance Problem

Piotr Berman and Nainan Kovoor
Computer Science Department, The Pennsylvania State University, University Park, PA 16804 USA

Panos M. Pardalos
Department of Industrial and Systems Engineering, University of Florida, Gainesville, FL 32611 USA

Abstract

We consider the problem of minimizing the Euclidean distance function on $\mathbb{R}^n$ subject to m equality constraints and upper and lower bounds (box constraints). We provide a parametric characterization in $\mathbb{R}^m$ of the family of solutions to this problem, thereby showing equivalence with a problem of search in an arrangement of hyperplanes in $\mathbb{R}^m$. We use this characterization and the technique for constructing arrangements due to Edelsbrunner, O'Rourke and Seidel to develop an exact algorithm for the problem. The algorithm is strongly polynomial running in time $\Theta(n^m)$ for each fixed m. We further develop an algorithm for the problem which uses the search scheme of Megiddo and Dyer to give a running time of $\Theta(n)$ for each fixed m.

Keywords: Complexity, least-distance problem, separable programming, arrangements of hyperplanes, polynomial time algorithms.

1 Introduction

We consider the following convex optimization problem, called the *least distance problem*, in which $V \in \mathbb{R}^{m \times n}, \mathbf{w} \in \mathbb{R}^m, \mathbf{l}, \mathbf{u} \in \mathbb{R}^n$:

$$
\begin{aligned}
\text{minimize} \quad & \mathbf{x} \cdot \mathbf{x} \\
\text{subject to} \quad & V\mathbf{x} = \mathbf{w} \\
& \mathbf{l} \le \mathbf{x} \le \mathbf{u}
\end{aligned}
\tag{1}
$$

Let $f : \mathbf{R}^n \to \mathbf{R}$ be a separable, differentiable and strictly convex function, i.e.

$$f(\mathbf{x}) = \sum_{j \in [1 \ldots n]} f_j(x_j)$$

where each $f_j : \mathbf{R} \to \mathbf{R}$ is a differentiable strictly convex function of one variable. Then we may consider the generalization of (1) to the following convex optimization problem, in which $V \in \mathbf{R}^{m \times n}, \mathbf{w} \in \mathbf{R}^m, \mathbf{l}, \mathbf{u} \in \mathbf{R}^n$:

$$
\begin{aligned}
\text{minimize} \quad & f(\mathbf{x}) \\
\text{subject to} \quad & V\mathbf{x} = \mathbf{w} \\
& \mathbf{l} \le \mathbf{x} \le \mathbf{u}
\end{aligned}
\tag{2}
$$

i.e. find $\mathbf{x} \in \mathbf{R}^n$ lying within a box (specified by the endpoints $\mathbf{l},\mathbf{u}$) and on the intersection of m hyperplanes (specified by $V, \mathbf{w}$), minimizing f. Let $P_f(V, \mathbf{w}, \mathbf{l}, \mathbf{u})$ denote the solution to (2) (provided it is feasible). By the strict convexity of f, $P_f(V, \mathbf{w}, \mathbf{l}, \mathbf{u})$ is uniquely defined (i.e. there is at most one minimum, local as well as global).

The problem is of interest because it belongs to a class of problems for which polynomial time algorithms have been discovered, but strongly polynomial algorithms are not known to exist. The least-distance problem can be solved in polynomial time using either the ellipsoid method (see [13]) or an interior point method (see [23]).

Previous work on the generalized problem has considered chiefly the case of a single equality constraint (the so-called resource-allocation problem). A survey of results is given by Ibaraki and Katoh [10]. For the special case of the least-distance problem with a single equality constraint, Helgason, Kennington and Lall [9] developed a parametric characterization of the family of solutions, and used it to obtain a strongly polynomial algorithm based on sorting and running in time $\Theta(n \log n)$. Subsequently Brucker [2], Calamai and Moré [3] and Pardalos and Kovoor [20] developed algorithms which used median-finding to achieve a running time of $\Theta(n)$.

Recently, Best and Tan [1] considered the case of $\|\cdot\|_2$ and two equality constraints, obtaining a strongly polynomial algorithm running in time $\Theta(n^2 \log n)$.

In this paper, we provide a parametric characterization of solutions of $P_f(V, \mathbf{w}, \mathbf{l}, \mathbf{u})$ as $\mathbf{w}$ varies. This characterization may be considered the natural extension of the characterization of Helgason, Kennington and Lall, and is related to (but more general than) results proved by Smith and Wolkowicz [21] and Li, Pardalos and Han [14].

We show that this parametric characterization results in a $\mathbf{R}^m$-map of all solutions in terms of the parameter. This map has an underlying combinatorial structure based on an arrangement of n pairs of parallel hyperplanes. The problem is thus transformed to one of finding the zero of a monotone vector field in R^n.

We exploit the combinatorial structure to obtain a family of exact algorithms for the least-distance problem which run in strongly polynomial time $\Theta(n^m)$ for each $m \in \mathbf{N}$, thus improving and generalizing previous results.

We also provide an algorithm for the least-distance problem which works in time $\Theta(n)$ for each fixed dimension m. This algorithm uses the multidimensional search technique developed by Megiddo [15] in connection with linear programming and refined by Dyer [21].

2 The Parametric Characterization and its Properties

We will now consider the conditions for a feasible point $\mathbf{x}$ to be a global minimum of $P_f(V, \mathbf{w}, \mathbf{l}, \mathbf{u})$, and use them to provide a parametric characterization of solutions of $P_f(V, \mathbf{w}, \mathbf{l}, \mathbf{u})$ as $\mathbf{w}$ varies. We first introduce some notation. Given $\mathbf{x}, \mathbf{l}, \mathbf{u} \in \mathbf{R}^n$, we define $[\mathbf{x}]_{\mathbf{l}}^{\mathbf{u}}$ to be $\mathbf{t}$ where $t_j = \min(\max(x_j, l_j), u_j)$ for each $j \in [1 \ldots n]$.

The Karush-Kuhn-Tucker conditions for $\mathbf{x} \in \mathbf{R}^n$ to be a local minimum of (2) are

$$
\begin{align}
V\mathbf{x} &= \mathbf{w} \tag{3} \\
\mathbf{l} - \mathbf{x} &\leq \mathbf{0} \tag{4} \\
\mathbf{x} - \mathbf{u} &\leq \mathbf{0} \tag{5} \\
\nabla f(\mathbf{x}) - V^T\boldsymbol{\lambda} - \boldsymbol{\mu} + \boldsymbol{\nu} &= \mathbf{0} \tag{6} \\
\boldsymbol{\mu} \cdot (\mathbf{l} - \mathbf{x}) &= \mathbf{0} \tag{7} \\
\boldsymbol{\nu} \cdot (\mathbf{x} - \mathbf{u}) &= \mathbf{0} \tag{8} \\
\boldsymbol{\mu} &\geq \mathbf{0} \tag{9} \\
\boldsymbol{\nu} &\geq \mathbf{0} \tag{10}
\end{align}
$$

where $\boldsymbol{\lambda} \in \mathbf{R}^m, \boldsymbol{\mu}, \boldsymbol{\nu} \in \mathbf{R}^n$ are the Lagrange multipliers associated with (3), (4), (5) respectively (since $\boldsymbol{\lambda}$ has components of unrestricted sign, it has been given a negative sign in (6), for convenience in the ensuing computations). Note that $\nabla f : \mathbf{R}^n \to \mathbf{R}^n$ is strict monotone increasing by the strict convexity of f, and hence $(\nabla f)^{-1} : \mathbf{R}^n \to \mathbf{R}^n$ is well-defined.

Theorem 1 *Let $f : \mathbf{R}^n \to \mathbf{R}$ be separable, differentiable and strictly convex. Then given $V \in \mathbf{R}^{m \times n}, \mathbf{l}, \mathbf{u} \in \mathbf{R}^n$:*

$$
\{P_f(V, \mathbf{w}, \mathbf{l}, \mathbf{u}) | \mathbf{w} \in \mathbf{R}^m\} = \{(\nabla f)^{-1}[V^T\mathbf{t}]_{\nabla f(\mathbf{l})}^{\nabla f(\mathbf{u})} | \mathbf{t} \in \mathbf{R}^m\} \tag{11}
$$

Proof.

We will prove (11) by two-way inclusion.

($\subseteq$) Suppose $\mathbf{x} \in \mathbf{R}^n$ is in the l.h.s. of (11). Then $\mathbf{x} = P_f(V, \mathbf{w}, \mathbf{l}, \mathbf{u})$ for some $\mathbf{w} \in \mathbf{R}^m$. Thus there exist $\boldsymbol{\lambda} \in \mathbf{R}^m, \boldsymbol{\mu}, \boldsymbol{\nu} \in \mathbf{R}^n$ satisfying the K-K-T conditions (3) – (10) together with $\mathbf{x}$.

From (6) we get

$$
V^T\boldsymbol{\lambda} = (\nabla f)(\mathbf{x}) - \boldsymbol{\mu} + \boldsymbol{\nu}
$$

and so for each $j \in [1 \ldots n]$,

$$\mathbf{V}_j \cdot \boldsymbol{\lambda} = (\nabla f)_j(x_j) - \mu_j + \nu_j \tag{12}$$

Now if $\mathbf{V}_j \cdot \boldsymbol{\lambda} < (\nabla f)_j(x_j)$ then $\mu_j > 0$ by (12), so $l_j - x_j = 0$ by (7). Thus

$$\mathbf{V}_j \cdot \boldsymbol{\lambda} < (\nabla f)_j(x_j) \;\;\Rightarrow\;\; x_j = l_j \tag{13}$$

and similarly

$$\mathbf{V}_j \cdot \boldsymbol{\lambda} > (\nabla f)_j(x_j) \;\;\Rightarrow\;\; x_j = u_j \tag{14}$$

Since $l_j \leq u_j$, we now have three cases to consider :

Case 1 : $\mathbf{V}_j \cdot \boldsymbol{\lambda} < (\nabla f)_j(l_j)$
Then $\mathbf{V}_j \cdot \boldsymbol{\lambda} < (\nabla f)_j(x_j)$ by (4) and the monotonicity of $(\nabla f)_j$, so $x_j = l_j$ by (13).

Case 2 : $(\nabla f)_j(l_j) \leq \mathbf{V}_j \cdot \boldsymbol{\lambda} \leq (\nabla f)_j(u_j)$
If now $\mathbf{V}_j \cdot \boldsymbol{\lambda} < (\nabla f)_j(x_j)$, then $x_j = l_j$ by (13), so $\mathbf{V}_j \cdot \boldsymbol{\lambda} \geq (\nabla f)_j(x_j)$, since $\mathbf{V}_j \cdot \boldsymbol{\lambda} \geq (\nabla f)_j(l_j)$ by assumption. This contradiction forces $\mathbf{V}_j \cdot \boldsymbol{\lambda} \geq (\nabla f)_j(x_j)$. Assuming that $\mathbf{V}_j \cdot \boldsymbol{\lambda} > (\nabla f)_j(x_j)$ similarly leads to a contradiction, so $\mathbf{V}_j \cdot \boldsymbol{\lambda} \leq (\nabla f)_j(x_j)$. Thus $\mathbf{V}_j \cdot \boldsymbol{\lambda} = (\nabla f)_j(x_j)$, so $x_j = (\nabla f)_j^{-1}(\mathbf{V}_j \cdot \boldsymbol{\lambda})$.

Case 3 : $\mathbf{V}_j \cdot \boldsymbol{\lambda} > (\nabla f)_j(u_j)$
Then $\mathbf{V}_j \cdot \boldsymbol{\lambda} > (\nabla f)_j(x_j)$ by (5) and the monotonicity of $(\nabla f)_j$, so $x_j = u_j$ by (14).

In all cases, we have $x_j = (\nabla f)_j^{-1}[\mathbf{V}_j \cdot \boldsymbol{\lambda}]_{(\nabla f)_j(l_j)}^{(\nabla f)_j(u_j)}$. Hence $\mathbf{x} = (\nabla f)^{-1}[V^T\boldsymbol{\lambda}]_{\nabla f(\mathbf{l})}^{\nabla f(\mathbf{u})}$ is in the r.h.s. of (11).

($\supseteq$) Suppose $\mathbf{x} \in \mathbf{R}^n$ is in the r.h.s. of (11). Then $\mathbf{x} = (\nabla f)^{-1}[V^T\mathbf{t}]_{\nabla f(\mathbf{l})}^{\nabla f(\mathbf{u})}$ for some $\mathbf{t} \in \mathbf{R}^m$. Let

$$
\begin{aligned}
\mathbf{w} &= V(\nabla f)^{-1}[V^T\mathbf{t}]_{\nabla f(\mathbf{l})}^{\nabla f(\mathbf{u})} \\
\boldsymbol{\lambda} &= \mathbf{t} \\
\boldsymbol{\mu} &= \mathbf{1} - (\nabla f)^{-1}[V^T\mathbf{t}]^{\nabla f(\mathbf{l})} \\
\boldsymbol{\nu} &= (\nabla f)^{-1}[V^T\mathbf{t}]_{\nabla f(\mathbf{u})} - \mathbf{u}
\end{aligned}
$$

We will show that the given $\mathbf{x}, \mathbf{w}, \boldsymbol{\lambda}, \boldsymbol{\mu}, \boldsymbol{\nu}$ satisfy the K-K-T conditions (3) – (10) for $P_f(V, \mathbf{w}, \mathbf{l}, \mathbf{u})$. Clearly $\mathbf{x}$ and $\mathbf{w}$ satisfy (3), $\mathbf{x}$ satisfies (4), (5), and $\boldsymbol{\mu}, \boldsymbol{\nu}$ satisfy (9), (10). For (6) – (8) we examine each $j \in [1 \ldots n]$. There are three cases to consider:

Case 1 : $\mathbf{V}_j \cdot \mathbf{t} < (\nabla f)_j(l_j)$
Then $l_j - (\nabla f)_j^{-1}(\mathbf{V}_j \cdot \mathbf{t}) > 0$, so $(\nabla f)_j^{-1}(\mathbf{V}_j \cdot \mathbf{t}) - u_j < 0$, since $l_j \leq u_j$.
Hence $x_j = l_j, \lambda = \mathbf{t}, \mu_j = l_j - (\nabla f)_j^{-1}(\mathbf{V}_j \cdot \mathbf{t}), \nu_j = 0$.

Case 2 : $(\nabla f)_j(l_j) \leq \mathbf{V}_j \cdot \mathbf{t} \leq (\nabla f)_j(u_j)$
Then $l_j - (\nabla f)_j^{-1}(\mathbf{V}_j \cdot \mathbf{t}) \leq 0$, $(\nabla f)_j^{-1}(\mathbf{V}_j \cdot \mathbf{t}) - u_j \leq 0$. Hence $x_j = (\nabla f)_j^{-1}(\mathbf{V}_j \cdot \mathbf{t}), \lambda = \mathbf{t}, \mu_j = \nu_j = 0$.

Case 3 : $\mathbf{V}_j \cdot \mathbf{t} > (\nabla f)_j(u_j)$
Then $(\nabla f)_j^{-1}(\mathbf{V}_j \cdot \mathbf{t}) - u_j > 0$, so $l_j - (\nabla f)_j^{-1}(\mathbf{V}_j \cdot \mathbf{t}) < 0$, since $l_j \leq u_j$.
Hence $x_j = u_j, \lambda = \mathbf{t}, \mu_j = 0, \nu_j = (\nabla f)_j^{-1}(\mathbf{V}_j \cdot \mathbf{t}) - u_j$.

In each case, $x_j, \lambda, \mu_j, \nu_j$ clearly satisfy (6) – (8). Thus $\mathbf{x}, \mathbf{w}, \lambda, \mu, \nu$ satisfy the K-K-T conditions, so $\mathbf{x} = P_f(V, \mathbf{w}, \mathbf{l}, \mathbf{u})$ is in the l.h.s. of (11).

∎

Motivated by the problem statement (2) and the parametric characterization (11), we define the functions $\mathbf{x} : \mathbf{R}^m \to \mathbf{R}^n, \mathbf{w} : \mathbf{R}^m \to \mathbf{R}^m$ by

$$\mathbf{x(t)} = (\nabla f)^{-1}[V^T \mathbf{t}]_{\nabla f(\mathbf{l})}^{\nabla f(\mathbf{u})} \tag{15}$$

$$\mathbf{w(t)} = V(\nabla f)^{-1}[V^T \mathbf{t}]_{\nabla f(\mathbf{l})}^{\nabla f(\mathbf{u})} \tag{16}$$

Then we obtain the following corollary which gives necessary and sufficient conditions for the existence of a solution to a particular instance of the problem :

Corollary 2 $\mathbf{x}^* \in \mathbf{R}^n, \mathbf{w}^* \in \mathbf{R}^m$ *satisfy* $\mathbf{x}^* = P_f(V, \mathbf{w}^*, \mathbf{l}, \mathbf{u})$ *iff there exists* $\mathbf{t}^* \in \mathbf{R}^m$ *such that*

$$\mathbf{x(t^*)} = \mathbf{x}^* \tag{17}$$

$$\mathbf{w(t^*)} = \mathbf{w}^* \tag{18}$$

PROOF. Immediate from (2) and (11). ∎

It follows from Corollary 2 that $\mathbf{x}^* = P_f(V, \mathbf{w}^*, \mathbf{l}, \mathbf{u})$ can be solved for $\mathbf{x}^*$ given $\mathbf{w}^*$, by first solving (18) for $\mathbf{t}^*$, and then calculating $\mathbf{x}^*$ using (17).

Let S be the set of solutions of (18), for a particular value of $\mathbf{w}^*$,

$$S = \{\mathbf{t} \in \mathbf{R}^m | \mathbf{w(t)} = \mathbf{w}^*\} \tag{19}$$

From (16), each component of $\mathbf{w(t)}$ is a linear combination of the same set of terms. Each such term $(\nabla f)_j^{-1}[\mathbf{V}_j \cdot \mathbf{t}]_{(\nabla f)_j(l_j)}^{(\nabla f)_j(u_j)}$ is a smooth function of $\mathbf{t}$ except on the pair of *break hyperplanes*

$$L_j = \{\mathbf{t} \in \mathbf{R}^m | \mathbf{V}_j \cdot \mathbf{t} = (\nabla f)_j(l_j)\}$$

$$U_j = \{\mathbf{t} \in \mathbf{R}^m | \mathbf{V}_j \cdot \mathbf{t} = (\nabla f)_j(u_j)\}$$

Let

$$\begin{aligned}
\mathcal{L} &= \{L_j | j \in [1 \cdots n]\} \\
\mathcal{U} &= \{U_j | j \in [1 \cdots n]\} \\
\mathcal{H} &= \mathcal{L} \cup \mathcal{U}
\end{aligned}$$

Define the *break set* $B_f(V, \mathbf{l}, \mathbf{u})$ by

$$B_f(V, \mathbf{l}, \mathbf{u}) = \bigcup_{H \in \mathcal{H}} H$$

Then $\mathbf{w}$ is a smooth function everywhere in $\mathbf{R}^m$ except in $B_f(V, \mathbf{l}, \mathbf{u})$. We will now investigate the combinatorial structure of the sets $B_f(V, \mathbf{l}, \mathbf{u})$ and $\mathbf{R}^m \setminus B_f(V, \mathbf{l}, \mathbf{u})$.

The $2n$ hyperplanes of $\mathcal{H}$ dissect $\mathbf{R}^m$ into connected sets of various dimensions. We call this dissection the *arrangement* $\mathcal{A}(\mathcal{H})$ *of* $\mathcal{H}$. In the sequel we assume familiarity with basic notions connected with arrangements. An introduction to arrangements can be found in Edelsbrunner [1987].

Clearly the set $\mathbf{R}^m \setminus B_f(V, \mathbf{l}, \mathbf{u})$ on which $\mathbf{w}$ is continuous is a (disjoint) union of cells of the arrangement,

$$\mathbf{R}^m \setminus B_f(V, \mathbf{l}, \mathbf{u}) = \bigcup_{c \in \mathcal{A}(\mathcal{H})} c$$

Further, each point of $\mathbf{R}^m$ is in the closure of some cell, so the closures of all the cells taken together cover all of $\mathbf{R}^m$,

$$\mathbf{R}^m = \bigcup_{c \in \mathcal{A}(\mathcal{H})} \text{cl } c \tag{20}$$

For each cell c of $\mathcal{A}(\mathcal{H})$ let $\mathbf{w}^c$ be the smooth function which is equal to $\mathbf{w}$ on c, let S_c be the intersection of S with the closure of c,

$$S_c = \text{cl } c \cap S \tag{21}$$

and let $T_{c,i} (i \in [1 \cdots m])$ be the sets on which the component equations of $\mathbf{w}^c(\mathbf{t}) = \mathbf{w}^*$ hold,

$$T_{c,i} = \{\mathbf{t} \in \mathbf{R}^m | w_i^c(\mathbf{t}) = w_i^*\} \tag{22}$$

If $w_i^c(\mathbf{t})$ is constant then $T_{c,i}$ is trivial (either null or $\mathbf{R}^m$). Otherwise T_c, i is a hypersurface of dimension $m - 1$.

It follows from (20) and (21) that

$$\bigcup_{c \in \mathcal{A}(\mathcal{H})} S_c = S \tag{23}$$

Further,

$$
\begin{aligned}
&\operatorname{cl} c \cap S \\
&= \{ \mathbf{t} \in \operatorname{cl} c \,|\, \mathbf{w}(\mathbf{t}) = \mathbf{w}^* \} \\
&= \{ \mathbf{t} \in \operatorname{cl} c \,|\, \mathbf{w}^c(\mathbf{t}) = \mathbf{w}^* \} \\
&= \operatorname{cl} c \cap \{ \mathbf{t} \in \mathbb{R}^m \,|\, \mathbf{w}^c(\mathbf{t}) = \mathbf{w}^* \}
\end{aligned}
$$

so

$$
S_c = \operatorname{cl} c \cap \bigcap_{i \in [1 \cdots m]} T_{c,i}
$$

Finally, define $S_{c,k}$ for each $c \in \mathcal{A}(\mathcal{H})$ and $k \in [0 \cdots m]$ by

$$
S_{c,k} = \operatorname{cl} c \cap \bigcap_{i \in [1 \cdots k]} T_{c,i}
$$

Then we obtain the recursive formulation

$$
\begin{aligned}
S_{c,0} &= \operatorname{cl} c \\
S_{c,k+1} &= S_{c,k} \cap T_{c,k+1} (k \in [0 \cdots m - 1])
\end{aligned}
\tag{24}
$$

Note that $S_{c,m} = S_c$. Also note that since $\operatorname{cl} c$ and $T_{c,k}(k \in [1 \cdots m])$ are closed, each of the sets $S_{c,k}$ is closed. Finally, it follows from the nature of the $T_{c,k}$s that each non-null $S_{c,k+1}$ is either equal to $S_{c,k}$ or has dimension one less than $S_{c,k}$.

For the special case of the least-distance problem P,

$$
\begin{aligned}
&\text{minimize} \quad \mathbf{x} \cdot \mathbf{x} \\
&\text{subject to} \quad V\mathbf{x} = \mathbf{w} \\
&\hphantom{\text{subject to} \quad} \mathbf{l} \leq \mathbf{x} \leq \mathbf{u}
\end{aligned}
\tag{25}
$$

we define the functions $\mathbf{x} : \mathbb{R}^m \to \mathbb{R}^n, \mathbf{w} : \mathbb{R}^m \to \mathbb{R}^m$ by

$$
\begin{aligned}
\mathbf{x}(\mathbf{t}) &= [V^T \mathbf{t}]_{\mathbf{l}}^{\mathbf{u}} \tag{26} \\
\mathbf{w}(\mathbf{t}) &= V[V^T \mathbf{t}]_{\mathbf{l}}^{\mathbf{u}} \tag{27}
\end{aligned}
$$

A result analogous to Corollary 2 then applies, so $\mathbf{x}^* \in \mathbb{R}^n, \mathbf{w}^* \in \mathbb{R}^m$ satisfy $\mathbf{x}^* = P(V, \mathbf{w}^*, \mathbf{l}, \mathbf{u})$ iff there exists $\mathbf{t}^* \in \mathbb{R}^m$ such that

$$
\begin{aligned}
\mathbf{x}(\mathbf{t}^*) &= \mathbf{x}^* \tag{28} \\
\mathbf{w}(\mathbf{t}^*) &= \mathbf{w}^* \tag{29}
\end{aligned}
$$

It follows that $\mathbf{x}^* = P(V, \mathbf{w}^*, \mathbf{l}, \mathbf{u})$ can be solved for $\mathbf{x}^*$ given $\mathbf{w}^*$, by first solving (29) for $\mathbf{t}^*$, and then calculating $\mathbf{x}^*$ using (28).

We now examine some of the properties of the function $\mathbf{w} : \mathbf{R}^m \to \mathbf{R}^m$

$$
\begin{aligned}
\mathbf{w(t)} \\
&= V[V^T \mathbf{t}]_1^{\mathbf{u}} \\
&= (\mathbf{V}_1 \ldots \mathbf{V}_n) \left[\begin{pmatrix} \mathbf{V}_1^T \\ \vdots \\ \mathbf{V}_n^T \end{pmatrix} \mathbf{t} \right]_1^{\mathbf{u}} \\
&= (\mathbf{V}_1 \ldots \mathbf{V}_n) \left[\begin{pmatrix} \mathbf{V}_1 \cdot \mathbf{t} \\ \vdots \\ \mathbf{V}_n \cdot \mathbf{t} \end{pmatrix} \right]_1^{\mathbf{u}} \\
&= \sum_{1 \le j \le n} [\mathbf{V}_j \cdot \mathbf{t}]_{l_j}^{u_j} \mathbf{V}_j \\
&= \sum_{1 \le j \le n} \mathbf{w}_j(\mathbf{t})
\end{aligned}
$$

where each $\mathbf{w}_j : \mathbf{R}^m \to \mathbf{R}^m$ is defined by

$$\mathbf{w}_j(\mathbf{t}) = [\mathbf{V}_j \cdot \mathbf{t}]_{l_j}^{u_j} \mathbf{V}_j$$

We will call a vector function $\mathbf{r} : \mathbf{R}^m \to \mathbf{R}^m$ a *ramp function* if it has the form

$$\mathbf{r(t)} = c[\mathbf{v} \cdot \mathbf{t}]_l^u \mathbf{v} + \mathbf{d}$$

for some $\mathbf{v}, \mathbf{d} \in \mathbf{R}^m, l, u, c \in \mathbf{R}$.

A vector function $\mathbf{w}_j : \mathbf{R}^m \to \mathbf{R}^m$ is said to be *monotone* if for all $\mathbf{t}_1, \mathbf{t}_2 \in \mathbf{R}^m$,

$$(\mathbf{f(t_1)} - \mathbf{f(t_2)}) \cdot (\mathbf{t}_1 - \mathbf{t}_2) \ge 0$$

In the case of ramp functions,

$$
\begin{aligned}
(\mathbf{r(t_1)} &- \mathbf{r}(t_2)) \cdot (\mathbf{t}_1 - \mathbf{t}_2) \\
&= (c[\mathbf{v} \cdot \mathbf{t}_1]_l^u \mathbf{v} - c[\mathbf{v} \cdot \mathbf{t}_2]_l^u \mathbf{v}) \cdot (\mathbf{t}_1 - \mathbf{t}_2) \\
&= c([\mathbf{v} \cdot \mathbf{t}_1]_l^u - [\mathbf{v} \cdot \mathbf{t}_2]_l^u)(\mathbf{v} \cdot \mathbf{t}_1 - \mathbf{v} \cdot \mathbf{t}_2)
\end{aligned}
$$

Since $[\mathbf{v} \cdot \mathbf{t}_1]_l^u - [\mathbf{v} \cdot \mathbf{t}_2]_l^u$ has the same sign as $\mathbf{v} \cdot \mathbf{t}_1 - \mathbf{v} \cdot \mathbf{t}_2$, it follows that $\mathbf{r}$ is monotone iff $c \ge 0$. In particular, each of the functions $\mathbf{w}_j$ is monotone. Thus the function $\mathbf{w}$, which is a sum of monotone ramp functions, is also monotone.

It is also easy to show that the projection onto a hyperplane of the restriction of a monotone ramp function to the hyperplane is also a monotone ramp function. Thus the projection of the restriction of a monotone ramp function on any affine subspace is also a monotone ramp function. Hence the projection of the restriction of $\mathbf{w}$ on any affine subspace is a sum of n monotone ramp functions.

Each ramp function $\mathbf{w}_j$ is bounded, continuous, and smooth except on the pair of *break hyperplanes*

$$
\begin{aligned}
L_j &= \{\mathbf{t} \in \mathbf{R}^m \mid \mathbf{V}_j \cdot \mathbf{t} = l_j\} \\
U_j &= \{\mathbf{t} \in \mathbf{R}^m \mid \mathbf{V}_j \cdot \mathbf{t} = u_j\}
\end{aligned}
$$

As before, we may define the sets of hyperplanes $\mathcal{L}, \mathcal{U}, \mathcal{H}$ and the break set $B(V, \mathbf{l}, \mathbf{u})$, such that $\mathbf{w}$ is a smooth function everywhere in $\mathbf{R}^m$ except in $B(V, \mathbf{l}, \mathbf{u})$. The $2n$ hyperplanes of $\mathcal{H}$ form an arrangement $\mathcal{A}(\mathcal{H})$. $\mathbf{w}$ is smooth on the closure of each cell of $\mathcal{A}(\mathcal{H})$.

We could easily solve for $\mathbf{t}^*$ in the equation $\mathbf{w}(\mathbf{t}^*) = \mathbf{w}^*$ if we knew in which cell of $\mathcal{A}(\mathcal{H})$ the solution lies, or equivalently, if we knew of the smooth function equal to $\mathbf{w}$ in that cell. We will follow the method of determining the smooth function.

3   An Algorithm Based on Construction of Arrangements

We will now develop an algorithm to solve (2). It follows from (23) that the set S of solutions $\mathbf{t}^*$ of (18) is nonempty iff S_c is nonempty for some cell c. Thus we may find a solution $\mathbf{x}^*$ to (2) if one exists, as follows :

1. Find a cell c of $\mathcal{A}(\mathcal{H})$ for which S_c is nonempty, or determine that there is no such cell, by examining each cell in turn.

2. If a nonempty S_c exists, find a point $\mathbf{t}^*$ in that S_c and compute a solution $\mathbf{x}^*$ to (2) using (17). If no nonempty S_c exists, report that the problem is infeasible.

Step 1 requires that we enumerate the cells of $\mathcal{A}(\mathcal{H})$ in some order. For each cell c we need to compute S_c. We will do this by constructing the sets cl $c = S_{c,0}, S_{c,1}, \ldots, S_{c,m} = S_c$ in sequence. We will begin with a representation of the set cl c and construct a representation of each successive $S_{c,k+1}$ from $S_{c,k}$ by intersecting it with the set $T_{c,k+1}$. Since the $T_{c,k}$s are determined by the corresponding $\mathbf{w}^c$s, we need to determine $\mathbf{w}^c$ for each cell c.

The smooth function $\mathbf{w}^c$ for a cell c is determined by c's position relative to the hyperplanes of $\mathcal{H}$. A straightforward method of determining $\mathbf{w}^c$s would compute the relative position anew for each cell of the arrangement. We will now describe a more efficient method. Define the function $\mathbf{y} : \mathbf{R}^m \to \mathbf{R}^n$ by

$$
\mathbf{y}(\mathbf{t}) = [V^T \mathbf{t}]_{\nabla f(\mathbf{l})}^{\nabla f(\mathbf{u})}
$$

Further, let $\mathbf{y}^c$ be the smooth (in fact, linear) function which is equal to $\mathbf{y}$ on cell c. Clearly

$$
\mathbf{w}^c(\mathbf{t}) = V(\nabla f)^{-1}\mathbf{y}^c(\mathbf{t})
$$

Each of the m components of $\mathbf{w}^c$ is a linear combination of functions of the n components of $\mathbf{y}^c$. We will maintain representations of both $\mathbf{w}^c$ and $\mathbf{y}^c$ for the cell currently being examined in Step 1.

Let c and d be adjacent cells of $\mathcal{A}(\mathcal{H})$ separated by a facet contained in either L_j or U_j. Then the linear functions $\mathbf{y}^c$ and $\mathbf{y}^d$ differ only by a linear expression in their jth components, and are identical in all other components. Thus if we enumerate the cells of $\mathcal{A}(\mathcal{H})$ by traversing its cell-adjacency graph, we may maintain representations of the current $\mathbf{w}^c$ and $\mathbf{y}^c$ by updating only a single component of $\mathbf{y}^c$ as we move from cell to cell.

We now describe the algorithm in greater detail. We will begin by constructing the face-incidence lattice of the arrangement, using the algorithm of Edelsbrunner, O'Rourke and Seidel [8]. Next we construct the cell-adjacency graph on the cells of the face-incidence lattice. Then we may execute Step 1. The loop through the cells is implemented by means of a traversal of the cell-adjacency graph. As we move from cell to cell, we update the functions $\mathbf{y}^c$ and $\mathbf{w}^c$ correspondingly. For each cell, we construct the sequence of $S_{c,k}$s beginning with $S_{c,0} = \text{cl } c$, for which we have a face-incidence lattice representation accessible (since we have been traversing the cell-adjacency graph). If the last element in the sequence is nonempty, we take any point $\mathbf{t}^*$ in it and compute $\mathbf{x}^*$ in Step 2 in a straightforward manner.

It now remains only to explain how to construct each $S_{c,k+1}$ from $S_{c,k}$. We first explain how to do this for the case of the least-distance problem. In this case $f(\mathbf{x}) = \frac{1}{2}\mathbf{x} \cdot \mathbf{x}$ and so $\nabla f(\mathbf{x}) = \mathbf{x}$. Hence each non-trivial $T_{c,k}$ is a hyperplane, and so $S_{c,0}, S_{c,1}, \ldots, S_{c,m}$ is a sequence of polytopes of non-increasing dimension. The face-incidence lattice of each polytope in the sequence can be constructed from that of the previous one by a method used by Edelsbrunner, O'Rourke and Seidel [8] as part of their construction of the face-incidence lattice of an arrangement. It is easy to see that each face of the polytope $S_{c,k+1}$ is the intersection of the hyperplane $T_{c,k+1}$ with a face of the polytope $S_{c,k}$. Hence $S_{c,k+1}$ has at most as many faces as $S_{c,k}$.

We now analyze the complexity of this algorithm for fixed m. The construction of the arrangement takes time $\Theta(n^m)$ for fixed m and the resulting structure has size $\Theta(n^m)$. Construction of the cell-adjacency graph also takes time $\Theta(n^m)$ since there are $\Theta(n^m)$ cells and faces. Traversal of the cell-adjacency graph takes time proportional to its size, i.e. $\Theta(n^m)$. As we move from cell to cell, only one component of the function $\mathbf{y}^c$ changes, so update of y^c takes $\Theta(1)$ for each traversal step. All of the m components of $\mathbf{w}^c$ will change correspondingly, but update of $\mathbf{w}^c$ takes time $\Theta(1)$ for fixed m. Thus over the entire traversal the updates take time $\Theta(n^m)$. Finally, construction of each polytope $S_{c,k+1}$ from $S_{c,k}$ takes time proportional to the size of $S_{c,k}$, and results in a polytope of no larger size, so construction of the sequence of polytopes $S_{c,0}, S_{c,1}, \ldots, S_{c,m}$ takes time proportional to the size of $S_{c,0}$. Hence construction of the sequences for all the cells takes time proportional to the sums of the sizes of the cells of the arrangement, and this is just the size of the arrangement, i.e. $\Theta(n^m)$. Thus the whole algorithm takes time $\Theta(n^m)$ for fixed m.

It is easy to see that the algorithm is strongly polynomial.

4 Queries and a Witness-Based Search Scheme

Let $\mathbf{w}$ be a sum of monotone ramp functions, and assume without loss of generality that we wish to find a zero $\mathbf{z}$ of $\mathbf{w}$, or show that a zero does not exist. We will accomplish this overall goal using the search scheme whose elementary steps are actions called enquiries. At the top level the algorithm makes a number of enquiries in m-dimensional space that depends only on m.

An *enquiry* with respect to a given hyperplane H indicates on which side of H the zero lies, or establishes that there is no zero. Denoting by $h(\mathbf{t})$ the value at $\mathbf{t}$ of the function associated with the hyperplane, the enquiry reports one of the following three conclusions:

- $h(\mathbf{t}) < 0 \Rightarrow \mathbf{w}(\mathbf{t}) \neq \mathbf{0}$

- $h(\mathbf{t}) > 0 \Rightarrow \mathbf{w}(\mathbf{t}) \neq \mathbf{0}$

- $\forall \mathbf{t} \ \ \mathbf{w}(\mathbf{t}) \neq \mathbf{0}$

A special case occurs when in the process of solving the enquiry we find a zero of $\mathbf{w}$ within hyperplane H; in this case, we just report that zero.

This is the rough outline of our algorithm. In one stage we solve a well-chosen set of enquiries. If one of them reports that $\mathbf{w}$ has no zero, (or finds a zero) we terminate. Otherwise, we have restricted possible locations of zeros of $\mathbf{w}$ to the closure of one cell c of the arrangement defined by the hyperplanes of enquiries. We simplify each of the monotone ramp functions by removing the break hyperplanes that are disjoint with the cell c. The way we choose the enquires assures that this step removes a constant fraction of break hyperplanes. Moreover, the new sum of monotone ramp functions has the same set of zeros, and we can start a new stage in respect to that simplified function. In the last stage, the number of break hyperplanes is so small that we can apply the algorithm from the previous section.

We now examine when the enquiry in respect to H may be solved by evaluation $\mathbf{w}$ in a single *witness point* $\mathbf{t}$ on H. If $\mathbf{w}(\mathbf{t}) = \mathbf{0}$, the enquiry is obviously solved. Othewise, let $\mathbf{z}$ be a zero of $\mathbf{w}$. By the monotonicity of $\mathbf{w}$,

$$(\mathbf{w}(\mathbf{t}) - \mathbf{w}(\mathbf{z})) \cdot (\mathbf{t} - \mathbf{z}) \geq 0$$

so

$$\mathbf{w}(\mathbf{t}) \cdot \mathbf{z} \leq \mathbf{w}(\mathbf{t}) \cdot \mathbf{t}$$

or

$$-\mathbf{w}(\mathbf{t}) \cdot (\mathbf{z} - \mathbf{t}) \geq 0$$

Thus evaluation of $\mathbf{w}$ at the point $\mathbf{t}$ provides a halfspace

$$\{\mathbf{y} \mid \mathbf{w}(\mathbf{t}) \cdot \mathbf{y} \leq \mathbf{w}(\mathbf{t}) \cdot \mathbf{t}\}$$

within which any zero $\mathbf{z}$ must lie. If this halfspace happens to be one of the two subspaces associated with the hyperplane H, $\mathbf{w}$ is a *witness point*. This in turn happens exactly when the component of $\mathbf{w}(\mathbf{t})$ that is tangential to H equals zero. Thus to find a witness point for the enquiry in respect to H we need to look for a zero of the tangent vector field on H. This problem can clearly be solved recursively.

Unfortunately, in some situations this tangent vector field does not possess a zero. The simplest example can be provided for $m = 2$. Let $\mathbf{w}(x, y)$ be a monotone linear funcion $(-y, x)$ and let H be defined by $h(x, y) = y + 1$. In this case, the tangent vector field is always equal to 1, and hence it has no zeros. However, we will show that in the absence of a *witness point* we can find a *witness ray* with equally good properties: evaluation of the witness takes time proportional to the number of break hyperplanes of the ramp functions and it resolves the enquiry.

Given a vector $\mathbf{a}$ and a unit vector $\widehat{\mathbf{b}}$, we define ray $(\mathbf{a}, \widehat{\mathbf{b}})$ as the set $\{\mathbf{a} + \alpha \widehat{\mathbf{b}} \mid \alpha \geq 0\}$. We define $\mathbf{w}(\mathbf{a}, \widehat{\mathbf{b}})$ as the limit

$$\lim_{\alpha \to +\infty} \mathbf{w}(\mathbf{a} + \alpha \widehat{\mathbf{b}})$$

(provided it exists). Similarly we denote by $\widehat{\mathbf{w}}(\mathbf{a}, \widehat{\mathbf{b}})$ the limiting unit vector

$$\lim_{\alpha \to +\infty} \widehat{\mathbf{w}}(\mathbf{a} + \alpha \widehat{\mathbf{b}})$$

Now we can formally define the witnesses that we will use to resolve our enquires. Because of the recursive nature of our algorithm, we will need to solve an enquiry in respect to arbitrary affine space A. We denote by $\mathbf{w}^A$ the projection onto A of the restriction $\mathbf{w}|_A$ of the function $\mathbf{w}$ to A. An enquiry in respect to A will be resolved by finding a witness for $\mathbf{w}^A$.

A *witness* for $\mathbf{w}^A$ (with respect to A) is defined to be either of the following:

- a *witness point* $\mathbf{a} \in A$ such that $\mathbf{w}^A(\mathbf{a}) = \mathbf{0}$ (i.e. a zero of $\mathbf{w}^A$)

- a *witness ray* $(\mathbf{a}, \widehat{\mathbf{b}}) \subset A$ such that $\widehat{\mathbf{w}^A}(\mathbf{a}, \widehat{\mathbf{b}}) = -\widehat{\mathbf{b}}$,

Note that when there is no witness point, the limit of the form $\widehat{\mathbf{w}^A}(\mathbf{a}, \widehat{\mathbf{b}})$ is well-defined, for then $\mathbf{w}^A(\mathbf{a} + \alpha \widehat{\mathbf{b}})$ is always non-zero.

We prove several geometric properties of witnesses for functions which are sums of monotone ramp functions. Lemma 3 shows that to evaluate a witness ray it suffices to evaluate a point. Theorem 4 shows that $\mathbf{w}^A$ must have a witness, but never a witness point together with a witness ray. Now assume that the affine subspace A is a hyperplane in another affine subspace B. Theorem 7 shows that if we get a witness for $\mathbf{w}^A$ and evaluate $\mathbf{w}^B$ on this witness, then either we get a witness for B, or we restrict the search for the witness for $\mathbf{w}^B$ to one of the halfspaces of B defined by A. Taken together, these properties justify our choice of the witness definition, and allow to obtain a linear time recursive algorithm for our problem.

The next lemma follows from the piecewise linearity and monotonicity of the class of functions of interest:

Lemma 3 *If $\mathbf{w}$ is a sum of monotone ramp functions and $(\mathbf{a}, \widehat{\mathbf{b}})$ is a witness ray for $\mathbf{w}^A$, then there exists $\alpha_0, \beta > 0$ such that for all $\alpha \geq \alpha_0$*

$$\mathbf{w}^A(\mathbf{a} + \alpha\widehat{\mathbf{b}}) = \mathbf{w}^A(\mathbf{a}, \widehat{\mathbf{b}}) = -\beta\widehat{\mathbf{b}}$$

PROOF.

Since $\mathbf{w}$ is a sum of monotone ramp functions, so is $\mathbf{w}^A$. Consider the arrangement $\mathcal{A}(\mathcal{H}^A)$ of break hyperplanes of the ramp functions of $\mathbf{w}^A$. There exists $\alpha_0 > 0$ and a cell $c \in \mathcal{A}(\mathcal{H}^A)$ such that for every $\alpha \geq \alpha_0$ the point $\mathbf{a} + \alpha\widehat{\mathbf{b}}$ is located in c. Recall that within c the function $\mathbf{w}^A$ is monotone and affine. Because it is affine, we have vectors $\mathbf{c}$ and $\mathbf{v}$ such that $\mathbf{w}^A(\mathbf{a} + \alpha\widehat{\mathbf{b}}) = \mathbf{c} + \alpha\mathbf{v}$ for all $\alpha \geq \alpha_0$. Because it is monotone, the product $\widehat{\mathbf{v}} \cdot \widehat{\mathbf{b}}$ is nonnegative.

Now we observe that $\mathbf{v} = 0$. Otherwise

$$\widehat{\mathbf{w}^A}(\mathbf{a}, \widehat{\mathbf{b}}) = \lim_{\alpha \to +\infty} \widehat{\mathbf{w}^A}(\mathbf{a} + \alpha\widehat{\mathbf{b}}) = \widehat{\mathbf{v}}.$$

Because $(\mathbf{a}, \widehat{\mathbf{b}})$ is a witness ray for $\mathbf{w}^A$, we have $\widehat{\mathbf{v}} \cdot \widehat{\mathbf{b}} = -\widehat{\mathbf{b}} \cdot \widehat{\mathbf{b}} = -1$, a contradiction, since this product is nonnegative.

This shows that for all $\alpha \geq \alpha_0$ the value $\mathbf{w}^A(\mathbf{a} + \alpha\widehat{\mathbf{b}})$ equals $\mathbf{c}$, and as a result $\mathbf{w}^A(\mathbf{a}, \widehat{\mathbf{b}})$ equals $\mathbf{c}$ as well. By the definition of a witness ray, $\mathbf{c}$ equals $-\beta\widehat{\mathbf{b}}$ for some positive β. ∎

As a consequence, $\mathbf{w}^A$ must have a finite nonzero value at a witness ray.

Note that our general goal is to find a zero of $\mathbf{w}$ or to verify that none exists. If we find a witness point for $\mathbf{w}$, we solve this problem in a positive fashion, as it is nothing else but a zero of $\mathbf{w}$. We will show that finding a witness ray for $\mathbf{w}$ shows that no zero of $\mathbf{w}$ exists. Moreover, this problem can always be solved by finding a witness. This result is summed up by the following theorem:

Theorem 4 *A sum of monotone ramp functions possesses either a witness point or a witness ray but not both.*

The proof of this theorem is immediate from Lemmas 5 and 6 which prove uniqueness of the type of witness and existence of some type of witness, respectively.

Lemma 5 *If $\mathbf{w}$ is a sum of monotone ramp functions and $(\mathbf{a}, \widehat{\mathbf{b}})$ is a witness ray for $\mathbf{w}^A$, then for every $\mathbf{x} \in A$, either $\mathbf{w}^A(\mathbf{x}) = \mathbf{w}^A(\mathbf{a}, \widehat{\mathbf{b}})$, or $\| \mathbf{w}^A(\mathbf{x}) \| > \| \mathbf{w}^A(\mathbf{a}, \widehat{\mathbf{b}}) \|$.*

PROOF. Assume to the contrary, i.e. $\mathbf{w}^A(\mathbf{a}, \widehat{\mathbf{b}}) = \mathbf{y}$, $\mathbf{w}^A(\mathbf{x}) = \mathbf{v}$, $\mathbf{y} \neq \mathbf{v}$ and $\| \mathbf{v} \| \leq \| \mathbf{y} \|$. In this case we have

$$(\mathbf{y} - \mathbf{v}) \cdot \mathbf{y} > 0$$

By Lemma 3 we have $\beta, \alpha_0 > 0$ such that $\mathbf{w}^A(\mathbf{a}, \widehat{\mathbf{b}}) = \mathbf{w}^A(\mathbf{a} + \alpha\widehat{\mathbf{b}}) = \mathbf{y} = -\beta\widehat{\mathbf{b}}$ for every $\alpha > \alpha_0$. By monotonicity of $\mathbf{w}^A$ we obtain the following for all $\alpha > \alpha_0$:

$$
\begin{aligned}
0 &\leq \frac{\beta}{\alpha}(\mathbf{w}^A(\mathbf{a} + \alpha\widehat{\mathbf{b}}) - \mathbf{w}^A(\mathbf{x})) \cdot (\mathbf{a} + \alpha\widehat{\mathbf{b}} - \mathbf{x}) \\
&= \frac{\beta}{\alpha}(\mathbf{y} - \mathbf{v}) \cdot (\mathbf{a} - \frac{\alpha}{\beta}\mathbf{y} - \mathbf{x}) \\
&= -(\mathbf{y} - \mathbf{v}) \cdot \mathbf{y} + \frac{\beta}{\alpha}(\mathbf{y} - \mathbf{v}) \cdot (\mathbf{a} - \mathbf{x})
\end{aligned}
$$

By going to the limit with α, we obtain $(\mathbf{y} - \mathbf{v}) \cdot \mathbf{y} \leq 0$, a contradiction. This contradiction shows that there is no witness point. $\blacksquare$

The preceding lemma shows that a witness ray cannot coexist with a witness point; otherwise we would have a non-zero value $\mathbf{y}$ of the witness ray, and the zero value of the witness point, obviously, $\| \mathbf{0} \| < \| \mathbf{y} \|$.

Our next proof makes use of the following well-known result, which is a corollary of the celebrated Brouwer fixed point theorem [18].

Result (Hairy ball theorem) *Every non-vanishing continuous vector field on B^{n+1} points normally inward at some point of S^n, and normally outward at some point of S^n.*

Here B^{n+1} is the $(n + 1)$-dimensional unit ball, and S^n is its surface.

Lemma 6 *A sum of monotone ramp functions must possess either a witness point or a witness ray.*

PROOF. Let $\mathbf{w} : \mathbf{R}^m \to \mathbf{R}^m$ be a sum of n monotone ramp functions, and assume that $\mathbf{w}$ possesses no finite witness, i.e. zero. Then $\widehat{\mathbf{w}}$ is defined throughout $\mathbf{R}^m$. We will prove the existence of a witness ray for $\mathbf{w}$.

Consider a sequence $B_j (j \in \mathbf{N})$ of balls centered at the origin and with radii increasing without bound. By the hairy ball theorem, there is at least one point $\mathbf{d}_j$ on the surface of each B_j at which $\widehat{\mathbf{w}}$ points normally inward. Thus we obtain a sequence $\mathbf{d}_j (j \in \mathbf{N})$ of points with norm increasing without bound, for each of which $\widehat{\mathbf{w}}(\mathbf{d}_j) = -\widehat{\mathbf{d}}_j$.

The break hyperplanes of the ramp functions partition $\mathbf{R}^m$ into a finite number of cells. Thus there exists some cell whose closure C contains an infinite subsequence $\mathbf{c}_j (j \in \mathbf{N})$ of the $\mathbf{d}_j$s. Since the $\mathbf{c}_j$s form an unbounded set, C must be an unbounded polytope.

Now consider the sequence $\widehat{\mathbf{c}}_j (j \in \mathbf{N})$. Since it takes values in a compact set (the unit sphere), there exists a subsequence $\mathbf{b}_j (j \in \mathbf{N})$ of $\mathbf{c}_j$s such that $\widehat{\mathbf{b}}_j (j \in \mathbf{N})$ converges to some limit, say $\widehat{\mathbf{b}}$. Define the sequences $\mathbf{e}_j, \beta_j (j \in \mathbf{N})$ by

$$
\mathbf{e}_j = \widehat{\mathbf{b}}_j - \widehat{\mathbf{b}}, \quad \beta_j = \| \mathbf{b}_j \|
$$

Then

$$\lim_{j\to\infty} \mathbf{e}_j = \mathbf{0}, \quad \lim_{j\to\infty} \beta_j = \infty \tag{30}$$

Further,

$$\mathbf{b}_j = \beta_j \widehat{\mathbf{b}}_j = \beta_j(\widehat{\mathbf{b}} + \mathbf{e}_j)$$

Within C, the function $\mathbf{w}$ reduces to a monotone affine function $\mathbf{w}^C$, i.e.

$$\mathbf{w}^C(\mathbf{t}) = A\mathbf{t} + \mathbf{a} \tag{31}$$

for some $A \in \mathbf{R}^{m\times m}$ positive semi-definite, $\mathbf{a} \in \mathbf{R}^m$. Then for all $j \in \mathbf{N}$,

$$
\begin{aligned}
\mathbf{w}^C(\mathbf{b}_j) &= \mathbf{w}^C(\beta_j(\widehat{\mathbf{b}} + \mathbf{e}_j)) \\
&= \beta_j A\widehat{\mathbf{b}} + \beta_j A\mathbf{e}_j + \mathbf{a} \\
&= \beta_j(A\widehat{\mathbf{b}} + A\mathbf{e}_j + \frac{\mathbf{a}}{\beta_j})
\end{aligned}
$$

so

$$\frac{\mathbf{w}^C(\mathbf{b}_j)}{\beta_j} = A\widehat{\mathbf{b}} + A\mathbf{e}_j + \frac{\mathbf{a}}{\beta_j}$$

It follows from (30) that the second and third terms above vanish when $j \to \infty$. Thus

$$\lim_{j\to\infty} \frac{\mathbf{w}^C(\mathbf{b}_j)}{\beta_j} = A\widehat{\mathbf{b}}$$

If $A\widehat{\mathbf{b}} \neq \mathbf{0}$ then the above relation implies that

$$-\widehat{\mathbf{b}} = \lim_{j\to\infty} \mathbf{w}^C(\mathbf{b}_j) = \widehat{A\mathbf{b}}$$

Comparing with (31) we obtain a contradiction to the assumption that $\mathbf{w}^C$ is monotone. Thus $A\widehat{\mathbf{b}} = 0$. Consequently $\mathbf{w}^C$ does not depend on the $\widehat{\mathbf{b}}$-coordinate.

We now claim that for any $\mathbf{a} \in C$ the ray $(\mathbf{a}, \widehat{\mathbf{b}})$ lies within C. Suppose this were false. Then there exists some halfspace H including C such that an infinite part of the ray $(\mathbf{a}, \widehat{\mathbf{b}})$ lies outside H. Say that $H = \{\mathbf{t} \mid \mathbf{h} \cdot \mathbf{t} \geq h\}$. Our assumption means that $\mathbf{h} \cdot \mathbf{a} \geq h$ while $\mathbf{h} \cdot (\mathbf{a} + \alpha\widehat{\mathbf{b}}) < h$ for some α. This occurs only if $\mathbf{h} \cdot \widehat{\mathbf{b}} < 0$. We can rescale $\mathbf{h}$ and h in such a way that $\mathbf{h} \cdot \widehat{\mathbf{b}} = -2$.

Now since $\lim_{j\to\infty} \widehat{\mathbf{b}}_j = \widehat{\mathbf{b}}$, we have $\mathbf{h} \cdot \widehat{\mathbf{b}}_j < -1$ for all sufficiently large j, moreover, for sufficiently large j we have $\| \mathbf{b}_j \| > h$. This in turn implies that

$$\mathbf{h} \cdot \mathbf{b}_j \ < \ -|h| \ \leq \ h$$

hence for all sufficiently large js $\mathbf{b}_j$ is not in H, and consequently in C, a contradiction.

Suppose that the sequence $\mathbf{w}(\mathbf{b}_k)(k \in \mathbf{N})$ is bounded in magnitude. The $\mathbf{w}(\mathbf{b}_k)$s take values in the image of the closed region C under $\mathbf{w}$. This is the image of a closed polytope under an affine linear function, and is consequently closed. Hence the limit

$\mathbf{w}_l$ of the $\mathbf{w}(\mathbf{b}_k)$s (with $\widehat{\mathbf{w}}_l = -\widehat{\mathbf{b}}$) occurs in the image of C, and therefore there is a point $\mathbf{d} \in C$ for which $\widehat{\mathbf{w}}(\mathbf{d}) = -\widehat{\mathbf{b}}$.

We already argued that the ray $(\mathbf{d}, \widehat{\mathbf{b}})$ is included in C and that within C the value of $\mathbf{w}^A$ does not depend on b-coordinate. Therefore $(\mathbf{d}, \widehat{\mathbf{b}})$ is the desired witness ray.

It only remains to show that the $\mathbf{w}(\mathbf{b}_k)$s are bounded. We prove it by contradiction. Suppose there exists a subsequence of the $\mathbf{b}_k$s with norm diverging to infinity. For $\mathbf{b}_k$s from this subsequence, we can show that $\widehat{\mathbf{b}_k - \mathbf{b}_1}$ converges to $\widehat{\mathbf{b}}$ while $\widehat{\mathbf{w}(\mathbf{b}_k) - \mathbf{w}(\mathbf{b}_1)}$ converges to $-\widehat{\mathbf{b}}$. Thus for sufficiently large k we have $(\mathbf{b}_k - \mathbf{b}_1) \cdot (\mathbf{w}(\mathbf{b}_k) - \mathbf{w}(\mathbf{b}_1)) < 0$, which contradicts the monotonicity of $\mathbf{w}$. ∎

5   A Linear Time Recursive Algorithm

Theorem 4 showed that our original problem can be solved by finding a witness for $\mathbf{w}$. Now, assume that an affine space B is a union of two open halfspaces B_- and B_+, and a hyperplane A. If there exists a witness point $\mathbf{z}$ for $\mathbf{w}^B$, then clearly $\mathbf{z} \notin B_-$ or $\mathbf{z} \notin B_+$. It is also easy to observe that if there exists a witness ray $(\mathbf{a}, \widehat{\mathbf{b}})$ for $\mathbf{w}^B$, then some witness ray $(\mathbf{a} + \alpha \widehat{\mathbf{b}}, \widehat{\mathbf{b}})$ is disjoint with B_- or with B_+. Therefore the enquiry problem—tell on which side of A is a witness point located or provide a witness for $\mathbf{w}^B$ contained in A—can be extended to witness rays. The next theorem shows that the extended enquiry problem can be always solved by finding a witness.

Let us say that a witness point $\mathbf{z}$ for $\mathbf{w}^A$ points to B_+ if $\mathbf{z} + \mathbf{w}^B(\mathbf{z}) \in B_+$. Similarly, we say that a witness ray $(\mathbf{a}, \widehat{\mathbf{b}})$ points to B_+ if for all sufficiently large αs we have

$$(\mathbf{a} + \alpha \widehat{\mathbf{b}}) + \mathbf{w}^B(\mathbf{a} + \alpha \widehat{\mathbf{b}}) \in B_+$$

Note, that if a witness for $\mathbf{w}^A$ does not point to either B_+, or B_-, then it provides a witness for $\mathbf{w}^B$, and thus solves the enquiry. The following theorem shows that in the remaining case the enquiry is solved as well.

Theorem 7 (Exclusion by a witness) *Assume that A, B are affine subspaces, $B - A$ has two connected components, B_- and B_+, and that there exists a witness for $\mathbf{w}^A$ that points to B_+. Then no witness for $\mathbf{w}^B$ is included in B_+.*

We divide the proof of the theorem into two lemmas, that deal with the following two cases: when $\mathbf{w}^A$ has a witness point, and when it has a witness ray.

Lemma 8 *If a witness point for $\mathbf{w}^A$ points to B_+ then B_+ contains no witnesses for $\mathbf{w}^B$.*

PROOF. Let $\mathbf{t}$ be the witness for $\mathbf{w}^A$ pointing to B_+ and let $\mathbf{s} = \mathbf{w}^B(\mathbf{t})$. Under our assumption, $B_+ = \{\mathbf{y} \in B \mid \mathbf{s} \cdot (\mathbf{y} - \mathbf{t}) > 0\}$. When we introduced the notion of the witness point at the beginning of this section, we have shown that this set cannot contain witness points of $\mathbf{w}^B$. Thus it remains to consider a witness ray $(\mathbf{u}, \hat{\mathbf{v}})$ of $\mathbf{w}^B$. By Lemma 3 we may assume without loss of generality that $\mathbf{w}^B$ is constant within this ray, in particular, for some $\beta > 0$ and every $\alpha \geq 0$

$$\mathbf{w}^B(\mathbf{u} + \alpha\hat{\mathbf{v}}) = -\beta\hat{\mathbf{v}}$$

By the monotonicity of $\mathbf{w}^B$, we have for any point $\mathbf{u} + \alpha\hat{\mathbf{v}}$ on the ray $(\mathbf{u}, \hat{\mathbf{v}})$

$$\begin{aligned}
0 &\leq (\mathbf{w}^B(\mathbf{u} + \alpha\hat{\mathbf{v}}) - \mathbf{w}^B(\mathbf{t})) \cdot (\mathbf{u} + \alpha\hat{\mathbf{v}} - \mathbf{t}) \\
&= (-\beta\hat{\mathbf{v}} - \mathbf{s}) \cdot (\mathbf{u} + \alpha\hat{\mathbf{v}} - \mathbf{t})
\end{aligned}$$

and by taking the limit with $\alpha \to \infty$

$$\begin{aligned}
0 &\leq (-\beta\hat{\mathbf{v}} - \mathbf{s}) \cdot \hat{\mathbf{v}} \\
\mathbf{s} \cdot \hat{\mathbf{v}} &\leq -\beta\hat{\mathbf{v}} \cdot \hat{\mathbf{v}} < 0
\end{aligned}$$

Because $\mathbf{s} \cdot \hat{\mathbf{v}} < 0$, the ray $(\mathbf{u}, \hat{\mathbf{v}})$ cannot be contained in B_+. $\blacksquare$

Lemma 9 *If a witness ray for $\mathbf{w}^A$ points to B_+ then B_+ contains no witnesses for $\mathbf{w}^B$.*

PROOF. By translating the coordinate system, we may assume that the assumed witness ray is of the form $(\mathbf{0}, \hat{\mathbf{t}})$. Because $\mathbf{w}$ is a piecewise linear function, there exists α_0 such that for all $\alpha \geq \alpha_0$ we evaluate $\mathbf{w}^B(\alpha\hat{\mathbf{t}})$ according to a single linear formula. By Lemma 3 and the fact, that this witness ray points to B_+, this formula can be written as follows

$$\mathbf{w}^B(\alpha\hat{\mathbf{t}}) = -\beta\hat{\mathbf{t}} + f(\alpha)\mathbf{s}$$

where β is a positive constant, f is a positive linear function (for $\alpha \geq \alpha_0$) and $\mathbf{s}$ a vector such that $B_+ = \{\mathbf{y} \mid \mathbf{s} \cdot \mathbf{y} > 0\}$.

Now, let $\mathbf{u} \in B$ be a witness point for $\mathbf{w}^B$, i.e. $\mathbf{w}^B(\mathbf{u}) = \mathbf{0}$. By applying the monotonicity of $\mathbf{w}^B$ for $\mathbf{u}$ and for any point $\alpha\hat{\mathbf{t}}(\alpha \geq \alpha_0)$ on the ray $(\mathbf{0}, \hat{\mathbf{t}})$ we obtain

$$\begin{aligned}
0 &\leq (\mathbf{w}^B(\alpha\hat{\mathbf{t}}) - \mathbf{w}^B(\mathbf{u})) \cdot (\alpha\hat{\mathbf{t}} - \mathbf{u}) \\
&= (-\beta\hat{\mathbf{t}} + f(\alpha)\mathbf{s}) \cdot (\alpha\hat{\mathbf{t}} - \mathbf{u}) \\
&= -\beta\alpha\hat{\mathbf{t}} \cdot \hat{\mathbf{t}} + \beta\hat{\mathbf{t}} \cdot \mathbf{u} - f(\alpha)\mathbf{s} \cdot \mathbf{u}
\end{aligned}$$

(note that $\mathbf{s} \cdot \hat{\mathbf{t}} = 0$). If $\lim_{\alpha \to +\infty} f(\alpha)\alpha^{-1} = 0$, then we have a contradiction: $0 \leq -\beta\hat{\mathbf{t}} \cdot \hat{\mathbf{t}}$. Thus assume that $\lim_{\alpha \to +\infty} f(\alpha)\alpha^{-1} = a > 0$. In this case, by taking the limit we obtain

$$\begin{aligned}
0 &\leq -\beta\hat{\mathbf{t}} \cdot \hat{\mathbf{t}} - a\mathbf{s} \cdot \mathbf{u} \\
\mathbf{s} \cdot \mathbf{u} &\leq -\frac{\beta}{a}\hat{\mathbf{t}} \cdot \hat{\mathbf{t}} < 0
\end{aligned}$$

hence $\mathbf{u} \in B_-$.

Next, let $(\mathbf{u}, \widehat{\mathbf{v}})$ be a witness ray for $\mathbf{w}^B$. By Lemma 3 we may assume that there exists $\gamma > 0$ such that for every $\alpha \geq 0$ we have $\mathbf{w}^B(\mathbf{u} + \alpha \widehat{\mathbf{v}}) = -\gamma \widehat{\mathbf{v}}$. As a preliminary step, we will show that $\widehat{\mathbf{t}} \cdot \widehat{\mathbf{v}}$ is negative.

By applying the monotonicity of $\mathbf{w}^B$ for $\mathbf{u}$ and for any point $\alpha\widehat{\mathbf{t}}(\alpha \geq \alpha_0)$ on the ray $(\mathbf{0}, \widehat{\mathbf{t}})$ we obtain

$$
\begin{aligned}
0 \;\leq\; & (\mathbf{w}^B(\alpha\widehat{\mathbf{t}}) - \mathbf{w}^B(\mathbf{u})) \cdot (\alpha\widehat{\mathbf{t}} - \mathbf{u}) \\
=\; & (-\beta\widehat{\mathbf{t}} + f(\alpha)\mathbf{s} - \gamma\widehat{\mathbf{v}}) \cdot (\alpha\widehat{\mathbf{t}} - \mathbf{u})
\end{aligned}
$$

By taking the limit

$$
\begin{aligned}
0 \;\leq\; & (-\beta\widehat{\mathbf{t}} + f(\alpha)\mathbf{s} - \gamma\widehat{\mathbf{v}}) \cdot \widehat{\mathbf{t}} \\
=\; & -\beta\widehat{\mathbf{t}} \cdot \widehat{\mathbf{t}} - \gamma\widehat{\mathbf{t}}\widehat{\mathbf{v}}) \\
\widehat{\mathbf{t}} \cdot \widehat{\mathbf{v}} \;\leq\; & -\frac{\beta}{\gamma}\widehat{\mathbf{t}} \cdot \widehat{\mathbf{t}} \;<\; 0
\end{aligned}
$$

Now, we apply the monotonicity of $\mathbf{w}^B$ for any pair of points $\alpha\widehat{\mathbf{t}}$ and $\mathbf{u} + \alpha\widehat{\mathbf{v}}$:

$$
0 \;\leq\; (\mathbf{w}^B(\alpha\widehat{\mathbf{t}}) - \mathbf{w}^B(\mathbf{u} + \alpha\widehat{\mathbf{v}})) \cdot (\alpha\widehat{\mathbf{t}} - \mathbf{u} - \alpha\widehat{\mathbf{v}}))
$$

By taking the limit

$$
\begin{aligned}
0 \;\leq\; & (\mathbf{w}^B(\alpha\widehat{\mathbf{t}}) - \mathbf{w}^B(\mathbf{u} + \alpha\widehat{\mathbf{v}})) \cdot (\widehat{\mathbf{t}} - \widehat{\mathbf{v}}) \\
=\; & (-\beta\widehat{\mathbf{t}} + \frac{f(\alpha)}{\alpha}\mathbf{s} + \gamma\widehat{\mathbf{v}}) \cdot (\widehat{\mathbf{t}} - \widehat{\mathbf{v}})
\end{aligned}
$$

Observe that $\widehat{\mathbf{t}} \cdot \widehat{\mathbf{v}} < 0$ implies that both $-\widehat{\mathbf{t}} \cdot (\widehat{\mathbf{t}} - \widehat{\mathbf{v}})$ and $\widehat{\mathbf{v}}(\widehat{\mathbf{t}} - \widehat{\mathbf{v}})$ are negative. Since the last expression is nonnegative, we must have

$$
\frac{f(\alpha)}{\alpha}\mathbf{s} \cdot (\widehat{\mathbf{t}} - \widehat{\mathbf{v}}) \;>\; 0
$$

and because $\mathbf{s} \cdot \widehat{\mathbf{t}} = 0$, this means that $\mathbf{s} \cdot \widehat{\mathbf{v}} < 0$. The latter implies

$$
\mathbf{s} \cdot (\mathbf{u} + \alpha\widehat{\mathbf{v}}) < 0
$$

for sufficiently large α, hence the ray $(\mathbf{u}, \widehat{\mathbf{v}})$ cannot be contained in B_+. $\blacksquare$

To solve the problem it suffices to determine the location of the witnesses (zeros) for $\mathbf{w}$ with respect to each of the defining hyperplanes. Our algorithm works in phases. In each phase we determine the position of the witnesses with respect to some fraction p of the defining hyperplanes. We may then remove these hyperplanes from the description of the problem, obtaining a modified function $\mathbf{w}'$ whose roots are the same as the roots of $\mathbf{w}$. We repeat phases until all hyperplanes have been removed from the description. At this point the location of the zeros is known with respect to all hyperplanes.

Each phase will use the search strategy of Megiddo [15] and Dyer [6]. Given some number k of active hyperplanes, we will decide on a set of enquiries, i.e. enquiry hyperplanes. The choice of enquiries will depend on the active hyperplanes, but their number is a function only of the m and not of k. As we explained, to solve an enquiry, we solve an identical problem but with a lower value of m (note that in the original setting three recursive subproblems are formed to solve an enquiry). After obtaining the answers for our enquiries, we either get an answer to the original problem, or we partially locate the solution, namely we identify the only cell of the arrangement that may contain a witness.

Note, that if the witnesses are located in more than one cell, then for one of the enquiries they are located on both sides of the respective hyperplane, say A. Our algorithm solves that enquiry by finding a witness for w^A, say $\mathbf{t}$. Because $\mathbf{t}$ cannot point to either of the sides of A, it must be a witness for w (see the discussion before Theorem 7).

Consequently, the interesting case is when we prove that all witnesses for the original problem lie in one closed cell C. Then we can drop all break hyperplanes that are not adjacent to that cell, and the simplified function w' will have the same set of witnesses (as we will show below). The method of choosing the enquiries guarantees that a finite fraction of break hyperplanes gets removed. The analysis of the running time may follow that of Dyer [6], except that in each phase we introduce three times less recursive subproblems, hence our procedure is somewhat faster (by factor 3^m), and surely linear in k. To close our argument, we remains to prove our claim that $\mathbf{w}$ and $\mathbf{w}'$ have the same set of witnesses.

First, notice that the witnesses for $\mathbf{w}$ are also the witnesses for $\mathbf{w}'$, because all of them are contained in the cell C in which the function did not change. Actually, there exist a positive distance ϵ, such that the function does not change not only within C, but also within distance ϵ from C. We consider two cases: first when our witness are witness points, and then when they are witness rays.

We start from the following lemma.

Lemma 10 *The set of zeros of a continuous monotone vector field is convex.*

PROOF. Let $\mathbf{f}$ be our vector field. It suffices to show that if $\mathbf{f}(\mathbf{0}) = \mathbf{f}(z) = \mathbf{0}, 0 < t < 1$, and $\mathbf{f}(t\mathbf{z}) = \mathbf{y}$, then $\mathbf{y} = \mathbf{0}$. By the monotonicity of $\mathbf{f}$

$$0 \leq \frac{1}{t}(\mathbf{f}(t\mathbf{z}) - \mathbf{f}(\mathbf{0})) \cdot (t\mathbf{z} - \mathbf{0}) = \mathbf{y} \cdot \mathbf{z}$$

$$0 \leq \frac{1}{1-t}(\mathbf{f}(t\mathbf{z}) - \mathbf{f}(\mathbf{z})) \cdot (t\mathbf{z} - \mathbf{z}) = -\mathbf{y} \cdot \mathbf{z}$$

Thus $\mathbf{y} \cdot \mathbf{z} = 0$. Now for the nonnegative values of ϵ we consider $\Delta(\epsilon)$ defined as

$$\Delta(\epsilon) = \mathbf{f}(t\mathbf{z} - \epsilon\mathbf{y}) - \mathbf{f}(t\mathbf{z})$$

By the monotonicity of $\mathbf{f}$

$$
\begin{aligned}
0 \;\le\;& \frac{1}{t}(\mathbf{f}(t\mathbf{z} - \epsilon\mathbf{y}) - \mathbf{f}(\mathbf{0})) \cdot (t\mathbf{z} - \epsilon\mathbf{y} - \mathbf{0}) \\
=\;& (\mathbf{y} + \Delta(\epsilon)) \cdot (\mathbf{z} - \frac{\epsilon}{t}\mathbf{y}) \\
0 \;\le\;& \frac{1}{1-t}(\mathbf{f}(t\mathbf{z} - \epsilon\mathbf{y}) - \mathbf{f}(\mathbf{z})) \cdot (t\mathbf{z} - \epsilon\mathbf{y} - \mathbf{z}) \\
=\;& (\mathbf{y} + \Delta(\epsilon)) \cdot (-\mathbf{z} - \frac{\epsilon}{1-t}\mathbf{y})
\end{aligned}
$$

By adding these two inequalities together and simplifying, we obtain $(\mathbf{y}+\Delta(\epsilon))\cdot\mathbf{y} \le 0$. Therefore for every positive ϵ we have $\|\,\Delta(\epsilon)\,\| \ge \|\,\mathbf{y}\,\|$. Because $\Delta(0) = \mathbf{0}$ and Δ is continuous, this implies $\mathbf{y} = \mathbf{0}$. $\blacksquare$

Now, suppose that $\mathbf{w}$ has a zero in the cell C. Because $\mathbf{w}'$ is monotone, by Lemma 5, it cannot have any witness rays. Would it have some "new" witness points, i.e. "new" zeros, they would had to be located in the distance larger than ϵ from C. However, the convex hull of old zeros and new zeros together must contain points outside C but arbitrarily close to C: for these points $\mathbf{w}'$ coinsides with $\mathbf{w}$, so these would be zeros of $\mathbf{w}$ outside C, a contradiction.

The following lemma shows that a very similar reasoning can be applied in the case when C contains a witness ray. We show that the locus of the witness rays coincides with the locus of one particular value of $\mathbf{w}$ (according to Lemma 5 this is the unique value that minimizes the norm).

Lemma 11 *Let $\mathbf{w}$ be a sum of monotone ramp fuctions and let $(\mathbf{a}, \widehat{\mathbf{b}})$ be a witness ray for $\mathbf{w}$. Let $\mathbf{y} = \mathbf{w}(\mathbf{a}, \widehat{\mathbf{b}})$. Then $(\mathbf{c}, \widehat{\mathbf{d}})$ is a witness ray iff $\widehat{\mathbf{d}} = \widehat{\mathbf{b}}$ and for some α_0 we have $\mathbf{w}(\mathbf{c} + \alpha_0\widehat{\mathbf{b}}) = \mathbf{y}$.*

PROOF.

Assume that $(\mathbf{c}, \widehat{\mathbf{d}})$ is a witness ray for $\mathbf{w}$. By Lemma 3, there exists α_0, β and γ such that for all $\alpha > \alpha_0$

$$
\begin{aligned}
\mathbf{w}(\mathbf{a} + \alpha\widehat{\mathbf{b}}) \;&=\; \mathbf{w}(\mathbf{a}, \widehat{\mathbf{b}}) \;=\; -\beta\widehat{\mathbf{b}} \;=\; \mathbf{y} \\
\mathbf{w}(\mathbf{c} + \alpha\widehat{\mathbf{d}}) \;&=\; \mathbf{w}(\mathbf{c}, \widehat{\mathbf{d}}) \;=\; -\gamma\widehat{\mathbf{d}}
\end{aligned}
$$

By Lemma 5, $-\beta\widehat{\mathbf{b}} = \gamma\widehat{\mathbf{d}}$. This concludes the proof of "only if".

Now assume that for some α_0 we have $\mathbf{w}(\mathbf{c} + \alpha_0) = \mathbf{y}$. If for every $\alpha \ge \alpha_0$ we have $\mathbf{w}(\mathbf{c} + \alpha) = \mathbf{y}$, $(\mathbf{c}, \widehat{\mathbf{d}})$ is a witness ray. Otherwise, there exists $\alpha > \alpha_0$ such that $\mathbf{w}(\mathbf{c} + \alpha) \ne \mathbf{y}$. Note tha Lemma 10 implies that the set $Y = \{\mathbf{x} \mid \mathbf{w}(\mathbf{x}) = \mathbf{y}\}$ is a convex closed polytope. Therefore there exists a halfspace h containing Y that the ray $(\mathbf{c}, \widehat{\mathbf{b}})$ exits H. But in that case all the rays with direction $\widehat{\mathbf{b}}$ exit H (and hence Y), including $(\mathbf{a}, \widehat{\mathbf{b}})$, contrary to the assumption that $\mathbf{w}(\mathbf{a}, \widehat{\mathbf{b}}) = \mathbf{y}$. $\blacksquare$

Now assume that cell C contains a witness ray, say $(\mathbf{a}, \hat{\mathbf{b}})$, where $\mathbf{w}(\mathbf{a}, \hat{\mathbf{b}}) = \mathbf{y}$. Suppose that $\mathbf{w}'$ has a witness ray disjoint with C. By the last lemma, we may assume that this ray is of the form $(\mathbf{c}, \hat{\mathbf{b}})$ and that $\mathbf{w}(\mathbf{c}, \hat{\mathbf{b}}) = \mathbf{w}(\mathbf{c}) = \mathbf{y}$. By Lemma 10, we know that for every $0 < t < 1$ the ray $(t\mathbf{a} + (1-t)\mathbf{c}, \hat{\mathbf{b}})$ is also a witness for the simplified function. It is clear, that for a certain value of t, one of such ray is within distance ϵ form C (at least, its final section), and hence it would be a witness ray for $\mathbf{w}$, disjoint with C, a contradiction.

6 Conclusions

In this paper, we considered the least-distance problem in any dimension. We obtained a parametric characterization of the solutions, thereby showing connections with a search problem in an arrangement of hyperplanes. We used this characterization to obtain an algorithm for the problem strongly polynomial for any fixed m. We further developed a search technique based on the notion of witnesses and used to obtain an algorithm for the problem linear for any fixed m.

It would be of interest to see if the arrangement based method could be extended easily to arbitrary convex separable functions.

References

[1] Michael J. Best and Ruo Yang Tan, An $O(n^2 \log n)$ strong polynomial algorithm for quadratic program with two equality constraints and lower and upper bounds, manuscript.

[2] Peter Brucker, An $O(n)$ algorithm for the quadratic knapsack problem, *Operations Research Letters* **3**(1984), 163–166.

[3] P. H. Calamai and J. J. Moré, Quasi-Newton updates with bounds, *SIAM Journal on Numerical Analysis* **24**(1987), 1434-1441.

[4] Bernard Chazelle, Leo J. Guibas, and D. T. Lee, The power of geometric duality, *Proc. 15th Annual ACM Symposium on Theory of Computing*, 217–225.

[5] M. E. Dyer, Linear time algorithms for two- and three-variable linear programs, *SIAM Journal of Computing* **13**(1984), 31–45.

[6] M. E. Dyer, On a multidimensional search technique and its application to the Euclidean one-center problem, *SIAM Journal of Computing* **15**(1986), 725—738.

[7] Herbert Edelsbrunner, *Algorithms in Combinatorial Geometry*, Springer Verlag (1987).

[8] Herbert Edelsbrunner, Joseph O'Rourke and Raimund Seidel, Constructing arrangements of lines and hyperplanes with applications, *SIAM Journal of Computing* bf 15(1986), 341–363.

[9] R. Helgason, J. Kennington, and H. Lall, A polynomial bounded algorithm for a singly constrained quadratic program, *Mathematical Programming* **18**(1980), 338–343.

[10] Toshihide Ibaraki and Naoki Katoh, *Resource allocation problems: algorithmic approaches*, The MIT Press, Cambridge, Massachusetts (1988).

[11] Nainan Kovoor and Panos M. Pardalos, Minimization of separable convex functions subject to equality and box constraints, Technical Report CS-91-16, Department of Computer Science, The Pennsylvania State University (June 1991).

[12] Nainan Kovoor and Panos M. Pardalos, A linear time algorithm for the least-distance problem, Technical Report CS-92-3, Department of Computer Science, The Pennsylvania State University (March 1992).

[13] M. K. Kozlov, S. P. Tarasov, and Leonid G. Khachiyan, Polynomial solvability of convex quadratic programming, *Soviet Math. Dokl.* **20**(1979), 1108-1111.

[14] Wu Li, Panos M. Pardalos, and C.-G. Han, Gauss-Seidel method for least distance problems, *Journal of Optimization Theory and Applications* **75**(1992), 487–500.

[15] Nimrod Megiddo, Linear programming in linear time when the dimension is fixed, *J. ACM* **31**(1984), 114–127.

[16] Nimrod Megiddo and Arie Tamir, Linear time algorithms for separable quadratic programming problems, IBM Research Report RJ 8501 (76722).

[17] C. A. Micchelli, P. W. Smith, J. Swetits, and J. D. Ward, Constrained L_p approximation, *Constructive Approximation* **1**(1985), 93-102.

[18] J. R. Munkres, Topology: A first course, *Prentice Hall* (1975).

[19] P.Orlik, and H.Terao, Arrangements of Hyperplanes, *Springer Verlag* (1992).

[20] Panos M. Pardalos and Nainan Kovoor, An algorithm for a singly constrained class of quadratic programs subject to upper and lower bounds, *Mathematical Programming* **46**(1990), 321–328.

[21] P. W. Smith and H. Wolkowicz, A nonlinear equation for linear programming, *Mathematical Programming* **34**(1986), 235–238.

[22] Stephen A.Vavasis, *Nonlinear Optimization: Complexity issues*, Oxford University Press (1991).

[23] Yinyu Ye and Edison Tse, An extension of Karmarkar's projective algorithm for convex quadratic programming, *Mathematical Programming* **44** (1989), 157-179.

Appendix

A finite set $\mathcal{H}$ of hyperplanes in $\mathbf{R}^d$ defines a dissection of $\mathbf{R}^d$ into connected sets of various dimensions. We call this dissection the *arrangement* $\mathcal{A}(\mathcal{H})$ of $\mathcal{H}$.

Given a vector $\boldsymbol{\eta} = (\eta_1, \eta_2, \dots, \eta_d) \in \mathbf{R}^d - \mathbf{0}$ and a number $\eta_0 \in \mathbf{R}$, we may define a hyperplane H and associated halfspaces H^-, H^+ by

$$\begin{aligned}
H &= \{\mathbf{x} \in \mathbf{R}^d \mid \boldsymbol{\eta} \cdot \mathbf{x} = \eta_0\} \\
H^- &= \{\mathbf{x} \in \mathbf{R}^d \mid \boldsymbol{\eta} \cdot \mathbf{x} < \eta_0\} \\
H^+ &= \{\mathbf{x} \in \mathbf{R}^d \mid \boldsymbol{\eta} \cdot \mathbf{x} > \eta_0\}
\end{aligned}$$

Clearly H, H^-, H^+ are disjoint and $H \cup H^- \cup H^+ = \mathbf{R}^d$.

We may now specify the location of a point relative to the set of hyperplanes $\mathcal{H} = \{H_1, H_2, \dots, H_n\}$. For a point $\mathbf{p}$ and $1 \leq j \leq n$, define

$$s_j(\mathbf{p}) = \begin{cases} -1 & \text{if } \mathbf{p} \in H_j^- \\ 0 & \text{if } \mathbf{p} \in H_j \\ +1 & \text{if } \mathbf{p} \in H_j^+ \end{cases}$$

The vector $\mathbf{s}(\mathbf{p}) = (s_1(\mathbf{p}), s_2(\mathbf{p}), \dots, s_n(\mathbf{p}))$ is called the *position vector* of $\mathbf{p}$.

Clearly there are at most 3^n possible position vectors; however, in general most of these will not occur. We say that points $\mathbf{p}$ and $\mathbf{q}$ *lie on the same face* if $\mathbf{s}(\mathbf{p}) = \mathbf{s}(\mathbf{q})$. The non-empty set of points with position vector $\mathbf{r}$ is called the *face* $f(\mathbf{r})$:

$$f(\mathbf{r}) = \{\mathbf{p} \in \mathbf{R}^d \mid \mathbf{s}(\mathbf{p}) = \mathbf{r}\}$$

The non-empty sets of this form are called the *faces* of the arrangement $\mathcal{A}(\mathcal{H})$. The position vector of a face $f(\mathbf{r}) = g$ is defined to be $\mathbf{r}$,

$$\mathbf{s}(f(\mathbf{r})) = \mathbf{r}$$

A face f is called a k-face if its dimension is k. Special names are used to denote k-faces for special values of k : a 0-face is called a *vertex*, a 1-face is called an *edge*, a $(d-1)$-face is called a *facet*, and a d-face is called a *cell*. A face f is said to be a *subface* of another face g if the dimension of f is one less than the dimension of g and f is contained in the boundary of g; it follows that $s_i(f) = 0$ unless $s_i(f) = s_i(g)$ for $1 \leq i \leq n$. If f is a subface of g, then we also say that f and g are *incident (upon each other)* or that they define an *incidence*.

An arrangement $\mathcal{A}(\mathcal{H})$ of $n \geq d$ hyperplanes is called *simple* if any d hyperplanes of $\mathcal{H}$ have a unique point in common and if any $d+1$ hyperplanes have no point in common. If $n < d$, we say that $\mathcal{A}(\mathcal{H})$ is simple if the common intersection of the n hyperplanes is a $(d-n)$-flat. For more details see [8], [7] and the recent monograph [19].

Complexity in Numerical Optimization, pp. 57-73
P.M. Pardalos, Editor
©1993 World Scientific Publishing Co.

Translational Cuts for Convex Minimization

James V. Burke
*Department of Mathematics, GN–50, University of Washington, Seattle, WA 98195
USA*

Allen A. Goldstein
*Department of Mathematics, GN–50, University of Washington, Seattle, WA 98195
USA*

Paul Tseng
*Department of Mathematics, GN–50, University of Washington, Seattle, WA 98195
USA*

Yinyu Ye
Department of Management Sciences, University of Iowa, Iowa City, IO 52242 USA

Abstract

We develop an iterative descent algorithm for minimizing the pointwise maximum of a finite collection of convex thrice-differentiable functions;
$$\min_x \{ F(x) := \max_{i=1,\ldots,n} f_i(x) \}.$$

The proposed algorithm begins each iteration with a number R and an inexact 'analytic center', x_R, of the lower level set $\{ x \in R^m : F(x) \leq R \}$; it then sets $R := (1 - \alpha)F(x_R) + \alpha R$, with α an arbitrarily chosen constant in $(0, 1)$, and recomputes x_R accordingly. The resulting sequence of inexact analytic centers is a descent sequence for F and it is shown that the F value along this sequence comes within ϵ of $\min_x F(x)$ after at most
$$2(1 - \alpha)^{-1}[n \ln(1/\epsilon) + \ln(\textstyle\prod_{i=1}^{n}(R_0 - f_i(x_0)))] + 3/2$$

iterations, where x_0 and R_0 are the initial values of x_R and R, respectively, and ϵ is the termination tolerance.

To recompute x_R after each update of R, we propose to use a global newton procedure of [3]. We show that, under a certain nondegeneracy assumption on F and assuming infinite precision arithmetic, the number of newton steps required to recompute x_R is at most a constant plus $\log_2 \log_2(1/\epsilon)$.

Keywords: Complexity, mini-max optimization, global Newton method, analytic center.

1 Introduction

Motivated by interior-point algorithms, several researchers recently studied ways of capturing the combinatorial structure of a convex polytope using a potential function (see [7], [8]). For example, it is shown in [8] that the max-potential of a convex polytope, where the polytope is given as
the intersection of halfspaces, is reduced by a factor of at least e, the base of the natural logarithm, when any of the hyperplanes defining the polytope is translated through the analytic center of the polytope [6]. In this paper we extend this result to convex sets that are given as the intersection of the lower level sets of a finite collection of convex thrice-differentiable functions; and, by using the extended result, we develop a translational-cuts algorithm for finding the minimum of the pointwise maximum of the given functions.

Let f_i, $i = 1, ..., n$, be given convex thrice-differentiable functions defined on the m-dimensional Euclidean space R^m. Let F be the function on R^m defined to be the pointwise maximum of the f_i's, that is,

$$F(x) := \max\{f_i(x) : i = 1, ..., n\}.$$

We are interested in the problem of finding a minimum point of F, i.e., an $x^* \in \mathrm{R}^m$ satisfying $F(x^*) = R^*$, where for notational convenience we let

$$R^* := \min_x F(x).$$

This is a classical problem in optimization [2]. For the moment we make only the following assumption on F.

Assumption A The set of minimum points of F, i.e., $\{x : F(x) = R^*\}$, is nonempty and compact.

Additional assumptions will be introduced as needed.

As an immediate consequence of the compactness assumption on the minimum points of F, we have that for every $R > R^*$, the lower level set

$$lev_R := \{x : F(x) < R\}$$

is nonempty and bounded. Then, the 'logarithmic potential' function ϕ_R defined by

$$\phi_R(x) := \sum_{i=1}^{n} \ln(R - f_i(x)),$$

which is strictly concave and differentiable on lev_R, attains its maximum on lev_R.

Following the standard practice, we call any maximum point of ϕ_R the *analytic center* of lev_R. As R approaches R^*, the analytic center of lev_R approaches the set of minimum points of F. This motivates an iterative method for finding a minimum point of F whereby, at each iteration, an analytic center of lev_R is computed and R is moved some fraction of the distance towards R^*. However, it is impractical to compute an analytic center exactly and, to resolve this difficulty, we introduce below a notion of an inexact analytic center.

Let $g : [0, \infty) \mapsto [0, \infty)$ be any continuous function satisfying $g(\theta) > 0$ for all $\theta > 0$ and $g(0) = 0$. Relative to g, we say that a point x_R in lev_R is an *inexact analytic center* of lev_R if

$$\|\nabla \phi_R(x_R)\| \leq \frac{\min\{R - F(x_R), 1\} g(R - R^*)}{R - R^*}. \tag{1}$$

Notice that, in contrast to previous notions of an inexact analytic center (see [6]), the preceding notion is based on the gradient of the potential function being 'small' (i.e., an inexact root of $\nabla \phi_R$). We have adopted this novel notion because the method which we will use to compute an inexact analytic center (for a fixed R) is based on reducing the norm of this gradient and the work is determined by the total reduction in norm. Suitable choices for g will be discussed shortly (see (3), (10)).

Below we formally describe our algorithm for finding a minimum point of F:

Translational-Cuts Algorithm.
0. Fix a parameter $\alpha \in (0,1)$ and a termination tolerance $\epsilon > 0$. Start with any $R_0 > R^*$ and any inexact center of lev_{R_0}. Let $R = R_0$ and go to Step 1.
1. Given an $R > R^*$ and an inexact analytic center of lev_R, say x_R, check to see if the termination criterion

$$F(x_R) \leq R^* + \epsilon$$

is met. If yes, stop the algorithm; otherwise go to Step 2.
2. Set

$$R' := (1 - \alpha)F(x_R) + \alpha R$$

and apply the global newton method of [3] to find an inexact center of $lev_{R'}$, say x'_R, with x_R as the starting point for the method. Go to Step 3.
3. Set

$$x_R := x'_R, \qquad R := R'$$

and return to Step 1.

A few words on the parameter α are in order. This parameter controls the decrease in the size of the lower level set or, equivalently, the amount by which the inequalities $f_i(x) \leq R$, $i = 1,...,n$, are translated (in going from R to R'). If $\alpha = 0$, then at least one inequality will be translated so to cut through the current inexact center x_R; if $\alpha = 1$, then no inequality will be translated; if $\alpha = 1/2$, then at least one inequality will translate halfway to x_R. The reason for introducing the parameter α is so that one can initiate the computation of the new inexact analytic center x'_R from the current inexact analytic center x_R.

The translational-cuts algorithm is most closely related to the 'large-step' non-parametric logarithmic barrier method of [4], applied to the following nonlinear programming formulation of $\min_x F(x)$:

$$(NLP) \quad \min \gamma$$
$$\text{subject to } f_i(x) - \gamma \leq 0, \ i = 1,...,n \ .$$

However, the method of [4] uses a different notion of inexact analytic center and requires a relative Lipschitz condition on $\nabla^2 f_i$ for all i. It also uses a different termination criterion and requires an objective value lower bound for initialization.

In the remainder of this paper, we analyze the complexity of the translational-cuts algorithm under additional assumptions on the f_i's (see Assumptions B–D). Notice that $F(x_R) < R$ and hence the value of R is monotonically decreasing in the algorithm. Also, although it appears that termination of the algorithm and the computation of an inexact center require knowledge of the optimal objective value R^*, we will show that, under mild assumptions on the f_i's, this difficulty can in fact be circumvented (see discussion at the end of Sections 2 and 3).

2 Iteration count for the translational-cuts algorithm

We will call each repetition of Steps 1-3 in the translational-cuts algorithm an iteration. In this section we estimate the number of iterations used in the algorithm.

By Assumption A, there exists a scalar $C_0 > 0$ such that

$$||x - x'|| \leq C_0 \quad \forall x, x' \in lev_{R_0}. \tag{2}$$

In the following lemma, we estimate the reduction in the potential after each iteration of the translational-cuts algorithm.

Lemma 2.1 *Fix an $\alpha \in (0,1)$. For any $R \in (R^*, R_0]$ satisfying $g(R-R^*)/(R-R^*) \leq (1-\alpha)/(2C_0)$, any inexact analytic center x of lev_R and any inexact analytic center x' of $lev_{R'}$, where $R' = (1-\alpha)F(x) + \alpha R$, we have*

$$\phi_{R'}(x') \leq \phi_R(x) - (1-\alpha)/2.$$

Proof Since x is an inexact analytic center of lev_R, we have from (1) and the definition of ϕ_R that

$$\left\| \sum_{i=1}^{n} \frac{\nabla f_i(x)}{R - f_i(x)} \right\| \leq \frac{g(R - R^*)}{R - R^*}.$$

Let $\bar{x}'$ be an analytic center of $lev_{R'}$. For each $i = 1, \cdots, n$, let $\beta_i = (R - R')/(R - f_i(x))$ and let $\beta = \sum_{i=1}^{n} \beta_i$. Observe that each β_i satisfies $0 < \beta_i \leq 1 - \alpha$ with at least one β_i equal to $1 - \alpha$. Thus, $\beta \geq 1 - \alpha$. Then, the convexity of the f_i's and the preceding relation yield:

$$
\begin{aligned}
\sum_{i=1}^{n} \frac{R' - f_i(\bar{x}')}{R - f_i(x)} &= \sum_{i=1}^{n} \frac{R - f_i(\bar{x}') - \beta_i(R - f_i(x))}{R - f_i(x)} \\
&= \sum_{i=1}^{n} \frac{-f_i(\bar{x}') + f_i(x) + (1 - \beta_i)(R - f_i(x))}{R - f_i(x)} \\
&= \sum_{i=1}^{n} \frac{-f_i(\bar{x}') + f_i(x)}{R - f_i(x)} + n - \beta \\
&\leq \sum_{i=1}^{n} \frac{-\nabla f_i(x)^T(\bar{x}' - x)}{R - f_i(x)} + n - \beta \\
&\leq \left\| \sum_{i=1}^{n} \frac{\nabla f_i(x)}{R - f_i(x)} \right\| \|\bar{x}' - x\| + n - \beta \\
&\leq \frac{g(R - R^*)}{R - R^*} \|\bar{x}' - x\| + n - \beta \\
&\leq \frac{g(R - R^*)}{R - R^*} C_0 + n - \beta \leq n - \beta/2,
\end{aligned}
$$

where the fourth inequality follows from (2) and the last inequality follows from the inequality $g(R - R^*)C_0/(R - R^*) \leq (1 - \alpha)/2 \leq \beta/2$. Dividing both sides by n and taking the logarithm, we obtain

$$\ln(1 - \frac{\beta}{2n}) \geq \ln\left(\frac{1}{n} \sum_{i=1}^{n} \frac{R' - f_i(\bar{x}')}{R - f_i(x)} \right)$$

$$\geq \frac{1}{n} \sum_{i=1}^{n} \ln\left(\frac{R' - f_i(\bar{x}')}{R - f_i(x)} \right) = \frac{1}{n}(\phi_{R'}(\bar{x}') - \phi_R(x)),$$

where the second inequality uses the concavity of the logarithm function. Upon observing that the lefthand side is bounded above by $-\beta/(2n)$, which in turn is bounded above by $-(1 - \alpha)/(2n)$, and then using $\phi_{R'}(\bar{x}') \geq \phi_{R'}(x')$ to bound the righthand side, the result follows. $\qquad\square$

By using Lemma 2.1, we can now estimate the number of iterations required to reduce the objective value $F(x_R)$ to within ϵ of the optimal objective value R^*.

Theorem 2.2 *Assume that g is chosen so that*

$$g(\theta)/\theta \leq (1-\alpha)/(4C_0) \quad \forall \theta \in (0, R_0 - R^*].\tag{3}$$

Then, the translational-cuts algorithm terminates in at most

$$\frac{2}{1-\alpha}[\phi_{R_0}(x_{R_0}) + n\ln(\frac{1}{\epsilon})] + \frac{3}{2}$$

iterations.

Proof Lemma 2.1 assures us that, after k iterations of the translational-cuts algorithm, we have

$$\phi_R(x_R) \leq \phi_{R_0}(x_{R_0}) - k(1-\alpha)/2.$$

Thus, for $k \geq 2[\phi_{R_0}(x_{R_0}) - n\ln\epsilon]/(1-\alpha) + 1/2$, we have

$$\phi_R(x_R) \leq n\ln\epsilon - \frac{1-\alpha}{4}.\tag{4}$$

Let $\bar{x}_R$ be any analytic center of lev_R. Then, using the concavity of ϕ_R and the fact that x_R is an inexact analytic center of lev_R yields

$$\begin{aligned}
\phi_R(\bar{x}_R) &\leq& \phi_R(x_R) + \nabla\phi_R(x_R)^T(\bar{x}_R - x_R)\\
&\leq& n\ln\epsilon - \frac{1-\alpha}{4} + \|\nabla\phi_R(x_R)\|\|\bar{x}_R - x_R\|\\
&\leq& n\ln\epsilon - \frac{1-\alpha}{4} + \frac{g(R-R^*)}{R-R^*}C_0\\
&\leq& n\ln\epsilon,
\end{aligned}$$

where the last inequality follows from (2) and (3). Therefore,

$$\phi_R(x^*) \leq \phi_R(\bar{x}_R) \leq n\ln\epsilon,$$

where x^* is any minimum point of F. Dividing both sides by n yields

$$\ln\epsilon \geq \frac{\phi_R(x^*)}{n} = \sum_{i=1}^{n}\frac{\ln(R - f_i(x^*))}{n} \geq \min_i \ln(R - f_i(x^*)) = \ln(R - R^*)$$

and thus $R \leq R^* + \epsilon$. Since $F(x_R) < R$ trivially, this completes the proof. $\qquad\square$

One possible choice for g that satisfies (3) is the linear function

$$g(\theta) = \theta(1-\alpha)/(4C_0).$$

With this choice, determining whether a point is an inexact analytic center does not require any knowledge of R^*.

Also, it can be seen from the analysis that a sufficient condition for terminating the translational-cuts algorithm is (4) and the iteration count given in Theorem 2.2 is that for satisfying this condition. Thus, we can instead use (4) as a termination condition. This condition has the advantage that it does not require knowledge of the optimal objective value R^*.

3 Work per iteration for the translational-cuts algorithm

In this section we estimate, under additional nondegeneracy and sharp-minimum assumptions (see Assumptions B–D), the work per iteration in the translational-cuts algorithm. This estimate, together with the iteration count obtained in the previous section, will then yield an overall complexity estimate for the translational-cuts algorithm.

The work per iteration is dominated by Step 2 in which the global newton method of [3] is applied to find an inexact analytic center of $lev_{R'}$, starting from x_R. Thus, our analysis will focus on estimating the work for this method.

We start our analysis by applying directly Claim 1 in [3] to obtain a preliminary estimate of the work in Step 2: Let R, R' and x_R be as in Step 2 and let

$$
\begin{aligned}
S_{R'} &:= \{\, x \in lev_{R'} \ : \ \|\nabla\phi_{R'}(x)\| \leq \|\nabla\phi_{R'}(x_R)\| \,\}, \\
\lambda_{R'} &:= \max_{x \in S_{R'}} \{\, \text{largest eigenvalue of } -\nabla^2\phi_{R'}(x) \,\}, \\
\mu_{R'} &:= \min_{x \in S_{R'}} \{\, \text{smallest eigenvalue of } -\nabla^2\phi_{R'}(x) \,\}, \\
L_{R'} &:= \max_{x \in S_{R'}} \|\nabla^3\phi_{R'}(x)\|.
\end{aligned}
$$

According to Claim 1 in [3], the number of steps required by the method of [3] to find an approximate root of $\nabla\phi_{R'}$ (i.e., reducing $\|\nabla\phi_{R'}\|$ to less than $\beta_{R'} := (\mu_{R'})^2/4L_{R'}$), starting from an inexact analytic center x_R of lev_R, is at most

$$
(\|\nabla\phi_{R'}(x_R)\| - \beta_{R'})/4\delta^2\beta_{R'}, \tag{5}
$$

where δ is chosen from the interval $(0, 1/\sqrt{8}]$. Moreover, the number of steps required by the method of [3] to find an inexact analytic center x'_R of $lev_{R'}$ (i.e., x'_R satisfies $\|\nabla\phi_{R'}(x'_R)\| \leq \min\{R'-F(x'_R), 1\}g(R'-R^*)/(R'-R^*))$, starting from an approximate root of $\nabla\phi_{R'}$, is at most

$$
\log_2\log_2\left(\frac{\mu_{R'}\lambda_{R'}(R'-R^*)}{L_{R'}\min\{R'-F(x'_R), 1\}g(R'-R^*)}\right). \tag{6}
$$

In the remainder of this section, we focus on estimating the four quantities:

$$
\|\nabla\phi_{R'}(x_R)\|, \ \lambda_{R'}, \ \mu_{R'}, \ L_{R'} \tag{7}
$$

in terms of the problem parameters. We will estimate the latter three quantities by replacing the set $S_{R'}$ in their definition with the following set

$$
T_{R'} := \left\{\, x \in lev_{R'} \ : \ \|\nabla\phi_{R'}(x)\| \leq \frac{2n\sqrt{n}B_1}{\alpha(R'-R^*)} \,\right\}, \tag{8}
$$

where B_1 is any positive scalar satisfying

$$\|\nabla f_i(x)\| \leq B_1 \quad \forall x \in lev_{R_0}, \ i = 1, ..., n \tag{9}$$

(B_1 exists by Assumption A).

Our first order of business is to show that $S_{R'} \subset T_{R'}$. To show this, we make the following simplifying assumption, namely, $-\nabla^2 \phi_R$ is positive definite uniformly over lev_R and over all $R \in (R^*, R_0]$. This assumption will also be needed for bounding $\mu_{R'}$ from below.

Assumption B There exists a scalar $\mu > 0$ such that

$$(\text{smallest eigenvalue of } -\nabla^2 \phi_R(x)) \geq \mu \quad \forall x \in lev_R, \ \forall R \in (R^*, R_0].$$

Assumption B is actually quite mild. For example, Assumption B holds when any one of the f_i's is strongly convex. Alternatively, in view of Assumption A, we can enforce that Assumption B holds by choosing a sufficiently large number L so that the box

$$\{ \, x \in \Re^m \mid x \leq (L, ..., L) \, \}$$

contains the set of minimum points of F, and then set

$$f_{n+j}(x_1, ..., x_m) := x_j - L - R_0, \quad j = 1, ..., m.$$

It is straightforward to verify that $\max_{i=1, m+n} f_i(x)$ has the same set of minimum points as $F(x)$ (so that Assumption A and Assumptions C and D, which are to come, hold for $f_1, ..., f_{n+m}$ whenever they hold for $f_1, ..., f_n$) and that Assumption B holds with $\phi_R(x)$ replaced by $\sum_{i=1}^{n+m} \ln(R - f_i(x))$.

We show below that $S_{R'} \subset T_{R'}$ through a sequence of three lemmas.

Lemma 3.1 *For any $R \in (R^*, R_0]$ and any analytic center $\bar{x}_R$ of lev_R, we have*

$$R - R^* \leq n(R - F(\bar{x}_R)).$$

Proof By definition of an analytic center, we have

$$0 = \nabla \phi_R(\bar{x}_R) = \sum_{i=1}^{n} \frac{\nabla f_i(\bar{x}_R)}{R - f_i(\bar{x}_R)},$$

so that

$$\sum_{i=1}^{n} \frac{R - f_i(\bar{x}_R) - \nabla f_i(\bar{x}_R)(x - \bar{x}_R)}{R - f_i(\bar{x}_R)} = n \quad \forall x \in lev_R.$$

For each i and each $x \in lev_R$, we have from the convexity of f_i that $R \geq f_i(x) \geq f_i(\bar{x}_R) + \nabla f_i(\bar{x}_R)(x - \bar{x}_R)$ and hence every term in the above sum is nonnegative. Thus, we conclude that

$$\frac{R - f_i(\bar{x}_R) - \nabla f_i(\bar{x}_R)(x - \bar{x}_R)}{R - f_i(\bar{x}_R)} \leq n \quad \forall x \in lev_R, \ i = 1, ..., n.$$

[This relation can also be inferred from observing that $\bar{x}_R$ is an analytic center of the polytope $\{x : \nabla f_i(\bar{x}_R)(x - \bar{x}_R) \leq R - f_i(\bar{x}_R), \ i = 1, ..., n\}$ and invoking a containing-ellipsoid property for this polytope.] Let x^* be any minimum point of F. Since $x^* \in lev_R$, we obtain, upon using the convexity of the f_i's and invoking the above relation with $x = x^*$, that

$$R - f_i(x^*) \leq R - f_i(\bar{x}_R) - \nabla f_i(\bar{x}_R)(x^* - \bar{x}_R) \leq n(R - f_i(\bar{x}_R)), \quad i = 1, ..., n.$$

Thus,

$$R - R^* \leq n(R - f_i(\bar{x}_R)), \quad i = 1, ..., n,$$

and the result readily follows. $\square$

Lemma 3.2 *Let Assumption B hold. Suppose that g is chosen so that*

$$g(\theta)/\theta \leq \mu/2B_1 \qquad \forall \theta \in (0, R_0 - R^*]. \tag{10}$$

Then, for any $R \in (R^*, R_0]$ and any inexact analytic center x_R of lev_R, we have

$$R - R^* \leq 2n(R - F(x_R)).$$

Proof Let $\bar{x}_R$ be an analytic center of lev_R. By using $\nabla \phi_R(\bar{x}_R) = 0$ and Assumption B and the concavity of $\phi_R(.)$, we have

$$(\mu/2)\|\bar{x}_R - x_R\|^2 \leq \phi_R(\bar{x}_R) - \phi_R(x_R) \leq \nabla \phi_R(x_R)(\bar{x}_R - x_R),$$

which implies

$$\|\bar{x}_R - x_R\| \leq \frac{2}{\mu}\|\nabla \phi_R(x_R)\|.$$

Also, for each $i \in \{1, ..., n\}$, we have

$$R - f_i(\bar{x}_R) - (R - f_i(x_R)) = \int_0^1 -\nabla f_i(x_R + t(\bar{x}_R - x_R))^T(\bar{x}_R - x_R)dt.$$

Upon combining the above two relations, we obtain

$$\left| \frac{R - f_i(\bar{x}_R)}{R - f_i(x_R)} - 1 \right| = \frac{|\int_0^1 \nabla f_i(x_R + t(\bar{x}_R - x_R))^T(\bar{x}_R - x_R)dt|}{R - f_i(x_R)}$$

$$\leq \frac{\int_0^1 \|\nabla f_i(x_R + t(\bar{x}_R - x_R))\|\|\bar{x}_R - x_R\|dt}{R - f_i(x_R)}$$

$$
\begin{aligned}
&\leq \frac{\int_0^1 B_1 \|\bar{x}_R - x_R\| dt}{R - f_i(x_R)} \\
&\leq \frac{2B_1 \|\nabla \phi_R(x_R)\|}{\mu(R - f_i(x_R))} \\
&\leq \frac{2B_1 g(R - R^*)(R - F(x_R))}{\mu(R - f_i(x_R))(R - R^*)} \\
&\leq \frac{2B_1 g(R - R^*)}{\mu(R - R^*)} \leq 1,
\end{aligned}
$$

where the second inequality follows from (9), the fourth inequality follows from (1), the fifth inequality follows from the relation $0 < R - F(x_R) \leq R - f_i(x_R)$, and the last inequality follows from (10). This inequality together with Lemma 3.1 yields

$$
2(R - f_i(x_R)) \geq R - f_i(\bar{x}_R) \geq \frac{R - R^*}{n}, \quad i = 1, ..., n,
$$

and the result readily follows. $\hfill\square$

Lemma 3.3 *Let Assumption B hold and suppose that g satisfies (10). Then, for any $\alpha \in (0,1)$, any $R \in (R^*, R_0]$, and any inexact analytic center x_R of lev_R, we have*

$$
\|\nabla \phi_{R'}(x_R)\| \leq \frac{2n\sqrt{n}B_1}{\alpha(R' - R^*)},
$$

where $R' := (1 - \alpha)F(x_R) + \alpha R$.

Proof From the definition of R', we have $R' - F(x_R) = \alpha(R - F(x_R))$ and so Lemma 3.2 yields

$$
R' - F(x_R) \geq \frac{\alpha}{2n}(R - R^*).
$$

Then, using the fact $x_R \in lev_{R_0}$ and (9), we conclude

$$
\begin{aligned}
\|\nabla \phi_{R'}(x_R)\| &= \left\| \sum_{i=1}^{n} \frac{\nabla f_i(x_R)}{R' - f_i(x_R)} \right\| \leq \frac{\sqrt{n}B_1}{R' - F(x_R)} \\
&\leq \frac{2n\sqrt{n}B_1}{\alpha(R - R^*)} \leq \frac{2n\sqrt{n}B_1}{\alpha(R' - R^*)},
\end{aligned}
$$

where the last inequality follows from $R \geq R'$. $\hfill\square$

Lemma 3.3 shows that $S_{R'} \subset T_{R'}$, provided that g is chosen to satisfy (10). We will now estimate the quantities (7) by replacing $S_{R'}$ in their definition with the larger set $T_{R'}$. To do this, we need to make the following assumption on F.

Assumption C The function F has a *strongly unique* minimum x^*, that is, $F(x^*) = R^*$ and there exists a scalar $\sigma > 0$ such that

$$
F(x) \geq R^* + \sigma \|x - x^*\| \quad \forall x.
$$

The notion of a strongly unique minimum (or 'sharp minimum' [5]) was first extensively studied in [1], where its connection to the convergence behavior of algorithms was reviewed. Notice that Assumption C superceeds Assumption A.

Let

$$I^* := \{i \in \{1, ..., n\} : f_i(x^*) = R^*\}. \tag{11}$$

A key part of our analysis lies in showing that (see (19))

$$\liminf_{R \downarrow 0} \min_{x \in T_R} \left\{ \frac{R - F(x)}{R - f_i(x)} \right\} > 0 \quad \forall i \in I^* .$$

To show this, we need to make one further assumption on F.

Assumption D For any set of nonnegative scalars λ_i, $i \in I^*$, satisfying

$$\sum_{i \in I^*} \lambda_i \nabla f_i(x^*) = 0 \quad \text{and} \quad \sum_{i \in I^*} \lambda_i = 1 , \tag{12}$$

there holds $\lambda_i > 0$ for all $i \in I^*$.

It can be seen that Assumption C implies $|I^*| \geq m + 1$ and that Assumption D implies $|I^*| \leq m + 1$. (To see the former, note that if $|I^*| \leq m$, there would exist a nonzero $d \in \mathbb{R}^m$ satisfying $\nabla f_i(x^*)^T d \leq 0$ for all $i \in I^*$, implying $\lim_{\theta \downarrow 0}(F(x^* + \theta d) - R^*)/\theta \leq 0$ and thus violating Assumption C.) Thus, Assumptions C and D together imply $|I^*| = m + 1$. The condition specified in Assumption D is a standard nondegeneracy condition used in the convergence analysis of algorithms for mini-max optimization. In fact, this condition corresponds to two separate conditions often employed in nonlinear programming: *linear independence of the active constraint gradients* and *strict complementary slackness*. To see this, recall that the problem $\min_x F(x)$ is equivalent to the nonlinear program (NLP). Then, Assumption D is equivalent to the assumption that the gradients of the constraint functions in (NLP) that are active at x^*, namely

$$\left\{ \begin{pmatrix} \nabla f_i(x^*) \\ -1 \end{pmatrix} : i \in I^* \right\} ,$$

are linearly independent and that there exists a set of strictly complementary Lagrange multipliers at x^*, i. e., there exist scalars $\lambda_i > 0$, $i \in I^*$, satisfying (12).

Lemma 3.4 *Suppose that Assumptions B through D hold. Fix any $\alpha \in (0,1)$. Then there exist scalars $C_1 > 0$ and $C_2 > 0$ satisfying*

$$R - F(x) \geq C_1(R - R^*) \qquad \forall x \in T_R,\ \forall R \in (R^*, R_0) \tag{13}$$
$$R - F(x) \geq C_2(R - f_i(x)) \qquad \forall x \in T_R,\ \forall R \in (R^*, R_0),\ \forall i \in I^*, \tag{14}$$

where T_R and I^* are given by (8) and (11), respectively.

Proof We will argue (13) by contradiction. Suppose that (13) does not hold for any $C_1 > 0$. Then, there would exist a sequence $\{(x^k, R^k)\}$ satisfying

$$F(x^k) < R^k < R_0 \text{ and } \|\nabla \phi_{R^k}(x^k)\| \le \frac{2n\sqrt{n}B_1}{\alpha(R^k - R^*)} \qquad \forall k, \tag{15}$$

and yet

$$\frac{R^k - F(x^k)}{R^k - R^*} \to 0. \tag{16}$$

These two relations together imply

$$(R^k - F(x^k)) \to 0 \tag{17}$$

and

$$\frac{\|\nabla \phi_{R^k}(x^k)\|}{\sum_{i=1}^n 1/(R^k - f_i(x^k))} \le \frac{\|\nabla \phi_{R^k}(x^k)\|}{1/(R^k - F(x^k))} \le \frac{2n\sqrt{n}B_1(R^k - F(x^k))}{\alpha(R^k - R^*)} \to 0 \tag{18}$$

as $k \to 0$. Let $((\lambda_1, ..., \lambda_n), R^\infty, x^\infty)$ be any cluster point of the sequence

$$\left\{ \frac{((1/(R^k - f_1(x^k)), ..., 1/(R^k - f_n(x^k))))}{\sum_{i=1}^n 1/(R^k - f_i(x^k))}, R^k, x^k \right\},$$

and let $I^\infty := \{i \in \{1, ..., n\} : f_i(x^\infty) = F(x^\infty)\}$. The relation (18) implies that

$$\sum_{i=1}^n \lambda_i \nabla f_i(x^\infty) = 0, \qquad \sum_{i=1}^n \lambda_i = 1, \qquad \lambda_i \ge 0, \ i = 1, ..., n.$$

Since $R^\infty \ge F(x^\infty) > f_i(x^\infty)$ for all $i \notin I^\infty$, we know from (17) that $\lambda_i = 0$ for all $i \notin I^\infty$. Therefore, x^∞ satisfies the sufficient conditions for a minimum point of F. Since x^* is the unique minimum point of F, this implies $x^\infty = x^*$ and $I^\infty = I^*$, so (17) yields $R^\infty = F(x^\infty) = R^*$ and Assumption D yields $\lambda_i > 0$ for all $i \in I^*$. Thus, we conclude that there exist positive scalars η_i, $i \in I^*$, such that

$$\frac{R^k - F(x^k)}{R^k - f_i(x^k)} > \eta_i \quad \forall i \in I^*, \ \forall k. \tag{19}$$

Fix any k. By using the concavity of ϕ_{R^k}, relation (15), and Assumption C, we have

$$\begin{aligned}
\phi_{R^k}(x^*) &\le \phi_{R^k}(x^k) + \langle \nabla \phi_{R^k}(x^k), x^* - x^k \rangle \\
&\le \phi_{R^k}(x^k) + \|\nabla \phi_{R^k}(x^k)\|\|x^* - x^k\| \\
&\le \phi_{R^k}(x^k) + \frac{2n\sqrt{n}B_1}{\alpha(R^k - R^*)}\|x^* - x^k\| \\
&\le \phi_{R^k}(x^k) + \frac{2n\sqrt{n}B_1}{\alpha\sigma}.
\end{aligned}$$

Upon combining this with (19), we conclude that

$$
\begin{aligned}
|I^*|\ln(R^k - R^*) \;+\; & \sum_{i\notin I^*}\ln(R^k - f_i(x^*)) \\
= \;& \phi_{R^k}(x^*) \\
\leq \;& \phi_{R^k}(x^k) + \frac{2n\sqrt{n}\,B_1}{\alpha\sigma} \\
= \;& \sum_{i\in I^*}\ln(R^k - f_i(x^k)) \\
& + \sum_{i\notin I^*}\ln(R^k - f_i(x^k)) + \frac{2n\sqrt{n}\,B_1}{\alpha\sigma} \\
\leq \;& |I^*|\ln(R^k - F(x^k)) - \sum_{i\in I^*}\ln(\eta_i) \\
& + \sum_{i\notin I^*}\ln(R^k - f_i(x^k)) + \frac{2n\sqrt{n}\,B_1}{\alpha\sigma}.
\end{aligned}
$$

Also, by continuity of the f_i's, there exists a scalar $\theta > 0$ such that

$$
R^k - f_i(x^k) \geq \theta \quad \forall i \notin I^* , \;\; \forall k.
$$

Upon combining the above two relations and using the fact that $\ln(\cdot)$ is an increasing function, we obtain that

$$
\begin{aligned}
|I^*|\ln(R^k - R^*) & + (n - |I^*|)\ln(\theta) \\
\leq \;\; |I^*|\ln(R^k - F(x^k)) & - \sum_{i\in I^*}\ln(\eta_i) + \sum_{i\notin I^*}\ln(B_2) + \frac{2n\sqrt{n}\,B_1}{\alpha\sigma},
\end{aligned}
$$

where B_2 is any positive scalar satisfying

$$
R_0 - f_i(x) \leq B_2 \qquad \forall i \in I^*, \; \forall x \in lev_{R_0}
$$

(such a B_2 exists by the compactness of lev_{R_0}). Dividing both sides by $|I^*|$ and then taking the exponential yields the inequality

$$
R^k - R^* \leq C'(R^k - F(x^k)) \quad \forall k,
$$

where C' is some suitable positive scalar. This contradicts (16) and thus (13) must hold for some $C_1 > 0$.

Finally, we show that there exists a scalar $C_2 > 0$ such that (14) holds. By the preceding argument, there exists a scalar $C_1 > 0$ such that (13) holds. Consider any $R \in (R^*, R_0)$ and any $x \in T_R$. For each $i \in I^*$, we have from $f_i(x^*) = R^*$ and the convexity of f_i that

$$
f_i(x) \geq R^* - B_1||x - x^*||,
$$

so Assumption C yields

$$R - f_i(x) \leq R - R^* + B_1\|x - x^*\|$$
$$\leq R - R^* + \frac{B_1}{\sigma}(R - R^*).$$

Combining this with (13) and we obtain

$$R - F(x) \geq \frac{C_1}{1 + B_1/\sigma}(R - f_i(x)) \qquad \forall i \in I^*.$$

Set $C_2 := C_1/(1 + B_1/\sigma)$. □

Armed with Lemmas 3.3 and 3.4, we can now proceed to bound, in terms of the problem parameters, the four quantities (7) used in the complexity estimates (5) and (6). In what follows, we assume that Assumptions B–D hold and that g is chosen to satisfy (10).

First, we bound from above $\lambda_{R'}$. For each R, straightforward calculation shows that

$$-\nabla^2\phi_R(x) = \sum_{i=1}^{n} \frac{\nabla f_i(x)\nabla f_i(x)^T}{(R - f_i(x))^2} + \sum_{i=1}^{n} \frac{\nabla^2 f_i(x)}{R - f_i(x)}.$$

Since lev_{R_0} is bounded (by Assumption A), it follows that the largest eigenvalue of $-\nabla^2\phi_R(x)$ is bounded above by some constant divided by $(R - F(x))^2$ for all $x \in T_R$ and all $R \in (R^*, R_0)$. By (13) in Lemma 3.4, the quantity $R - F(x)$ is bounded below by a constant times $R - R^*$; hence we conclude that (also using the fact $S_{R'} \subset T_{R'}$)

$$\lambda_{R'} \leq C_3/(R' - R^*)^2,$$

for some scalar $C_3 > 0$.

Next, we bound from below $\mu_{R'}$. By Assumption C, the smallest eigenvalue of

$$\sum_{i \in I^*} \nabla f_i(x)\nabla f_i(x)^T$$

is bounded below by some positive constant, uniformly over all x in some neighborhood of x^*. (Otherwise there would exist a $d \in \mathbf{R}^m$ satisfying $\nabla f_i(x^*)^T d = 0$ for all $i \in I^*$, implying $\lim_{\theta\downarrow 0}(F(x^* + \theta d) - R^*)/\theta = 0$ and thus violating Assumption C.) This observation together with (14) in Lemma 3.4 implies that the smallest eigenvalue of

$$\sum_{i \in I^*} \frac{\nabla f_i(x)\nabla f_i(x)^T}{(R - f_i(x))^2}$$

is bounded below by some positive constant divided by $(R - F(x))^2$, uniformly over all x in T_R and all R sufficiently close to R^*. Since $F(x) \geq R^*$, we can replace $F(x)$ in the preceding bound by R^*. Summarizing the above results, we see that the smallest eigenvalue of $-\nabla^2 \phi_R$ is bounded below by some positive scalar C_4 divided by $(R - R^*)^2$, uniformly over T_R and over all R sufficiently close to R^*. In view of Assumption B (and by taking C_4 sufficiently small if necessary), we can extend this bound to hold over T_R and over all $R < R_0$. Since $S_{R'} \subset T_{R'}$, this yields

$$\mu_{R'} \geq C_4/(R' - R^*)^2.$$

Finally, we determine a suitable value for $L_{R'}$. For each R, straightforward calculation shows that

$$\|\nabla^3 \phi_R(x)\| \leq 3 \sum_{i=1}^{n} \frac{\|\nabla f_i(x)\| \|\nabla^2 f_i(x)\|}{(R - f_i(x))^2} + \sum_{i=1}^{n} \frac{\|\nabla^3 f_i(x)\|}{R - f_i(x)} + 2 \sum_{i=1}^{n} \frac{\|\nabla^2 f_i(x)\|^3}{(R - f_i(x))^3}.$$

Since lev_{R_0} is bounded, this implies that $\|\nabla^3 \phi_R(x)\|$ is bounded above by some constant divided by $(R - F(x))^3$, uniformly over all $x \in T_R$ and all $R < R_0$. By (13) in Lemma 3.4, we can in turn bound $R - F(x)$ from below by $C_1(R - R^*)$. Since $S_{R'} \subset T_{R'}$, this implies that we can take

$$L_{R'} := C_5/(R' - R^*)^3$$

for some suitable scalar C_5.

Thus

$$\beta_{R'} = \frac{(\mu_{R'})^2}{4 L_{R'}} \geq \frac{(C_4/(R' - R^*)^2)^2}{4(C_5/(R' - R^*)^3)} = \frac{(C_4)^2}{4 C_5 (R' - R^*)},$$

which together with Lemma 3.3 yields that the quantity in (5) is bounded above by some positive constant independent of $R' - R^*$. Similarly, using the trivial observation $\mu_{R'} \leq \lambda_{R'}$, we have that the quantity in (6) is bounded above by

$$\log_2 \log_2 \left(\frac{(C_3)^2}{C_5 \min\{R' - F(x'_R), 1\} g(R' - R^*)} \right).$$

Since x'_R is an inexact analytic of $lev_{R'}$, we can apply (13) with x and R replaced by x'_R and $R'' := (1 - \alpha)F(x'_R) + \alpha R'$, respectively, to obtain

$$R' - F(x'_R) \geq R'' - F(x'_R) \geq C_1(R'' - R^*) \geq \alpha C_1(R' - R^*).$$

Combining the preceding observations and we obtain the following key complexity estimate for Step 2 of the translational-cuts algorithm.

Theorem 3.5 *Let Assumptions B–D hold and assume that g is chosen to satisfy (10). Then, the total number of steps required by the method of [3] to find an inexact analytic center of $lev_{R'}$, starting from an inexact analytic center x_R of lev_R (and with $R' := (1 - \alpha)F(x_R) + \alpha R$), is at most a constant plus $\log_2 \log_2(1/(R' - R^*) g(R' - R^*))$.*

Suppose furthermore that g is an increasing function. Then, since the termination criterion in Step 1 is not met each time we visit Step 2, we know that $R' - R^* = (1 - \alpha)(F(x_R) - R^*) + \alpha(R - R^*) \geq F(x_R) - R^* > \epsilon$; hence the complexity estimate in Theorem 3.5 is in turn bounded above by some constant plus $\log_2 \log_2(1/\epsilon g(\epsilon))$.

Finally, we note that a similar estimate can be made if (4) is used as the termination criterion. In particular, observe that

$$\phi_R(x_R) \leq (n - 1)\ln(B_2) + \ln(R - F(x_R))$$

for any $R \in (R^*, R_0)$ and any x_R in lev_R, where B_2 is defined as in Lemma 3.4. Thus, if in addition (4) is not satisfied (so that $\phi_R(x_R) > n\ln\epsilon - (1 - \alpha)/4$), we would have

$$R - R^* \geq R - F(x_R) \geq \frac{\epsilon^n}{e^{(1-\alpha)/4}(B_2)^{n-1}}$$

and so $R' - R^*$ (which is bounded below by $\alpha(R - R^*)$) is bounded below by some scalar C_6 times ϵ^n. Thus, in this case (and assuming that g is an increasing function), the total number of steps required by the method of [3] to find an inexact analytic center of $lev_{R'}$, starting from an inexact analytic center x_R of lev_R, is at most a constant plus $\log_2 \log_2(1/\epsilon^n g(C_6\epsilon^n))$.

References

[1] L. Cromme, *Strong uniqueness*, Numerische Mathematik, 29 (1978), 179–193.

[2] V.F. Demyanov and V.N. Malozemov, *Introduction to Minimax*, John Wiley and Sons, New York, 1974.

[3] A.A. Goldstein, *A global Newton method*, Applied Geometry and Discrete Mathematics, 4 (1991), 301-307.

[4] D.D. Hertog, C. Roos, and T. Terlaky, *A large-step analytic center method for a class of smooth convex programming problems*, SIAM Journal on Optimization, 2 (1992), 55-70.

[5] B.T. Polyak, *Sharp minima*, Institute of Control Sciences Lecture Notes, Moscow (1979); Presented at the IIASA Workshop on Generalized Lagrangians and Their Applications, Laxenburg (1979).

[6] G. Sonnevend, *An "analytic center" for polyhedrons and new classes of global algorithms for linear (smooth, convex) programming*, System Modelling and Optimization: Proceedings of the 12th IFIP–Conference, Budapest (1985); Eds. A. Prekopa, J. Szelezsan and B. Strazicky, Lecture Notes in Control and Information Sciences, 84 (1986), 866–876.

[7] P.M. Vaidya, *A new algorithm for minimizing convex functions over convex sets*, in Proceedings of 30th Annual IEEE Symposium on the Foundations of Computer Science (IEEE Computer Society Press, Los Alamitos, 1989), 338-343.

[8] Y. Ye, *A Combinatorial property of the analytic centers of polytopes*, Manuscript, University of Iowa, Iowa City (1989).

Complexity in Numerical Optimization, pp. 74-87
P.M. Pardalos, Editor
©1993 World Scientific Publishing Co.

Maximizing Concave Functions in Fixed Dimension

Edith Cohen[1]
AT&T Bell Laboratories, Murray Hill, NJ 07928 USA

Nimrod Megiddo
*IBM Research, Almaden Research Center, San Jose, CA 95120-6099 USA, and
School of Mathematical Sciences, Tel Aviv University, Tel Aviv, Israel.*

Abstract

In [3, 7, 2] the authors introduced a technique which enabled them to solve
the parametric minimum cycle problem with a fixed number of parameters in
strongly polynomial time. In the current paper[2] we present this technique as
a general tool. In order to allow for an independent reading of this paper, we
repeat some of the definitions and propositions given in [3, 7, 2]. Some proofs
are not repeated, however, and instead we supply the interested reader with
appropriate pointers.

Suppose $Q \subset R^d$ is a convex set given as an intersection of k halfspaces, and let
$g : Q \to R$ be a concave function that is computable by a piecewise affine algorithm (i.e., roughly, an algorithm that performs only multiplications by scalars,
additions, and comparisons of intermediate values which depend on the input).
Assume that such an algorithm $\mathcal{A}$ is given and the maximal number of operations required by $\mathcal{A}$ on any input (i.e., point in Q) is T. We show that under
these assumptions, for any fixed d, the function g can be maximized in a number of operations polynomial in k and T. We also present a general framework
for parametric extensions of problems where this technique can be used to obtain strongly polynomial algorithms. Norton, Plotkin, and Tardos [13] applied
a similar scheme and presented additional applications.

---

[1]Research done while the author was at Stanford University and IBM Almaden Research Center.
Research partially supported by NSF PYI Grant CCR-8858097, and ONR Contract N00014-88-K-
0166.

[2]See [4, 2] for an earlier version.

1 Introduction

A *convex optimization problem* is a problem of maximizing a concave function g over a convex set $S \subset R^d$. Equivalently, we can consider minimizing a convex function. We consider the problem of maximizing a concave function g, where the dimension of the space is fixed. We assume that g is provided to us by a piecewise affine algorithm (see Definition 2.2) that can evaluate g at any given point.

The results of this paper can be extended easily to the case where the range of g is R^ℓ for any $\ell \geq 1$. We then define the notions of maximum and concavity of g with respect to the lexicographic order as follows. We say that a function $g : Q \subset R^d \to R^\ell$ is *concave* with respect to the lexicographic order $\leq_{\text{lex}}$ if for every $\alpha \in [0, 1]$ and $\boldsymbol{x}, \boldsymbol{y} \in Q$,

$$\alpha g(\boldsymbol{x}) + (1 - \alpha)g(\boldsymbol{y}) \leq_{\text{lex}} g(\alpha \boldsymbol{x} + (1 - \alpha)\boldsymbol{y}) .$$

Applications where the range of g is R^2 were given in [5, 8].

In Section 2 we define the problem. In Section 3 we introduce the subproblem of hyperplane queries, which is essential for the design of our algorithm. In Section 4 we discuss the multi-dimensional search technique which we utilize for improving our time bounds. In Section 5 we introduce the optimization algorithm. In Sections 6 and 7 we prove the correctness and analyze the time complexity of the algorithm. In Section 8 we discuss applications of the technique introduced here to obtain strongly polynomial time algorithms for parametric extensions of other problems.

2 Preliminaries

Let R_1^d denote the set of vectors $\boldsymbol{\lambda} = (\lambda_1, \ldots, \lambda_d)^T \in R^d$ such that $\lambda_d = 1$. For $\boldsymbol{\lambda} \in R^{d-1}$, denote by $\hat{\boldsymbol{\lambda}} \in R_1^d$ the vector $\hat{\boldsymbol{\lambda}} = (\boldsymbol{\lambda}, 1)$. For a halfspace F, denote by ∂F the boundary of F, i.e., ∂F is a hyperplane.

Definition 2.1 For a finite set $C \subset R^{d+1}$, denote by $L_C : R^{d+1} \to R$ the minimum envelope of the linear functions that correspond to the vectors in C, i.e.,

$$L_C(\boldsymbol{\lambda}) = \min_{\boldsymbol{c} \in C} \boldsymbol{c}^T \boldsymbol{\lambda} .$$

Denote by $\hat{L}_C : R^d \to R$ the function given by

$$\hat{L}_C(\boldsymbol{\lambda}) = L_C(\hat{\boldsymbol{\lambda}}) .$$

The vectors in $C \subset R^{d+1}$ are called the *pieces* of $\hat{L}_C$. For a piece $\boldsymbol{c} \in C$ and a vector $\boldsymbol{\beta} \in R^d$ such that $\boldsymbol{c}^T \hat{\boldsymbol{\beta}} = \hat{L}_C(\boldsymbol{\beta})$, we say that $\boldsymbol{c}$ is *active* at $\boldsymbol{\beta}$.

Definition 2.2 For a function $g : Q \to R$, where $Q \subset R^d$:

   i. Denote by Λ_g (or Λ, for brevity) the set of maximizers of g. The set Λ may be empty.

ii. An algorithm that computes the function g (i.e., for $\lambda \in \mathcal{Q}$ returns the value $g(\lambda)$ and otherwise stops or returns an arbitrary value) is called *piecewise affine*, if the operations it performs on intermediate values that depend on the input vector are restricted to additions, multiplications by constants, comparisons, and copies.

iii. For a piecewise affine algorithm $\mathcal{A}$, denote by $T(\mathcal{A})$ and $C(\mathcal{A})$ the maximum number of operations and the maximum number of comparisons, respectively, performed by $\mathcal{A}$. We assume that this numbers are finite.

iv. If $g = \hat{L}_{\mathcal{C}}$ for some $\mathcal{C} \subset R^{d+1}$, we say that $g' = \hat{L}_{\mathcal{C}'}$ is a *weak approximation* of g, if the set of pieces of g' is a subset of the set of pieces of g ($\mathcal{C}' \subset \mathcal{C}$), and the affine hulls[3] aff Λ_g and aff $\Lambda_{g'}$ are equal. The function $g' = \hat{L}_{\mathcal{C}'}$ is a *minimal* weak approximation of g, if there is no $\mathcal{C}'' \subset \mathcal{C}'$ such that $\hat{L}_{\mathcal{C}''}$ is a weak approximation of g.

Remark 2.3 Suppose that $\mathcal{A}$ is a piecewise affine algorithm. Consider the computation tree (i.e., the tree consisting of all possible computation paths) of $\mathcal{A}$. Observe that all the intermediate values along a computation path, including the final output, can be expressed as linear functions (i.e., are of the form $a^T \hat{\lambda}$) of the input vector. These linear functions can be easily computed and maintained during a single execution of the algorithm. These linear functions map the input vectors whose computation path coincides so far with that same path to the corresponding value. Moreover, the linear function which corresponds to the final output at a single execution is a piece which is active at the input vector.

Remark 2.4 Suppose $g : \mathcal{Q} \subset R^d \to R$ is concave and computable by a piecewise affine algorithm $\mathcal{A}$. It is easy to see that there exists a finite set $\mathcal{C} \subset R^{d+1}$ such that g coincides with $\hat{L}_{\mathcal{C}}$.

Definition 2.5 Suppose $\mathcal{Q} = F_1 \cap \cdots \cap F_k$ is the intersection of k closed halfspaces and $g : \mathcal{Q} \to R$, $g = \hat{L}_{\mathcal{C}}$, is concave and computable by a piecewise affine algorithm.

i. For $\beta \in R^d$, denote

$$\mathcal{Q}_\beta = \bigcap \{F_i \ : \ \beta \in \partial F_i\} \ .$$

Denote by $g_\beta : \mathcal{Q}_\beta \subset R^d \to R$ the function whose pieces are all the pieces of g which are active at β,

$$g_\beta(\lambda) = \min_{c \in \mathcal{C}} \{c^T \hat{\lambda} \ : \ c^T \hat{\beta} = g(\beta)\} \ .$$

Note that $g_\beta = \hat{L}_{\mathcal{C}'}$ where $\mathcal{C}' = \{c \in \mathcal{C} \mid c^T \hat{\beta} = g(\beta)\}$. See Figure 1 for an example. Later, we describe an algorithm for evaluating g_β.

---

[3]The affine hull aff S of a set S is the smallest flat that contains S.

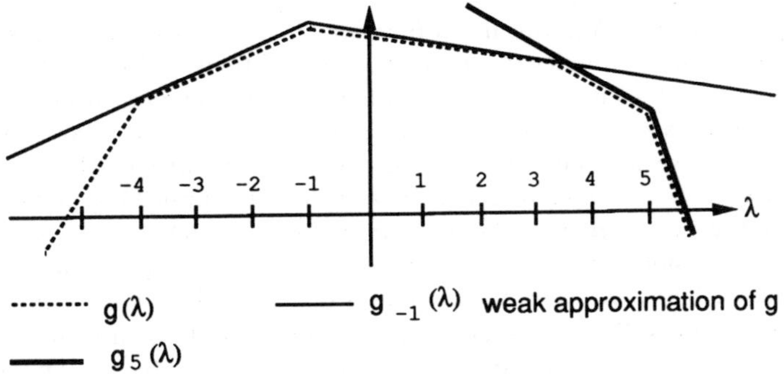

Figure 1: Examples of restrictions of g

ii. Suppose an ℓ-dimensional flat $S \subset R^d$, is represented as a set of solutions of a linear system of equations. There exists an affine mapping M from R^ℓ onto S, which can be computed in $O(d^3)$ time. Denote $Q^S = \{\lambda \in R^\ell \mid M(\lambda) \in Q\}$, and define $g^S : Q^S \to R$ by $g^S = g \circ M$.

Proposition 2.6 *Let $\mathcal{A}$ be a piecewise affine algorithm for evaluating $g : Q \to R$, where $Q \subset R^d$ is given as the intersection of k halfspaces. By modifying $\mathcal{A}$ we can obtain the following piecewise affine algorithms:*

i. *For any given vector $\beta \in Q$, we obtain an algorithm $\mathcal{A}_\beta$ for evaluating g_β so that $T(\mathcal{A}_\beta) = T(\mathcal{A}) + dC(\mathcal{A})$ and $C(\mathcal{A}_\beta) = C(\mathcal{A})$.*

ii. *For any ℓ-dimensional flat $S \subset R^d$, represented as the set of solutions of a system of linear equations, we obtain an algorithm $\mathcal{A}^S$ for evaluating g^S so that $T(\mathcal{A}^S) = T(\mathcal{A}) + O(\ell d)$ and $C(\mathcal{A}^S) = C(\mathcal{A})$.*

Proof: Part ii is straightforward, since we can choose the algorithm $\mathcal{A}^S$ to be a composition of the appropriate affine mapping and $\mathcal{A}$. We discuss the construction of the algorithm $\mathcal{A}_\beta$ for part i. Consider an input vector $\lambda \in Q_\beta$. Let $\epsilon > 0$ be such that for all ϵ' $(0 < \epsilon' \leq \epsilon)$, $\beta + \epsilon'(\lambda - \beta) \in Q$, and the set of pieces of g which are active at $\beta + \epsilon'(\lambda - \beta)$ is equal to the set of pieces which are active at $\beta + \epsilon(\lambda - \beta)$. It is immediate to see that such an ϵ always exists. It follows from the definition that $g_\beta(\lambda)$ is the value of λ at the linear pieces of g which are active at $\beta + \epsilon(\lambda - \beta)$. The algorithm $\mathcal{A}_\beta$, when executed with an input λ, follows the computation path of $\mathcal{A}$ which corresponds to the input $\beta + \epsilon(\lambda - \beta)$. The algorithm computes the linear functions associated with the intermediate values of this path (see Remark 2.3). Recall that the linear function which corresponds to the final value is a piece of g which is active at $\beta + \epsilon(\lambda - \beta)$. Hence, the value of $g_\beta(\lambda)$ is obtained by substituting λ in this linear function. In order to follow the desired run of $\mathcal{A}$, the algorithm $\mathcal{A}_\beta$ mimics the work of $\mathcal{A}$ on additions and multiplications by scalars, keeping track of linear functions rather than just numerical values. When the run of $\mathcal{A}$ reaches a comparison (branching point), $\mathcal{A}_\beta$ does as follows. Without loss of generality we assume that the branching is according to the sign of the linear function $a^T \hat{x}$. In order to decide what to do at a branching point, $\mathcal{A}_\beta$ has to determine the sign of $a^T \hat{x}$ at the point $x = \beta + \epsilon(\lambda - \beta)$. Since ϵ is not given explicitly, the decision cannot be made directly by substitution. The decision is made as follows. The algorithm first computes $\alpha = a^T \hat{\beta}$. If $\alpha \neq 0$, then obviously for any vector y, for sufficiently small number $\epsilon > 0$ $a^T(\widehat{\beta + \epsilon y})$ has the same sign as α. In particular this holds for $y = \lambda - \beta$ and the sign is detected. Otherwise, if $\alpha = 0$, it follows that $a^T(\widehat{\beta + \epsilon(\lambda - \beta)}) = \epsilon a^T \hat{\lambda}$. Hence $a^T \hat{\lambda}$ has the same sign as $a^T(\widehat{\beta + \epsilon(\lambda - \beta)})$. It remains to compute the sign of $a^T \hat{\lambda}$ and branch accordingly. It is easy to verify that $\mathcal{A}_\beta$ evaluates the function g_β for any vector $\lambda \in \mathcal{Q}_\beta$, and performs the stated number of operations. $\blacksquare$

Proposition 2.7 *If $\beta \in \Lambda_g$ then g_β is a weak approximation of g (see [7, 2]) for a proof).*

The goal is to solve the following problem:

Problem 2.8 The input of this problem consists of a polyhedron $\mathcal{Q} = F_1 \cap \cdots \cap F_k$, given as the intersection of k closed halfspaces and a piecewise affine algorithm $\mathcal{A}$ for evaluating a concave function $g : \mathcal{Q} \to R$. Decide whether or not g is bounded. If so, then find a $\lambda^* \in \mathrm{rel\,int}\,\Lambda$. We refer to the following as the "optional" part of the problem: If g is bounded, then find a subset $\mathcal{C}$ of the set of pieces of g, such that $\hat{L}_\mathcal{C}$ is a minimal weak approximation of g, and $|\mathcal{C}| \leq 2d$.

The set $\mathcal{C}$ may be viewed as a *certificate* for the fact that the maximum of the function g does not exceed $g(\lambda^*)$. In the current paper we do not discuss the details of solving the optional part of the problem. See [7, 2] for an existence proof and an algorithm which finds such a set.

 We propose an algorithm for Problem 2.8. In any fixed dimension d, the total number of operations performed by this algorithm is bounded by a polynomial in $T(\mathcal{A})$ and k. The algorithm is based on solving instances of a subproblem, which we call *hyperplane query*. For a given hyperplane H_0, decide on which side of H_0 the function g is either unbounded or attains its maximum. A procedure for hyperplane queries is called an *oracle*. Obviously, an oracle can be utilized to perform a binary search over the polyhedron $\mathcal{Q}$. However, in order to attain an exact solution within time bounds that depend only on d, T, and k, we use the oracle in a more sophisticated way. The number of hyperplane queries needed by the algorithm, and hence the number of oracle calls, is bounded by the number of comparisons performed by $\mathcal{A}$.

We later discuss applying the multi-dimensional search technique, what allows us to do even better. When $\mathcal{A}$ is a parallel algorithm, we may reduce the number of queries corresponding to a single parallel step to a polylogarithm of the number of concurrently performed comparisons.

The function g is a concave piecewise linear mapping. Concave functions have the property that it can be effectively decided which side of a given hyperplane H_0 contains the maximum of the function. If the domain of g does not intersect H_0, then the answer is the side of H_0 which contains the domain of g. Otherwise, the decision can be made by considering a neighborhood of the maximum of the function relative to H_0, searching for a direction of ascent from that point. This principle is explained in detail in [12].

For a hyperplane $H_0 \subset R^d$, we wish to decide on which side of H_0 the set rel int Λ lies. By solving a linear program with d variables and $k+1$ constraints, we determine whether or not $H_0 \cap Q = \emptyset$, and if so, we determine which side of H_0 contains Q. It follows from [12] that this can be done in $O(k)$ time. If $H_0 \cap Q \neq \emptyset$, then the oracle problem solves the original problem, when g is restricted to H_0. If g is unbounded on H_0 the oracle reveals that. If $\Lambda = \emptyset$, or if rel int Λ is either contained in H_0 or extends into both sides of H_0 (i.e., $H_0 \cap$ rel int $\Lambda \neq \emptyset$), then we find $\lambda \in H_0 \cap$ rel int Λ and the oracle will actually solve Problem 2.8.

Problem 2.9 Given are a set $Q = F_1 \cap \cdots \cap F_k$, a piecewise affine algorithm $\mathcal{A}$ which evaluates a concave function $g : Q \to R$, and a hyperplane H_0 in R^d. Do as follows:

i. If $Q \cap H_0 = \emptyset$, recognize which of the two halfspaces determined by H_0 contains Q. Otherwise,

ii. recognize whether or not g is bounded on H_0. If it is, then

iii. find $\lambda \in H_0 \cap$ rel int Λ if such λ exists, and solve Problem 2.8 relative to g. Otherwise, if $H_0 \cap$ rel int $\Lambda = \emptyset$, then

iv. recognize which of the two halfspaces determined by H_0 has either a nonempty intersection with rel int Λ, or has g unbounded on it.

A procedure for solving Problem 2.9 will be called an *oracle* and the hyperplane H_0 will be called the *query hyperplane*. Problem 2.8 is solved by running a modification of the algorithm $\mathcal{A}$, where additions and multiplications are replaced by vector operations and comparisons are replaced by hyperplane queries. Problem 2.9 is solved by three recursive calls to instances of Problem 2.8 of the form $(\mathcal{A}^H, Q^H)$, $(\mathcal{A}_\beta^H, Q_\beta^H)$, where $\beta \in R^d$, and H is a hyperplane (see Definitions 2.5 and 2.6). Note that these algorithms compute, respectively, the functions $g^H : Q^H \to R$, $g_\beta^H : Q_\beta^H \to R$, where Q^H and Q_β^H are subsets of R^{d-1}. Hence, the recursive calls are made to instances of lower dimension.

In Section 5 we propose Algorithm 5.2 for Problem 2.8. The algorithm executes calls to the oracle problem (Problem 2.9) relative to g. An algorithm for the oracle problem is given in Section 3. A call to the oracle is costly. Therefore, one wishes to solve many hyperplane queries with a small number of oracle calls. In Section 4 we discuss the multi-dimensional search technique (introduced in [12] and later refined in [1, 9]).

3   Hyperplane queries

For a hyperplane $H \subset R^d$, we solve Problem 2.9 for g relative to H.

Theorem 3.1 *Problem 2.9 can be reduced to the problem of solving three instances of Problem 2.8 on functions defined on an intersection of at most k closed halfspaces in R^{d-1}. The time complexity of the additional computation is $O(d^3)$.*

Proof: We solve Problem 2.8 with the function g^H, where $H = \{\boldsymbol{\lambda} \in R^d \mid \boldsymbol{a}^T \hat{\boldsymbol{\lambda}} = \alpha\}$. If g is unbounded on H, then this fact is detected; otherwise, suppose $\boldsymbol{\lambda}^{(0)}$ is in the relative interior of the set of maximizers of $g(\boldsymbol{\lambda})$ subject to $\boldsymbol{\lambda} \in H$, and we get the collection $\mathcal{C}^{(0)}$. Let $t^{(0)} = g(\boldsymbol{\lambda}^{(0)})$. We wish to recognize whether $\boldsymbol{\lambda}^{(0)}$ is also a relative interior point of the set of global maxima (i.e., relative to R^d). If not, then we wish to decide whether for all $\boldsymbol{\lambda}^* \in R^d$ such that $g(\boldsymbol{\lambda}^*) \geq g(\boldsymbol{\lambda}^{(0)})$, necessarily $\boldsymbol{a}^T \boldsymbol{\lambda}^{*\prime} > \alpha$, or whether for all of them $\boldsymbol{a}^T \boldsymbol{\lambda}^{*\prime} < \alpha$. These are the two possible cases. Consider the function $g_{\lambda^{(0)}}$. We solve Problem 2.8 on two restrictions of $g_{\lambda^{(0)}}$ to hyperplanes (see Proposition 2.6), where in one case it is restricted to $H^{(1)} = \{\boldsymbol{\lambda} \mid \boldsymbol{a}^T \hat{\boldsymbol{\lambda}} = \alpha - 1\}$, and in the other to $H^{(-1)} = \{\boldsymbol{\lambda} \mid \boldsymbol{a}^T \hat{\boldsymbol{\lambda}} = \alpha + 1\}$. Note that the domains $\mathcal{Q}_{\lambda^{(0)}}^{H^{(\delta)}} (\delta \in \{-1, 1\})$ are $(d-1)$-dimensional. Denote the respective optimal values of $g_{\lambda^{(0)}}^{H^{(\delta)}}$ by $t^{(\delta)}$ $(\delta \in \{-1, 1\})$, and let $\mathcal{C}^\delta$ be the respective minimal weak approximations. Only one of the optimal values $t^{(1)}, t^{(-1)}$ can be greater than $t^{(0)}$. If this is the case, or if one of $t^{(1)}, t^{(-1)}$ equals $t^{(0)}$ and the other is smaller, then the side of the hyperplane that contains $\operatorname{rel int} \Lambda$ is determined. Otherwise, if both values are less than or both values are equal to $t^{(0)}$, then $t^{(0)}$ is the global optimal value. In the latter case $\boldsymbol{\lambda}^{(0)} \in \operatorname{rel int} \Lambda$. It follows from Proposition 2.7 that the pieces of g active in a minimal weak approximation have the value $t^{(0)}$ at $\boldsymbol{\lambda}^{(0)}$. Thus, a minimal weak approximation of the function $g_{\lambda^{(0)}}$ is a minimal weak approximation of g. It follows from analysis done in [7, 2] that by using $O(d^3)$ operations we can construct a minimal weak approximation of $g_{\lambda^{(0)}}$. Furthermore, the number of pieces involved in a minimal weak approximation is at most $2d$. $\blacksquare$

As an example, consider an application of the algorithm described in the proof to decide on which side of the hyperplane $H = \{2\}$ the function $g(\boldsymbol{\lambda}) = \min\{\boldsymbol{\lambda}/5 + 2, -4\boldsymbol{\lambda} + 12.5\}$ is maximized (see Figure 2). Note that maximizing a function $f : R \to R$ on a hyperplane corresponds to evaluating it at a single point. Therefore, the maximum value of g on H is 2.4. The algorithm considers the restriction $g_2 = \boldsymbol{\lambda}/5 + 2$,

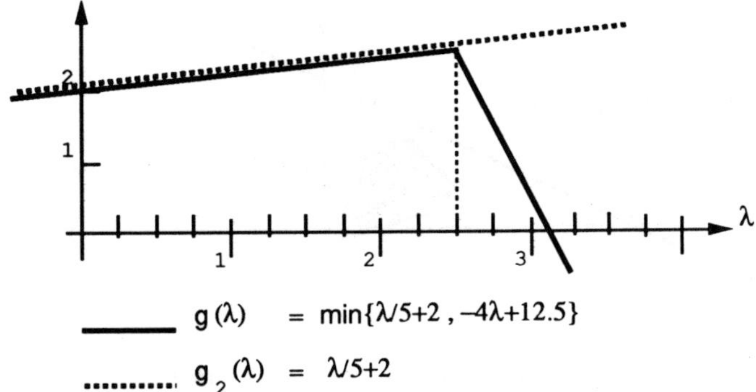

Figure 2: Example: hyperplane query at $H = \{2\}$

and maximizes it on the hyperplanes $H^{(1)} = \{1\}$ and $H^{(-1)} = \{3\}$. The corresponding maxima are $t^{(1)} = 2.2$ and $t^{(-1)} = 2.6$, and hence, the algorithm concludes that the maximizers of g are contained in the halfspace $\{\lambda \in R | \lambda > 2\}$. Observe that this conclusion could not have been made if the algorithm considered the values of g, rather than the values of the restriction g_2, at the hyperplanes $\{1\}$ and $\{3\}$.

4　Employing multi-dimensional search

The definitions and propositions stated in this section appeared in [3, 7, 2]. They are presented here to allow for an independent reading of this paper. For proofs, the reader is referred to [3, 7, 2]. The multi-dimensional search problem was defined and used in [12] for solving linear programming problems in fixed dimension. In this section we employ it to achieve better time bounds.

Definition 4.1 We define a partial order on $R^d \setminus \{0\}$, relative to a concave function $g : Q \to R$, where $Q \subset R^{d-1}$ is a nonempty polyhedral set. For any pair of distinct vectors $a_1, a_2 \in R^d$, denote

$$H = H(a_1, a_2) = \{\lambda \in R^{d-1} : a_1^T \hat{\lambda} = a_2^T \hat{\lambda}\} \ .$$

If g is unbounded on $H(a_1, a_2)$ or if $H(a_1, a_2) \cap \mathrm{rel\,int}\, \Lambda \neq \emptyset$, then we write $a_1 <>_\Lambda a_2$. Otherwise, g can be unbounded on at most one of the open halfspaces determined by H, and also $\mathrm{rel\,int}\, \Lambda$ can intersect at most one of these open halfspaces. If g is undefined on H (i.e., $Q \cap H = \emptyset$), then Q is contained in one of these halfspaces. We denote $a_1 <_\Lambda a_2$ (respectively, $a_1 >_\Lambda a_2$) if there exists a $\hat{\lambda} \in \mathrm{rel\,int}\, \Lambda$ such that $a_1^T \hat{\lambda} < a_2^T \hat{\lambda}$ (respectively, $a_1^T \hat{\lambda} > a_2^T \hat{\lambda}$), in which case the same holds for all these $\hat{\lambda}$'s, or if g is unbounded on the halfspace determined by the inequality $a_1^T \hat{\lambda} < a_2^T \hat{\lambda}$

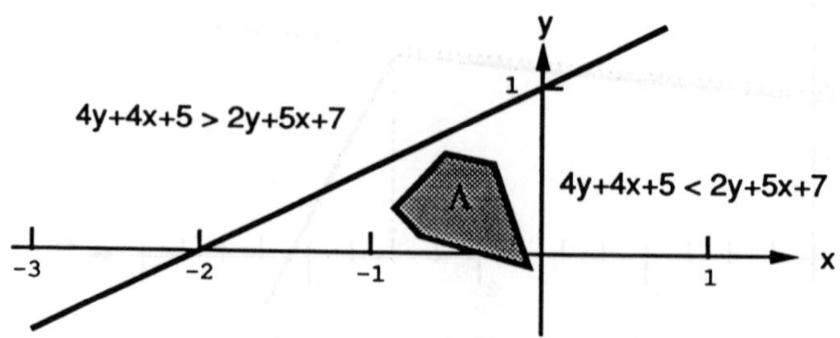

Figure 3: An example where $(4, 4, 5) <_\Lambda (2, 5, 7)$

(respectively, $a_1^T \hat{\lambda} < a_2^T \hat{\lambda}$). See Figure 3 for an example. We also use the notation $<_P$ for a similar partial order relative to any set P.

Problem 4.2 Suppose we are given finite sets $A_1, \ldots, A_r$ of nonzero vectors, where $A_i = \{a_1^i, \ldots, a_{s_i}^i\}$ $(a_j^i \in R^d)$ and $s = \sum s_i$. The goal is either to find a minimal element, with respect to the partial order $<_\Lambda$, in each of the sets A_i, or (if we encounter two incomparable elements) to reduce the problem to a lower dimension. More specifically, we need to do either one of the following:

   i. Find a collection of closed halfspaces whose intersection P contains rel int Λ, and indices $1 \leq m_i \leq s_i$ $(i = 1, \ldots, r)$ as follows. For every $1 \leq i \leq r$ and every $1 \leq j \leq s_i,\ j \neq m_i$, we have $a_{m_i}^i <_\Lambda a_j^i$ and $a_{m_i}^i <_P a_j^i$.

   ii. Find a hyperplane H such that either g is unbounded on H or $H \cap$ rel int $\Lambda \neq \emptyset$.

Proposition 4.3 *Problem 4.2 can be solved using $O(\gamma(d-1)\log s)$ oracle calls plus additional computation which can be performed in either*

   i. $O(\gamma(d-1)\log^2 s)$ parallel time on $O(s)$ processors, or

   ii. $O(\gamma(d-1)s\log s)$ sequential time.

The function $\gamma(d)$ arises from the multi-dimensional search [12]. It follows from [1, 9] that $\gamma(d) = 3^{O(d^2)}$.

5   The algorithm

The algorithm described below solves Problem 2.8. It finds a vector $\lambda^* \in$ rel int Λ, unless g is unbounded. It also returns a collection $\mathcal{C}$ of pieces of g whose minimum envelope $\hat{L}_\mathcal{C}$ is a minimal weak approximation of g. The number of vectors in $\mathcal{C}$ is at most $2d$.

Definition 5.1 For a piecewise affine algorithm $\mathcal{A}$, we define the corresponding *lifted computation*. The lifted computation is a run of the algorithm on a set of inputs. The computation is done symbolically on linear functions instead of on scalars. It follows the path on the computation tree of $\mathcal{A}$ that corresponds to input vectors which are in Λ. The additions and scalar multiplications are trivially generalized to operations on linear functions. When a comparison is done between $f_1^T \hat{\lambda}$ and $f_2^T \hat{\lambda}$, it is resolved according to the partial order $<_\Lambda$. We compute the hyperplane $H(f_1, f_2)$ and solve Problem 2.9 (hyperplane query) relative to H. The hyperplane query decides whether or not the vectors are comparable. If they are, it decides whether $f_1 <_\Lambda f_2$. If $f_1 <>_\Lambda f_2$, then the lifted computation halts since an oracle call resulted in a solution to Problem 2.8. Otherwise, the resolved hyperplane query tells us which of the halfspaces defined by $H(f_1, f_2)$ contains the set $\mathrm{rel\,int}\,\Lambda$, and the comparison is resolved.

Sets of independent comparisons performed by $\mathcal{A}$ correspond to sets of independent hyperplane queries. Recall from Section 4 that a set of independent hyperplane queries can be solved by performing a logarithmic number of "oracle" calls. The lifted computation maintains a set $\mathcal{H}$ of closed halfspaces which is initially empty. Whenever an oracle call is executed the resulting halfspace is added to $\mathcal{H}$.

Algorithm 5.2 [Find a vector $\lambda \in \mathrm{rel\,int}\,\Lambda_g$]

Step 1. Run the lifted computation, collecting into $\mathcal{H}$ all the halfspaces resulting from oracle calls where comparisons are resolved. If the computation halts, then some comparison is not resolved but a global solution is found, so stop. Otherwise, denote by $m = (m_1, \ldots, m_{d+1})^T \in R^{d+1}$ the piece of g that corresponds to the computation path followed.

Step 2. Denote by P the intersection of the halfspaces in $\mathcal{H}$.

    i. Compute $\lambda^* \in \mathrm{rel\,int}\,P \cap Q$. This amounts to a linear programming problem with d variables and $|\mathcal{H}|$ constraints, and hence it can be solved in $O(|\mathcal{H}|)$ sequential time [12]. Note that the size of $\mathcal{H}$ is bounded by the number of oracle calls.

    ii. If $\hat{L}_{\{m\}}$ is not constant on R^d, that is, not all of $m_1, m_2, \ldots, m_d$ equal zero, then g is unbounded. Otherwise,

    iii. consider $g(\lambda^*) = m_{d+1}$. The function $\hat{L}_{\{m\}}$ is a weak approximation of g, and $P = \Lambda$. Hence, $\lambda^* \in \mathrm{rel\,int}\,\Lambda$. Output λ^* and $\mathcal{C} = \{m\}$.

6   Correctness

If an oracle call results in a solution during Step 1 of Algorithm 5.2, then correctness follows by induction on the dimension. We now assume that no oracle call resulted in a solution during Step 1. In this case, a collection $\mathcal{H}$ of closed halfspaces is obtained. Recall that if an oracle call on a hyperplane H did not result in a solution, then the halfspace F returned has the following properties: (i) if the function g is bounded then $\Lambda \subset F$ but $\Lambda \not\subset H$, (ii) if the function g is unbounded, then it must be bounded on the hyperplane H, and unbounded on the halfspace F. Let P be the polyhedron $P = \bigcap_{F \in \mathcal{H}} F$. It follows that if g is bounded then $P \supset \Lambda$, and if g is unbounded then it must be bounded outside and on the boundary of P. Note that P must be of full dimension ($\dim P = d$), for if not, then it must be contained in one of the query hyperplanes, which contradicts the previous statement.

Observe that for all pairs a_1, a_2 of vectors compared by the lifted computation, one of the following must hold: either $a_1 <_\Lambda a_2$ and $a_1 <_P a_2$, or $a_2 <_\Lambda a_1$ and $a_2 <_P a_1$. The latter is obvious when we call the oracle to resolve each hyperplane query, and it is easy to see that it still holds when we employ the multi-dimensional search technique (see Problem 4.2 and Proposition 4.3) and solve these hyperplane queries by a smaller number of oracle calls. Thus, the piece m (the maximizer) found by the lifted computation must satisfy $m <_\Lambda c$ and hence $m <_P c$ for all pieces c of g. It follows that $g(\lambda) = m^T \hat{\lambda}$ for all $\lambda \in P$. Thus, g is unbounded if and only if $\hat{L}_{\{m\}}$ is not constant, and the correctness of step ii follows. To show the correctness of step iii assume that $\hat{L}_{\{m\}}$ is constant, and thus $g = m_{d+1}$ for all $\lambda \in P$. Since $P \supseteq \Lambda$ we have $P = \Lambda$. It follows that $\lambda^* \in \text{rel int}\,\Lambda$, $\text{aff}\,\Lambda = R^d$, and $\hat{L}_{\{m\}}$ is a minimal weak approximation of g.

7   Complexity

Consider the algorithm $\mathcal{A}$. Suppose that the $C(\mathcal{A})$ comparisons performed by $\mathcal{A}$ can be divided into r phases, where C_i independent comparisons are performed during phase i ($i = 1, \ldots, r$). It follows from Proposition 4.3, that the lifted computation can be implemented in such a way that it performs $\gamma(d) \sum_{i=1}^{r} \lceil \log C_i \rceil$ oracle calls. It follows from Theorem 3.1 that each oracle call involves three recursive calls to instances of Problem 2.8 of lower dimension. The piecewise affine algorithms that correspond to these instances have the same number of comparisons as $\mathcal{A}$, divided into phases in the same way, and $O(d)$ times more operations. Thus, the total number of operations needed for the lifted computation is

$$d^3 \gamma(d) k T(\mathcal{A}) \left(\sum_{i=1}^{r} \lceil \log C_i \rceil \right)^d .$$

The number of parallel phases needed in the above computation is bounded by the product of the number of phases of the algorithm $\mathcal{A}$ with $\sum_{i=1}^{r} \lceil \log C_i \rceil)^d$. If the algorithm $\mathcal{A}$ is inherently sequential, then the total number of operations is $O(k T(\mathcal{A}) C(\mathcal{A})^d)$.

8 Parametric extensions of problems

The technique described in this paper was employed in [3, 7] to get algorithms for the parametric extensions of the minimum cycle and the minimum cycle-mean problems. This technique can be applied to a variety of other problems, where we consider a strongly polynomial algorithm for a problem and obtain a strongly polynomial algorithm for a parametric extension of the problem (when the number of parameters is fixed). We state the conditions where this technique is applicable and present applications.

Definition 8.1 [Parametric extensions]

i. A *problem* $S : \mathcal{P} \to R$ is a mapping from a set $\mathcal{P}$ of instances into the set of real numbers. We say that $S(P)$ is the *solution* of the problem for the instance $P \in \mathcal{P}$. Suppose that every instance $P \in \mathcal{P}$ has a *size* $\|P\|$ associated with it. The size of an instance is not necessarily defined to be the number of bits in its representation. It may be any natural parameter (for example, the number of edges in a weighted graph).

ii. Let $\mathcal{A}$ be an algorithm that computes $S(P)$. Denote by $T_{\mathcal{A}}(P)$ the number of elementary operations the algorithm performs on the instance P. The algorithm $\mathcal{A}$ is *polynomial* if $T_{\mathcal{A}}(P) = O(p(\|P\|))$ for some polynomial $p(\bullet)$.

iii. A *d-parametric extension* $P^d = (\mathcal{M}, \mathcal{Q})$ of $\mathcal{P}$ is defined as follows, where $\mathcal{Q} \subset R^d$ is a polyhedron given as an intersection of k halfspaces, and $\mathcal{M} : \mathcal{Q} \to \mathcal{P}$ is a mapping from points $\boldsymbol{\lambda} \in \mathcal{Q}$ to instances of $\mathcal{P}$. The extension P^d corresponds to a subset of instances $\{\mathcal{M}(\boldsymbol{\lambda}) \mid \boldsymbol{\lambda} \in \mathcal{Q}\} \subset \mathcal{P}$. We refer to $\mathcal{M}(\boldsymbol{\lambda}) \in \mathcal{P}$ as the instance of $\mathcal{P}$ *induced* by $\boldsymbol{\lambda}$. For an extension P^d, we define $g : \mathcal{Q} \to R$ as a mapping from vectors $\boldsymbol{\lambda} \in \mathcal{Q}$ to the solution of the corresponding induced instance $g(\boldsymbol{\lambda}) = S(\mathcal{M}(\boldsymbol{\lambda}))$. A *solution of the parametric extension* P^d is defined as follows. Consider the maximum of $g(\boldsymbol{\lambda})$. If it is finite, a solution consists of the maximum and a vector $\boldsymbol{\lambda} \in R^d$ that belongs to the relative interior of the set of vectors which maximize S. Formally, if $\mathcal{Q}$ is empty or if $S(\mathcal{M}(\boldsymbol{\lambda}))$ is unbounded on $\mathcal{Q}$, these facts are recognized. Otherwise, a pair $(m, \boldsymbol{\lambda}^*) \in R \times R^d$, where $m = \max_{\boldsymbol{\lambda} \in \mathcal{Q}} g(\boldsymbol{\lambda})$, and $\boldsymbol{\lambda}^* \in \mathrm{rel\,int}\{\boldsymbol{\lambda} \mid g(\boldsymbol{\lambda}) = m\}$ is computed. We denote $T = \max_{\boldsymbol{\lambda} \in \mathcal{Q}} T_{\mathcal{A}}(\mathcal{M}(\boldsymbol{\lambda}))$.

Theorem 8.2 *Let $S : \mathcal{P} \to R$ be a problem in the sense of Definition 8.1. Let $\mathcal{A}$ be an algorithm that evaluates S, and let $P^d = (\mathcal{M}, \mathcal{Q})$ (where $|\mathcal{Q}| = k$) be a corresponding parametric extension. We assume that*

i. the function g is concave,

ii. the mapping $\mathcal{M}$ is computable by a piecewise affine algorithm $\mathcal{A}_{\mathcal{M}}$ (see Definition 2.2) in less than T operations, and *iii. the combined algorithm which computes an instance $\mathcal{A}_{\mathcal{M}}(\lambda) \in \mathcal{P}$ and applies $\mathcal{A}$ to $\mathcal{A}_{\mathcal{M}}(\lambda)$, is piecewise affine.*

Denote by C the maximum (over $\lambda \in Q$) number of comparisons performed by the combined algorithm. Suppose the comparisons can be divided into r sets of sizes $C_1, \ldots, C_r$ ($C = \sum_{i=1}^{r} C_i$) such that the algorithm runs in r phases, where C_i independent comparisons are performed in phase i.

Under these conditions, the d-parametric extension P^d can be solved using

$$\beta(d) kT (\sum_{i=1}^{r} \lceil \log C_i \rceil)^d \text{ operations, where } \beta(d) = 3^{O(d^2)} .$$

Remark 8.3 In the above formulation we defined a problem as a mapping into the set of real numbers $S : \mathcal{P} \to R$. The results generalize to cases where the range of S is R^ℓ for $\ell > 1$ and the notions of maximum and concavity of g are defined with respect to the lexicographic order as discussed in the introduction.

Below we present some applications of Theorem 8.2. Additional applications were found by Norton, Plotkin, and Tardos [13].

Adding variables to LP's with two variables per inequality   Linear programming problems with at most two variables in each constraint and in the objective function were shown to have a strongly polynomial time algorithm by Megiddo [11]. Lueker, Megiddo and Ramachandran [10] gave a polylogarithmic time parallel algorithm for the problem which uses a quasipolynomial number of processors. The best known time bounds for the problem were given in [6, 2]. Cosares, using nested parametrization, extended Megiddo's strong polynomiality result to allow objective functions which have a fixed number of nonzero coefficients. This result can be further extended to include the following. For a fixed d, we consider linear programming problems as above, but we allow certain d additional variables to appear anywhere in the constraints and in the objective function without being "counted." This problem is a d-parameter extension of the two variables per constraint problem, where the "parameters" are the d additional variables. For each choice of values for the parameters we have a corresponding induced system with two variable per constraint. It is easy to verify that the conditions of Theorem 8.2 hold. Hence, this class of problems also has a strongly polynomial time algorithm, and a polylogarithmic time parallel algorithm which uses a quasipolynomial number of processors.

Parametric flow problems   Theorem 8.2 was applied in [5, 8] to generate strongly polynomial algorithms for parametric flow problems with a fixed number of parameters and to some constrained flow problems with a fixed number of additional constraints. Complementing results showing the P-completeness of these problems when the number of parameters is not fixed, were also given.

References

[1] K. L. Clarkson. Linear programming in $O(n \times 3^{d^2})$ time. *Information Processing Let.*, 22:21–27, 1986.

[2] E. Cohen. *Combinatorial Algorithms for Optimization Problems*. PhD thesis, Department of Computer Science, Stanford University, Stanford, Ca., 1991.

[3] E. Cohen and N. Megiddo. Strongly polynomial and NC algorithms for detecting cycles in dynamic graphs. In *Proc. 21st Annual ACM Symposium on Theory of Computing*, pages 523–534. ACM, 1989.

[4] E. Cohen and N. Megiddo. Maximizing concave functions in fixed dimension. Technical Report RJ 7656 (71103), IBM Almaden Research Center, San Jose, CA 95120-6099, August 1990.

[5] E. Cohen and N. Megiddo. Complexity analysis and algorithms for some flow problems. In *Proc. 2nd ACM-SIAM Symposium on Discrete Algorithms*, pages 120–130. ACM-SIAM, 1991.

[6] E. Cohen and N. Megiddo. Improved algorithms for linear inequalities with two variables per inequality. In *Proc. 23rd Annual ACM Symposium on Theory of Computing*, pages 145–155. ACM, 1991.

[7] E. Cohen and N. Megiddo. Strongly polynomial time and NC algorithms for detecting cycles in periodic graphs. *J. Assoc. Comput. Mach.*, 1992. To appear.

[8] E. Cohen and N. Megiddo. Complexity analysis and algorithms for some flow problems. *Algorithmica*, 1993. To appear.

[9] M. E. Dyer. On a multidimensional search technique and its application to the Euclidean one-center problem. *SIAM J. Comput.*, 15:725–738, 1986.

[10] G. S. Lueker, N. Megiddo, and V. Ramachandran. Linear programming with two variables per inequality in poly log time. *SIAM J. Comput.*, 19(6):1000–1010, 1990.

[11] N. Megiddo. Towards a genuinely polynomial algorithm for linear programming. *SIAM J. Comput.*, 12:347–353, 1983.

[12] N. Megiddo. Linear programming in linear time when the dimension is fixed. *J. Assoc. Comput. Mach.*, 31:114–127, 1984.

[13] C. H. Norton, S. A. Plotkin, and É. Tardos. Using separation algorithms in fixed dimension. In *Proc. 1st ACM-SIAM Symposium on Discrete Algorithms*, pages 377–387. ACM-SIAM, 1990.

Complexity in Numerical Optimization, pp. 88-106
P.M. Pardalos, Editor
©1993 World Scientific Publishing Co.

Approximating the Steiner Minimum Tree

Ding-Zhu Du[1]
*Computer Science Department, University of Minnesota, Minneapolis, MN 55455,
USA and Institute of Applied Mathematics, Academia Sinica, Beijing 100080, PRC*

Abstract

Finding a shortest network interconnecting a given set of points in a metric
space is called the Steiner tree problem. This problem is known to be NP-
hard in several metric spaces, including euclidean metric and rectilinear metric.
Thus, the study on approximation solutions for this problem is very important.
In this paper, we survey recent developments on this topic.

1   Introduction

The Steiner tree problem is a classic intractable problem with many applications in
the design of computer circuits, long-distance telephone lines, or mail routing, etc.
Given a set of points in a metric space, the problem is to find a shortest network
interconnecting the points in the set. Such a shortest network is called a *Steiner minimum tree* on the point set. The Steiner tree problem can be seen as a generalization
of Fermat's problem. Three hundred years ago, Fermat proposed a problem of finding
a point to minimize the total distance from this points to three given points in the
euclidean plane; its solution is exactly the Steiner minimum tree on the three points.
The general form of the Steiner tree problem was proposed by Jarnik and Kossler [34]
in 1934. During the last two years, several important developments on the Steiner
tree problem have been made. They are all connected to approximating the Steiner
minimum tree. In this paper, we survey those developments with emphasis on the
analysis of approximation performance ratios.

---

[1]Support in part by the NSF under grant CCR-9208913

Let P be a set of n points in a metric space. A tree interconnecting points in P is called a *Steiner tree* (ST) on P if all leaves in the tree belong to P. The Steiner minimum tree (SMT) is the shortest Steiner tree. A Steiner tree may contain some vertices not in P. Such vertices are called *Steiner points* (S-point) while vertices in P are called *regular points*. Computing the SMT is an NP-hard problem in euclidean metric, in rectilinear metric, and in graphs [28, 27, 36, 25]. A *spanning tree* on P is a Steiner tree without Steiner points. A shortest spanning tree on P is called a *minimum spanning tree* on P. The minimum spanning tree can be computed fast. For example, it needs only $O(n \log n)$ time in the euclidean plane [8]. Thus, the minimum spanning tree can be seen as a polynomial-time approximation of the SMT. Currently, polynomial-time heuristics for the SMT [2, 3, 24, 23, 35, 37, 38, 39, 40, 43, 49, 50, 52] can be classified into two classes, MST-based heuristics and Sub-SMT-based heuristics. In the next three sections, we discuss the performance ratios of the minimum spanning tree and these two types of heuristics, respectively, where the performance ratio is the largest lower bound for the ratio of lengths between the SMT and the approximation solution for the same set of points.

2 Minimum spanning trees

The minimum spanning tree is a bounded heuristic of the SMT. In fact, in any mertic space, for any point set P,

$$L_m(P) \leq 2L_s(P)$$

where $L_m(P)$ and $L_s(P)$ are the lengths of the minimum spanning tree and the SMT on P, respectively [36]. The performance ratio of the minimum spanning tree, defined by

$$\rho(S) = \inf_{P \subset S} \frac{L_s(P)}{L_m(P)},$$

is called the *Steiner ratio* in the metric space S.

Determine the Steiner ratio in various metric spaces is a traditional combinatorial problem. Hwang [32] showed that the Steiner ratio in the rectilinear plane is 2/3. Upon many efforts [10, 11, 20, 44], Du and Hwang [21] proved the truth of Gilbert-Pollak conjecture [29] that the Steiner ratio in the Euclidean plane is $\sqrt{3}/2$. Gilbert and Pollak [29] also suggested that the Steiner ratio in the n-dimensional Euclidean space is achieved by the n-dimensional regular simplex. Smith [47] showed, with a computer proof, that this suggestion is wrong for $3 \leq n \leq 9$. Recently, Du and Smith [22] found a simple argument which shows that for $n \geq 3$, Gilbert and Pollak's suggestion is wrong.

The argument is based on a property of Euclidean SMTs: Each Steiner point has degree exactly three and each angle at a Steiner point is exactly $120°$. The main idea is as follows.

Let us first look at the case $n = 3$. Consider four unit regular simplices (i.e., regular simplices of unit edge length) S_1, S_2, S_3, S_4 (see Figure 1) whose vertices are

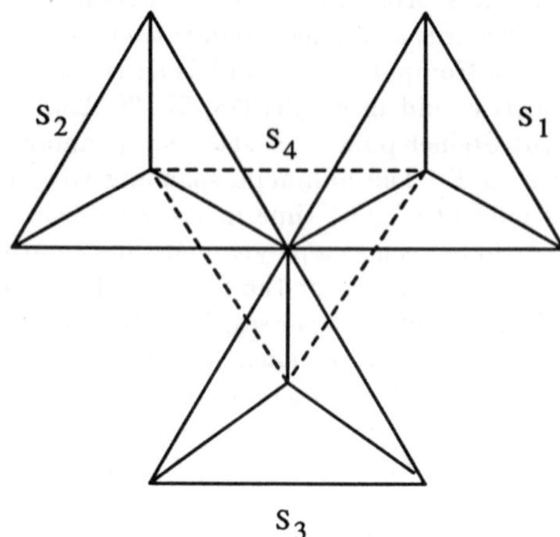

Figure 1: The four simplices in the 3-dimensional space.

defined as follows:

$$
\begin{array}{llll}
S_1: & (0,0,0), & (1,0,0), & (\tfrac{1}{2}, \tfrac{\sqrt{3}}{2}, 0), & (\tfrac{1}{2}, \tfrac{\sqrt{3}}{6}, \tfrac{\sqrt{3}}{3}); \\
S_2: & (0,0,0), & (-1,0,0), & (-\tfrac{1}{2}\tfrac{\sqrt{3}}{2}, 0), & (-\tfrac{1}{2}, \tfrac{\sqrt{3}}{6}, \tfrac{\sqrt{3}}{3}); \\
S_3: & (0,0,0), & (\tfrac{1}{2}, \tfrac{\sqrt{3}}{2}, 0), & (-\tfrac{1}{2}, \tfrac{\sqrt{3}}{2}, 0), & (0, -\tfrac{\sqrt{3}}{3}, \tfrac{\sqrt{3}}{3}); \\
S_4: & (0,0,0), & (\tfrac{1}{2}, \tfrac{\sqrt{3}}{6}, -\tfrac{\sqrt{3}}{3}), & (-\tfrac{1}{2}, \tfrac{\sqrt{3}}{6}, -\tfrac{\sqrt{3}}{3}), & (0, \tfrac{\sqrt{3}}{3}, -\tfrac{\sqrt{3}}{3}).
\end{array}
$$

Let P be the union of vertex sets of the four simplices. It is easy to verify that every two points in P has the distance at least one. In fact, these 12 points around the original point $o(0,0,0)$ are chosen from Kepler's spherical packing in which 12 unit balls around the unit ball at the point o [31]; the 12 points are centers of the 12 unit balls. Since every pair of points in P has distance at least one, we see that

$$
L_m(P) = 12 = \sum_{i=1}^{4} L_m(S_i).
$$

Suppose that the Gilbert-Pollak suggestion is true for $n = 3$. Then

$$
\rho(\mathbf{R}^3) = \frac{L_s(S_i)}{L_m(S_i)} \text{ for } i = 1, 2, 3, 4.
$$

Let T be the union of the SMT's on S_i for $i = 1, 2, 3, 4$. Then the length of T equals $\rho(\mathbf{R}^3)L_m(P) \le L_s(P)$. Thus, T is an SMT on P. However, the point o in T

has degree four, contradicting the property of the Euclidean SMT. From the above argument, it is easy to see that to disprove the Gilbert-Pollak's suggestion in the n-dimensional space, it suffices to find four n-dimensional unit regular simplices such that

(a) they have a vertex in common and

(b) the distance between any pair of vertices is at least one.

In the n-dimensional Euclidean space, there are actually $n + 1$ n-dimensional unit regular simplices satisfying the above conditions. To give a simple construction for these $n + 1$ regular simplices, consider an n-dimensional subspace in the $(n + 1)$-dimensional space, determined by the equation

$$x_1 + x_2 + \cdots + x_{n+1} = 0.$$

The $n + 1$ n-dimensional regular simplices S_i for $i = 1, \cdots, n + 1$ in this subspace are as follows:

> S_i consists of $n+1$ vertices. One is the origin o; in each of other n vertices, the ith component is $\sqrt{2}/2$ and other components are all 0's but exactly one $-\sqrt{2}/2$.

Du and Hwang's proof for the Gilbert-Pollak conjecture suggests a new approach to prove the lower bound, especially to determine the Steiner ratio. In their approach, the central part is a minimax theorem about minimizing the maximum value of several concave functions over a polytope as follows.

Theorem 1 (Du-Hwang Minimax Theorem) *Let $f(x) = \max_{i \in I} g_i(x)$ where I is a finite index set and $g_i(x)$ is a continuous, concave function over a polytope X. Then the minimum value of $f(x)$ over the polytope X is achieved at some* critical *point, namely, a point satisfying the following property:*

> *(*) There exists an extreme subset[2] Y of X such that $x \in Y$ and the index set $M(x)$ $(= \{i \mid f(x) = g_i(x)\})$ is maximal over Y.*

The Steiner ratio problem is first transformed to a minimax problem such that

> $g_i(x)=$ (the length of a Steiner tree) - (the Steiner ratio)·(the length of a spanning tree with graph structure i)

where x is a vector whose components are edge-lengths of the Steiner tree. The minimax theorem reduces the minimax problem to the problem of finding the minimax value of the concave functions at critical points. Then each critical point is transformed back to an input set of points with special geometric structure; it is a subset

---

[2]A subset Y is an extreme subset of X if for any two points x and y in X, $\alpha x + (1 - \alpha)y \in Y$ for some $0 < \alpha < 1$ implies $x, y \in Y$.

of a lattice formed by equilateral triangles. This special structure enables us to verify the conjecture corresponding to the non-negativeness of minimax value of the concave functions.

Clearly, in order to use the minimax approach to solve a problem, three questions should be addressed:

(1) How can the problem be transformed to a minimax problem meeting the condition that the functions are concave?

(2) How can the critical geometric structure be determined ?

(3) How can the function value on the critical structure be verified?

Developping techniques for answering these three questions will enable us to solve more open problems. In fact, many problems can be transformed into the minimax problem. However, usually, one of the above three questions is difficult to solve.

Using this approach, Gao, Du and Graham [26] recently determined the tight lower bound for the Steiner ratio in normed planes as follows.

Theorem 2 *In any normed plane, the Steiner ratio is at least 2/3.*

This resolves a conjecture made by Clieslik [12] and Du, Gao, Graham, Liu and Wan [17] independently. The main part of their work is to overcome the difficulty in answering question (3) for this particular problem. In fact, Du and Hwang [21] already gave solutions for questions (1) and (2). Let us sketch their proof in the following.

An SMT is *full* if every regular point is a leaf. In a plane with differentiable and strictly convex norm, every full SMT consists of three sets of parallel segments [17]. A tree is called a *3-regular tree* if every vertex which is not a leaf has degree three. So, in a plane with strictly convex and differentiable norm, every full SMT is a 3-regular tree. This implies that every full SMT for more than three points must have a local structure either in Figure 2(a) or in Figure 2(b) [16].

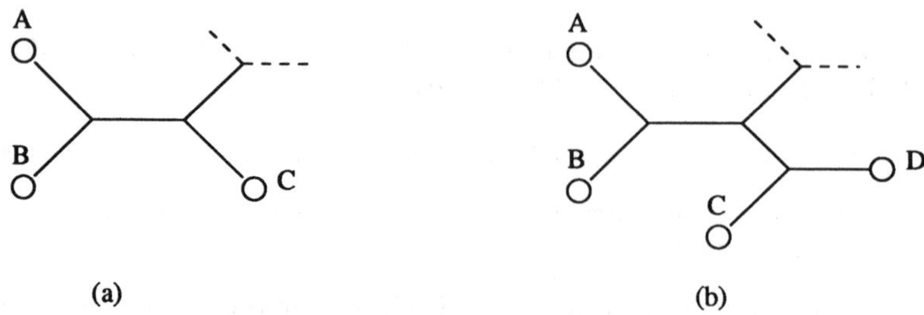

Figure 2: Local structures of full SMT.

Consider a full SMT T in a plane with strictly convex and differentiable norm. If we inflate the edge of T to have an ϵ width, then T is a polygonal region with a boundary. Two regular points are called *adjacent* if they are consecutive on the boundary. For example, in Figure 2, B and C are adjacent regular points of A. Each regular point A has exactly two adjacent regular points. One can be reached from A going along T in clockwise direction. The other one can be reached from A going along T in counterclockwise direction. These two paths are called *clockwise* path and *counterclockwise* path, respectively. A path is a *monotone* path if it is either a clockwise path or a counterclockwise path.

Connecting all pairs of adjacent regular points by straight line segment forms a polygon, called the *characteristic area* of T. A spanning tree lying inside of the characteristic area is called *inner* spanning tree. The minimum inner spanning tree is the shortest one among inner spanning trees. A point set P is called a *critical set* if there is a minimum Steiner tree T for P such that the union of minimum inner spanning tree with respect to T divides the characteristic area of T into equilateral triangles. Based on these equilateral triangles, we can draw a lattice consisting of equilateral triangle cells. The vertices on this lattice are called *lattice points*.

Since two similar sets has the same ratio of lengths between the minimum Steiner tree and the minimum spanning tree, we may consider only the critical set with equilateral triangles of edges of unit length. Clearly, for every critical set, a minimum inner spanning tree is also a minimum spanning tree; its length is $n - 1$ where n is the number of regular points. Also, two adjacent regular points have distance one.

Denote

$$\rho_n = \min_{|P|=n} \frac{L_s(P)}{L_m(P)}.$$

If $\rho_{n-1} > \rho_n$, then n is called a *stair* number. Du and Hwang [21] proved the following.

Theorem 3 *In a plane with a strictly convex and differentiable norm, if n is a stair number, then ρ_n is achieved by some critical set.*

To prove Theorem 2, first note that the norm can be assumed to be differentiable and strictly convex because every norm can be approximated by a sequence of differentiable and strictly convex norms. By contradiction, assume that Theorem 2 is false. Let n be the minimum number such that $\rho < 2/3$. Then n is a stair number. Thus, there exists a critical set P of n points such that

$$\frac{L_s(P)}{L_m(P)} = \rho_n < \frac{2}{3}.$$

that is,

$$L_s(P) < \frac{2}{3} L_m(P) = \frac{2}{3}(n - 1).$$

Let T be an SMT on P with a critical set P. Then the following properties of T are established.

(1) T is a full Steiner tree and every edge of T has length less than $2/3$.

(2) Let T_1 be a 3-regular subtree of T. Suppose T_1 has f leaves. Then

$$\ell(T_1) < \frac{2}{3}(f - 1).$$

(3) Let S_1 be a Steiner point in T, adjacent to two regular points A and B. Then S_1 can be legally moved to either A or B, but not both. S_1 can be *legally* moved to A if the subpath $S_1 S_2 S_3$ of the path from A to its adjacent regular point C other than B can be parallelly moved along $S_1 A$ to A without disconnecting A and C. (See Figure 3.)

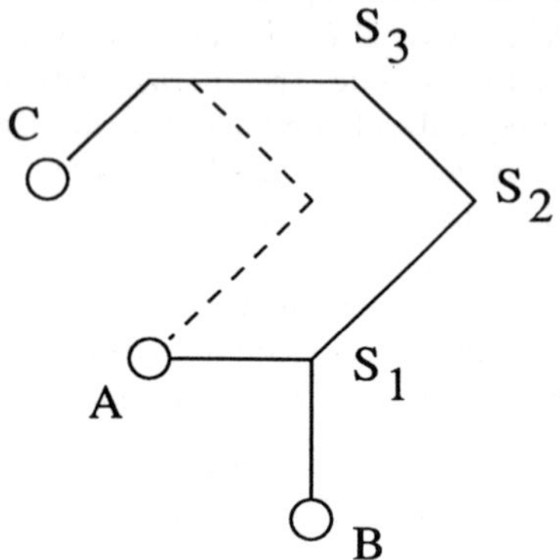

Figure 3: Legal movement.

These properties enable them to find a contradiction in the local structures as shown in Figure 2.

Since Hwang's result on the Steiner ratio in the rectilinear plane is a corollary of Theorem 2, this proof gives a new way to study the Steiner ratio in the rectilinear metric space. Graham and Hwang [30] conjectured that in the n-dimensional rectilinear space, the Steiner ratio is $n/(2n - 1)$. Whether the new technique can be useful to attack this conjecture or not is unclear right now.

The tight upper bound $\sqrt{3}/2$ for the Steiner ratio in normed planes is also conjectured by Clieslik [12] and Du, Gao, Graham, Liu and Wan [17] independently. Du, Gao, Graham, Liu and Wan [17] proved an upper bound of 0.8686. Although this is very close to $\sqrt{3}/2 = 0.866$, it still needs new efforts. For fundamental knowledge about SMT in normed planes, the reader is referred to [1, 13, 41, 45].

3 MST-based heuristics

For MST-based heuristics, the basic strategy is on improvements of minimum spanning trees. There are two types of strategies. The first one is that after constructing

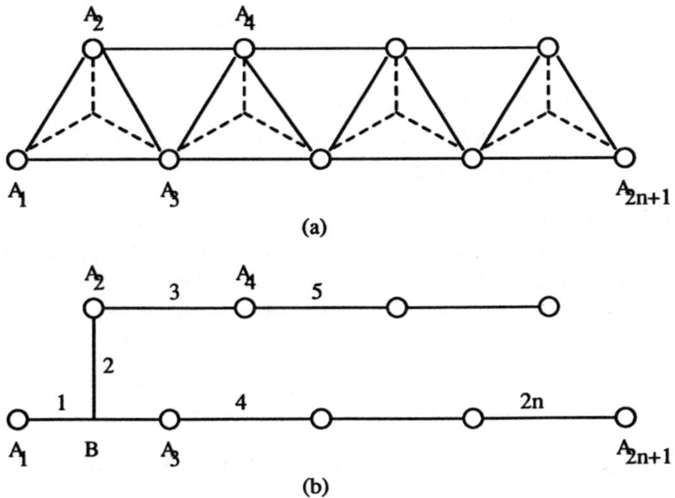

Figure 4: Counterexample.

a minimum spanning tree, do some improvements on it. Strategies of this type can be found in EI-Arbi[24], Kou, Markowsky, and Berman [38], and Bharath-Kumar and Jaffe [6]. The second type of strategies is that improvements are inserted into the algorithm for generating minimum spanning trees. For example, Kruskal's algorithm constructs a minimum spanning tree recursively by connecting the shortest edge, between original points, which does not generate a cycle. In a heuristic based on Kruskal's algorithm, the shortest edge is replaced by the shortest distance between the current trees constructed in the previous steps. Strategies of the second type can be found in Wu, Widmayer, and Wong [50] and Bern [4]. Since the minimum spanning tree can be constructed very fast, MST-based heuristics are usually fast heuristics. Bern [4] and Hwang and Yao [33] proved that some heuristics in this class can have average performance ratio larger than the Steiner ratio. However, none of them has been found to have the worst-case performance ratio larger than the Steiner ratio.

It was a long-standing open problem whether there exists a polynomial-time heuristic with performance ratio larger than $\sqrt{3}/2$. Shor and Smith [48] proposed a greedy construction as follows: Start with all n points, regarded as a forest of n 1-node trees. At any stage, add the shortest possible line segment to the current forest, which cause two trees to merge. Continue until the forest is completely merged into a single tree. They conjectured that this greedy heuristic has performance ratio larger than $\frac{\sqrt{3}}{2}$. However, Du [19] found the following counterexample for their conjecture.

Consider a union of $2n - 1$ equilateral triangles $\triangle A_1 A_2 A_3$, $\triangle A_2 A_3 A_4$, $\cdots$, $\triangle A_{2n-1} A_{2n} A_{2n+1}$ with edges of unit length as shown in Figure 4 (a). Let ϵ be a

small positive number. One disturbs the union by increasing the edge-lengths
such that

$$d(A_1, A_2) = 1 + \epsilon, \quad d(A_1, A_3) = 1,$$
$$d(A_i, A_{i+1}) = d(A_i, A_{i+2}) = 1 + i\epsilon \quad \text{for } 2 \le i \le 2n - 1, \text{ and}$$
$$d(A_{2n}, A_{2n+1}) = 1 + 2n\epsilon,$$

where $d(A, B)$ is the Euclidean distance between points A and B. Let $P_{n,\epsilon} = \{A_1, A_2, \cdots, A_{2n+1}\}$. According to the greedy construction of Shor and Smith, one should first connect $A_1 A_3$ and secondly, connect A_2 to the edge $A_1 A_3$ by the shortest line segment $A_2 B$. Then one has to connect $A_2 A_4$, $A_3 A_5$, $\cdots$, $A_{2n} A_{2n+1}$ (see Figure 1(b)). Therefore, the length of the greedy tree for the point set P is

$$L_g(P_{n,\epsilon}) = 2n - 1 + \frac{\sqrt{3}}{2} + O(\epsilon).$$

Moreover, it is easy to see that the length of the SMT for these $2n + 1$ points is

$$L_s(P_{n,\epsilon}) = n\sqrt{3} + O(\epsilon).$$

Thus, the ratio of the two lengths is

$$\frac{L_s(P_{n,\epsilon})}{L_g(P_{n,\epsilon})} = \frac{n\sqrt{3} + O(\epsilon)}{2n - 1 + \frac{\sqrt{3}}{2} + O(\epsilon)}.$$

As ϵ goes to zero and n goes to infinity, this ratio converges to $\frac{\sqrt{3}}{2}$. Hence,

$$\inf_P \frac{L_s(P)}{L_g(P)} \le \frac{\sqrt{3}}{2}.$$

Clearly, $L_g(P) \le L_m(P)$ for any set P. By the result of Du and Hwang [21],

$$\frac{L_s(P)}{L_g(P)} \ge \frac{L_s(P)}{L_m(P)} \ge \frac{\sqrt{3}}{2}.$$

Therefore,

$$\inf_P \frac{L_s(P)}{L_g(P)} = \frac{\sqrt{3}}{2}.$$

Shor and Smith [48] also proposed the following construction: Start with the line segment defined by the closest pair of two points, regarded as a 2-node, 1-edge tree. Extend the current tree by drawing the shortest possible line segment which joins the current tree to an unattached point. Continue in this manner until all points are joined. The above argument also proves that for this construction, the performance ratio is still $\frac{\sqrt{3}}{2}$.

Du [16] proposed the following conjecture: Consider a set P of n points A_1, A_2, $\cdots$, A_n in the Euclidean plane. For any permutation σ of 1, 2, $\cdots$, n, let T_σ denote the tree constructed as follows: Start with the edge $A_{\sigma(1)}A_{\sigma(2)}$. At the ith step, extend the current tree by drawing the shortest line segment from $A_{\sigma(i+2)}$ to the tree. Let $L_{op}(P)$ denote the minimum length of T_σ over all permutations. Then

$$\frac{L_s(P)}{L_{op}(P)} \geq \frac{2\sqrt{3}}{2+\sqrt{3}} = 0.92820\cdots.$$

The lower bound is achieved by the three vertices of an equilateral triangle. This conjecture can be verified for three points as follows.

Let $a = d(B,C)$, $b = d(C,A)$, and $c = d(A,B)$. Let h_a, h_b, and h_c be the lengths of three altitudes on BC, CA, and AB, respectively. Then the following two facts can be proved.

(A) If $a \leq \min(b,c)$, then $a + h_a \leq \min(b + h_b, c + h_c)$.

(B) If $\angle B \geq \frac{\pi}{2}$, then $b + h_b \geq a + c$.

To show (A), note that $h_b \leq a$ and $h_c \leq a$. So, $h_b \leq b$ and $h_c \leq c$. Moreover, $h_a \leq b$ and $h_a \leq c$. Thus,

$$\max(a, h_a) \leq b = \max(b, h_b) \text{ and } \max(a, h_a) \leq c = \max(c, h_c).$$

Note that $ah_a = bh_b = ch_c$. Define $f(x) = x + \frac{ah_a}{x}$. Then $f(x)$ is increasing for $x \geq \sqrt{ah_a}$. Thus,

$$a + h_a = f(\max(a, h_a)) \leq f(b) = b + h_b$$

and

$$a + h_a = f(\max(a, h_a)) \leq f(c) = c + h_c.$$

To show (B), let $\angle A = \alpha$ and $\angle C = \beta$. Then $c = h_b \sec \alpha$, $a = h_b \csc \beta$, and $b = h_b(\cot \alpha + \cot \beta)$. Thus, $b + h_b \geq a + c$ is equivalent to

$$\sec \alpha + \csc \beta \leq 1 + \cot \alpha + \cot \beta,$$

that is,

$$\frac{1 - \cos \alpha}{\sin \alpha} + \frac{1 - \cos \beta}{\sin \beta} \leq 1.$$

Simplifying it, one obtains

$$\tan \frac{\alpha}{2} + \tan \frac{\beta}{2} \leq 1.$$

This inequality is proved by using an inequality in [51] as follows: Let $x_1 \leq x_2 \leq x_3 \leq x_4$ and $x_1 + x_4 = x_2 + x_3$. If f is a convex function, then

$$f(x_1) + f(x_4) \geq f(x_2) + f(x_3).$$

Note that $\tan x$ is an increasing convex function on $[0, \frac{\pi}{2}]$. In fact, $(\tan x)' = \sec^2 x$ and $(\tan x)'' = \tan x \sec^2 x$. Moreover, $\alpha + \beta \leq \frac{\pi}{2}$. Thus,

$$
\begin{aligned}
& \tan \frac{\alpha}{2} + \tan \frac{\beta}{2} \\
\leq\ & \tan \frac{\alpha}{2} + \tan(\frac{\pi}{4} - \frac{\alpha}{2}) \\
\leq\ & \tan 0 + \tan \frac{\pi}{4} \\
=\ & 1.
\end{aligned}
$$

Using facts (A) and (B), one can prove that if P is a set of three points, then

$$L_g(P) = L_{op}(P).$$

In fact, the proof can be done in the following two cases.

Case 1. The triangle ABC has no angle larger than $\frac{\pi}{2}$. In this case, $L(T_{CAB}) = a + h_a$, $L(T_{ABC}) = b + h_b$, and $L(T_{BCA}) = c + h_c$ where $L(T)$ denotes the total length of a tree T. Thus, by Lemma 1,

$$L_{op}(A, B, C) = \min(L(T_{CAB}), L(T_{ABC}), L(T_{BCA})) = a + h_a.$$

Clearly, according to the construction of Shor and Smith,

$$L_g(A, B, C) = a + h_a = L_{op}(A, B, C).$$

Case 2. The triangle has an inner angle larger than $\frac{\pi}{2}$. Without loss of generality, assume $\angle B > \frac{\pi}{2}$. Then $L(T_{ABC}) = b + h_b$ and $L(T_{BCA}) = L(T_{CAB}) = a + c$. Thus,

$$L_{op}(A, B, C) = \min(b + h_b, a + c) = a + c.$$

Clearly, the greedy tree constructed by the method of Shor and Smith is $T_{BCA} = T_{CAB}$. Thus, $L_{op}(A, B, C) = L_g(A, B, C)$.

The Du's conjecture for three points is a corollary of the equation $L_{op}(A, B, C) = L_g(A, B, C)$ because Cole [14] proved already that

$$\frac{L_S(A, B, C)}{L_g(A, B, C)} \geq \frac{2\sqrt{3}}{2 + \sqrt{3}}.$$

Recently, this conjecture for four points has been verified by Cole [15].

If the minimum T_σ could be computed in polynomial time, then it would be a good approximation for the SMT. However, since the number of permutation is exponential, the minimum T_σ is unlikely polynomial-time computable. To reduce the computational complexity, one possible way is to choose a small subset S of permutations or a single permutation σ and use $\min_{\sigma \in S} T_\sigma$ or T_σ to approximate the SMT. How does one make a good choice? It is a topic for further research.

4   Sub-SMT-based heuristics

For sub-SMT-based heuristics, the heuristics in [3, 23, 52] are typical examples. These heuristics are based on constructing SMTs for at most k points. Those Steiner minimum trees are sub-SMTs for the heuristics. It has been found that the heuristics of this type can have the performance ratio larger than the Steiner ratio.

There are two types of strategies for using the sub-SMT. The first one is to grow a tree recursively by utilizing a variation of the procedure of constructing the sub-SMT. Strategy of this type can be found in Kahang [35]. The second one is to partition the set of original points into subsets such that each subset has such a sub-SMT and the union of sub-SMT's for all subsets forms a Steiner tree for whole point set. For the second type of strategies, an important question is how to find an optimal partition. For sub-SMTs of at most k points, finding such an optimal partition is also an intractable problem. Zelikovsky [52] gave a polynomial-time heuristic with performance ratio 6/11 for the Steiner tree problem in graphs by successfully using a near-optimal partition generated by a greedy algorithm instead of the optimal partition. The running time of this type of heuristics depends on the running time of constructing sub-SMT's and the algorithm for computing the partition. In Zelikovsky's algorithm, the latter takes a lot of time.

Since Zelikovsky's work, component-size bounded heuristics have attracted a lot of attentions. An ST is called a *k-size ST* if it can be decomposed at regular points into smaller trees each of which contains at most k regular points. A *minimum k-size ST* is a k-size ST with the minimum total edge-length. The *k-Steiner ratio* is defined to be

$$\rho^{(k)}(S) = \inf_{P \subset S} \frac{L_s(P)}{L_s^k(P)}$$

where $L_s^k(P)$ is the length of a minimum k-size ST on P. This way, a spanning tree is a 2-size ST and $\rho_2 = \rho$.

Berman and Ramaiyer [3] obtained a polynomial-time heuristic with performance ratio larger than 2/3 for the SMT in the rectilinear plane. They also improved Zelikovsky's result by giving the following.

Theorem 4 *If the Steiner minimum tree for k points can be constructed in polynomial-time in a metric space S, then there exists polynomial-time heuristic for the Steiner minimum tree in the metric S, with performance ratio*

$$(\rho^{(2)}(S)^{-1} - \sum_{i=3}^{k} \frac{\rho^{(i-1)}(S)^{-1} - \rho^{(i)}(S)^{-1}}{i-1})^{-1}.$$

Note that $\rho^{(k)}(S)$ is nondecreasing as k increases. From Theorem 4, it is easy to see that to show the existence of a polynomial-time heuristic for the Euclidean SMT, it suffices to prove that for some $k \geq 3$, $\rho^{(k)}(\mathbf{R}^2) > \sqrt{3}/2$. Du, Zhang, and Feng [23] proved the following lower bound for the k-Steiner ratios.

Theorem 5 *In any metric space S, $\rho^{(k)}(S) \geq \lfloor \log_2 k \rfloor / (\lfloor \log_2 k \rfloor + 1)$.*

Recently, Du [18] has obtained an interesting result as follows.

Theorem 6

$$\inf_S \rho^{(3)}(S) = \frac{3}{5}$$

where S is over all metric spaces.

The part $\inf_S \rho^{(3)}(S) \geq 3/5$ was obtained by Zelikovsky [52]. (Note that $\rho^{(2)} \geq 1/2$. Thus, the lower bound of $3/5$ for $\rho^{(3)}$ results in a polynomial-time heuristic with performance ratio $6/11$ by Theorem 4.) The main contribution of Du [18] is the proof of $\inf_S \rho^{(3)}(S) \leq 3/5$. To do so, he considered a balanced binary tree T of $n+1$ levels; each edge has length one. Let S be the metric space consisting of all vertices of T; the distance $d(x,y)$ between two vertices x and y of T is defined to be the length of the path between them in T. Suppose that the given set P consists of all leaves of T. Let T^* be a minimum 3-size ST on P. Consider a full component of size 3. Let s be the Steiner point and x_1, x_2, and x_3 three regular points in the component. Then three paths from s to x_1, x_2, and x_3, respectively, must be disjoint in T. Otherwise, one is able to shorten T^* by repacing s with another point, contradicting the optimality of T^*. From the disjointness of the three paths, it is easy to see that in this component, s must connect to two regular points, say x_1 and x_2, with the same distance and to the third regular point x_3 with the longer distance (see Figure 5). Define $p(s) = d(s, x_1)$.

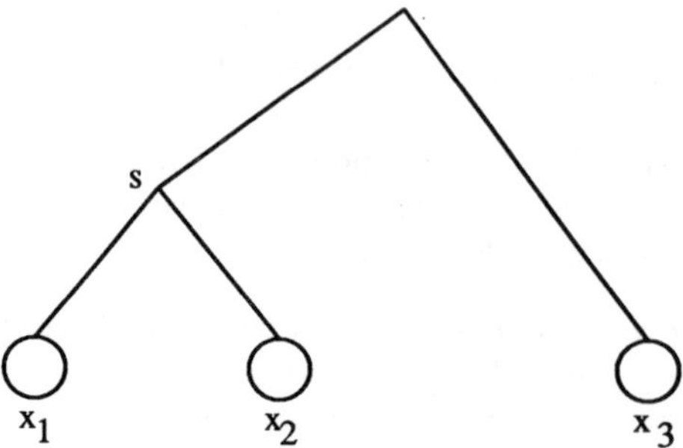

Figure 5: A full component of size 3.

Note that if one replaces this full component by connecting $x_1 x_2$ and $x_2 x_3$, then the total length increases exactly by $p(s)$. Thus,

$$L_s^3(P) \geq L_m(P) - \sum_{s \in V_s} p(s).$$

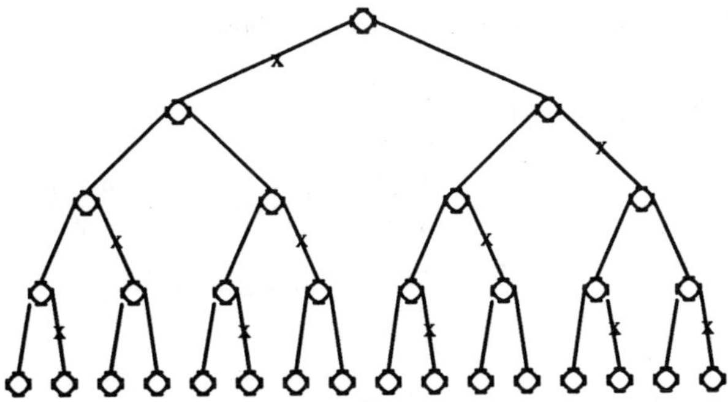

Figure 6: The matching H.

where V_s is the set of all Steiner points of T^*. He further proved the following facts about V_s.

(1) For any $s, s' \in V_s$, s cannot be a child of s'.

(2) There exists a minimum 3-size ST T^* such that for every pair $s, s' \in V_s$, s and s' have different fathers.

From these two facts, it is known that there exists a minimum 3-size ST T^* such that all Steiner points of T^* are independent in G, i.e., no edge of G is between two Steiner points of T^*. Note that no Steiner point can be the root or a leaf of T and every internal point other than the root must belong to two triangles in G; each triangle is formed by a father and his two children. Now, one constructs G' from G as follows: G' takes all triangles of G as vertices. An edge e_s between two vertices u and v in G' exists if and only if the corresponding two triangles in G have a vertex s in common. Clearly, the mapping $s \to e_s$ is a one-to-one and onto correspondence between the set of internal vertices of T other than the root and the edge set of G'. The set S of all Steiner points of T^* corresponds to a matching in G'. For each edge e_s in G', assign a weight $p(s)$ to e_s. It is well-known that a matching is maximum if and only if for any maximal alternating path, the total weight of edges in the matching is not less than the total weight of others. By this condition, it is easy to verify that the matching H, which consists of one edge at the first level, one edge at the second level, three edges at the third level, five edges at the fourth level, and so on, is the maximum weight matching of G' (see Figure 6). Let $f(k)$ denote the number of edges of the matching H at the kth level. Then $f(k) + f(k+1) = 2^k$. Thus,

$$\sum_{k=1}^{h} f(k) = \begin{cases} 2^1 + 2^3 + \cdots + 2^{h-1} & \text{if } h \text{ is even} \\ 1 + 2^2 + 2^4 + \cdots + 2^{h-1} & \text{if } h \text{ is odd} \end{cases}$$
$$= \begin{cases} \frac{1}{3} \cdot (2^{h+1} - 2) & \text{if } h \text{ is even} \\ \frac{1}{3} \cdot (2^{h+1} - 1) & \text{if } h \text{ is odd.} \end{cases}$$

Note that G' has $n-1$ edge-levels and that any edge e_s in the kth level has weight

$p(s) = n - k$. Thus, the total weight of H is as follows.

$$\sum_{k=1}^{n-1} f(k)(n-k) = \sum_{h=1}^{n-1}\sum_{k=1}^{h} f(k)$$

$$= \frac{1}{3}(\sum_{h=1}^{n-1} 2^{h+1} - (n-1) - \lfloor\frac{n-1}{2}\rfloor)$$

$$= \frac{1}{3}(2^{n+1} - n - 2 - \lfloor\frac{n+1}{2}\rfloor).$$

Hence,

$$L_s^3(P) \geq L_m(P) - \frac{1}{3}(2^{n+1} - 2 - n - \lfloor\frac{n+1}{2}\rfloor).$$

Moreover, by Kruskal's method, it is easy to see that the minimum spanning tree on P can be constructed as follows: For every pair of brothers x and y, connect a leaf of T_x to a leaf of T_y where T_x is the subtree rooted at x. Thus,

$$L_m(P) = 2^n + 2 \cdot 2^{n-1} + 3 \cdot 2^{n-2} + \cdots + (n-1) \cdot 2 = 2^{n+2} - 2(n+1).$$

It follows that

$$L_s^3(P) \geq \frac{1}{3}(5 \cdot 2^{n+1} - 5n - 4 + \lfloor\frac{n+1}{2}\rfloor).$$

Note that

$$L_s(P) = 2^{n+1} - 2.$$

Therefore

$$\rho^{(3)}(S) = \frac{L_s(P)}{L_s^3(P)} \leq \frac{2^{n+1} - 2}{\frac{1}{3}(5 \cdot 2^{n+1} - 5n - 4 + \lfloor\frac{n+1}{2}\rfloor)}.$$

Letting $n \to \infty$, one obtains $\inf_S \rho^{(3)}(S) \leq \frac{3}{5}$.

Du [18] conjectured that

$$\inf_S \rho^{(2^k)}(S) = \frac{k}{k+1}.$$

References

[1] M. Alfaro, M. Conger, K. Hodges, A. Levy, R. Kochar, L. Kuklinski, Z. Mahmood and K. von Haam, "The structure of singularities of Φ-minimizing networks in R^2", *Pac. J. Math.*, 149 (1991) 201-210.

[2] J.E. Beasley, A heuristic for the euclidean and rectilinear Steiner problems, Technical Report of Management School, Imperial College, London, 1989.

[3] P. Berman and V. Ramaiye, "An approximation algorithm for the Steiner tree problem," manuscript.

[4] M. Bern, "Two probabilistic results on rectilinear Steiner trees," *Proceedings of the 18th Annual ACM Symposium on Theory of Computing* (1986) 433-441.

[5] M. Bern and Plassmann, "The Steiner problem with edge lengths 1 and 2," *Information Processing Letters* (1989) 171-176.

[6] K. Bharath-Kumar and J.M. Jaffe, "Routing to multipe destinations in computer networks", *IEEE Trans. Communications*, COM-31 (1983) 343-351.

[7] G.D. Chakerian and M.A. Ghandehari, "The Fermat problem in Minkowski space", *Geometriae Dedicata* 17 (1985) 227-238.

[8] D. Cheriton and R.E. Tarjan, "Finding minimum spanning trees", *SIAM J. Comput.* 5 (1976) 724-742.

[9] F.R.K. Chung and E.N. Gilbert, "Steiner trees for the regular simplex," *Bull. Inst. Math. Acad. Sinica* 4 (1976) 313-325.

[10] F.R.K. Chung and R.L. Graham, "A new bound for euclidean Steiner minimum trees," *Annual New York Academy of Sciences*, 440 (1985) 328-346.

[11] F.R.K. Chung and F.K. Hwang, "A lower bound for the Steiner tree problem," *SIAM Journal of Applied Mathematics*, 34 (1978) 27-36.

[12] D. Cieslik, "The Steiner ratio of Banach-Minkowski planes", *Contemporary Methods in Graph Theory* (ed. R. Bodendiek), (BI-Wissenschatteverlag, Mannheim, 1990) 231-248.

[13] E.J. Cockayne, "On the Steiner problem", *Canad. Math. Bull.* 10 (1967) 431-450.

[14] T. Cole, "A problem in Steiner networks", manuscript.

[15] T. Cole, "Some results on greedy algorithm conjectures", to appear in *Discrete Applied Mathematics*.

[16] D.Z. Du, "On Steiner ratio conjectures", *Annals of Operations Research*, 33 (1991) 437-449.

[17] , D.Z. Du, B. Gao, R.L. Graham, Z.-C. Liu, and P.-J. Wan, "Minimum Steiner trees in normed planes", to appear in *Discrete and Computational Geometry*.

[18] D.Z. Du, "On component-size bounded Steiner trees", to appear in *Discrete Applied Mathematics*.

[19] D.Z. Du, "On greedy heuristics for Steiner minimum trees", TR-92. Computer Science Department, University of Minnesota.

[20] D.Z. Du and F.K. Hwang, "A new bound for the Steiner ratio," *Transactions of American Mathematical Society*, 278 (1983) 137-148.

[21] D.Z. Du and F.K. Hwang, "An approach for proving lower bounds: solution of Gilbert-Pollak's conjecture on Steiner ratio," *Proceedings of the 31th Annual Symposium on Foundations of Computer Science* (1990) 76-85.

[22] D.Z. Du and W.D. Smith, "Three disproofs for Gilbert-Pollak conjecture on the Steiner ratio in high dimensional Euclidean space", manuscript.

[23] D.Z. Du, Y.-J. Zhang, and Q. Feng, "On better heuristic for Euclidean Steiner minimum trees," *Proceedings of 32nd FOCS*, (1991) 431-439.

[24] C. El-Arbi, "Une heuristique pour le probleme de l'Arbre de Steiner", *RAIRO* 12 (1978) 207-212.

[25] L.R. Foulds and R.L. Graham, "The Steiner problem in phylogeny is NP-complete," *Advances in Applied Mathematics*, 3 (1982) 43-49.

[26] B. Gao, D.Z. Du, and R.L. Graham, "Tight lower bound for the Steiner ratio in normed planes", manuscript.

[27] M.R. Garey, R.L. Graham and D.S. Johnson, "The complexity of computing Steiner minimal trees," *SIAM Journal of Applied Mathematics*, 32 (1977) 835-859.

[28] M.R. Garey and D.S. Johnson, "The rectilinear Steiner tree problem is NP-complete," *SIAM Journal of Applied Mathematics*, 32 (1977) 826-834.

[29] E.N. Gilbert and H.O. Pollak, "Steiner minimal trees," *SIAM Journal of Applied Mathematics*, 16 (1968) 1-29.

[30] R.L. Graham and F.K. Hwang, "Remarks on Steiner minimal trees," *Bulletin of Institute of Mathematics, Academia Sinica*, 4 (1976) 177-182.

[31] W.-Y. Hsiang, "On the density of sphere packing in E^3 (I)", manuscript.

[32] F.K. Hwang, "On Steiner minimal trees with rectilinear distance," *SIAM Journal of Applied Mathematics*, 30 (1976) 104-114.

[33] F.K. Hwang and Y.C. Yao, "Comments on Bern's probabilistic results on rectilinear Steiner trees," *Algorithmica* 5 (1990) 591-598.

[34] V. Jamik and O. Kossler, "O minimalnich gratech obsahujicich n danych bodu" (in Czech), *Casopis Pesk. Mat. Fyr.*, 63 (1934) 223-235.

[35] A. Kahng and G. Robins, "A new family of Steiner tree heuristics with good performance: the iterated 1-Steiner approach", *Tech.Rep., UCLA CS Dept.*, 1990.

[36] R.M. Karp, "Reducibility among combinatorial problems," in R.E. Miller and J.W. Tatcher (ed.), *Complexity of Computer Computation* (Plenum Press, New York, 1972) 85-103.

[37] P. Korthonen, "An algorithm for transforming a spanning tree into a Steiner tree", *Survey of Mathematical Programming (2)*, (North-Holland, 1979) 349-357.

[38] L. Kou and K. Makki, "An even faster approximation algorithm for the Steiner tree problem in graph", *Congressus Numerantium* 59 (1987) 147-154.

[39] L. Kou, G. Markowsky and L. Berman, "A fast algorithm for Steiner trees", *Acta Informatica* 15 (1981) 141-145.

[40] A.J. Levin, "Algorithm for the shortest connection of a group of vertices", *Sovirt Math. Dokl.* 12 (1971) 1477-1481.

[41] Z.C. Liu and D.Z. Du, "On Steiner minimal trees with L_p distance", *Algorithmica* (1991).

[42] H.O. Pollak, "Some remarks on the Steiner problem," *Journal of Combinatorial Theory, Ser.A*, 24 (1978) 278-295.

[43] D. Richards, "Fast heuristic algorithms for rectilinear Steiner trees", *Algorithmica*, 4 (1989) 191-207.

[44] J.H. Rubinstein and D.A. Thomas, "The Steiner ratio conjecture for six points," to appear in *J. Combinatoria Theory, Ser.A*.

[45] J.M. Smith and M. Gross, "Steiner minimal trees and urban service networks", *J. Socio. Econ. Plng.*, 16 (1982) 21-38.

[46] J.M. Smith and J.S. Liebman, "Steiner trees, Steiner circuits and interference problem in building design", *Engineering Optimization*, 4 (1979) 15-36.

[47] W.D. Smith, "How to find Steiner minimal trees in euclidean d-space", *Algorithmica* (1991).

[48] W.D. Smith and P.W. Shor, "Steiner tree problems", *Algorithmica*, 7 (1992) 329-332.

[49] A. Suzuki and M. Iri, "A heuristic method for euclidean Steiner problem as a geographical optimization problem", *Asia-Pacific Journal of Operations Research*, 3 (1986) 106-122.

[50] Y.F. Wu, P. Widmayer, and C.K. Wong, "A faster approximation algorithm for the Steiner problem in graphs", *Acta Informatica*, 23 (1986) 223-229.

[51] Hong Yi, Yang Hongcang and Du Dingzhu, "An inequality for convex functions", *Kexue Tongbao* 27 (1982) 901-904.

[52] A.Z. Zelikovsky, "The 11/6-approximation algorithm for the Steiner problem on networks", manuscript.

Complexity in Numerical Optimization, pp. 107-127
P.M. Pardalos, Editor
©1993 World Scientific Publishing Co.

The Complexity of Allocating Resources in Parallel: Upper and Lower Bounds

Eric J. Friedman
Department of IEOR, University of California, Berkeley, CA 94720

Abstract

We examine the differences in complexity for distributed and non-distributed algorithms by focusing on the example of resource allocation. We construct lower bounds for these problems which are valid in many models of computation. We then present an efficient primal-dual algorithm for allocating one or two resources and describe a method for constructing one for an arbitrary number of resources that is easily implemented in parallel.

Keywords: Complexity, convex programming, lower bounds, resource allocation, global optimization, local optimization.

1   Introduction

The general (finite dimensional) convex programming problem consists of minimizing a convex function f over a convex set $G \subset \Re^n$. In general computing the solution of this problem exactly is not possible. Thus any reasonable theory of complexity for solving convex problems must allow for approximate solutions.

Given some measure of accuracy we can then consider the computational complexity of an algorithm that guarantees to be at least ϵ-accurate. This will be the running time of the algorithm as a function of the size of the problem and the accuracy obtained. Note that we must also define the *size* of a convex program.

In this paper we carefully define these quantities, and discuss why several natural measures are not useful. Then we discuss lower bounds for complexity in both serial and distributed computation. As a concrete example, we consider the problem of resource allocation, since it arises naturally in a distributed setting and demonstrates the relation between serial and distributed optimization.

The paper is organized as follows. Section 2 describes a formalism for general convex optimization and fundamental lower bounds. In section 3 we specialize to resource allocation and derive lower bounds for both serial and parallel computation. Section 4 describes standard (primal) algorithms for this problem, which do not distribute efficiently. In section 5 we remedy this deficiency by constructing a primal-dual algorithm that distributes naturally. This algorithm requires the solution of a difficult geometrical problem that we explicitly solve for the case of two (or fewer) resources. The extension of this algorithm to more than two resources is still an open problem, which we believe can be solved using techniques similar to those we develop for the two resources case. We conclude in section 6 with some comments on complexity theory for nonlinear programming.

Note that our proofs in the main part of this paper are motivated by pedagogical concerns; however, in the appendix, our proof of the primal–dual algorithm is quite technical and contains only the most important elements of the proof.

1.1   Notation

In this paper we will define $e = (1, 1, \ldots, 1)$ and e_i to be the unit vector with all zeros except for a one in the ith component. Note that we will not explicitly state the dimension of these vectors as this will be apparent from the context. Also, we will not distinguish between row and column vectors, and will denote the standard inner product between two vectors x, y as $x \cdot y$.

We will make use of the standard p-norms

$$\|x\|_p = (\sum_i |x_i|^p)^{1/p}.$$

For example $\|x\|_1 = \sum_i |x_i|$, $\|x\|_\infty = \max_i |x_i|$, and $\|x\|_2 = \sqrt{x \cdot x}$.

We will also often use the notation $x \in \Re^{nm}$ where a typical element of this vector is x_i^j for $1 \leq i \leq n$, $1 \leq j \leq m$. We will use the shorthand $x_i \in \Re^m$, and $x^j \in \Re^n$. Note that $\sum_i x_i$ is a sum of vectors and is itself in $\Re^m$.

2   General Convex Optimization

2.1   Convex Programs

The general convex program is:

$$\min\{f(x) \mid x \in G\} \quad \text{(GCP)}$$

where f is a convex function and $G \subset \Re^n$ is convex. Since f can be an arbitrary convex function, we will assume that f is given by an oracle that can return either function values or components of a supporting hyperplane[1] to f at x. Typically G is defined by a set of functions, $G = \{x \mid g_j(x) \leq 0\}$ for $j \in \{1, 2, \ldots, m\}$.

The case where the g_j's are general can be quite complicated. Even finding a feasible point $x_f \in G$ can be very difficult when the volume of G is small compared to the volume of the region in which we know G must lie. When the $g_i(x)$'s are linear we know that G must be contained in a ball centered at the origin of radius $2^{O(L)}$ and must have volume greater than $2^{-O(L)}$, where L is the encoding length of the polytope. (See, e.g. [14].) Thus the volume ratio of these two regions is $2^{-O(L)}$ which allows for the polynomial solution of linear programming as in Khachian's algorithm [8]. However, when the $g_i(x)$'s are nonlinear polynomials (e.g. quadratic) the ratio can be doubly exponential in m [3]. This suggests (but does not prove) that finding a feasible point $x \in G$ for nonlinear constraints is quite difficult, since it would require an algorithm which converges at least quadratically, while as we will discuss later, this is not possible in general. In this paper we will consider only the case where all the g_j's are linear.

Consider the convex program with linear constraints:

$$\min\{f(x) \mid Ax \leq b\} \quad \text{(CPLC)}$$

where A is an $m \times n$ matrix, $x \in \Re^n$ and $b \in \Re^m$. (We assume that the feasible set is nonempty.) The Lipschitz constant of $f(x)$ (in the Euclidean metric) is defined as

$$\Lambda(f) = \max_{x, x'} \frac{|f(x) - f(x')|}{\|x - x'\|_2}$$

which we will use as a measure of the size of $f(x)$. Also, we will let L be the size of (A, b), which we define in the standard manner [14, Chapter 8].

There are two main candidates for the measure of accuracy. The first is based on function value. Let x^* be the optimal solution to CPLC and define the error of a solution x to be

$$\epsilon(x) = \begin{cases} |f(x) - f(x^*)| & \text{if } x \in G \\ \infty & \text{otherwise} \end{cases}$$

The second is defined in terms of coordinates.

$$\epsilon_c(x) = \begin{cases} \|x - x^*\| & \text{if } x \in G \\ \infty & \text{otherwise} \end{cases}$$

However, as we will see later, this second measure is not useful in general.

Now we define the complexity of an algorithm for CPLC to be the worst case running time of an algorithm that guarantees $\epsilon(x_o) \leq \epsilon$, where x_o is the approximate solution found by the algorithm, as a function of L, ϵ, and Λ.

---

[1] A supporting hyperplane to f at x_0 can be described as a vector c such that $f(y) \geq f(x) + c \cdot (y - x)$ for all $y \in G$.

Note that we are working in a real number model of computing [10, 2, 1]. That is, we assume that we can manipulate real numbers, and do any of the following operations in unit time: addition, subtraction, multiplication, division, and comparison. Furthermore, for ease of exposition, we will also assume that we can compute arbitrary roots $\sqrt[p]{x}$ for $x \geq 0$ in $O(1)$ computations.[2]

The assumption of real number computation is not critical to our results. In fact, any of the algorithms that we discuss could be implemented in the standard Turing machine model of integer computation.

2.2   Lower Bounds for Complexity

It is useful to consider the complexity of an algorithm as arising from a sequential game in which the algorithm is playing against an adversarial oracle. At each query the oracle can give any reply, subject to the condition that there exists some function that is consistent with its entire history of replies. This viewpoint can be used to simplify the discussion of lower bounds. Thus we can imagine that the oracle can at each query change the function f in any way that does not invalidate any of its previous replies to queries by the algorithm.

Our first result is based on work by Nemirovsky and Yudin [12]. It implies that the use of coordinates as the measure of accuracy is not useful, since there is no finite algorithm that can guarantee accuracy of this type.

Theorem 2.1 (Nemirovsky and Yudin) *Consider the simple 2-dimensional convex problem*

$$\min\{f(x,y) \mid 0 \leq x \leq 1,\ 0 \leq y \leq 1\}$$

where $\Lambda(f) \leq 1$. Then there is no finite algorithm which can compute an (x_o, y_0) such that $\epsilon_c(x_o) \leq 1/2$

Proof: The adversary's strategy is very simple. It begins with some function that is independent of y, $\hat{f}(x,y) = g(x)$ which it uses to answer all of the algorithm's queries. Now when the algorithm terminates after a finite number of queries it must choose some point as its solution, say (x_o, y_o). Now the adversary will just choose some function f which agrees with $\hat{f}$ at all the points that the algorithm has queried, but has its optimum at (x^*, y^*) where y^* is far from y_o, and x^* is the minimum of g.

For example assume that $y_o \leq 1/2$. Let G_q be the set of points for which the algorithm has queried the oracle. Note that this set must be finite, and we can assume that $x^* \notin P_x(G_q)$, where P_x is the projection onto the first coordinate. Now the adversary defines the function to be $f(x,y) = \max[\hat{f}(x,y), g(\hat{x}) - (g(\hat{x}) - g(x^*))y]$

---

[2]These can be computed using a bisection method or Newton's method very quickly to high accuracy. For the algorithms we present in this paper the finite accuracy of these solutions will not significantly affect the algorithms' complexity. Note that we are not assuming that we can solve algebraic equations rapidly. This is in general a very difficult problem. (See e.g. [15, 17])

where $\hat{x} = \mathrm{argmin}_{x \in P_x(G_q)} g(x)$. Note that this function is convex, consistent with all of the adversary's answers and has its minimum at $(x^*, 1)$. $\Box$

Thus, if we were to use ϵ_c as our measure of accuracy then the whole theory of convex optimization would be trivial – no algorithms would be possible!

However, in apparent contradiction to this result, this measure has been used in the analysis of algorithms for convex problems [6, 7]. However, in these papers it was assumed that $f(x)$ is separable. Thus $f(x) = \sum_i f_i(x_i)$ where each x_i is a scalar. In this case the level sets of the objective function are well behaved and avoid the problems that may arise in general objective functions. Note that this structure breaks down for nonseparable functions. (Even if f_i only depends on two variables then no finite algorithm can guarantee solution in coordinates, as shown in our proof of Theorem 1.) Thus, this measure cannot be extended to more general problems, and the natural measures of complexity associated with it do not lead to an interesting theory of complexity for convex optimization.

For the remainder of this paper we will measure the accuracy of a solution in terms of objective value using $\epsilon(x)$. We now describe the basic complexity results for this measure.

Define the simple convex program to be

$$\min\{f(x) \mid \forall i,\ 0 \leq x_i \leq b_i\} \quad \text{(SCP)}$$

which is just the minimization of a convex function over a hypercube. Now we show that using very simple scaling arguments we can reduce the study of SCP to that of a convex function with Lipschitz constant 1 over the unit hypercube

$$\min\{f(x) \mid \forall i,\ 0 \leq x_i \leq 1\} \quad \text{(UCP)}$$

where we require that $\Lambda(f) \leq 1$.

Theorem 2.2 *Let $T_1(n, \epsilon)$ be the complexity of some algorithm for solving UCP, then there exists an algorithm for SCP that has complexity*

$$T(n, \Lambda, b, \epsilon) = T_1(n, \frac{\epsilon}{\Lambda \, \|b\|_\infty}).$$

Let $L_1(n, \epsilon)$ be a lower bound for solving UCP, then

$$L(n, \Lambda, b, \epsilon) = L_1(n, \frac{\epsilon}{\Lambda \, \|b\|_{-\infty}})$$

is a lower bound for SCP, where $\|b\|_{-\infty} = \min_j |b_j|$. (Note that as ϵ decreases the complexity must increase.)

Proof: Given some problem in SCP we solve a scaled version of it

$$\min \frac{f(x \times b)}{\Lambda \|b\|_\infty} \mid \forall i,\ 0 \leq x_i \leq 1$$

where $x \times b$ is shorthand for the vector with components $x_i b_i$. Note that this problem is in UCP, since the Lipschitz constant of the objective function is less than 1. Now if we solve this problem to accuracy $\epsilon/(\Lambda||b||_\infty)$ then we have solved SCP to accuracy ϵ. The second part of the theorem follows from a similar argument. $\square$

Thus we can study the problem UCP when computing lower bounds or designing algorithms for SCP. Now we present a basic lower bound.

Theorem 2.3 (Nemirovsky and Yudin) *The complexity of solving UCP is at least*

$$\Omega(n \log \frac{1}{\epsilon})$$

Proof: This result is due to Nemirovsky and Yudin [12], where a very general result for general convex programs is given. As their result and proof is quite complicated we will give a significantly simpler proof for this special case, than the one found in their book.

The basic idea of the proof is to consider a function that is defined by

$$f_H(x) = \max_{h \in H} h(x)$$

where each h is a linear function of the form $a \cdot x + b$. Let

$$H_0 = \{h \mid h(x) = -e_i \cdot x, \ 1 \leq i \leq n\} \bigcup \{h \mid h(x) = e_i \cdot x - 1, \ 1 \leq i \leq n\}.$$

Note that f_{H_0} has Lipschitz constant equal to 1 and has value 0 for all points on the boundary of the unit hypercube.

We will prove the lower bound as follows. The adversary begins with f_{H_0} defined on the unit hypercube, C_0. After n queries by the algorithm the adversary will be using some function f_{H_1} which is just a scaled version of f_{H_0} on a hypercube, C_1, with length 1/4 edges. Formally, there exists an affine transformation $T(x) = x_0 + 4x$ such that $T(C_1) = T(C_0)$ and for all $x \in C_1$ we have that $f_{H_1}(x) = f_{H_0}(T(x))$. Noting that f_{H_1} has Lipshitz constant 1/4 on C_1, we see that after n steps the algorithm is facing exactly the same problem as it was initially, but with everything scaled down. Continuing this process we see that after Nn queries to the oracle the algorithm is facing a scaled version of the initial problem, where everything has been scaled dowm by $(1/4)^N$. Thus, the algorithm is converging linearly (at best), which implies that it requires at least $\Omega(n \log 1/\epsilon)$ query points.

First we show that if the algorithm is faced with f_{H_0} on C_0 then it cannot produce a solution with accuracy better than 1/32 without making any queries.

Lemma 2.4 *Assume the algorithm knows the function f_{H_0} only on the boundary of the unit hypercube. Then there exists an adversary such that it cannot produce a solution with accuracy better than 1/32.*

Proof: Assume the algorithm chooses some point $\hat{x}$. Assume that $\hat{x}_1 \leq 1/2$. (The other case is similar.) Define $H' = H_0 \cup \{-e_1 \cdot x/4 - 1/8\}$. Now $f_{H'}(\hat{x}) \geq -1/4$ and $f_{H'}(e/2 + e_i/8) \leq -9/32$. Therefore $|f_{H'}(\hat{x}) - f_{H'}(x^*)| \geq 1/32$. $\diamond$

Lemma 2.5 *The adversary can play against any algorithm such that the following is true: It begins with the function f_{H_0} over the unit hypercube C_0. After n queries by the algorithm, the adversary is using the function f_{H_1}, where f_{H_1} is just a scaled version (as described above) of f_{H_0} on the hypercube C_1 with edge length $1/4$ whose interior does not contain any query points. The function f_{H_1} has Lipschitz constant $1/4$ on C_1.*

Proof: Begin with $N = \{1, 2, \ldots, n\}$. Let C be the interior of the unit hypercube and $H = H_0$. The algorithm makes a query about some point x and the adversary plays in the following manner.

1. Choose some $j \in N$ at random and set $N = N - \{j\}$.

2. If $x_j > 1/2$

    (a) Set $H := H \cup \{-e_1 \cdot x/4 - 3/16\} \cup \{e_1 \cdot x/4 - 1/8\}$.

    (b) Let $C := C \cap \{y \mid 1/4 \leq y_j < 1/2\}$.

3. else

    (a) Set $H := H \cup \{-e_1 \cdot x/4 - 1/8\} \cup \{e_1 \cdot x/4 - 3/16\}$.

    (b) Let $C := C \cap \{y \mid 1/2 \leq y_j \leq 3/4\}$.

4. Reply to the query with f_H.

After n queries C and H satisfy the conditions of the lemma. $\diamond$

These lemmas show that after every n operations the best the algorithm can do it to face a scaled version of the original problem[3] which proves the theorem. $\square$

Note that this lower bound is quite robust to changes in our model of computation, since it relies solely on properties of convex functions.[4]

---

[3]Note that in step 1 it is not necessary for the adversary to choose j at random; however, an extension of this proof shows that the algorithm must take $\Omega(n)$ computations per iteration, on average. Thus we could get a slightly better lower bound of $\Omega(n^2 \log(1/\epsilon))$ for the complexity. However, the techniques needed to prove this are beyond the scope of this paper.

[4]It is interesting to note that in these proofs the adversary chooses piecewise linear functions. Thus it is easy to see that superlinearly convergent algorithms such as Newton's method must fail, since Newton's method relies on second derivatives, which are all zero in this case and supply no information about the function.

2.3   A Comment on Measures of Problem Size

In our formulation we let the Lipschitz constant be one of our measures of problem size. However, for general convex optimization the Lipschitz constant Λ as a measure of the problem size may not be useful, since for a typical convex function, it becomes unbounded at infinity. Thus we must use the value of the Lipschitz constant over some bounded region. However, the feasible region can be quite large $2^{O(L)}$ or quite small $2^{-O(L)}$. Thus for many problems, the Lipschitz constant might be quite large in this large region, but of reasonable size in a smaller feasible region, for a specific problem. Thus, this measure of size is inadequate for the general problem GCLC and other definitions are necessary. One possible measure is discussed in [4] where under this measure, we construct strongly polynomial algorithms for a large class of problems.

However, the resource allocation problem has a very natural structure which makes the Lipschitz constant a useful and productive measure of problem size as we will demonstrate later.

This emphasizes a very important aspect of defining complexity for resource allocation problems. The definitions and measures of complexity may only make sense for a certain class of problems. For example, as mentioned earlier, the use of coordinates as a measure of accuracy may be interesting in the study of convex separable problems, but is not useful for nonseparable problems.

Also, further restrictions on the class of convex functions allowed can also effect the success or failure of certain types of algorithms. For example, interior point algorithms may be used to solve convex programs when restrictions are imposed on the class of allowable functions. One of the most general conditions which allow for such methods to work is that of self-concordancy [13, 5]. However, this condition is difficult to interpret. Thus, in this paper we will allow for general convex functions, which accounts for our neglect of interior point methods, since they are not useful for solving this class of problems. However, due to their effectiveness in practice, it would be useful to find natural conditions under which interior point methods are effective, and to compute upper and lower bounds for the complexity of problems in this class of problems.

3   Resource Allocation

Resource allocation is a fundamental problem in economics. It is also important in distributed computing and the running of large organizations. In each of these problems there is a resource (food, memory, capital) that must be divided up among many processors (consumers, processes, accounting centers) in an efficient manner. The resource allocation problem can be formulated as follows,

$$\max\{f(x) = \sum_i f_i(x_i) \mid \sum_i x_i \le b, \ x_i^j \ge 0\} \quad (RA)$$

where $0 \leq b \in \Re^m$, $0 \leq x_i \in \Re^m$ and $i \in \{1, 2, \cdots, n\}$. Here x_i represents the resources allocated to agent i, and m is the total number of types of resources. Assuming that resources are goods we require that f_i is nondecreasing in its arguments. Also, in order to be consistent with the literature on resource allocation we consider the maximization problem where each f_i is concave. Combining these conditions on f_i we see that the maximum of the Lipschitz constant of f_i occurs in a neighborhood of 0, and thus in order to compute the size of f we need only consider a small neighborhood of the origin, thus avoiding the difficulties we encountered with the general problem.

We are still unable to measure the accuracy in coordinates, since there is no bounded (even exponential) algorithm using this measure, except in the case of a single resource. However, by measuring the accuracy in terms of function value, we can construct polynomial algorithms for this problem,

For this problem the matrix A is very simple; thus we just need to consider the size of b in our analysis as $size(A) = O(n)$. Also, we will let $\Lambda = \max_i \Lambda(f_i)$, which can be at most $\sqrt{n}$ smaller than the Lipschitz constant of f. We will be interested in algorithms whose running times are bounded by a polynomial in $n, m, \log(\Lambda)$, and $\log(\|b\|_\infty)$.

3.1 Distributed Resource Allocation

The resource allocation problem has a very natural distributed structure. Typically, each 'utility function' f_i belongs to a specific agent, who has some computational ability. Thus we imagine that we have n processors, and the information about f_i is located at processor i. Now, if these processors are connected up on a reasonable network, then we would like to construct an efficient distributed algorithm for this problem. For concreteness we will assume that the processors, or agents, are connected via a 'Butterfly Network', as shown in Figure 1. Note that many useful computations can be performed on a Butterfly Network in $O(\log n)$ time, such as adding n numbers or performing matrix operations on small matrices [9].

For simplicity we will consider the time an algorithm uses for computation and communication on the network, with the assumption that both times are similar. In this setting an efficient algorithm should be polynomial in $\log n$ as in standard parallel algorithms. Note that our definition of distributed algorithm is based on the structure of the problem, in contrast with many parallel algorithms in which a very large number of processors are used.

3.2 Lower Bounds

Our method for constructing lower bounds is based on an idea used by Hochbaum [6]. Very simply, we remove the interdependencies that occur in the problem due to the resource constraints. This can be accomplished by adding a 'sink' agent.

Thus we consider the class of resource allocation problems with i) $f_n(x_n) = e \cdot x_n/2 - m/2$ and ii) for which we are given that $x_i^* \leq e/n$, for $1 \leq i < n$.

Now, the problem is completely separable[5], and each subproblem is independent. Thus, we can solve the entire problem in parallel in the time required to solve a single subproblem. Also, solving the problem on a serial computer requires n times the time required to solve a subproblem. Note that we will compute lower bounds for the case of $\Lambda = ||b||_\infty = 1$ since this implies lower bounds for the general case due to Theorem 2.2.

Theorem 3.1 *The resource allocation problem has a lower bound on complexity of*

$$\Omega(m \log \frac{1}{n\epsilon})$$

in parallel and

$$\Omega(nm \log \frac{1}{n\epsilon})$$

in serial.

Proof: Note that the problem:

$$\max\{\sum_{i=1}^{n} f_i(x_i) \mid \sum_i x_i \leq e, \ x_i \leq e/m \ \forall 1 \leq i < m\}$$

is equivalent to

$$\max\{\sum_{i=1}^{n-1} (f_i(x_i) - e \cdot x_i/2) \mid x_i \leq e/m, \ \forall i < n.\}$$

Now, this problem is completely separable and each subproblem is independent. Also we can choose $(f_i(x_i) - e \cdot x_i/2)$ to be an arbitrary convex function with Lipschitz constant less than $1/2$, and then f_i will satisfy our requirements for the objective function of RA. Now, if we solve any of the subproblems with accuracy worse than ϵ then we will not have solved the entire problem with accuracy ϵ thus the time to solve the independent subproblems to accuracy ϵ gives us a lower bound on the time to solve RA. Applying Theorems 2.2 and 2.3 completes the proof. $\square$

4   Primal Algorithms

The basic ideas for constructing polynomial algorithms for convex problems are described in Nemirovsky and Yudin [12]. These are the only methods known for approximately minimizing arbitrary convex functions in a finite number of computations.

---

[5]Note that for the purpose of computing lower bounds we can restrict the class of possible functions in any way, since this can not increase the complexity of the problem.

All other methods, of which we are aware, require the functions to satisfy certain regularity conditions. For example, interior point methods require some consistency of the second derivatives of the objective function. This is necessary, as the methods depend strongly on the second derivatives, and could not operate correctly if this 'information' were rapidly changing. Thus, we are left with the methods first introduced by Nemirovsky and Yudin, general convex bisection methods and the method of Mirror Descent.

4.1 Convex Bisection Methods

The basic idea of bisection methods is as follows. Assume the solution of a convex minimization problem is initially contained within some convex body S_0. Find some point $x \in S_0$ and compute a separating hyperplane to the objective function at x. This hyperplane defines a halfspace $H = \{x \mid a \cdot x \geq b\}$ for which any point $y \notin H$ implies that $f(y) \leq f(x)$. Thus we know that the solution must lie within $S_0 \cap H$. Thus we set $S_1 \supseteq S_0 \cap H$ where we choose S_1 to have less volume than S_0 (on average), and to be simple enough that we do not spend too much time computing it. We then continue this procedure computing $S_1, S_2, \ldots$ etc. If the volume of S is decreasing sufficiently rapidly, on average, then our algorithm will converge to the solution, since the volume of the set $\{x \mid |f(x) - f(x^*)| \leq \epsilon\}$ is at least $n^{-n/2}(\epsilon/\Lambda)^n$.

A well known example of this method is the ellipsoid method. In this case we always take S_k to be an ellipsoid, and after each iteration we find a new ellipsoid S_{k+1} that contains $S_k \cap H$. This method requires $O(n^2 \log 1/\epsilon)$ iterations and $O(n^2)$ computations per iteration to compute the ellipsoid. Note that the number of iterations is $O(n)$ more than the lower bound we discussed in the previous section. This is due to the loss of volume, when we expand the set $S_k \cap H$ to the ellipsoid S_{k+1}.

In [18] Vaidya gives an algorithm that uses polytopes instead of ellipsoids to obtain an algorithm which requires $O(n \log 1/\epsilon)$ iterations. However, the complications involved in manipulating a large polytope require $O(n^{2.38})$ operations[6] per iteration, making it slightly more efficient than the ellipsoid algorithm in a theoretical sense. It may also be significantly better for actual implementation [18].

4.2 Mirror Descent Methods

For low accuracy (less than n iterations or $\epsilon^2 \geq 1/n$) Nemirovsky and Yudin developed the method of Mirror Descent. This algorithm, which is a generalization of gradient descent to general metric spaces, (with precisely determined step sizes) requires $O(\log n/\epsilon^2)$ iterations with $O(n)$ operations per iteration. This method may also be useful in practice [11].

---

[6]Note that the exponent 2.38 comes from fast matrix multiplications.

4.3   Application to Resource Allocation

The only primal methods in the literature that we are aware of for resource allocation require us to ignore much of the special structure in the problem. We can apply Vaidya's Algorithm or Mirror Descent directly to the resource allocation problem to get an algorithm that solves resource allocation in

$$O(\min[(mn)^{3.38}\log\frac{mn}{\epsilon}, mn\frac{\log mn}{\epsilon^2}])$$

operations in serial, and these can be reduced to

$$O(\log n \min[mn\log\frac{mn}{\epsilon}, \frac{\log mn}{\epsilon^2}])$$

in parallel. Thus for high accuracy we cannot lose the (at least linear) dependence on n or the slow convergence $(1/\epsilon^2)$ in parallel.

In the next section we develop an algorithm which is logarithmic in n and $1/\epsilon$. This requires the use of a primal-dual approach in which we use the structure of the problem to reduce it to the repeated solution of smaller subproblems.

5   Primal–Dual (Price Based) Algorithms

The primal-dual algorithm is based on the following formulation of SRA

$$\min_p \max_x \mathcal{L}(x,p) \quad \text{(PDRA)}$$

where $\mathcal{L}(x,p) = f(x) - p \cdot s(x)$ and $s(x) = \sum_i x_i - e$. The fundamental theorem of duality in convex programming [16] says that if (x^*, p^*) is the optimal solution to PDRA then x^* is the optimal (and therefore feasible) solution to RA. Note that $x^* = x(p^*)$ where $x(p) = \text{argmax}_x \mathcal{L}(x,p)$.

5.1   Solving the Dual of Resource Allocation

The primal-dual formulation allows us to use the following procedure for solving RA in parallel. Note that $\mathcal{L}(x,p)$ is separable in x, $\mathcal{L}(x,p) = \sum_i \mathcal{L}_i(x_i,p)$. Now define $F_i(p) = \max_{x_i} \mathcal{L}_i(x_i,p)$ and $F(p) = \sum_i F_i(p)$. Thus the problem of solving the resource allocation problem can be distributed as follows, where we will think of the processors as having two distinct functions. The first is that of independent consumers. In this capacity each processor can independently solve the consumption problem, $F_i(p) = \max_{x_i} \mathcal{L}(x_i,p)$ which has only m coordinates. This can be done in $O(m^{3.38}\log(nm)/\epsilon)$ time to produce an $O(\epsilon^4/nm^4)$-accurate solution by using Vaidya's Algorithm. Thus since we can add up the F_i in $O(\log n)$ time, we can compute the value of $F(p)$ to $O(\epsilon^4/m^4)$ accuracy in $O(m^{3.38}\log mn/\epsilon)$ distributed time.

Next we think of the processors as a group, in which they must minimize $F(p)$. However, they must use only function values because supporting hyperplanes computed directly from the inexact solutions of $\max_x \mathcal{L}(x,p)$ may be very inaccurate. This can be accomplished through a randomized scheme for approximating gradients that requires $O(m)$ function values. Note that a deterministic method could fail due to 'kinks' in the function. As a group, the processors can apply Vaidya's algorithm to minimize $F(p)$. Thus, we can get a good solution to the min-max problem in the following sense.

Definition 5.1 *A solution $(\hat{x}, \hat{p})$ is an ϵ accurate solution of PDRA if:*

1. $F(\hat{p}) - F(p^*) \leq \epsilon/2$.

2. $F(\hat{p}) - \mathcal{L}(\hat{x}, \hat{p}) \leq \epsilon/2$.

Now we can compute a good solution of PDRA.

Theorem 5.1 *For any $\delta > 0$ the algorithm, described above, will find an ϵ accurate solution of PDRA with probability greater than $1 - \delta$.*
The distributed running time of this algorithm is

$$O(m^{5.38}(\log \frac{nm}{\epsilon})^2).$$

Proof: See appendix.

Surprisingly, this solution is **not** necessarily a good solution to RA since $\hat{x}$ may be arbitrarily far from feasibility! This is due to the approximations used in the solution and is intimately related to the fact that we cannot solve convex problems accurately in terms of coordinates. Thus we must use a post-processing step in order to find a good feasible solution.

5.2 Constructing a Feasible Solution

The problem of constructing a feasible solution in a distributed manner is quite complicated. Most straightforward methods destroy the separability of PDRA, and thus cannot be implemented in parallel. Thus we must find a method that allows the problem to remain separated.

The method we sketch below follows from two important observations:

Lemma 5.2   • *A processor can compute $\{\max c \cdot x_i \mid \ |\mathcal{L}(x_i, \hat{p}) - \mathcal{L}(\hat{x}_i, \hat{p})| \leq \epsilon\}$ to $O(\epsilon/n)$ accuracy in $O(m^{3.38} \log n/\epsilon)$ time.*

• *If x and y satisfy $|\mathcal{L}(x, \hat{p}) - F(\hat{p})| \leq \epsilon$, $|\mathcal{L}(y, \hat{p}) - F(\hat{p})| \leq \epsilon$, and if z is a convex combination of x and y, then $|\mathcal{L}(z, \hat{p}) - F(\hat{p})| \leq \epsilon$.*

Proof: The first part follows from the application of Vaidya's algorithm. The second part follows by noting that $\mathcal{L}(z,\hat{p}) \leq \max[\mathcal{L}(x,\hat{p}), \mathcal{L}(y,\hat{p})]$ by the concavity of $\mathcal{L}(x,\hat{p})$ in x and the optimality of $x(\hat{p})$ which implies that $\mathcal{L}(z,\hat{p}) \leq \mathcal{L}(x(\hat{p}),\hat{p})$. $\diamond$

Thus if we can find two (or more) good solutions of PDRA whose convex hull contains a feasible solution, then that solution will be a good solution to RA and approximately solve the problem. Note that we can easily construct many good solutions to PDRA, the difficult part is to guarantee that their convex hull will contain a feasible solution.

This we do in two steps. First we prove that there is indeed a nearly feasible solution in a neighborhood of the solution. Then we describe a method for finding this nearly feasible solution[7] as the convex hull of several good solutions for the case $m = 2$. We believe that this procedure can be generalized to arbitrary m, but do not pursue this question here, as the geometry becomes much more complicated for $m > 2$.

Theorem 5.3 *Let $(\hat{x},\hat{p})$ be an ϵ^2/n accurate solution to PDRA in the sense of Theorem 5.1. Let*

$$S_i = \{x_i \mid |\mathcal{L}(x_i,\hat{p}) - \mathcal{L}(\hat{x}_i,\hat{p})| \leq \epsilon^2/2\}$$

and $S = S_1 \times \cdots \times S_n$. Then there exists some $x \in S$ such that $\|\sum_i x_i - e\|_2 \leq \epsilon/2$.

Proof: See appendix.

The basic idea of the algorithm is to parametrize the search directions c by their direction θ. We will then perform a binary search in θ in order to 'trap' a feasible solution.

First we need a few definitions. Let $c(\theta) = (\cos\theta, \sin\theta)$,

$$x_i(\theta) = \mathrm{argmax}_{x_i}\{c(\theta) \cdot x_i \mid x_i \in S\}$$

and $\hat{s}(\theta) = s(x(\theta))$. Recall that $s(x) = \sum_i x_i - e$. Note that if we know $c(\theta_l)$ and $c(\theta_h)$ we can compute $c(\theta_m) = (c(\theta_l) + c(\theta_h))/\|c(\theta_l) + c(\theta_h)\|_2$, where $\theta_m = (\theta_l + \theta_h)/2$, except in the initial case when $\theta_l = 0$ and $\theta_h = \pi$, in which case $\theta_m = \pm\pi/2$ and $c(\theta_m) = \pm(0,1)$.

Now given $s_l = \hat{s}(\theta_l)$, $s_h = \hat{s}(\theta_h)$, and $s_m = \hat{s}(\theta_m)$ we see that they form a triangle. Now either the feasible point $(0,0)$ is contained in this triangle or it lies above one of the lines $\overline{s_l s_m}$ or $\overline{s_m s_h}$. (See figure 2.) Now if it lies in the triangle we are done, since the three points x_l, x_m, x_h contain a feasible solution in their convex hull. Otherwise we continue the bisection algorithm by using the two points whose line has 'trapped' the feasible solution and taking the point halfway between them (in θ) and continuing the process. This procedure ends when either the point has been found in a triangle,

---

[7] Note that there are several layers of approximation in the feasibility algorithm. In the following discussion most of the quantities discussed are approximate, even though we do not explicitly state this, in order not to interrupt the flow of the argument. We show in the appendix that these approximations do not significantly effect the algorithm.

or $\theta_h - \theta_m \leq \epsilon$. In the latter case, the region containing the feasible solution must be quite narrow, and thus a feasible solution must lie quite close to the line $\overline{x_l x_j}$. (See Figure 3.) In this case we have a nearly feasible solution, which we can simply scale, $x^j \to x^j/(s^j(x) + 1)$, to find a feasible solution without a significant loss of accuracy.

Theorem 5.4 *The feasibility algorithm (as described above) computes an ϵ-accurate solution in at most $O(\log \frac{1}{\epsilon})$ iterations.*

Proof: See appendix.

Thus, for $m = 2$, we can combine the two algorithms and compute a feasible, ϵ-accurate solution.

Theorem 5.5 *Combining the two algorithms above we can compute an ϵ-accurate solution to RA (for $m = 2$) in $O((\log \frac{n}{\epsilon})^2)$ distributed time.*

Proof: Note that the first stage dominates the complexity analysis.

Thus for $m = 2$ we have an efficient distributed algorithm for resource allocation. We believe that with a straightforward extension of the feasibility algorithm we can also solve the general problem efficiently.

6 Conclusions

In this paper we have described a model of complexity for convex optimization. These ideas have been motivated by the well studied theory of complexity for discrete optimization, and recent formalizations of real number computation.

Using very general arguments we have constructed lower bounds for convex problems, both for serial and distributed algorithms. These bounds are very robust and hold in most common models of computation. This is because they are based on very general ideas such as scaling and information, and not specific properties of the computational model.

We have also developed an efficient distributed primal-dual algorithm for allocating two resources. We believe that we can extend this algorithm to the case with an arbitrary number of resources without a significant increase in complexity.

The study of complexity for convex optimization is still in its infancy. The development of complexity measures is quite delicate. Any such theory must strike a balance between generality and constructibility. For example, while quite natural, the use of coordinates as a measure of accuracy does not lead to a useful theory. Also, the restrictions on objective functions are very important, as they can drastically alter the nature of the problem, and its complexity. These topics require further study, but may lead to a useful theory for the construction of algorithms for solving real-word convex problems.

7   Acknowledgements

I would like to thank Ross Baldick and Adam Landsberg for useful comments and
suggestions.

A   Appendix: Proof for the Primal–Dual Algorithm

In this appendix we give the proof for the Primal-Dual algorithm. Our presentation
is compact and contains only the most important elements of the proof.

A.1   Theorem 5.1

We can approximate a supporting hyperplane to $F(p)$ at p with the following procedure:

1. Given T and τ, choose $\hat{p}$ at random from the box $\{\hat{p} \mid |\hat{p}_j - p_j| \leq T\}$.

2. Let

$$c = \sum_{j=1}^{m} \frac{F(\hat{p} + \tau e_i) - F(\hat{p})}{\tau}.$$

   Then c is an approximation to the subgradient of F at p.

Lemma A.1 *For any $\hat{\delta}$ there exist T and τ such that with probability greater than*
$1 - \hat{\delta}$, c computed by the above procedure satisfies $\forall p' \in G$,

$$F(p') \geq F(\hat{p}) + c \cdot (p' - \hat{p}) - \epsilon/2$$

$$F(p) \leq F(\hat{p}) + c \cdot (p - \hat{p}) + \epsilon/2$$

assuming that the values of F used by the procedure are $O(\epsilon^4/(n\sqrt{m}))$ accurate. (Thus
this is a good approximation to a supporting hyperplane.)

Proof: See [12]. $\diamond$

Lemma A.2 *Vaidya's algorithm in stage one succeeds with probability greater than*
$1 - \delta$.

Proof: The probability that the approximation of the supporting hyperplane is not
accurate at each iteration is less than $\hat{\delta}$. Since there are $O(m \log(1/\epsilon))$ iterations,
the probability of having a failure at any stage during the entire process is less than
$O(m \log(1/\epsilon)\hat{\delta})$. Thus we may choose $\hat{\delta} = O(\delta/(m \log 1/\epsilon))$. $\diamond$

   Thus using the gradient computed by the method described above, we can use
Vaidya's algorithm to produce an ϵ-accurate solution, proving the theorem. $\square$

A.2 Theorem 5.3

We prove this theorem through a sequence of lemmas in which we will assume that f is differentiable. It is easy to show that any convex function can be approximated to arbitrary accuracy by a C^∞ convex function, which is sufficient for our purposes. Let $\hat{\epsilon} = \epsilon/8n$.

Lemma A.3 *Given p is an $\hat{\epsilon}^2$ accurate solution of $F(p)$, there exists a p' such that $\|p - p'\|_2 \le \hat{\epsilon}$ and $\|\nabla F(p')\|_2 \le \hat{\epsilon}$.*

Proof: Consider the path generated by

$$\frac{d\gamma(t)}{dt} = \frac{\nabla F(\gamma(t))}{\|\nabla F(p(t))\|_2} \quad \text{s.t.} \quad \gamma(0) = p$$

which is a path from p to p^* where t is the arclength of the path from $\gamma(0)$ to $\gamma(t)$. Note that $\gamma(l) = p^*$ where l is the length of the path γ.

Now

$$F(\hat{p}) - F(p^*) = \int_0^l \nabla F(p(t)) \cdot \frac{d\gamma(t)}{dt} dt = \int_0^l \|\nabla F(\gamma(t))\|_2 \, dt$$

which is just the line integral of ∇F.

By assumption $F(\hat{p}) - F(p^*) \le \epsilon^2$. Thus the following must hold

$$\int_0^{\hat{\epsilon}} \|\nabla F(\gamma(x))\|_2 \, d\gamma \le \hat{\epsilon}^2$$

as the integrand is non-negative.

Now assume that the theorem is false. This implies that for all $x \le \hat{\epsilon}$ we must have that $\|\nabla F(\gamma(x))\|_2 > \hat{\epsilon}$ as the Euclidean distance $\|\gamma(0) - \gamma(\hat{\epsilon})\|_2$ is less than the distance along γ. However, this implies that

$$\int_0^{\hat{\epsilon}} \|\nabla F(\gamma(x))\|_2 \, d\gamma > \hat{\epsilon}^2$$

Since

$$\int_0^{\hat{\epsilon}} \|\nabla F(\gamma(x))\|_2 \, d\gamma < \hat{\epsilon} \min_{0 \le x \le \hat{\epsilon}} \|\nabla F(\gamma(x))\|_2$$

thus providing a contradiction and proving the theorem. $\diamond$

However, the gradient of F is related to feasibility.

Lemma A.4 *The feasibility, $s(x(p)) = \nabla F(p)$.*

Proof: This follows from the well known envelope theorem,

$$\frac{dF(p)}{dp^j} = \frac{\partial F(p)}{\partial p^j} = s^j(x(p))$$

and is easy to compute directly. $\diamond$

Now we prove the lemma.

Lemma A.5 *Given $\hat{p}$, $\hat{x}$ is an $\hat{\epsilon}^2$ accurate solution of $\mathcal{L}(x,p)$, there exists a feasible x such that $|\mathcal{L}(\hat{x}_i, \hat{p}) - \mathcal{L}(x_i, \hat{p})| \leq 4\hat{\epsilon}$.*

Proof: From the previous lemma there exists a p' such that $\|s(x(p'))\|_2 \leq \epsilon$ then letting $x^j = x^j(p')/(s^j(p') + 1)$ gives a x that is feasible and Noting that

$$\|x - x(p')\|_2 \leq 2\hat{\epsilon}$$

implies that

$$|\mathcal{L}(x_i(p'), p') - \mathcal{L}(x_i, \hat{p})| \leq 2\hat{\epsilon}$$

by the Lipschitz constant of $\mathcal{L}_i$. Now by the Lipschitz constant of F_i for changing p we know that

$$|F_i(\hat{p}) - F_i(p')| \leq \hat{\epsilon}$$

Combining these inequalities produces the required result. $\diamond$

A.3   Theorem 5.4

In order to prove that the algorithm constructs a feasible solution we note that Stage 2 actually performs a binary search in θ, and the only ways it can halt are:

1. A feasible solution x is found as the convex combination of three points x_l, x_m, x_h.

2. A solution x with $\|s(x)\|_2 \leq \hat{\epsilon}$ is the convex combination of two points x_l, x_h.

3. $\theta_l - \theta_h \leq 2\pi/\hat{\epsilon}$.

In the first case Lemma 5.2 shows that the convex combination of x_l, x_m, x_h that gives a feasible solution is $\hat{\epsilon}$-accurate.

From the previous lemma we see that for case 2 we can simply round the solution to feasibility.

Lemma A.6 *In case two the allocation defined by $x^j = x^j/(s^j(x) + 1)$ satisfies $|\mathcal{L}(x, \hat{p}) - F(\hat{p})| \leq 2\hat{\epsilon}$.*

Proof: Note that x_j is feasible. Also $\|x' - x\|_2 \leq \epsilon$ since $\|s(x)\|_2 \leq \hat{\epsilon}$. Thus by the Lipschitz continuity of $\mathcal{L}$ we get the desired result. $\diamond$

Now we will show that case 3 is actually equivalent to case 2 since $|\theta_l - \theta_h| \leq 2\pi\hat{\epsilon}$ implies that there exists a solution x which is the convex combination of $x(\theta_l)$ and $x(\theta_h)$ and $\|s(x)\|_2 \leq \hat{\epsilon}$.

Lemma A.7 *Assume that the feasibility algorithm received the exact solutions $\hat{s}(\theta)$ at each iteration. Then if $|\theta_l - \theta_h| \leq \hat{\epsilon}$ then $(0,0)$ must be within $\hat{\epsilon}$ of the line $\overline{s_l s_h}$.*

Proof: This follows from elementary geometry and the fact that the feasible solution $(0,0)$ must be contained in the region bounded by $\overline{s_l s_h}$ by construction. Also we know that $c(\theta_l) \cdot s_l \geq 0$ and $c(\theta_m) \cdot s_m \geq 0$ since these (s_l, s_h) are the maxima of their respective c's. This region forms a triangle with angle $\alpha = \pi - \hat{\epsilon}$ since $\hat{\epsilon} = |\theta_h - \theta_l|$. (See Figure 3.)

Since we know that $\|s_l - s_h\|_2 \leq 2$, the height of the triangle must be less than $2\sin \hat{\epsilon}/2$ which is less than $\hat{\epsilon}$. Thus since the point $(0,0)$ is contained within this triangle, we see that the distance from $(0,0)$ to $\overline{s_l s_h}$ must be less than $\hat{\epsilon}$. $\diamond$

However, the feasibility algorithm does not receive exact solutions from the processors. The solutions from the agents are only accurate to $\hat{\epsilon}^2$. In fact the solutions $\hat{s}(\theta)$ can be arbitrarily far from the true solutions $\hat{s}(\theta)$. However, as the next lemma shows, these solutions are actually close to the real solution for a slightly different θ. Thus we can imagine that the values of θ are uncertain. However, this does not effect the binary search if these errors in θ are less than $\hat{\epsilon}$.

Lemma A.8 *Assume that $|c(\theta) \cdot (s - \hat{s}(\theta))| \leq \hat{\epsilon}^2$. Then there exists a θ' such that $|\theta - \theta'| \leq 2\pi\hat{\epsilon}$ and $\|s - \hat{s}(\theta')\|_2 \leq \hat{\epsilon}$.*

Proof: As $|c(\theta) \cdot (s - \hat{s}(\theta))| \leq \hat{\epsilon}^2$ we see that there exists a $\hat{t}$ such that $\left\|s - \hat{s}(\hat{t})\right\|_2 \leq \hat{\epsilon}$. Let $\gamma(x)$ be the boundary of the feasible region with $\gamma(0) = \hat{s}(\theta)$ and $\gamma(\hat{t}) = \hat{s}(\hat{t})$ and assume that γ is parametrized by arc length.

We will now show that there exists a θ' such that $|\theta' - \theta| \leq \hat{\epsilon}$ while $\left\|\hat{s}(\theta') - \hat{s}(\hat{t})\right\|_2 \leq \hat{\epsilon}$. Combining these inequalities produces the theorem.

First note that at $\hat{s}(\theta)$ the normal to γ must point in the direction defined by θ. Note that

$$c(\theta) \cdot (\hat{s}(\theta) - \hat{s}(\beta)) = \int_0^{x(\beta)} \tan(\theta(u) - \theta)du \leq \int_0^{x(\beta)} 2/\pi \, (\theta(u) - \theta)du$$

where $x(\beta)$ is the smallest x such that $\theta(x) \leq \beta$.

Now assume the lemma is false and that all $\theta(x)$ for $|x - x(\hat{t})|$ satisfy $|\theta(x) - \theta| > \hat{\epsilon}$. Then

$$c(\theta) \cdot (\hat{s}(\theta(x)) - \hat{s}(\theta)) > \hat{\epsilon}^2$$

by the previous equation. However this contradicts the fact that $\hat{s}$ is $\hat{\epsilon}^2$-accurate proving the lemma. $\diamond$

Thus the algorithm works even with the inexact computations done by the processors. $\square$

References

[1] L. Blum. Lectures on a theory of computation and complexity over the reals (or an arbitrary ring). Technical Report TR–89–065, International Computer Science Institute, Berkeley, CA, December 1989.

[2] L. Blum, M. Shub, and S. Smale. On a theory of computation over the real numbers: *NP*-completeness, recursive functions, and universal machines. *Bulletin of the ACM*, 21(1):1–46, 1988.

[3] J. Canny. *The Complexity of Robot Motion Planning*. ACM doctoral dissertation award. MIT Press, Cambridge, Mass., 1987.

[4] E. J. Friedman. A strongly polynomial algorithm for convex programs with combinatorial constraints and resource allocation. Technical Report UCB/ERL/IGCT M92/6, Interdisciplinary Group on Coordination Theory, Electronics Research Laboratory, University of California, Berkeley, CA 94720, January 1992.

[5] D. Den Hertog. *Interior Point Approach to Linear, Quadratic, and Convex Programming*. PhD thesis, Technische Universiteit Delft, 1992.

[6] D. S. Hochbaum. Lower and upper bounds for the allocation problem and other nonlinear optimization problems. Manuscript, School of Business Administration and Industrial Engineering and Operations Research, University of California, Berkeley, CA 94720. September 1989, Revised December 1990.

[7] D. S. Hochbaum and J. G. Shanthikumar. Convex separable optimization is not much harder than linear optimization. *Journal of the ACM*, 37(4):843–862, October 1990.

[8] L.G. Khachian. A polynomial algorithm for linear programming. *Soviet Math. Doklady*, 20:191–4, 1979.

[9] F. T. Leighton. *Introduction to parallel algorithms and architectures*. Morgan-Kaufman, 1991.

[10] N. Megiddo. Towards a genuinely polynomial algorithm for linear programming. *SIAM J. Comp.*, 12(2):347–59, 1983.

[11] A.S. Nemirovsky and I.A. Teslya. Optimizing the selection of load resistors in gate-array chips. *Engineering Cybernetics*, 24(6), 1986.

[12] A.S. Nemirovsky and D.B. Yudin. *Problem Complexity and Method Efficiency in Optimization*. Wiley, New York, 1983.

[13] Y.E. Nesterov and A.S. Nemirovsky. Self-concordant functions and polynomial time methods in convex programming. Technical report, Central Economic and Mathematical Institute, USSR Academy of Science, Moscow, USSR, 1989.

[14] C. H. Papadimitriou and K. Steiglitz. *Combinatorial Optimization: Algorithms and Complexity*. Prentice-Hall, Inc., Englewood Cliffs, New Jersey, 1982.

[15] J. Renegar. On the computational complexity and geometry of the 1st order theory of the reals: Parts 1,2, and 3. *Journal of Symbolic Computation*, 13(3):255–352, 1992.

[16] R. T. Rockafellar. *Convex Analysis*. Princeton University Press, Princeton, New Jersey, 1970.

[17] M. Shub and S. Smale. The complexity of Bezout's theorem. parts I, II, and III. Unpublished Manuscript, 1992.

[18] P.M. Vaidya. A new algorithm for minimizing convex functions over convex sets. In *30th Annual Symposim on the Foundations of Computer Science*, pages 332–337. IEEE, October 1989.

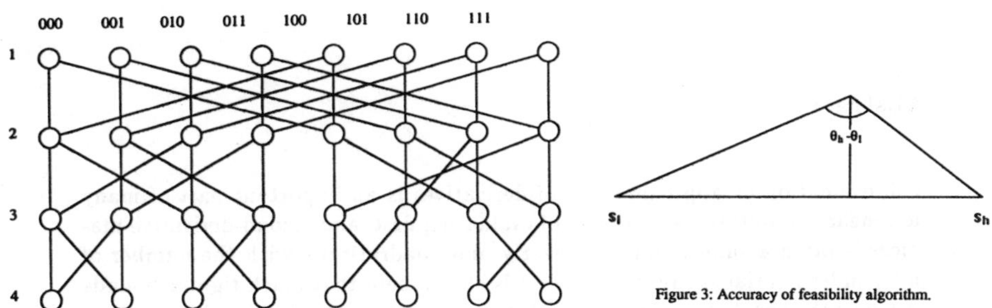

Figure 1 : Processors organized on a Butterfly network.

Figure 3: Accuracy of feasibility algorithm.

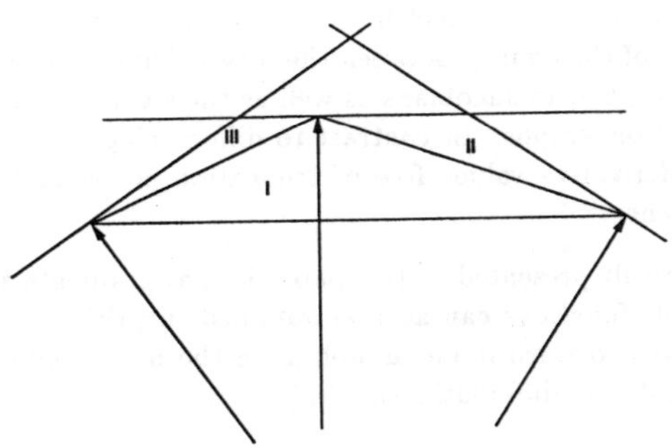

Figure 2: Geometry of feasibility algorithm.

Complexity in Numerical Optimization, pp. 128-162
P.M. Pardalos, Editor
©1993 World Scientific Publishing Co.

Some Bounds on the Complexity of Gradients, Jacobians, and Hessians [*]

Andreas Griewank
*Mathematics and Computer Science Division, Argonne National Laboratory,
Argonne, IL 60439 USA*

Abstract

The evaluation or approximation of derivatives is an important part of many
nonlinear computations. The cost of evaluating first- and second-derivative matrices is often assumed to grow linearly and quadratically with the number of
independent variables, respectively. It is shown here that much tighter bounds
can be achieved through the exploitation of partial function- and argument-
separability in combination with the forward and reverse mode of computational, or automatic, differentiation.

The new separability concepts facilitate the reduction of chromatic numbers
and maximal row lengths, which determine the complexity of the Curtis-Powell-
Reid and Newsam-Ramsdell schemes for estimating sparse derivative matrices.
Because of the duality between the forward and reverse modes these techniques
can be applied to Jacobians as well as their transposes and the associated row-
intersection graphs. In contrast to differencing, computational differentiation
yields derivative values free of truncation errors and without any parameter
dependence.

A key result presented in this paper is that gradients and Hessians of partially
separable functions can also be obtained surprisingly cheaply in the easily implemented forward mode as well as in the more sophisticated reverse mode of
computational differentiation.

---

[*]This work was supported by the Office of Scientific Computing, U.S. Department of Energy,
under Contract W-31-109-Eng-38.

1 Introduction

Most algorithms for the numerical solution of nonlinear programming problems involve the gradient of the objective function, the Jacobian of the active constraints, and the Hessian of the Lagrangian function. Some of this derivative information may be used only implicitly or in projected form. However, since optimizers are locally characterized by the KKT conditions in terms of objective and constrained gradients, evaluation errors in these first derivatives directly limit the solution accuracy. Moreover, except for algebraically simple test functions and some large, sparse problems, the computational effort of evaluating or approximating the required derivative information may dominate the cost of the numerical linear algebra and other algorithmic overhead.

As in numerical ordinary differential equations and other nonlinear computational fields, designers of optimization methods have usually assumed that derivatives are hard to come by and that their provision belongs to the realm of the user. This is sometimes called the black box model [21], where the optimization algorithm relies exclusively on a subroutine for evaluating objectives and gradients at given arguments, usually with unspecified precision. Unless the user provides additional codes for gradient and Jacobian evaluation, the optimization method must either resort to divided difference approximations or model the optimization problem on the basis of function values alone. The latter approach is rarely advantageous since it typically results in slow and unreliable convergence.

1.1 The Computational Model

In this paper we develop complexity estimates for gradients, Jacobians, and Hessians of functions that are defined by computer programs as compositions of arithmetic operations and intrinsic functions. We will assume that these elementary functions are performed in floating-point arithmetic with fixed precision and that their computational cost is independent of the argument. This scenario applies to the majority of practical optimization calculations, but not to symbolic, interval, or variable-precision computations. The overall temporal complexity is measured by simply counting the number of elementary operations and the number of random or sequential memory accesses. Here, it is assumed that the memory hierarchy can be split into a randomly accessed "core" storage and a sequentially accessed "disk" storage. Because of the increasing lag between processor and memory speed, the distinction between various memory access patterns becomes increasingly important not only on supercomputers but also on modern workstations.

The process of calculating overall derivatives vectors and matrices from the "local" derivatives of the elementary functions has become known as *automatic differentiation* [19],[13] but might be better called *computational differentiation*. Depending on the order in which the chain rule is applied, one obtains the forward, reverse, or mixed mode of computational differentiation. As was shown in [12], even a sophisticated implementation of the reverse mode of computational differentiation involves a logarithmic increase in the storage requirement, but the extra data can be stored and accessed sequentially in successive forward and reverse sweeps. In contrast, this nearly perfect data locality is lost if one tries to minimize the operations count for Jacobian evaluations by the general vertex elimination scheme described in [14]. Moreover, the combinatorial task of finding an elimination ordering that absolutely minimizes the operations count is conjectured to be NP hard. Therefore, we confine ourselves in this paper to sequential schemes for which the randomly accessed core memory can be a priori restricted to a small multiple of that used by the original function evaluation.

Several other practical aspects of relative efficiency are very hard to quantify, even in an informal discussion of computational complexity. In our context this applies, for example, to the alternative of compilable derivative code versus more interpretive derivative-evaluation schemes. A related issue is the relative efficiency of dynamically sparse vector operations with indirect addressing versus dense vector operations on contiguous arrays. While it is relatively easy to generate compilable code for the forward mode of automatic differentiation (see, e.g., [2]), all current implementations of the reverse mode incur a large number of procedure calls or other interpretive overheads, unless the original evaluation source is very restricted. Therefore, we will emphasize ways of reducing the temporal complexity of the forward mode by exploiting partial separability and other structure, even when the basic reverse mode has a lower operations count, as is the case for gradients.

Fortunately, much of the excellent research that has been conducted regarding the estimation of sparse Jacobians and Hessians by differencing (see, e.g., [10], [9], [18], [20]) carries over to computational differentiation. The main difference is that, instead of approximating Jacobian vector products by divided differences, one obtains them without any truncation errors by the forward mode of automatic differentiation. Moreover, since the reverse mode yields vector Jacobian products accurately and efficiently, one can exploit the sparsity structure not only columnwise but also rowwise. Therefore the well-known coloring techniques can be applied either to the column- or the row-intersection graph. Hence our complexity bounds involve the maximal clique size and the chromatic number of these undirected graphs. We have delayed until the central section of this paper the introduction of the directed *computational graph* that is associated with function evaluation programs. Until then, the results are developed in a more familiar matrix-vector notation. Concepts such as operations counts and intermediate variable will be used in an intuitive way until they are defined rigorously in Section 3.

The paper is organized as follows. In the remainder of the introduction we review the derivative requirements in optimization calculations and the basic properties of the forward and reverse modes of computational differentiation in combination with the CPR (Curtis-Powell-Reid) and NR (Newsam-Ramsdell) approaches to estimating sparse derivative matrices. In Subsections 2.1 and 2.2 we discuss whether and how *partial function-separability* and a dual concept of *partial argument-separability* can be used to express the original function as contraction of larger systems, whose Jacobians or their transposes have shorter row lengths and lower chromatic numbers. In Subsection 2.3 we discuss the tearing of functions and arguments, which amounts to a forced function or argument separation that entails a certain duplication of intermediate calculations. The resulting complexities are closely related to those of dynamically sparse implementations of the forward and reverse modes. In Subsection 2.4 we discuss the relation to multicoloring and in Subsection 2.5 we examine a generic binary interaction example. In Section 3 we provide a rigorous foundation and generalization of all previous developments in terms of computational graphs. In Section 4 we compile and discuss complexity bounds for the various methods and decompositions. The paper concludes with a brief summary in Section 5.

1.2   Derivative Requirements in Nonlinear Optimization

As a focal point of our investigation we consider the constrained optimization problem

$$\min w^T x \quad \text{s.t.} \quad f(x) = 0 \quad \text{and} \quad \underline{x} \leq x \leq \bar{x}$$

with $w \in \mathbb{R}^n$ some fixed-price vector and $f(x) : \mathbb{R}^n \to \mathbb{R}^m$ at least twice continuously differentiable. Without loss of generality we have assumed that the objective and the inequality constraints are linear; hence, all nontrivial derivatives occur in the equality constraints. As a consequence of this normalization, some components of x are likely to be slack variables; thus, the corresponding derivatives of f have a special structure as well. Sparsity in the Jacobian $J(x) \equiv J$ and the second-derivative tensor f'' will be a major concern throughout this paper. However, we will not make use of the observation that one need not evaluate rows of J that correspond to "safely" inactive constraints because their slack variable value is very large. Similarly, one could theoretically skip columns of J corresponding to "nonbasic" variables x_i that are certain to remain fixed at their lower or upper bound $\underline{x}_i$ or $\bar{x}_i$ over the current iteration.

Typically, one step of an iterative optimization algorithm proceeds as follows. First, the scalar objective $w^T x$ and the constraint vector $f(x)$ are evaluated at a new trial point x, which may be assumed feasible with respect to the bound constraints. Occasionally, the combination of values $(w^T x, f(x))$ may be deemed unacceptable, and the algorithm backtracks to a previous iterate. Otherwise, one evaluates the Jacobian $J(x)$ and computes a matrix S, whose $n - m$ columns span the null-space of $J(x)$. This is usually done by factoring $J(x)$, for example, into the product of a lower triangular or orthogonal

problem involves (besides the objective gradient w) the constraint Jacobian J and some information about the Hessian of the Lagrangian

$$L(x, u) \equiv w^T x + \sum_{i=1}^{m} u_i f_i(x).$$

Here the Lagrange multipliers u_i are usually estimates from the previous iteration. For one-step quadratic convergence a Newton-like algorithm must know the one-sided projection

$$H \equiv \nabla_x^2 L(x, u) S \quad \text{with} \quad J S = 0 \in \mathbb{R}^{m \times (n-m)} \quad .$$

For two-step quadratic convergence it suffices to evaluate the two-sided projection $S^T H \in \mathbb{R}^{p \times p}$ with $p = n - m$; but as we will show, this is not necessarily a saving with regard to computational differentiation. Moreover, it is important to recognize that the one-side projection onto the columns of S can be built into the differentiation process so that the full Hessian need never be formed. This implicitly projected approach is likely to be efficient on problems with a comparatively small degree of freedom, p.

1.3   Basic Properties of the Forward and Reverse Modes

Rather than considering $f(x)$ just as a mathematical mapping, we will assume from now on that it is defined by a procedural (sub)program for its evaluation at any given argument $x \in \mathbb{R}^n$. For simplicity we will exclude the possibility that the program takes different branches for various values of x or that some intrinsic function such as square root is evaluated at a point of nondifferentiability. Then one can by hand or automatically (i.e., with the help of some software package) extend the original program to a code that also evaluates the Jacobian-matrix product

$$J(x) S \quad \text{for} \quad S \in \mathbb{R}^{n \times p} \tag{1.1}$$

"analytically," that is, without incurring any truncation error. The choice of the *seed matrix* S provides the user with a great deal of flexibility at run time. For example, one might choose $S = I$ with $p = n$ to obtain the full Jacobian J in one sweep. Alternatively, one may have $p = 1$ over several sweeps and let $S = [s]$ range over a sequence of column vectors $s \in \mathbb{R}^n$, possibly as part of an iterative Newton-step calculation. Under realistic assumptions on the computational model, we will establish in Proposition 1 of Section 4 the bounds

$$\text{OPS}\{f, JS\} \quad \leq \quad (2 + 3p)\, \text{OPS}\{f\} \tag{1.2}$$
$$\text{RAM}\{f, JS\} \quad \leq \quad (1 + p)\, \text{RAM}\{f\} \quad . \tag{1.3}$$

Here, OPS denotes a conventional operations count, and RAM represents the number of randomly accessed storage locations required by the function or derivative evaluation program. On a serial machine the run time will be proportional to the operations count, and we will therefore refer to $\text{OPS}\{f, \text{task}\}/\text{OPS}\{f\}$ as the run-time ratio for some additional computational task related to f.

For $p = 1$ the operations count and storage requirement may grow fivefold and twofold, respectively. This situation is somewhat troubling, since a divided difference of the form $[f(x + \varepsilon s) - f(x)]/\varepsilon$ yields an approximation to Js at the cost of one extra function evaluation and without any increase in storage. Here we have assumed that f has already been evaluated at the base point x. Fortunately, the factor 3 in (1.2) is quite pessimistic. On the other hand, the leading constant 2 does not take account of some of the overhead in the derivative code. Empirically it was found [1] that, when the forward mode is implemented as compilable code and p is in the double digits, then the resulting run-time ratio is typically between half and twice the divided difference ratio $(1 + p)$.

The up to $(1 + p)$-fold increase in memory stems from the fact that a p-vector of derivatives is associated with most or all scalar variables in the original evaluation program. To limit this increase, one could alternatively evaluate one or a few derivative components at a time, but that would increase the accumulated run time significantly, since certain overhead costs are incurred repeatedly. On the other hand, when p reaches into the hundreds, it will probably be better to split the columns of S into groups in order to reduce the number of page faults in accessing intermediate-derivative vectors. Provided these p vectors are allocated as contiguous arrays, the number of memory accesses in the sense of pointer dereferencing does not really grow $(1 + p)$-fold, but merely doubles. On the other hand, if they are stored and manipulated as sparse vectors, the indirect addressing overhead goes up, but the storage requirement may be significantly reduced.

As a first approximation one may view the forward mode as truncation- and parameter-free equivalent of divided differences. The relationship is close enough that the CPR [10] and the NR [18] approaches for estimating sparse Jacobians J from some product $J S$ with $p < n$ can be applied virtually unchanged. In the CPR approach the rows of the seed matrix S are Cartesian p vectors, whereas in the NR approach S is usually chosen as a Vandermonde matrix. The potential ill-conditioning of the latter choice may be of less concern here since the projected Jacobian $J S$ is obtained essentially to working accuracy. The number p of columns required for the NR method is simply the row-length $\rho(J)$, that is, the maximal number of nonzeros in any row of the Jacobian. The corresponding lower bound for the CPR approach is the chromatic number $\chi(J)$ of the column-intersection graph of the Jacobian. This undirected graph has one node for each column of J, with two of them being connected by an edge exactly if the corresponding column pair shares a nonzero in at least one row. Otherwise the two columns are said to be structurally orthogonal. The row length and chromatic number satisfy the trivial relation

$$\rho(J) \leq \chi(J) \quad ,$$

which means that the seed matrix for the CPR method has at least as many columns as the minimum needed for the NR approach. However, to improve the conditioning of the linear systems in the latter approach, one may prefer to define S using complex roots of unity, which effectively doubles p to $2\rho(J)$.

From a mathematical point of view, the reverse mode is much more interesting since it has completely different complexity properties compared with differencing.

Instead of multiplying J from the right by a seed matrix S, whose columns represent directions in the domain of f, the Jacobian is now multiplied from the right by a matrix W^T, whose rows represent linear functionals on the range of f. Formally, the reverse mode yields

$$W^T J(x) \quad \text{for} \quad W^T \in \mathbb{R}^{q \times m} \quad .$$

At first sight the corresponding complexity estimates are also neatly transposed in that, for some integer $r \geq 1$,

$$\text{OPS}\{f, W^T J\} \quad \leq \quad (1 + r + 3q)\,\text{OPS}\{f\} \tag{1.4}$$

$$\text{RAM}\{f, W^T J\} \quad \leq \quad (1 + q)\,\text{RAM}\{f\} \quad . \tag{1.5}$$

However, there is a crucial difference, namely, that the reverse mode requires the ability to run the evaluation process for f step by step backward. This requirement explains the terminology *reverse mode,* or top-down method, which is preferable to the term *backward differentiation,* a label that invites confusion with a well-established class of methods for the numerical solution of stiff differential equations. The "naive" program reversal based on a full execution trace of the forward evaluation requires temporary storage proportional to $\text{OPS}\{f\}$. Fortunately, the generation and utilization of the trace data occur strictly sequentially in opposite order, so that it makes sense to quantify this storage requirement as SAM (for Sequentially **A**ccessed **M**emory). Hence we obtain the basic bound

$$\text{SAM}\{f, \overleftarrow{f}\} \quad = \quad \mathcal{O}(\text{OPS}\{f\}) \quad , \tag{1.6}$$

where $\overleftarrow{f}$ denotes the reversal of the evaluation process for f. Because of the strictly sequential access pattern, this potentially very large data set can be stored on external mass storage devices. When the ratio

$$h\{f\} \quad \equiv \quad \text{OPS}\{f\}\,/\,\text{RAM}\{f\} \tag{1.7}$$

is small or of moderate size, the proportionality relation (1.6) may not be worrisome. This situation applies, for example, for any three-dimensional composite structure without long-range interaction between its components (e.g., static models of buildings or mechanical devices and discretizations of differential operators). However, especially on explicitly time-dependent problems, the ratio $h\{f\}$ may be so large that (1.6) represents an unacceptable increase in memory requirement. This severe limitation of the basic reverse mode can be remedied as follows.

Rather than generating and storing the full execution trace in one piece, one can break it into slices that are (re)generated several times from snapshots taken at judiciously selected checkpoints. A detailed analysis of this recursive reverse mode in [12] shows that there exists a constant c such that for all integers $r \geq 1$, the reversal costs can be limited according to

$$\text{OPS}\{\overleftarrow{f}\} \quad \leq \quad r\,\text{OPS}\{f\} \tag{1.8}$$

$$\text{SAM}\{\overleftarrow{f}\} \quad = \quad c\,\text{RAM}\{f\}\,\sqrt[r]{h\{f\}} \quad . \tag{1.9}$$

As an alternative to accepting a constant increase in the operations count and an algebraic growth of the memory requirement with respect to h, one can also limit both increases to order $\log h$. In this paper we will use the algebraic option for the program reversal, whose operations count (1.8) was already included in (1.4). The total complexity of this reverse mode variant is therefore described by the three equations (1.4), (1.5), and (1.9). Despite these rather tight results, the practical reversal of a sizable program is a difficult problem, and no implementation achieving these bounds is currently available.

The key difference between (1.4) and the earlier (1.2) is that the operations count for the reverse mode depends on the number of dependent rather than independent variables. In particular, one obtains gradients (where $m = 1 = q$) at a small multiple of the cost of evaluating the underlying functions. Here, and later, we neglect the additional SAM requirement (1.9), which grows very slowly if r is sizable (say, greater than 5). For vector-valued functions, f, one can apply the CPR or NR approach to the rows rather than the columns of the Jacobian J, so that their respective complexities are determined by

$$\rho(J^T) \leq \chi(J^T) \ .$$

It is sometimes mistakenly concluded that Jacobians also have essentially the same complexity as the underlying vector function since each row is a gradient of the corresponding function component. Applying (1.4) with $q = 1$ to each component separately, we have indeed

$$\text{OPS}\{f, I\,J\} \leq 5 \sum_{i=1}^{m} \text{OPS}\{f_i\} \ . \tag{1.10}$$

However, the sum of the right hand side can be almost m times larger than $\text{OPS}\{f\}$, as a result of the multiple usage of common intermediates in the simultaneous evaluation of all function components. For example, this will be the case if $m = n$ and f is of the form

$$f_i(x) \quad = \quad x_i * crunch(x_1, x_2, \ldots, x_n),$$

with $crunch(x)$ being a computationally very expensive scalar-valued function. On the other hand, the evaluation of any two components $f_i(x)$ and $f_j(x)$ may not involve any common intermediates, so that the right-hand side of (1.10) is indeed exactly equal to $5\,\text{OPS}\{f\}$. Since this can easily happen even when J is dense, one would certainly wish to do better than applying the standard forward or reverse mode. In fact, this case would be ideal for the more general elimination procedure described in [14], whose optimal application is conjectured to be an NP-hard combinatorial problem. On the downside, even if greedy heuristics are used, the resulting Markowitz-like procedure requires RAM of order $\text{OPS}\{f\}$, which appears to be a serious drawback on larger problems. Therefore, we introduce column- and row-splitting techniques that sometimes yield similar reductions in the operations count without ever violating the storage bounds (1.3) or (1.5), and (1.9).

2 Generalizations of Partial Separability

Rather than applying computational differentiation techniques to calculate the Jacobian J directly, we can first strip it of linear premultiplier and postmultipliers in order to obtain an even sparser central part that contains all nonlinearities. For the purposes of computational differentiation, the aim is to reduce the maximal number of zeros per row or column and the chromatic number of the row- or column-intersection graphs, which determine the cost of the forward and reverse mode, respectively. Especially for large-scale problems, one can expect that not only the Jacobian J but especially the derivative tensor $f'' \in \mathbb{R}^{m \times n \times n}$ is quite sparse. As observed in [15], any scalar function $h \in C(\mathbb{R}^n)$ whose Hessian is sparse can be decomposed into a sum

$$ h(x) \quad = \quad \sum_k h_k(P_k x) \quad , $$

where the projection P_k picks out the subset of components in x on which h_k depends in a nontrivial fashion. Whenever a Hessian entry $\partial^2 h / \partial x_i \partial x_j$ vanishes identically, each P_k annihilates one of the Cartesian basis vectors e_i and e_j or both.

It is very important to understand that this partial separability property does not imply that h is best evaluated by evaluating each additive term h_k separately. For example, we could have a function of the form

$$ \begin{aligned} h_1(x_1, \ldots, x_{n-1}) &= x_1 \cdot crunch(x_2, \ldots, x_{n-1}) \\ h_2(x_2, \ldots, x_n) &= crunch(x_2, \ldots, x_{n-1}) \cdot x_n, \end{aligned} $$

with $crunch(x_2, \ldots, x_{n-1})$ a computationally intensive and nonseparable common intermediate. Then $h = h_1 + h_2$ is partially separable because x_1 and x_n do not interact in a nonlinear fashion, so that the $(1, n)$ and $(n, 1)$ entries of the Hessian vanish. However, evaluating h_1 and h_2 separately is clearly not a good idea since it would entail computing $crunch(x_2, \ldots, x_{n-1})$ twice. Instead one should rewrite $h(x)$ as

$$ h(x) = (1, 1)\, H(x) \quad \text{with} \quad H(x) \equiv (h_1(x), h_2(x))^T, $$

so that the gradient of h is given by

$$ \nabla h(x) \quad = \quad (1, 1)\, H'(x) \quad . $$

The Jacobian of the vector function $H : \mathbb{R}^n \to \mathbb{R}^2$ is sparse and can therefore be evaluated at reduced cost. More generally, we can use the following generalization of partial separability.

2.1 Partial Function Separability and Row Splitting

We will call the vector function f *partially function-separable* whenever one of its components is partially separable in the usual sense and can therefore be split into two. The corresponding row of the Jacobian is then also split into two so that the number of nonzeros in either of the new rows cannot be greater than the number of nonzeros in the original row. On the other hand, unless the new rows are structurally orthogonal, there is at least one column in which the number of nonzeros grows by one, whereas it is nondecreasing in all others. Hence we find that row splitting is likely to decrease the row length but increase the column length. Similarly, the number of edges in the column-intersection graph cannot grow, whereas the number of nodes and edges in the row-intersection graph must go up. However, the chromatic number of the row intersection can go down, as can be seen in the following example.

Consider a vector function $f : \mathbb{R}^3 \to \mathbb{R}^3$ whose Jacobian has a vanishing diagonal and whose first component function is partially separable. Then the first component can be split into two, and we obtain an vector function $\hat{f} : \mathbb{R}^3 \to \mathbb{R}^4$ such that the Jacobians J and $\hat{J} \equiv \hat{f}'$ have the sparsity pattern

$$J = \begin{bmatrix} 0 & \times & \times \\ \times & 0 & \times \\ \times & \times & 0 \end{bmatrix} \quad \Rightarrow \quad \hat{J} = \begin{bmatrix} 0 & \times & 0 \\ 0 & 0 & \times \\ \times & 0 & \times \\ \times & \times & 0 \end{bmatrix} \quad .$$

The associated row-intersection graphs take the following simple forms:

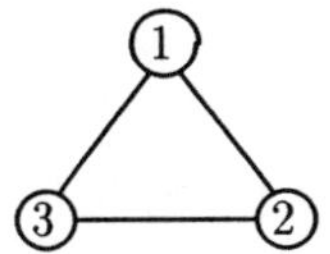 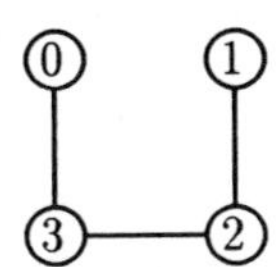

The column-intersection graphs are both identical to the row-intersection graph of J, so that

$$\chi(\hat{J}) \; = \; 2 \; < \; 3 \; = \; \chi(J) = \chi(J^T) = \chi(\hat{J}^T) \quad .$$

In this example, row splitting makes no difference for the forward mode but is beneficial for the CPR approach in the reverse mode. The corresponding weight matrix would be simply

$$W^T = \begin{bmatrix} 1 & 0 & 1 & 0 \\ 0 & 1 & 0 & 1 \end{bmatrix} \quad .$$

We will see later that the dual process of column splitting is sometimes beneficial for the CPR approach in the forward mode.

If the row-splitting process is carried out as far as possible, one obtains a representation of the form

$$f(x) \; = \; B\hat{f}(x) \quad \text{with} \quad \hat{f} : \mathbb{R}^m \to \mathbb{R}^{\hat{m}} \quad \text{and} \quad \mathrm{OPS}\{\hat{f}\} \; = \; \mathrm{OPS}\{f\} \quad , \qquad (2.1)$$

where each column of the matrix $B \in \mathbb{R}^{m \times \hat{m}}$ is a Cartesian basis vector. Here and throughout the paper we consider the computational cost of merging the components of $\hat{f}$ to those of f by addition as negligible. Applying the arguments given above, by induction one finds that

$$\rho(\hat{J}) \leq \rho(J) \quad , \quad \chi(\hat{J}) \leq \chi(J) \quad , \quad \rho(\hat{J}^T) \geq \rho(J^T) \quad . \tag{2.2}$$

If the first two inequalities hold strictly, it is advantageous in the forward mode first to evaluate $\hat{J}$ and then to multiply it by B in order to obtain $J = B\hat{J}$. It is natural to ask whether one can split off a similar linear factor on the right in order to make the column- rather than the row-incidence graph sparser. In the next subsection, we show how to do this by splitting independent variables, or arguments, rather than dependent variables, or functions. First, however, we will end this subsection with an observation that will be useful regarding the complexity of second derivatives.

For $\hat{J}$ to be nonseparable it is necessary that the nonzeros in the Hessians of each component function $\hat{f}_i$ form a dense square block. Otherwise, $\hat{f}_i$ would still be partially separable and could be split in at least two smaller component functions (increasing $\hat{m}$ by one in the process). Hence we have the following lemma.

Lemma 1 *If $\hat{f}(x) : \mathbb{R}^n \to \mathbb{R}^{\hat{m}}$ is not partially function-separable, its column-intersection graph is identical to the incidence graph G of the symmetric Hessian $\nabla_x^2[\hat{u}^T \hat{f}(x)]$ for a generic multiplier vector $\hat{u} \in \mathbb{R}^{\hat{m}}$.*

Using the terminology of Coleman and Cai [6], we find for the path and cyclic chromatic numbers $\chi_\pi(G)$ and $\chi_0(G)$

$$\rho(\hat{J}) = \chi(G) \leq \chi_0(G) \leq \chi_\pi(G) \leq \chi(G^2) \quad . \tag{2.3}$$

Here G^2 is the column-intersection graph of the Hessian, which does not reflect its symmetry. The chromatic numbers $\chi_\pi(G)$ and $\chi_0(G)$ determine how many gradient evaluations are required to estimate the Hessian by differencing with direct and indirect substitution, respectively. The same techniques have been applied in combination with computational differentiation, which yields Hessian vector products without any truncation errors. Again one may employ either the purely forward mode or a combination with the reverse mode, which has a lower operations count but requires more storage, as shown in Proposition 2 of Subsection 4.2. In any case it is possible to obtain second-derivative information directly from the evaluation program for f, without asking the user to supply a gradient code.

2.2 Partial Argument-Separability and Column Splitting

The row splitting described above proceeded by identifying one dependent variable $y_k = f_k(x)$ and a corresponding set of independent variables $\{x_i\}_{i \in \mathcal{I}}$ that do not interact nonlinearly in the evaluation of y_k. By *not interacting nonlinearly* we mean that $\partial y_k / \partial x_i \not\equiv 0$ for all $i \in \mathcal{I}$ but that

$$\partial^{|\mathcal{I}|} y_k \left/ \prod_{i \in \mathcal{I}} \partial x_i \right. \equiv 0 \quad ,$$

where $|\mathcal{S}|$ denotes the cardinality of a set $\mathcal{S}$. This effect must occur in particular when y_k is calculated as a sum of intermediate values each of which is constant with respect to at least one x_i with $i \in \mathcal{I}$. In the simplest case two independent variables x_i, x_j and the dependent variable y_k form a triplet (x_i, x_j, y_k) that is *disconnected* in the sense that at least one of the three dependencies between them and any intermediate variable is trivial.

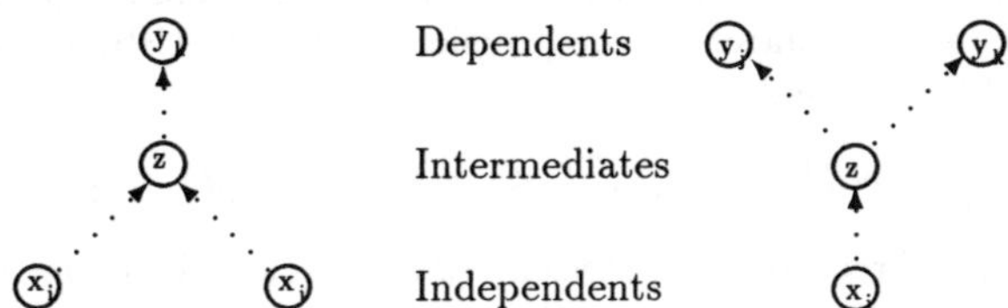

Two triplets that are connected by some intermediate variable z

Similarly, there may be triplets (x_i, y_j, y_k) consisting of one independent x_i and two dependent variables y_j, y_k that are disconnected in the sense that no intermediate that depends on x_i can impact both y_j and y_k. Then we may create a duplicate $x_{i-1/2}$ of x_i and replace for all intermediates that impact y_k their functional dependence on x_i by exactly the same dependence on $x_{i-1/2}$. For example, if $\exp(x_i)$ enters into the evaluation of y_j, and $\sin(x_i)$ enters into the evaluation of y_k but not y_j we can replace the definition of the second intermediate by $\sin(x_{i-1/2})$. Thus we have increased the number of independents by one and obtained a function

$$\check{f}(x_1, x_2, \ldots, x_{i-1}, x_{i-1/2}, x_i, \ldots, x_n) \ : \ \mathbb{R}^{n+1} \to \mathbb{R}^n$$

so that

$$f(x) = \check{f}(\tilde{I}_i x) \quad \text{and} \quad \text{OPS}\{\check{f}\} = \text{OPS}\{f\} \quad ,$$

where $\tilde{I}_i \in \mathbb{R}^{(n+1)\times n}$ is obtained from the $(n+1) \times (n+1)$ identity matrix by adding the i-th and $(i+1)$-st column together. The operations count for $\check{f}$ is the same as that for f even if the value for $x_{i-1/2}$ is chosen differently from x_i. This follows from the assumed disconnectedness of the triplet (x_i, y_j, y_k). For example, in the situation discussed above, exp and sin are no longer necessarily evaluated at the same argument, but the kind and number of such intrinsic functions remain exactly unchanged. Since by the chain rule

$$J(x) = \check{J}(I_i x)\, \tilde{I}_i \quad \text{with} \quad \check{J} = (\check{f})',$$

we see that the i-th column of J has indeed been split.

More generally, we will call x_i separable and f partially argument-separable, if there exists a subset of dependent variables $\{y_k\}_{k \in \mathcal{K}}$ such that $\partial y_k / \partial x_i \not\equiv 0$ for all $k \in \mathcal{K}$ but each intermediate that depends on x_i impacts only a proper subset of the $\{y_k\}_{k \in \mathcal{K}}$. For each of these subsets we may then make a copy of x_i and redefine the dependencies accordingly. The process of column splitting can be continued until none of the independents is separable any more. The resulting decomposition has the form

$$f(x) = \check{f}(Ax) \quad \text{with} \quad \check{f} : \mathbb{R}^{\check{n}} \to \mathbb{R}^n \quad \text{and} \quad \text{OPS}\{\check{f}\} = \text{OPS}\{f\} \quad ,$$

where all rows of the matrix $A \in \mathbb{R}^{\check{n} \times n}$ are Cartesian basis vectors.

Comparing the relation $J = \hat{J} A$ with (1.1), we see that column splitting is formally the opposite process of the Jacobian compression used in the CPR approach. As exact transposition of (2.2) we find

$$\rho(\check{J}^T) \leq \rho(J^T) \quad , \quad \chi(\check{J}^T) \leq \chi(J^T) \quad , \quad \rho(\check{J}) \geq \rho(J) \quad , \tag{2.4}$$

but there is no general relationship between $\chi(\check{J})$ and $\chi(J)$. Hence one might generally expect that column splitting improves the situation primarily for the reverse mode. However, as discussed in the Subsection 2.4 the splitting of independents may also be used to implement CPR with so-called multicoloring in the forward mode. But first, let us conclude this section by looking at an ideal situation for column splitting with regard to the reverse mode.

Consider a vector function f where each component f_k is evaluated by a separate computation so that the ratio

$$\check{\gamma}\{f\} = \left[\sum_{k=1}^{m} \text{OPS}\{f_k\} \right] \Big/ \text{OPS}\{f\} \tag{2.5}$$

is exactly equal to one. This situation occurs frequently in ordinary differential equations when the right hand side is given as a set of algebraic expressions. Suppose the k-th of these right hand sides depends nontrivially on $n_k \leq n$ of the variables. Then we can define $\check{f} : \mathbb{R}^{\check{n}} \to \mathbb{R}^m$ with

$$\check{n} = \sum_{i=1}^{n} n_i$$

by making for all component functions f_k one replica of each x_i that they depend on. The resulting extended Jacobian has the form

$$\check{J} = \begin{bmatrix} \underbrace{\times \cdots \cdot \times}_{n_1} & \underbrace{0 \cdots 0}_{n_2} & \cdots\cdots\cdots & \underbrace{0 \cdots\cdot 0}_{n_{m-1}} & \underbrace{0 \cdot 0}_{n_m} \\ 0 \cdots\cdot 0 & \times \cdots \times & \cdots\cdots\cdots & 0 \cdots\cdot 0 & 0 \cdot 0 \\ \cdots\cdots & \cdots\cdots & \cdots\cdots\cdots & \cdots\cdots & \cdots \\ 0 \cdots\cdot 0 & 0 \cdots 0 & \cdots\cdots\cdots & \times \cdots\cdot\times & 0 \cdot 0 \\ 0 \cdots\cdot 0 & 0 \cdots 0 & \cdots\cdots\cdots & 0 \cdots\cdot 0 & \times \cdot \times \end{bmatrix} \quad . \tag{2.6}$$

The nonzeros in $\check{J}$ are exactly the same as the nonvanishing entries of J, but now only one nontrivial entry occurs in each column, so that

$$\rho(\check{J}^T) = 1 = \chi(\check{J}^T) \quad .$$

Consequently, one reverse sweep with $q = 1$ yields $\check{J}$ and thus its row contraction, J. In other words, we consider each component function separately and evaluate its gradient in the scalar reverse mode. Obviously, this ideal example has an extreme degree of argument separability. Nevertheless, one can expect that significant savings are still possible if many arguments can be split with respect to most of the functions. For example, it can be seen that, if f_i and f_j share common intermediates only if $|i-j|$ is less than some width b, then $\check{f}$ can be defined such that $\rho(\check{J}^T) \leq b$.

2.3 Decomposition by Tearing of Rows and Columns

One may ask whether the complete splitting described above cannot be applied even when the ratio $\check{\gamma}\{f\}$ defined in (2.5) is greater than one. This will be the case if and only if certain intermediate values are shared in the joint computation of f and must therefore be calculated repeatedly if its components f_k are evaluated separately. To indicate that this process involves some losses in efficiency, we will refer to it as tearing of columns or rows. To emphasize the contrast, we will sometimes refer to the row and column splittings discussed in Subsections 2.1 and 2.2 as *exact*. If each column is torn into as many copies as it contains nonzeros, we obtain a function $\check{F}$ with $\mathrm{OPS}\{\check{F}\}/\mathrm{OPS}\{f\} = \check{\gamma}\{f\}$, as defined in (2.5), and a Jacobian of the structure (2.6). This complete column tearing of f amounts to considering each of its component as a completely separate function, and we have $J = \check{F}' A$ with $A^T = (I, \ldots, I) \in \mathbb{R}^{n \times n \cdot n}$. Then the reverse mode yields $\check{F}'$ with the operations count

$$
\begin{aligned}
\mathrm{OPS}\{I\check{F}'\} &\leq (r+4)\,\mathrm{OPS}\{\check{F}\} \\
&= (r+4)\,\check{\gamma}\,\{f\}\mathrm{OPS}\{f\} \\
&\leq (r+4)\,\rho\{\check{J}^T\}\,\mathrm{OPS}\{f\} \quad ,
\end{aligned}
$$

where the last inequality follows from the relation $\check{\gamma}\{f\} \leq \rho\{\check{J}^T\}$, which will be established in Lemma 3 of Section 3.3. By comparison with (1.4) for $q \geq \rho\{J^T\}$ we see that complete column tearing can theoretically not be worse and is likely to have a lower operations count than the reverse NR approach applied directly to $\check{J}$. Moreover, there may also be a benefit for the CPR approach in the forward mode.

In contrast to exact argument splitting, complete column tearing keeps the number of nonzeros in all rows constant. Moreover, the column-intersection graph becomes the union of m disconnected cliques, which contain at most $\rho(J) = \max\{n_k\}$ elements. Hence we have

$$
\chi(\check{F}') = \rho(\check{F}') = \rho(J) \leq \chi(J) \quad ,
$$

so that in the forward CPR approach

$$
\begin{aligned}
\mathrm{OPS}\{\check{F}'I\} &\leq [1 + 3\chi(\check{F}')]\mathrm{OPS}\{\check{F}\} \\
&= \check{\gamma}\{f\}\,[1 + 3\rho(J)]\mathrm{OPS}\{f\} \\
&\leq [\rho(J^T) + 3\,\rho(J^T)\rho(J)]\mathrm{OPS}\{f\} \quad .
\end{aligned}
$$

If the gap between $\chi(J)$ and $\rho(J)$ is sufficiently large, this approach may be more efficient than straightforward CRP. However, in practice it may be hard to extract from the evaluation program for f exactly those calculations needed to obtain one particular dependent value y_k without carrying out any unnecessary calculations. As we will see, essentially the same operations count can be obtained by a dynamically sparse version of the reverse mode.

The "transposed" concept of *row tearing* is somewhat less intuitive, and the resulting vector function depends in particular on the current argument. Suppose we define for each x_i the univariate function

$$f_x^{(i)} \equiv f(\bar{x}_1, \ldots, \bar{x}_{i-1}, x_i, \bar{x}_{i+1} \ldots, \bar{x}_n) \quad , \tag{2.7}$$

where $\bar{x}_j$ represents x_j held constant at its current value. Let $m_i \leq m$ be the number of component functions that depend nontrivially on x_i in that $\partial f_k/\partial x_i \not\equiv 0$. Then we may omit the $m - m_i$ constant components so that $f_x^{(i)} : \mathbb{R} \to \mathbb{R}^{m_i}$. For each i we may theoretically evaluate and store all intermediates occurring in the evaluation of f that do not depend on x_i. These operations need not be counted in the effort for subsequently evaluating $f_x^{(i)}$, so we obtain a generally lower complexity

$$\mathrm{OPS}\{f_x^{(i)}\} \leq \mathrm{OPS}\{f\} \quad .$$

This is exactly the complexity that would arise during differencing if $f(x + \varepsilon e_i)$ could be evaluated without redoing the parts of the evaluation at the base points x that are unaffected by the increment ε in x_i. If each f_k is fully separable (i.e., the sum of univariate scalar functions), then the ratio

$$\hat{\gamma}\{f\} = \left[\sum_{i=1}^{n} \mathrm{OPS}\{f_x^{(i)}\}\right] \bigg/ \mathrm{OPS}\{f\} \leq n \tag{2.8}$$

is equal to one. In the presence of joint intermediates, $\hat{\gamma}\{f\}$ may be any number between 1 and n. In any case we can combine the $f^{(i)}$ to the function

$$\hat{F} \equiv [(f^{(1)})^T, (f^{(2)})^T, \ldots, (f^{(n)})^T]^T \ : \ \mathbb{R}^n \to \mathbb{R}^{\hat{m}} \quad ,$$

where $\hat{m} = \sum_i m_i$. The associated Jacobian $\hat{F}'$ has the transposed structure of (2.6), so that now

$$\rho(\hat{F}') = 1 = \chi(\hat{J}) \quad .$$

Hence the forward mode yields J with the complexity

$$\begin{aligned} \mathrm{OPS}\{\hat{F}'e\} &\leq 5\,\mathrm{OPS}\{\hat{F}\} \\ &= 5\,\hat{\gamma}\{f\}\,\mathrm{OPS}\{f\} \\ &\leq 5\,\rho\{\hat{J}\}\,\mathrm{OPS}\{f\} \quad , \end{aligned}$$

where the last inequality follows from the relation $\hat{\gamma}\{f\} \leq \rho(\hat{J})$, which will be established in Section 3.3.

The superscripts were chosen such that going from f to $\hat{f}$ by row splitting **and** on to $\hat{F}$ by tearing is generally beneficial for the forward mode, whereas successively increasing the number of columns by going from f to $\check{f}$ and $\check{F}$ is generally beneficial for the reverse mode. The superscripts of the corresponding ratios $\hat{\gamma}$ and $\check{\gamma}$ can also be memorized as representing the average size of predecessor sets and successor sets in the computational graph, respectively. This relation is established in Subsection 3.3. In view of the Jacobian structure (2.6), we may also refer to $\check{f}$ and $\check{F}$ as *horizontal* expansions of f. Analogously $\hat{f}$ and $\hat{F}$ may be called *vertical* expansions of f. Conversely we may refer to f as horizontal and vertical contraction of $\check{f}$ and $\check{F}$ or $\hat{f}$ and $\hat{F}$, respectively.

2.4   Relation to Multicoloring

It has often been observed [20] that for a partitioned vector function

$$f(x) = \begin{bmatrix} f^{(1)}(x) \\ f^{(2)}(x) \end{bmatrix} \quad : \quad \mathbb{R}^n \to \mathbb{R}^{m_1 + m_2}$$

the partial Jacobians $J^{(i)} = \left(f^{(i)}\right)'$ for $i = 1, 2$ may satisfy

$$\chi(J) > \chi(J^{(1)}) + \chi(J^{(2)}) \quad .$$

Then the CPR approach should be applied to obtain $J^{(1)}$ and $J^{(2)}$ separately, which is more economical even if

$$\mathrm{OPS}\{f\} \approx \mathrm{OPS}\{f^{(i)}\} \quad \text{for} \quad i = 1, 2$$

because of the presence of many common intermediates in evaluating $f^{(1)}$ and $f^{(2)}$. Formally, this can be interpreted as tearing all columns of J into two copies so that

$$\check{F}\begin{pmatrix} x^{(1)} \\ x^{(2)} \end{pmatrix} = \begin{bmatrix} f^{(1)}(x^{(1)}) \\ f^{(2)}(x^{(2)}) \end{bmatrix} \quad : \quad \mathbb{R}^{2n} \to \mathbb{R}^m$$

with $x^{(1)}$ and $x^{(2)}$ duplicates of x. Hence we have the representation

$$f(x) = \check{F}(Ax) \quad \text{with} \quad A^T = [I, I]^T \in \mathbb{R}^{n \times 2n} \quad .$$

The extended Jacobian takes the form

$$\check{F}' = \begin{bmatrix} J^{(1)} & 0 \\ 0 & J^{(2)} \end{bmatrix} \quad ,$$

an expansion that has also been considered in [20]. Since the column-intersection graph for $\check{J}$ is the disconnected union of the two graphs associated with $J^{(1)}$ and $J^{(2)}$, its chromatic number is given by

$$\chi(\check{J}) = \max\{\chi(J^{(1)}), \chi(J^{(2)})\} \quad .$$

Hence the CPR approach yields $\check{J}$ and thus J at a total cost not exceeding

$$\chi(J^{(1)})\,\mathrm{OPS}\{f^{(1)}\} + \chi(J^{(2)})\,\mathrm{OPS}\{f^{(2)}\} \leq \max\{\chi(J^{(1)}), \chi(J^{(2)})\}[\mathrm{OPS}\{f^{(1)}\} + \mathrm{OPS}\{f^{(2)}\}]$$

In general, one has to split f into more than two subvectors in order to get the individual $\chi(J^{(i)})$ down to a small number. It then becomes important to ask how the sum of the complexities $\mathrm{OPS}\{f^{(i)}\}$ grows with the number of parts. If there are none or few common intermediates, it will be essentially equal to $\mathrm{OPS}\{f\}$; but if most intermediates are shared by all dependents, the ratio (2.5) can grow proportionally to the number of partitions.

For the NR approach in the forward mode, row partitioning makes no sense since the maximal row length satisfies

$$\rho(J) \;=\; \max\{\rho(J^{(1)}), \rho(J^{(2)})\}$$

and therefore cannot be reduced at all.

2.5   A Binary Interaction Example

Suppose we have an unconstrained problem with an objective function of the form

$$f(x) \;\equiv\; \sum_{1\leq i<j\leq n} f_{ij}(x_i, x_j),$$

where all element functions $f_{ij} : \mathbb{R}^2 \to \mathbb{R}$ are nonseparable. For example, the variables x_i could represent the coordinates of atoms that are aligned along one coordinate axis. The energy of such an arrangement is often modeled as the sum of $n(n+1)/2$ pairwise interactions f_{ij}. This situation has alos been examined in some detail as the *exponential* example in [20].

Now suppose we wish to compute the gradient of f, given a computer program for its evaluation. To illustrate the crucial role of common intermediates, we write each f_{ij} in the form

$$f_{ij}(x_i, x_j) = \tilde{f}_{ij}(prep_i(x_i), prep_j(x_j)) \quad ,$$

where $\tilde{f}_{ij} : \mathbb{R}^2 \to \mathbb{R}$ like f_{ij}, and the n univariate functions $prep_i$ somehow prepare the variables x_i for their involvement in the f_{ij}. Since $m = 1 = \rho(J^T)$, the reverse mode yields the gradient of f at no more than $r + 4$ times the operations count of f itself. However, suppose we wish to evaluate the gradient and Hessian in the forward mode, possibly to avoid the increase in memory requirement and interpretive overhead, or simply because no suitable software for the reverse mode is available.

Using the obvious partial separability, we may rewrite

$$f(x) \;=\; e^T \hat{f}(x) \quad \text{with} \quad \hat{f}(x) \;=\; (f_{ij}(x))_{1\leq i<j\leq n},$$

where $e^T = A$ is the vector of $n(n+1)/2$ ones and the ordering of the components in $\hat{f}$ does not matter. Because any two columns of $\hat{J}$ share nonzeros in (exactly) one row, the column intersection graph is a clique and therefore only allows the trivial coloring with $\chi(\hat{J}) = n$. Hence the CPR technique for estimating the gradient of f via the Jacobian of $\hat{f}$ would require n full evaluations of $\hat{f}$.

To avoid this unacceptable complexity, we consider a complete column tearing, where each contribution f_{ij} is considered as an independent function from all the others. Then the work ratio defined in (2.5) is given by

$$\check{\gamma}\{\hat{f}\} \;=\; \frac{\sum_{1 \leq i < j \leq n}\left[\mathrm{OPS}\{\tilde{f}_{ij}\} + \mathrm{OPS}\{prep_i\} + \mathrm{OPS}\{prep_j\}\right]}{\sum_{1 \leq i \leq n} \mathrm{OPS}\{prep_i\} + \sum_{1 \leq i < j \leq n} \mathrm{OPS}\{\tilde{f}_{ij}\}}$$

$$\;=\; 1 + \frac{(n-2)\sum_{1 \leq i \leq n} \mathrm{OPS}\{prep_i\}}{\sum_{1 \leq i \leq n} \mathrm{OPS}\{prep_i\} + \sum_{1 \leq i < j \leq n} \mathrm{OPS}\{\tilde{f}_{ij}\}} \;.$$

If the preparatory functions are mere identities $pred_i(x_i) = x_i$, then $\hat{f}$ is fully argument separable, and we have $\check{\gamma}\{\hat{f}\} = 1$. The complete tearing is still advantageous if the preparatory functions $prep_i$ are cheap compared to the actual binary interactions $\tilde{f}_{ij}$. On the other hand, if these univariate transformations are expensive, then one may get an n-fold increase in complexity, which amounts to the same as differencing the gradient of each f_{ij} separately.

One can also apply a complete row tearing to $\hat{f}$. Since each independent impacts $(n-2)$ components of $\hat{f}$ the ratio $\hat{\gamma}$ defined in (2.8) is given by

$$\hat{\gamma}\{\hat{f}\} \;=\; \frac{\sum_{1 \leq i \leq n}\left[\mathrm{OPS}\{prep_i\} + \sum_{j \neq i} \mathrm{OPS}\{\tilde{f}_{ij}\}\right]}{\sum_{1 \leq i \leq n} \mathrm{OPS}\{prep_i\} + \sum_{1 \leq i < j \leq n} \mathrm{OPS}\{\tilde{f}_{ij}\}}$$

$$\;=\; 1 + \frac{\sum_{1 \leq i < j \leq n} \mathrm{OPS}\{\tilde{f}_{ij}\}}{\sum_{1 \leq i \leq n} \mathrm{OPS}\{prep_i\} + \sum_{1 \leq i < j \leq n} \mathrm{OPS}\{\tilde{f}_{ij}\}} \;.$$

The row tearing is efficient if the preparatory functions $prep_i$ dominate the computational cost, in which case the column tearing is not so advantageous. If these costs are significant but not dominant both tearings may result in an unacceptable complexity for the CPR approach in the forward mode. In this particular example one may then utilize a mixture between the two in the following way.

First one can order the components of $\hat{f}$ such that the top quarter corresponds to element functions f_{ij} with $1 \leq i < j \leq n/2$ and the second quarter to those with $n/2 < i < j \leq n$. Then the rows in the remaining bottom half correspond to element functions f_{ij} with $1 \leq i \leq n/2 < j \leq n$. The chromatic number of this $(n^2/2) \times n$ or $(n^2-1)/2 \times n$ matrix is 2, since the first $n/2$ and the last $n - n/2$ columns form groups that are pairwise structurally orthogonal. Hence we can estimate the bottom part using two function evaluations or, equivalently, Jacobian-vector products. The two top quarters represent copies of the original problem with the number of independents cut in half. Hence they can be decomposed recursively in the same way; and since the corresponding columns are structurally orthogonal, each evaluation for the top quarter can be combined with an evaluation for the second quarter. In this way the whole Jacobian can be computed by using only $2\log_2 n$ Jacobian-vector evaluations.

Using the method of Newsam and Ramsdell [18], one can do even better. Let s be any n-vector whose components σ_i are all different from each other. Then the two Jacobian-vector products

$$\frac{\partial \hat{f}(x + \alpha e + \beta s)}{\partial(\alpha, \beta)}\bigg|_{\alpha=0=\beta} = \hat{J}(x)[e, s] \in \mathbb{R}^{\hat{m} \times 2}$$

completely determine $\hat{J}(x)$ since the nonzeros in the row corresponding to f_{ij} form a small vector-matrix product of the form

$$\frac{\partial f_{ij}(x + \alpha e + \beta s)}{\partial(\alpha, \beta)}\bigg|_{\alpha=0=\beta} = \left[\frac{\partial f_{ij}}{\partial x_i}, \frac{\partial f_{ij}}{\partial x_j}\right] \begin{pmatrix} 1 & \sigma_k \\ 1 & \sigma_l \end{pmatrix}$$

for some index pair $k \neq l$. Since by assumption $\sigma_k \neq \sigma_l$, each 2×2 matrix is invertible, and the two nonzero components of ∇f_{ij} can thus be easily recovered. While uniformly spaced σ_k would be acceptable in this case, more sophisticated choices such as roots of unity may be required to make the linear systems reasonably well conditioned when the maximal row length is larger.

Differentiating once more, one obtains with $x(\alpha, \beta) \equiv x + \alpha e + \beta s$

$$\left[\begin{array}{cc} \frac{\partial^2}{\partial \alpha^2} f_{ij}(x(\alpha, \beta)) & \frac{\partial^2}{\partial \alpha \partial \beta} f_{ij}(x(\alpha, \beta)) \\ \frac{\partial^2}{\partial \beta \partial \alpha} f_{ij}(x(\alpha, \beta)) & \frac{\partial^2}{\partial \beta^2} f_{ij}(x(\alpha, \beta)) \end{array}\right]_0 = \begin{pmatrix} 1 & 1 \\ \sigma_k & \sigma_l \end{pmatrix} \left[\begin{array}{cc} \frac{\partial^2 f_{ij}}{\partial x_i^2} & \frac{\partial^2 f_{ij}}{\partial x_i \partial x_j} \\ \frac{\partial^2 f_{ij}}{\partial x_j \partial x_i} & \frac{\partial^2 f_{ij}}{\partial x_j^2} \end{array}\right] \begin{pmatrix} 1 & \sigma_k \\ 1 & \sigma_l \end{pmatrix}.$$

Thus we see that all Hessians $\nabla^2 f_{ij}$ can be computed from the $n \times 2 \times 2$ derivative tensor of $f(x + \alpha e + \beta s)$ with respect to α and β.

It is often optimistically assumed that for most square sparsity patterns, not only the difficult to compute chromatic number itself but also a heuristically computed coloring number lies within a factor of two of the maximal row length. It seems unlikely that this property holds for the vertical expansions of partially separable functions, which may have very long columns and thus highly connected intersection graphs. Therefore we presume that the method of Newsam and Ramsdell deserves further investigation in this context. Also, one might hope that sometimes the column intersection graph of the transposed $\hat{J}$ is less connected, even though it is likely to have more nodes. However, on the scalar example above, each column has exactly $n - 1$ nonzero entries so that the reverse mode is not cheaper if one wishes to calculate the whole Jacobian J. Nevertheless, it is very efficient for calculating the accumulated gradient

$$\nabla f(x) = e^T \hat{J}(x) \, ,$$

or the gradient of a Lagrangian, where the vector of ones e would be replaced by a vector of Lagrange multipliers.

3   The Evaluation Program and Its Complexity

In this third section we will describe the computational graph for a given vector function f and develop several complexity estimates at the elemental level. We will also define the following key characteristics of f and its Jacobian J.

$\mathrm{RAM}\{f\}$	Bound on the the maximal number of live variables.
$\mathrm{OPS}\{f\}$	Sum of all elemental operations counts.
$\chi(\hat{J})$	Chromatic number of the incidence graph of the generic Lagrangian.
$\rho(J)$	Maximal number of independents that impact any dependent variable.
$\rho(\hat{J})$	Maximal number of independents that impact any intermediate variable.
$\hat{\gamma}\{f\}$	Average number of independents on which an intermediate depends.
$\rho(J^T)$	Maximal number of dependent variables impacted by any independent.
$\rho(\check{J}^T)$	Maximal number of dependent variables impacted by any intermediate.
$\check{\gamma}\{f\}$	Average number of dependents impacted by an intermediate variable.

In computational practice, a vector function $f : \mathbb{R}^n \to \mathbb{R}^m$ is evaluated by a sequence of scalar assignments

$$v_j = \varphi_i(v_i)_{i \to j} \quad , \tag{3.1}$$

where the variables v_j may be real or complex. The relation $i \to j$ between integer indices means that v_j *depends directly* on v_i because the latter is an argument of the *elementary function* $\varphi_j \in \mathcal{C}^2$. We will assume that the partial ordering induced by this dependency relation is compatible with the natural ordering of the integers, so that

$$j \to i \Longrightarrow j < i \quad . \tag{3.2}$$

Obviously there may be several such *topological* orderings of the *computational graph* with the vertices v_j. The edges are defined by the dependence relation $\to$, and the vertices can be labeled with the elementary function φ_j. Without loss of generality we may require that the first n variables v_i represent the *independent* variables and the last m represent the *dependent variables*. Then the function $y = f(x)$ can evaluated by executing the elementary assignments (3.1) in one big loop with j ranging from 1 to $o + m$. Here the nonnegative integer o represents the number of intermediate variables, which we expect to be much larger than both n and m for seriously nonlinear problems. Thus we can combine the variables v_i into the three vectors

$$
\begin{aligned}
x &\equiv (v_{1-n}, \ldots \quad \ldots, v_{-1}, v_0) \quad (\textit{independent}) \\
z &\equiv (v_1, v_2, \ldots \quad \ldots, v_o) \quad (\textit{intermediate}) \\
y &\equiv (v_{o+1}, \ldots \quad \ldots, v_{o+m}) \quad (\textit{dependent})
\end{aligned}
\quad ,
$$

so that $x_i \equiv v_{i-n}$ and $y_k \equiv v_{o+k}$. The independent and dependent variables represent the roots and leaves of the computational graph. To make partial function-separability as defined in Subsection 2.1 a special case of partial separability in the usual sense, we have to impose a *final summation* condition, which can always be achieved by minor modifications of the graph. We state this requirement formally.

Final Summation Condition

Elementary function of the additive form

$$
v_j = \varphi_j(v_i)_{i \to j} = \sum_{i \to j} v_i \tag{3.3}
$$

must occur for all $i > o$ but cannot occur for any j with $v_j \to y_k$ for some $k \le m$. The cost of the final summations is considered negligible.

In other words, each final elementary function must be a (possibly unary) addition, but none of the immediate predecessors may be obtained as a sum. When a final elementary function φ_{o+k} is multivariate and nonadditive, we may simply relabel v_{o+k} as an intermediate and use a unary summation to make copy that serves as the k-th dependent variable. On the other hand, immediate predecessors of dependents that are themselves obtained as sums can be eliminated by merging the two summations. The "transposed" condition on the independent variables is that no identical copies are made so that none of the φ_j with $x_i \to v_j$ for some x_i may be the identity function.

3.1   Counting Contemporaries, Ancestors, and Descendants

The v_j considered here are mathematical variables rather than memory locations on the computer. Since the storage requirement would otherwise be at least o, we must allow that v_j overwrites some v_i, provided it is certain that

$$
i \to k \implies k < j \quad , \tag{3.4}
$$

which means that v_i can no longer occur as an argument once v_j has been computed. In the remainder we will denote by RAM$\{f\}$ an upper bound on the number of *live* variables v_j that must be in storage at any one time during the evaluation process. For Fortran programs, a suitable bound can be determined at compile time since there is no dynamic storage allocation.

We will use the notation

$$i \rightsquigarrow j \quad \text{or} \quad v_i \rightsquigarrow v_j$$

to indicate that v_j depends on v_i either directly in that $i \rightarrow j$ or indirectly through intermediates. In graph theoretical terms $i \rightsquigarrow j$ means that a directed path connects the vertices v_i and v_j. For independent and dependent variables we will often write

$$x_i \rightsquigarrow v_j \quad \text{and} \quad v_j \rightsquigarrow y_k$$

in lieu of $i \rightsquigarrow (j - n)$ or $j \rightsquigarrow (o + k)$, respectively. For all v_j we may define the sets

$$X(v_j) \equiv \{x_i : x_i \rightsquigarrow v_j\} \quad \text{and} \quad Y(v_j) \equiv \{y_k : v_j \rightsquigarrow y_k\} \tag{3.5}$$

of independent ancestors and dependent descendants, respectively. By subsuming constants into the definition of elementary functions and eliminating unnecessary calculations, one can ensure that the sets $X(v_j)$ and $Y(v_j)$ are nonempty for all intermediates. In other words, all intermediates lie on a path from an independent to a dependent variable, so that for all $1 \leq j \leq o$

$$x_i \rightsquigarrow v_j \rightsquigarrow y_k \quad \text{for some} \quad x_i \quad \text{and} \quad y_k \quad . \tag{3.6}$$

The set $X(y_k)$ contains exactly the independent variables on which y_k may be nontrivially dependent. Similarly, $Y(x_i)$ contains exactly those dependent variables that are nontrivially dependent on x_i. Denoting by $|\mathcal{S}|$ the cardinality of a set $\mathcal{S}$ one finds that the maximal row and column lengths of the Jacobian J are given by

$$\rho\{J\} = \max_{k \leq m} |X(y_k)| \quad \text{and} \quad \rho\{J^T\} = \max_{i \leq n} |Y(x_i)| \quad . \tag{3.7}$$

The concepts of function- and argument-separability used in Subsections 2.1 and 2.2 can now be reintroduced rigorously as follows.

3.2 Function and Argument Splitting on the Graph

Function separability

A dependent variable y_k is called separable if for all $1 \leq j \leq o$

$$v_j \rightsquigarrow y_k \quad \Rightarrow \quad x_i \not\rightsquigarrow v_j \quad \text{for} \quad \text{some} \quad x_i \rightsquigarrow y_k \quad ,$$

in which case f is called partially function-separable.

This condition is equivalent to the property

$$v_j \rightarrow y_k \quad \Rightarrow \quad X(v_j) \subset X(y_k) \quad , \tag{3.8}$$

where the symbol $\subset$ excludes equality. Now let X_{kl} for $l = 1, \ldots, \bar{l}$ be a numbering of all $\bar{l}$ distinct subsets of the form $X(v_j)$ with $v_j \rightarrow y_k$. Then we can split y_k into $\bar{l}$ copies y_{kl} defined by

$$y_{kl} = \sum_{X_{kl}=X(v_j)} v_j \ . \tag{3.9}$$

After renumbering some vertices and updating the dependencies accordingly, one has now obtained a computational graph for a function $\hat{f} : \mathbb{R}^n \rightarrow \mathbb{R}^{\hat{m}}$ with $\hat{m} = m + \bar{l} - 1$ as originally introduced in Section 2.1. From now on we will assume that $\hat{f}$ is maximal, that is, that it has been obtained from f by performing all possible function splittings.

Argument separability

An independent variable x_i is called separable if for all $1 \leq j \leq o$

$$x_i \rightsquigarrow v_j \quad \Rightarrow \quad v_j \not\rightsquigarrow y_k \quad \text{for} \quad \text{some} \quad x_i \rightsquigarrow y_k \ ,$$

in which case f is called partially argument-separable.

This condition is equivalent to the property

$$x_i \rightarrow v_j \quad \Rightarrow \quad Y(v_j) \subset Y(x_i) \ , \tag{3.10}$$

where the symbol $\subset$ still excludes equality. In exact analogy to the function-separable case discussed above we may now number $\bar{l}$ sets $Y_{il} = Y(v_j)$ for some $v_j \leftarrow x_i$ and replace the original assignment $v_j = \phi_j(\ldots, x_{i-1}, x_i, x_{i+1} \ldots)$ by

$$v_j \equiv \phi_j(\ldots, x_{i-1}, x_{kl}, x_{i+1}, \ldots) \quad \text{if} \quad Y_{il} = Y(v_j) \ . \tag{3.11}$$

Whenever all $x_{il} = x_i$ the values of the direct successors $v_j \leftarrow x_i$ as well as all later intermediates are obviously the same. Therefore, the horizontal expansion $\check{f} : \mathbb{R}^{\check{n}} \rightarrow \mathbb{R}^m$ defined by the new graph with $\check{n} = n + \bar{l} - 1$ has the properties described in Section 2.2. From now on we will assume that $\check{f}$ is maximal, that is, that it has been obtained from f by performing all possible argument splittings.

Now we can characterize the row and column lengths of the Jacobians $\hat{f}$ and $\check{f}$ directly in terms of the computational graph.

Lemma 2

Under the final summation condition we have for the maximal expansions $\hat{f}$ and $\check{f}$

$$\rho(\hat{J}) \quad = \quad \max_{1 \leq j \leq o} |X(v_j)| \ \leq \ \rho(J) \ ,$$

$$\rho\{\check{J}^T\} \quad = \quad \max_{1 \leq j \leq o} |Y(v_j)| \ \leq \ \rho(J^T) \ ,$$

where the $X(v_j)$ and $Y(v_j)$ are defined in (3.5).

Proof. Since we have assumed that all intermediates impact at least one dependent of f, the same is true for $\hat{f}$. Excluding the possibility of accidental cancellation, we must therefore have $\rho\{\hat{J}\} \geq |X(v_j|$ for all $1 \leq j \leq o$. Now suppose that the gradient $\nabla \hat{y}_k$ of some component $\hat{y}_k$ of $\hat{f}$ has more nonzeros than any of the preceding intermediates v_j. Then that $\hat{y}_k$ is in fact separable, which contradicts the definition of $\hat{f}$. Thus we must have equality as asserted, and the bound by $\rho(J)$ is an immediate consequence of (3.7). The second assertion follows analogously. ∎

Finally, we note that the row- and column splitting processes reinforce rather than obstruct each other. More specifically, when some independent x_i is split, all separable dependents y_k maintain that property even if $X(y_k)$ contains x_i and may therefore be enlarged. Similarly, row splittings cannot reduce the number of separable arguments, which can be used for subsequent column splittings. Hence, there must be a function

$$f^\circ \ : \ \mathbb{R}^{\check{n}} \to \mathbb{R}^{\hat{m}}$$

such that

$$f(x) \ = \ Bf^\circ(Ax) \quad \text{and} \quad \text{OPS}\{f^\circ\} \ = \ \text{OPS}\{f\} \quad ,$$

where $A \in \mathbb{R}^n \to \mathbb{R}^{\check{n}}$ and $B \in \mathbb{R}^m \to \mathbb{R}^{\hat{m}}$ as before. We can identify the previously discussed vertical and horizontal expansions of f as

$$\hat{f}(x) \ = \ f^\circ(Ax) \quad \text{and} \quad \check{f}(\check{x}) \ = \ Bf^\circ(\check{x}) \quad ,$$

where $\check{x} \in \mathbb{R}^{\check{n}}$ is the replicated variable vector. Obviously the only difference between the computational graphs for $\hat{f}, \check{f}, f^\circ$, and the original f is in the leaves, roots and the way the first layers of intermediates v_j (with $x_i \to v_j$ for some i) are defined as elementary functions of the intermediates. Without ambiguity we may therefore denote the dependents of $\hat{f}$ and f° by $\hat{y} = (\hat{y}_k)_{1 \leq k \leq \hat{m}}$ and the independents of $\check{f}$ and f° by $\check{x} = (\check{x}_i)_{1 \leq i \leq \check{n}}$, respectively. The relation between these vectors and the original independents and dependents is simply $y = B\hat{y}$ and $\check{x} = Ax$. The definition of the first layer will be clear from the context.

Generally speaking, we can expect that the Jacobian of f° is much larger and sparser than that of the original f. Because of (2.2) and (2.4) it is clear that the NR approach in the forward and reverse modes is best applied to the $\hat{f}$ and $\check{f}$, respectively. This is no longer true when the factors A and B are allowed to be general linear transformations. Such further generalization is useful in cases where the evaluation of f involves linear functionals like the average $e^T x$ of the independent or dependent variables.

3.3   Elemental Complexity Assumptions

Our main restriction on the elementary functions φ_i is that the partial derivatives

$$c_{ij} \equiv \begin{cases} \partial\varphi_i/\partial v_j & \text{if} \quad j \to i \\ 0 & \text{if} \quad j \not\to i \end{cases}$$

and

$$c_{ijk} \equiv \begin{cases} \partial^2\varphi_i/\partial v_j \partial v_k & \text{if} \quad j \to i \quad \text{and} \quad k \to i \\ 0 & \text{if} \quad j \not\to i \quad \text{or} \quad k \not\to i \end{cases}$$

are well defined and easily evaluated at all arguments of interest. This is certainly the case when the evaluation program is written in a standard high-level language such as Fortran or C. Then the compiler breaks down the evaluation into a sequence of arithmetic operations and intrinsic function calls.

For some purposes it is advantageous to view more complex computational units as elementary building blocks. This approach has the advantage of reducing the interpretive overhead and facilitates some local preaccumulation of derivatives. For example, in the source translator ADIFOR, right-hand sides of assignments are treated as *elementary functions*, whose gradients are computed by the reverse mode in the form of compilable code. This compile-time differentiation can be easily generalized to function and subroutines, especially if their code is tight in that it does not contain variable dimensions or loop lengths. It has proven very efficient for the evaluation of first derivatives. Unfortunately, the trade-offs are more complicated, if one also wishes to compute second or higher derivatives. However, it is clear that linear or bilinear vector-vector and matrix-vector operations should be treated as *elementary functions*, since their first and higher derivatives are easy to store and manipulate, with many of them vanishing altogether. For notational simplicity we will continue to assume that all elementary functions are scalar valued, but we allow the number of local independents $|\{i : i \to j\}|$ to be arbitrarily large.

Our key assumption is that the cost for computing the first and second derivatives of each φ_j is no more than twice that of computing ϕ_j by itself, so that

$$\text{OPS}\{\varphi_j, \nabla\varphi_j, \nabla^2\varphi_j\} \leq 2\,\text{OPS}\{\varphi_j\} \quad . \tag{3.12}$$

In fact, this bound is quite pessimistic, since for all linear and bilinear operations the derivatives come virtually for free, and for most intrinsic functions the first two derivatives are easily obtained from the function itself. For sinusoidal functions the bound appears to be sharp, but even there sin and cos are often evaluated in pairs anyway, in which case no extra derivative evaluations are required in theory. In practice, such savings could be realized only if the automatic differentiation tool did some compiler-like dependency analysis and optimization.

The temporal complexity measure OPS{} may account not only for arithmetic operations but also for memory accesses. Naturally, we cannot distinguish the access costs to different levels of the memory hierarchy and will assume exact additivity so that (at least on a serial machine)

$$\mathrm{OPS}\{f\} \equiv \sum_{j=1}^{o} \mathrm{OPS}\{\varphi_j\} \quad , \tag{3.13}$$

where we have again assumed that the cost of the final summations φ_{o+k} for $k = 1 \ldots m$ is negligible. Apart from generating the derivatives $\nabla\varphi_j$ and $\nabla^2\varphi_j$, we must also consider the cost of incorporating them into the chain rule. The elementary operations addition and subtraction play a special role, because all first derivatives are 1 or -1 and all second derivatives vanish identically. In these cases no multiplications are required to multiply the local gradients or Hessians by vectors or matrices. In general, we assume that the effort of forming an inner product of the gradient $\nabla\varphi_j$ with a compatible vector, or multiplying the Hessian $\nabla^2\varphi_j$ from the left and right by two vectors, or incrementing a given vector by a multiple of $\nabla\varphi_j$, is bounded according to

$$\max\left\{\mathrm{OPS}\{(\nabla\varphi_j)^T w\}, \mathrm{OPS}\{u^T \nabla^2\varphi_j w\}, \mathrm{OPS}\{+\omega\nabla\varphi_j\}\right\} \leq 3\,\mathrm{OPS}\{\varphi_j\} \quad , \tag{3.14}$$

where the $+$ sign indicates that adding the result to a give vector is considered an integral part of the calculation. If a multiplication is no more expensive than an addition, the bound is sharp for the multiplication operator $v_j = \varphi_j(v_1, v_2) = v_1 \cdot v_2$, where $(\nabla\varphi_j)^T w = v_2 w_1 + v_1 w_2$ and $u^T \nabla^2 \phi w = u_1 w_2 + u_2 w_1$.

Let us finally perform individual operations counts for the component functions f_k and the univariate functions $f_x^{(i)}$ defined in Subsection 2.3. After discounting all elementary functions φ_j that have no impact on a given f_k, we obtain the operations count

$$\mathrm{OPS}\{f_k\} = \sum_{j \rightsquigarrow k} \mathrm{OPS}\{\varphi_j\} \quad . \tag{3.15}$$

Similarly, the (re)evaluation of $f_x^{(i)}$ requires only the calculation of the elementary functions that depend on x_i, so that

$$\mathrm{OPS}\{f_x^{(i)}\} = \sum_{i \rightsquigarrow j} \mathrm{OPS}\{\varphi_j\} \quad , \tag{3.16}$$

where the subscript x indicates that the definition of $f_x^{(i)}$ depends on the "current" point x viewed as a constant. Substituting these expressions into the definitions (2.8) and (2.5), we obtain the following result.

Lemma 3 *The complexity ratios defined in (2.5) and (2.8) satisfy*

$$\check{\gamma}\{f\} \leq \rho(\check{J}^T) \quad \text{and} \quad \hat{\gamma}\{f\} \leq \rho(\hat{J}) \quad .$$

Proof. Summing (3.15) and (3.16) over k and i, we obtain with $Y(v_j)$ and $X(v_j)$ as defined in (3.5) by changing the order of summation:

$$\check{\gamma}\{f\}\,\mathrm{OPS}\{f\} \;=\; \sum_{j=1}^{o}\left(\sum_{j\leadsto k}\mathrm{OPS}\{\varphi_j\}\right)$$

$$=\; \sum_{j=1}^{o}|Y(v_j)|\,\mathrm{OPS}\{\varphi_j\} \;\leq\; \rho(\check{J}^T)\,\mathrm{OPS}\{f\}\quad,$$

where the last inequality follows from (3.7). By interchanging rows and columns we find similarly

$$\hat{\gamma}\{f\}\mathrm{OPS}\{f\} \;=\; \sum_{j=1}^{o}\sum_{i\leadsto j}\mathrm{OPS}\{\varphi_j\}$$

$$=\; \sum_{j=1}^{o}|X(v_j)|\,\mathrm{OPS}\{\varphi_j\} \;\leq\; \rho(\hat{J})\,\mathrm{OPS}\{f\}\quad,$$

which completes the proof. ∎

In this thirs section we have shown that there exist constant matrices $B \in \mathbb{R}^{m\times\hat{m}}$ and $A \in \mathbb{R}^{\hat{n}\times n}$ (whose rows and columns are Cartesian basis vectors, respectively) such that

$$f(x) \;=\; B\hat{f}(x) \;=\; \check{f}(Ax) \;=\; Bf^{\circ}(Ax)$$

and consequently

$$J(x) \;=\; B\,\hat{J}(x) \;=\; \check{J}(Ax)\,A \;=\; B\,J^{\circ}(Ax)\,A\quad.$$

Since pre- and post-multiplication by A and B involve only additions, we neglect these costs and assume that

$$\mathrm{OPS}\{f\} = \mathrm{OPS}\{\check{f}\} = \mathrm{OPS}\{\hat{f}\} = \mathrm{OPS}\{f^{\circ}\}$$

and

$$\mathrm{OPS}\{J\} \leq \min\left\{\mathrm{OPS}\{\check{J}\}, \mathrm{OPS}\{\hat{J}\}, \mathrm{OPS}\{J^{\circ}\}\right\}.$$

In other words, we view J as a free by-product of any method for calculating $\hat{J}, \check{J}$, or J°. The same assumption will be made regarding the evaluation of second-derivative matrices or tensors.

4 Results and Discussion

In this final section we formulate rigorous bounds on the complexity of evaluating first and second derivatives of a vector function f in various ways. Similar bounds have been derived repeatedly in the automatic differentiation literature (see, e.g., [16], [17], and [4] as recent references).

4.1 First and Second Derivatives in the Forward Mode

Suppose the independent variables x are considered as linear functions

$$x(d) \equiv x + Sd$$

of the *differentiation parameter* vector $d \in \mathbb{R}^p$. We will refer to $S \in \mathbb{R}^{n \times p}$ as the seed matrix, which may vary from the $n \times n$ identity to a single-direction vector. Then all intermediates v_j have associated gradients and Hessians

$$\nabla_S v_j \in \mathbb{R}^p \quad \text{and} \quad \nabla_S^2 v_j \in \mathbb{R}^{p \times p} \quad .$$

Starting from $\nabla_S x_i = e_i^T S$ and $\nabla_S^2 x_i = 0$, one can propagate these derivatives forward by the chain rules

$$\nabla_S v_j = \sum_{i \to j} c_{ji} \nabla_S v_i \tag{4.1}$$

and

$$\nabla_S^2 v_j = \sum_{i \to j} \left[c_{ji} \nabla_S^2 v_i + \nabla_S v_i \sum_{k \to j} c_{jik} (\nabla_S v_k)^T \right] \quad . \tag{4.2}$$

At the end one obtains the *reduced* gradients

$$\nabla_S \hat{y}_k = \left. \frac{\partial f_k(x + Sd)}{\partial d} \right|_{d=0} = \nabla \hat{f}_k(x) S$$

and the two-sided projected Hessian

$$\nabla_S^2 y_k = \frac{\partial^2 \hat{f}_k(x + Sd)}{\partial d^2} = S^T \nabla^2 \hat{f}_k S \quad .$$

Now we obtain from the elemental complexity assumptions in the preceding section the following result.

Proposition 1 *The forward propagation of first and second derivatives with respect to p differentiation parameters can be achieved with*

$$\begin{aligned}
\mathrm{OPS}\{f, \hat{J} S\} &\leq (2 + 3p) \, \mathrm{OPS}\{f\} \\
\mathrm{OPS}\{f, \hat{J} S, S^T \hat{f}_k'' S \text{ for } k = 1 \ldots \hat{m}\} &\leq [2 + 3p(p + 1)] \, \mathrm{OPS}\{f\}
\end{aligned}$$

operations. The corresponding memory requirement is bounded by

$$\begin{aligned}
\mathrm{RAM}\{f, \hat{J} S\} &\leq (1 + p) \, \mathrm{RAM}\{f\} \\
\mathrm{RAM}\{f, \hat{J} S, S^T \hat{f}_i'' S \text{ for } k = 1 \ldots \hat{m}\} &\leq (1 + p)(2 + p)/2 \, \mathrm{RAM}\{f\} \quad .
\end{aligned}$$

Proof. First let us note that the bounds on the randomly accessed memory reflect the fact that the p vector $\nabla_S \varphi_j$ and possibly also the symmetric $p \times p$ matrix $\nabla_S^2 \varphi_j$ must be kept for each scalar variable v_j. These gradient and Hessian objects can also be overwritten when v_j itself is overwritten by a new value.

Equation (4.1) can be interpreted as the multiplication of the row vector $\nabla \varphi_j$ by a matrix with p columns from the right. Hence the computational effort consists of exactly p inner products between $\nabla \varphi_j$ and a generally dense vector w. Thus we derive from (3.12), (3.14) and (4.1)

$$\text{OPS}\{f, \hat{J} S\} = \sum_{j=1}^{o} \left[\text{OPS}\{\varphi_j, \nabla \varphi_j\} + p\,\text{OPS}\{(\nabla \varphi_j)^T w\} \right] \tag{4.3}$$

$$\leq \sum_{j=1}^{o} (2 + 3p)\,\text{OPS}\{\varphi_j\} = (2 + 3p)\,\text{OPS}\{f\} \quad . \tag{4.4}$$

Similarly we see that (4.2) requires the computation of $p(p+1)/2$ inner products in $\nabla \varphi_j$ and exactly the same number of quadratic forms, $u^T \nabla^2 \varphi_j w$, so that by (4.2) and again (3.14)

$$\text{OPS}\{f, \hat{J} S, S^T \hat{f}_k'' S \ k \leq \hat{m}\}$$
$$= \sum_{j=1}^{o} \left[\text{OPS}\{\varphi_j, \nabla \varphi_j, \nabla^2 \varphi_j\} + \frac{1}{2} p(p+1) \left(\text{OPS}\{(\nabla \varphi_j)^T w\} + \text{OPS}\{u^T \nabla^2 \varphi_j w\} \right) \right]$$
$$\leq \sum_{j=1}^{o} [2 + 3p(p+1)]\,\text{OPS}\{\varphi_j\} = [2 + 3p(p+1)]\,\text{OPS}\{f\} \quad ,$$

which completes the proof. ∎

A key advantage of the forward mode is that no extra sequentiallly accessed storage (SAM) is required and that sweeps of various order can be carried out simultaneously with the function evaluation by compilable code. If J and J' are dense, they can be calculated from one forward sweep with $p = n$ parameters. Alternatively, one can use slicing to obtain the Jacobian f' or the collection of Hessians f'' over several sweeps with S obtained from a partitioning of the identity matrix. For Jacobians the temporal complexity is strictly additive, but for Hessians the operations count may grow by a factor of two as a result of slicing [3]. In the constrained optimization case, one only needs projections of the objective and constraint Hessians to the range space of S anyway.

Even when f is neither function- nor argument-separable and J is dense, it is quite likely that the ratio $\hat{\gamma}$ defined in (2.8) is significantly smaller than n. Then the Jacobian J could theoretically be calculated more efficiently as a contraction of the vertically expanded Jacobian $\hat{F}'$. The difficulty with this approach is that one can, in general, not easily separate the calculations for (re)evaluating the various functions

$f^{(i)}$ by themselves. A similar effect is achieved if one performs the recursion (4.1) with $S = I$ and hence $p = n$, but with $\nabla_S v_j$ and $\nabla_S^2 v_j$ stored and manipulated as sparse vectors and matrices, respectively. Since each $\nabla_S v_j$ has at most $|X(v_j)|$ nonzero entries, we obtain the following corollary.

Corollary 1
If the forward propagation of first and second derivatives is carried out using the sparsity of the gradients $\nabla_S v_j$ and Hessians $\nabla_S^2 v_j$, then the operations count is

$$\begin{aligned}
\mathrm{OPS}\{f, \hat{J}\} &\leq (2 + 3\hat{\gamma}\{f\})\,\mathrm{OPS}\{f\} \\
\mathrm{OPS}\{f, \hat{J}, \hat{f}_k'' \text{ for } k = 1 \ldots \hat{m}\} &\leq [2 + 3\hat{\gamma}\{f\}(\rho(\hat{J}) + 1)]\,\mathrm{OPS}\{f\} \quad,
\end{aligned}$$

and the corresponding RAM requirements are bounded as in Prop. 1 with $p = \rho(\hat{J})$.

Proof. The first inequality follows by definition of $\hat{\gamma}$ from (4.3) with p on the right-hand side replaced by $|X(v_j)|$. To prove the second inequality, we first note that the nonzeros of each Hessian $\nabla_S^2 v_j$ form a nonzero square submatrix of order $|X(v_j)|$, so that on the right-hand side of (4.5) the factor p can also be replaced by $|X(v_j)|$. Hence we have instead

$$\begin{aligned}
\mathrm{OPS}\{f, JS, S^T \hat{f}_k'' S, \ k \leq \hat{m}\} &= \sum_{j=1}^{o} [2 + 3\,|X(v_j)|(|X(v_j)| + 1)]\,\mathrm{OPS}\{\varphi_j\} \\
&\leq \left[2 + 3\hat{\gamma}\{f\}(\rho(\hat{J}) + 1)\right]\,\mathrm{OPS}\{f\} \quad,
\end{aligned}$$

where we have used Lemma 3 to bound the second factor $|X(v_j)|$. ∎

Since $\hat{\gamma} \leq \rho(\hat{J}) \leq \chi(\hat{J})$, it is clear that the sparse forward mode yields the lowest operations count followed by NR where we may choose $p = \rho(\hat{J})$ and CPR with $p \geq \chi(\hat{J})$. However, the NR and CPR methods may actually have a lower run-time, since on most computing platforms, vectors of fixed length p can be accessed and manipulated much faster than dynamically sparse vectors with a comparable number of nonzeros on average. In comparing the NR and CPR methods, we have so far ignored the fact that the former scheme requires the solution of $\hat{m}$ linear Vandermonde systems. According to [11] this adds

$$\sum_{k=1}^{\hat{m}} 2.5\hat{m}\,|X(\hat{y}_k)|^2 \leq 2.5\hat{m}\rho(\hat{J})^2$$

floating-point operations to complexity. As pointed out in [18], the conditioning of these linear systems can be improved by defining the Vandermonde matrix S using only $\chi(\hat{J})$ distinct real abscissas or defining them as complex roots of unity if the chromatic number is still too large. In the latter case, since all $\nabla_S v_j$ are complex, the arithmetic cost exactly doubles, because no complex multiplications or divisions are required.

4.2   First Derivatives in the Reverse Mode

In this subsection we first consider the complexity bounds for evaluating first derivatives in the reverse mode. Given the weight matrix $W^T \in \mathbb{R}^{q \times n}$, we may associate with each intermediate variable v_j the *adjoint* vector

$$\Delta^W v_j \equiv \frac{\partial W^T y}{\partial v_j} = W^T \frac{\partial y}{\partial v_j} \in \mathbb{R}^q \quad , \tag{4.5}$$

where all v_k with $v_j \not\rightsquigarrow v_k$ are held constant with respect to the differentiation. It is well known that the $\Delta^W v_j$ satisfy the backward recurrence

$$\Delta^W v_j = \sum_{v_j \rightsquigarrow v_k} c_{kj} \, \Delta^W v_k \quad , \tag{4.6}$$

which can be executed only if the elementary partials $c_{kj} = \partial \phi_k / \partial x_j$ can be provided in reverse order, namely, for $k = o, o-1, \ldots, 1$. As we have mentioned in Subsection 1.3, it was shown in [12] that this program reversal can be performed at the computational costs (1.8) and (1.9) for some integer r, which determines a trade-off between temporal and spatial complexity. At the end of the reverse computation one obtains the adjoint vectors

$$\Delta^W \check{x}_i = W^T \frac{\partial \check{f}(\check{x})}{\partial \check{x}_i} = W^T \hat{J} e_i \quad .$$

The complexity of this reverse sweep is bounded by the following result

Proposition 2
The reverse differentiation of the q functions $W^T \check{f} : \mathbb{R}^{\check{n}} \to \mathbb{R}^q$ with respect to the expanded independents $\check{x}$ can be achieved at the costs

$$\begin{aligned}
\text{OPS}\{f, W^T \check{J}\} &\leq (r+1+3q)\text{OPS}\{f\} \\
\text{RAM}\{f, W^T \check{J}\} &\leq (1+q)\text{RAM}\{f\} \\
\text{SAM}\{f, W^T \check{J}\} &\leq c\,\text{RAM}\{f\}\sqrt[r]{h\{f\}} \quad ,
\end{aligned}$$

where $h\{f\} \equiv \text{OPS}\{f\}/\text{RAM}\{f\}$ as before.

Proof. The RAM requirement follows from the need to store an adjoint q-vector $\Delta^W v_j$ for each variable that is live during the reverse sweep. Using the third inequality implicit in (3.14), we find that the backward propagation of the q vectors $\Delta^W v_j$ according to (4.6) requires also no more than $3q\,\text{OPS}\{\varphi_k\}$ operations per intermediate node. Together with the cost for evaluating $\nabla\varphi_j$ and that for reversing the program as described in [12], this yields the operations count as well as the SAM requirement. ∎

If one wishes to obtain the whole Jacobian $\check{J}$ in order to compute $J = A\check{J}$, one may use the NR approach with $q = \rho(\check{J}^T)$ and W a Vandermonde matrix or the

CPR approach with $q \leq \chi(\check{J}^T)$ and W a $0-1$ matrix. Similarly, one can also employ a dynamically sparse reverse mode with $W = I$ for which q is effectively replaced by $\check{\gamma}\{f\}$ as defined (2.5). The advantages and disadvantages of these three alternatives are essentially the same as in the forward mode. Again the operations count is highest for CPR and lowest for the dynamically sparse procedure, which does, however, involve more overhead. The NR approach may again suffer from poor conditioning unless the matrix W is chosen carefully, possibly using a coloring or complex roots of unity.

4.3 Combinations of Forward and Reverse Sweeps

In Proposition 1 we have shown that the full second-derivative tensor, $\hat{f}''$, and thus its contraction, f'', can be obtained at a complexity that grows quadratically with $p = \rho(\hat{J})$ or $p = \chi(\hat{J})$, depending on whether one uses the NR or CPR approach. It is interesting to note that, if one were to use CPR in the forward mode to evaluate the gradient of a scalar function f and then to use directional derivatives of this vector function ∇f in an indirect substitution method as described and analyzed in [6], then by (2.3) the complexity would be proportional to $\mathrm{OPS}\{f\}$ times

$$\chi(G)\chi_0(G) \geq \chi^2(G) = \chi^2(\hat{J}),$$

where G is the incidence graph of $\nabla^2 f$, which coincides by Lemma 1 with the column-intersection graph of the expanded Jacobian $\hat{J}$. Consequently, even indirect substitution on a gradient that is evaluated in the forward mode is likely to be less efficient than the calculation of the Hessian by differentiating $\hat{f}$ twice in the forward mode. Even lower complexities can be achieved if the forward and reverse modes are combined (see, for example, [5]). By combining Propositions 1 and 2 we obtain our final result.

Corollary 2
With $f : \mathbb{R}^n \to \mathbb{R}^m$ as before, $u \in \mathbb{R}^m$ a vector of Lagrange multipliers, and $S \in \mathbb{R}^{n \times p}$ the one-sided projection

$$\nabla^2 L(x)\, S = \sum_{i=1}^{\hat{m}} \hat{u}_i \nabla^2 \hat{f}_i S$$

can be calculated as the complexity

$$
\begin{aligned}
\mathrm{OPS}\{\nabla^2 L(x)\, S\} &\leq (4+r)(2+3p)\mathrm{OPS}\{f\} \\
\mathrm{RAM}\{\nabla^2 L(x)\, S\} &\leq 2(1+p)\mathrm{RAM}\{f\} \\
\mathrm{SAM}\{\nabla^2 L(x)\, S\} &\leq c(1+p)\,\mathrm{RAM}\{f\}\sqrt[r]{h\{f\}} \quad ,
\end{aligned}
$$

where c may be larger by a factor of 3 compared with Proposition 2.

Proof. This result can be achieved by first evaluating the vector function $g(x) \equiv J(x)S : \mathbb{R}^n \to \mathbb{R}^p$. According to Proposition 1 the forward mode yields these values with an operations count no greater than $(2 + 3p)\text{OPS}\{f\}$ and a RAM requirement no greater than $(1 + p)\text{RAM}\{f\}$. Hence the ratio between the operations count and the RAM requirement grows by a factor less than 3. Applying Proposition 2 to this calculation with $W = u$ and $q = 1$, we pick up another factor of $r+4$ for the operations count, a factor of 2 for the RAM requirement, and a factor less than $(1 + p)\sqrt[r]{3}$ for the SAM requirement. ∎

Since the number n of independents does not occur in the bounds of Corollary 2, we see that the complexity of the one-sided projected Hessian of the Lagrangian depends only on the degrees of freedom $p = n - m$ in a constrained optimization problem. It also appears that the cheapest way of obtaining the two-sided projection is to multiply the one-sided projection by S. Further cost reductions might be achievable if one exploits sparsity of $\nabla^2 LS$. The columns of S must span the null-space of J and are often defined on the basis of an LU or QR factorization of J. It would appear that these choices may be far from optimal regarding the sparsity of $\nabla^2 LS$, since they tend to introduce dense rows into S and consequently $\nabla^2 LS$. This question deserves further investigation.

5   Summary and Conclusion

In this paper we have shown how function separability and the new concept of argument separability can be exploited to yield first and second derivatives by the forward or reverse mode of computational differentiation with surprisingly low complexity.

The ideal case of function separability is that of a partially separable objective function f, whose gradient and Hessian can be obtained in the forward mode at a complexity of $\rho(\hat{J})$ and $\rho^2(\hat{J})$, respectively. Here $\rho(\hat{J})$ represents the maximal number of variables that are truly intertwined in a nonlinear fashion during the evaluation of f. The ideal case of argument separability is that of a vector function f, whose components f_k are evaluated completely separately from each other. Then one may apply the reverse mode to the horizontal expansion $\check{f}$ and obtain the full Jacobian $J = \check{J}A$ at no more than five times the cost of evaluating f itself. It is likely that substantial savings can be realized in mixed cases, but the implementation in a computational differentiation tool is a nontrivial task.

Rather than just considering additive decompositions with 0-1 matrices A and B, one can generalize the separability concepts, so that arbitrary linear pre-factors B and post-factors A are removed from the given vector functions to facilitate more efficient differentiation on the remaining nonlinear part.

Acknowledgments

The author had the benefit of extended discussions with Jorge Moré and Brett Averick, and he is greatly indebted to Christian Bischof and Paul Plassmann for their careful reading of the first draft.

References

[1] B. M. AVERICK, J. J. MORÉ, C. H. BISCHOF, A. CARLE, AND A. GRIEWANK, *Computing large sparse Jacobian matrices using automatic differentiation*, Preprint MCS-P348-0193, Mathematics and Computer Science Division, Argonne National Laboratory, Argonne, Illinois, 1993.

[2] C. BISCHOF, A. CARLE, G. CORLISS, A. GRIEWANK, AND P. HOVLAND, *ADIFOR: Generating derivative codes from Fortran programs*, Scientific Programming, 1 (1992), pp. 1–29.

[3] C. BISCHOF, G. CORLISS, AND A. GRIEWANK, *Computing second- and higher-order derivatives through univariate Taylor series*, Preprint MCS-P296-0392, Mathematics and Computer Science Division, Argonne National Laboratory, Argonne, Illinois, 1992.

[4] B. CHRISTIANSON, *Automatic Hessians by reverse accumulation*, IMA J. of Numerical Analysis, 12 (1992) pp. 135–150 .

[5] B. CHRISTIANSON, *Reverse accumulation and accurate rounding error estimates for Taylor series coefficients*, Optimization Methods and Software, 1 (1992), pp. 81–94.

[6] T. F. COLEMAN AND JIN-YI CAI, *The cyclic coloring problem and estimation of sparse Hessain matrices*, SIAM J. Alg. Disc. Meth., 7 (1986), pp. 221-235.

[7] T. F. COLEMAN, B. S. GARBOW, AND J. J. MORÉ, *Fortran subroutines for estimating sparse Jacobian matrices*, ACM Trans. Math. Software, 10 (1984), pp. 346–347.

[8] ——, *Software for estimating sparse Jacobian matrices*, ACM Trans. Math. Software, 10 (1984), pp. 329–345.

[9] T. F. COLEMAN AND J. J. MORÉ, *Estimation of sparse Jacobian matrices and graph coloring problems*, SIAM J. Numer. Anal., 20 (1983), pp. 187–209.

[10] A. R. CURTIS, M. J. D. POWELL, AND J. K. REID, *On the estimation of sparse Jacobian matrices*, J. Inst. Math. Appl., 13 (1974), pp. 117–119.

[11] G. H. GOLUB, AND C. F. VAN LOAN, *Matrix Computations*, second edition, The Johns Hopkins University Press, Baltimore (1988)

[12] A. GRIEWANK, *Achieving logarithmic growth of temporal and spatial complexity in reverse automatic differentiation*, Optimization Methods and Software, 1 (1992), pp. 35–54.

[13] A. GRIEWANK AND G. F. CORLISS, eds., *Automatic Differentiation of Algorithms: Theory, Implementation, and Application*, SIAM, Philadelphia, 1991.

[14] A. GRIEWANK AND S. REESE, *On the calculation of Jacobian matrices by the Markowitz rule*, in **Automatic Differentiation of Algorithms: Theory, Implementation**, and Application (A. Griewank and G. Corliss, eds.), SIAM, Philadelphia, 1991, pp. 126–135.

[15] A. GRIEWANK AND PH.L. TOINT *On the unconstrained optimization of partially separable objective functions*, in **Nonlinear Optimization 1981** (M. J. D. Powell, ed.), Academic Press, London, 1981, pp. 301–312.

[16] MASAO IRI, *History of automatic differentiation and rounding estimation*, in Automatic Differentiation of Algorithms: Theory, Implementation, and Application, A. Griewank and G. Corliss, eds., SIAM, Philadelphia, 1991, pp. 1–16

[17] R. D. NEIDINGER, *An efficient method for the numerical evaluation of partial derivatives of arbitrary order*, ACM Trans. Math. Software, 18(1992), pp. 159–173 .

[18] G. N. NEWSAM AND J. D. RAMSDELL, *Estimation of sparse Jacobian matrices*, SIAM J. Alg. Disc. Meth., 4 (1983), pp. 404–417 .

[19] L. B. RALL, *Automatic Differentiation: Techniques and Applications*, **Lecture Notes in Computer Science**, Vol. 120, Springer-Verlag, Berlin, 1981.

[20] TROND STEIHAUG AND A. K. M. SHAHADAT HOSSAIN, *Graph coloring and the estimation of sparse Jacobian matrices using row and column partitioning*, Report 72, Department of Informatics, University of Bergen, 1992.

[21] STEPHEN A. VAVASIS, *Nonlinear Optimization, Complexity Issues*, Oxford University Press, Oxford, 1991.

Complexity in Numerical Optimization, pp. 163-179
P.M. Pardalos, Editor
©1993 World Scientific Publishing Co.

Complexity Issues in Nonconvex Network Flow Problems

Geoffrey M. Guisewite
HRB Systems, State College, PA 16804 USA

Panos M. Pardalos
Department of Industrial and Systems Engineering, University of Florida, Gainesville, FL 32611 USA

Abstract

Nonconvex-cost network flow problems are known to be NP-hard. However, additional insight into the complexity of this class of problems can be obtained by considering network flow formulations of various NP-complete problems. In this paper we summarize current complexity results for the uncapacitated version of this problem. We also provide new results for the capacitated case. These results indicate that capacities on the arc flows add additional complexity, and local search is more difficult for the capacitated case than for the uncapacitated case.

Keywords: Complexity, concave-cost network flow, capacitated, global optimization, local optimization.

1 Introduction

The general nonconvex network flow problem can be stated formally as follows:

Given a directed graph $G = (N_G, A_G)$ consisting of a set N_G of n nodes and a set A_G of m ordered pairs of distinct nodes called arcs, coupled with a n-vector $d = (d_i)$ (demand vector) and a cost function for each arc, $c_{ij}(x_{ij})$, then solve

$$global\ min \sum_{(i,j)\in A_G} c_{ij}(x_{ij})$$

subject to

$$\sum_{(k,i)\in A_G} x_{ki} - \sum_{(i,k)\in A_G} x_{ik} = d_i, \forall i \in N_G \tag{1}$$

and

$$0 \leq a_{ij} \leq x_{ij} \leq b_{ij}, (i,j) \in A_G. \tag{2}$$

The x_{ij} represent units of flow on arc (i,j). All constraints and demands are assumed to be integral. A consistent system satisfies $\sum_{i=1}^n d_i = 0$, which states that total source demand equals total sink demand. The constraints in (1) are called the conservation of flow equations. The constraints in (2) are called capacity constraints on the arc flows. Nodes with $d_i < 0$ correspond to sinks. Nodes with $d_i > 0$ correspond to sources. The problem is uncapacitated if $a_{ij} = 0$ and $b_{ij} = \infty$, $\forall (i,j) \in A_G$.

The nonconvex cost case arises from nonconvex functions for the arc costs, $c_{ij}(x_{ij})$. The case where all arc cost functions are concave gives rise to the minimum concave-cost network flow problem (MCNFP). MCNFP falls into the category of constrained concave global optimization problems. As a result, it possesses the property that if a feasible solution exists, then an optimal solution occurs at some vertex of the convex polyhedron defined by (1) and (2), [13]. And, because of the concavity of the objective function, local optimality does not imply global optimality. This implies that the classical approaches of nonlinear programming will find only a local optimum. These two properties combined indicate that finding the solution to a MCNFP instance corresponds to searching the set of vertices of the polyhedron defined by (1) and (2) for the overall minimum. It is well known that almost every nonconvex function can be expressed as the sum of a convex and a concave function. Since the source of nonconvexity is due to the concave part of the objective function, we restrict ourselves here to the study of concave-cost network flow probleems.

Any network flow satisfying both constraints (1) and (2) is called a feasible flow. The set of all feasible flows is called the feasible region. A feasible flow is an extreme flow if it is not the convex combination of any other feasible flows. This corresponds to a vertex of the convex polyhedron defined by the linear constraints. Extreme flows correspond to basic feasible solutions in the simplex tableau [1]. If the objective function has a finite global minimum on the feasible region, then there is an extreme flow that is an optimal flow [2]. Extreme flows have been characterized for possible exploitation in solving MCNFP. For the uncapacitated case, a flow is extremal if it contains no positive loops [1, 21]. Here a positive loop is defined as follows: A path is a sequence of nodes $n_1, n_2, \ldots, n_k$ such that (n_i, n_{i+1}) or (n_{i+1}, n_i) is in A_G. A loop is a path with $n_1 = n_k$. A positive loop is a loop with all arcs (i,j) satisfying $x_{ij} > 0$. For the capacitated problem a positive loop is a loop with all arcs (i,j) satisfying $x_{ij} \neq 0$ and $x_{ij} \neq b_{ij}$ [4]. This property of extremal flows implies that for a single-source, uncapacitated (SSU) network (only one $d_i > 0$) an extreme flow is an arborescence, and for an uncapacitated single-source, single-sink network an extreme flow is a shortest weighted path [21]. The single-source uncapacitated version of MCNFP is denoted by SSU MCNFP.

MCNFPs arise naturally in a number of application areas including production planning, transportation and communication network design, facilities location, and VLSI design. Each application area gives rise to problems with distinct features for both the objective function and the underlying network. In each case, the concave objective functions originate from, for example, start-up costs, discounts and economies of scale.

In this paper, we address the complexity of MCNFP. In Section 2 we summarize recent results for the uncapacitated case. In Section 3 we present new results for the capacitated case.

2 The Uncapacitated Case

Most nonconvex optimization problems fall into the realm of NP-hard problems [14, 15]. MCNFP has been specifically shown to be NP-hard even when the arc weights are constant [13] and the underlying graph is bipartite [6], or the ratio of the fixed-charge to the linear charge is constant [12]. These proofs of NP-hardness for MCNFP all involve fixed costs $(f_{ij}(0) = 0, f_{ij}(x > 0) = w_{ij})$. In this section we present uncapacitated network formulations for several NP-complete problems including Subset Sum, 3-Dimensional Matching, and 3-Satisfiability. These results were originally presented in [7, 8] and are restated here for comparison to the capacitated results in Section 3. These results indicate the complexity of the general MCNFP, the single-source uncapacitated case of MCNFP, and the complexity of local search for MCNFP.

2.1 Results for the General MCNFP

The following polynomial time transformation demonstrates that the general uncapacitated version of MCNFP is NP-hard for cases involving costs other than fixed-charge. Consider the Subset Sum problem that is known to be NP-complete [6]:

Instance: Finite set D, size $s(d) \in Z^+$, $\forall d \in D$, positive integer B.

Question: Is there a subset $D' \subseteq D$ such that $\sum_{d \in D'} s(d) = B$?

We construct the following flow problem with $N = |D|$:

1. Create N source vertices, V_i, with source flow equal to $s(i)$.

2. Create two sink vertices S_B and $S_{\overline{B}}$ with flow requirements B and $\sum_{d \in D} s(d) - B$.

3. Create an arc from each source to each sink, (V_d, S_B) and $(V_d, S_{\overline{B}}), \forall d \in D$.

4. The arcs are uncapacitated.

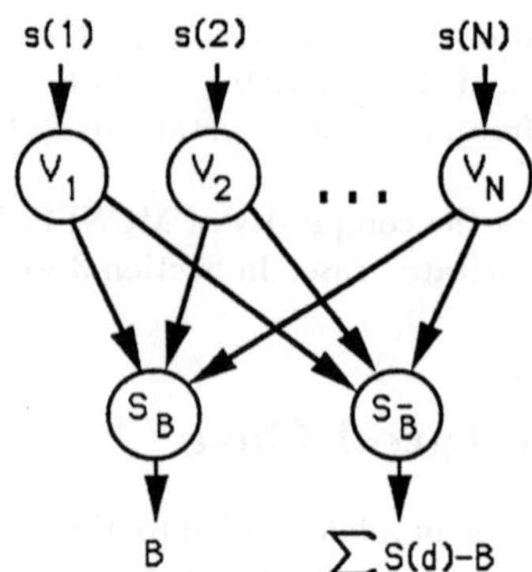

Figure 1: Network Resulting from Subset Sum Transformation

The resulting network is pictured in Figure 1. Consider the following cost functions:

1. Start-up costs (assume the nodes are labelled 1,2,..., $N + 2$); this corresponds to the fixed-charge case: $c_{ij}(0) = 0$, $c_{ij}(x) = a + (b * x)$, $a > 0$.

2. Cost functions satisfying: $c_{ij}(0) = 0$, $c_{ij}(x + y) < c_{ij}(x) + c_{ij}(y)$, $x, y > 0$. This case includes all strictly concave functions, and a number of nonconcave, nonconvex functions.

For case 1, consider when all arcs have cost 1 for nonzero flow. Then the optimal cost of the network problem is N if and only if the Subset Sum instance has answer *yes*. In general, if the cost on each arc is $c_{ij}(x) = a + bx$, then the optimal cost of the network problem is $(N * a) + (b * \sum_{i=1}^{N} s(i))$ if and only if the Subset Sum instance has answer *yes*. In both cases the result follows immediately by noting that if any flow is split, then we incur additional startup costs.

For case 2, consider where the cost of (V_i, S_B) and $(V_i, S_{\overline{B}})$ are identical and satisfy the specified constraint. Here we find that the optimal cost of the network problem is $\sum_{i=1}^{N} c_{V_i,S_B}(s(i))$ if and only if the Subset Sum instance has answer *yes*. This follows from the observation that if no split of the flow occurs, the cost is as specified above and the arcs of the flow establish the solution to the Subset Sum problem. If any split occurs, then the cost increases. For the case with no split:

$$\text{FLOW COST} = \sum_{i=1}^{N} c_{V_i,S_B}(x_{i,S_B}) + \sum_{i=1}^{N} c_{V_i,S_{\overline{B}}}(x_{i,S_{\overline{B}}})$$
$$= \sum_{i=1}^{N} c_{V_i,S_B}(s(i))$$

$$= \sum_{i=1}^{N} c_{V_i,S_{\overline{B}}}(s(i)).$$

For the case with the flow split at V_m:

$$\begin{aligned}
\text{FLOW COST} &= \sum_{i=1}^{N} c_{V_i,S_B}(x_{i,S_B}) + \sum_{i=1}^{N} c_{V_i,S_{\overline{B}}}(x_{i,S_{\overline{B}}}) \\
&= \left(\sum_{i=1,i\neq m}^{N} c_{V_i,S_B}(s(i)) \right) + c_{V_m,S_B}(x_{m,S_B}) + c_{V_m,S_{\overline{B}}}(x_{m,S_{\overline{B}}}) \\
&> \left(\sum_{i=1}^{N} c_{V_i,S_B}(s(i)) \right).
\end{aligned}$$

Inspection of the above network flow problem reveals that when all arcs have cost 1 for nonzero flow, it is trivial to find a flow with cost one more than the optimal flow cost (i.e. force flows from V_i to S_B until this flow is met). This provides a simple example where enumerative search techniques, such as branch-and-bound, remain inefficient even if a very good initial upper bound for the flow cost is obtainable. In addition, any optimal solution for the above problem involves a feasible flow with the minimum number of active (nonzero flow) edges. This indicates the problem of computing a feasible flow with this property is NP-hard. Also, the underlying network in the above transformation is bipartite. This indicates that the general concave-cost network flow problem is NP-hard when restricted to bipartite networks, even for cases which do not involve fixed-charges on edge flows.

2.2 Complexity for the SSU Case

The general single-source uncapacitated version (SSU) of MCNFP is known to be NP-hard. This follows immediately from this class of problems containing the Steiner Tree in Graphs problem [6, 13]. This result is based on arc costs that correspond to fixed arc weights.

The following transformation demonstrates that SSU MCNFP is NP-hard for cases involving objective functions other than the fixed-charge case. Consider the 3 Dimensional Matching (3DM) problem that is known to be NP-complete [6]:

Instance: Set $M \subseteq W \times X \times Y$, where W, X, and Y are disjoint sets having the same number q of elements.

Question: Does M contain a matching, i.e. a subset $M' \subseteq M$ such that $|M'| = q$ and no two elements of M' agree in any coordinate?

We construct the following flow problem:

1. Create a single source vertex S with source flow $3 * q$.

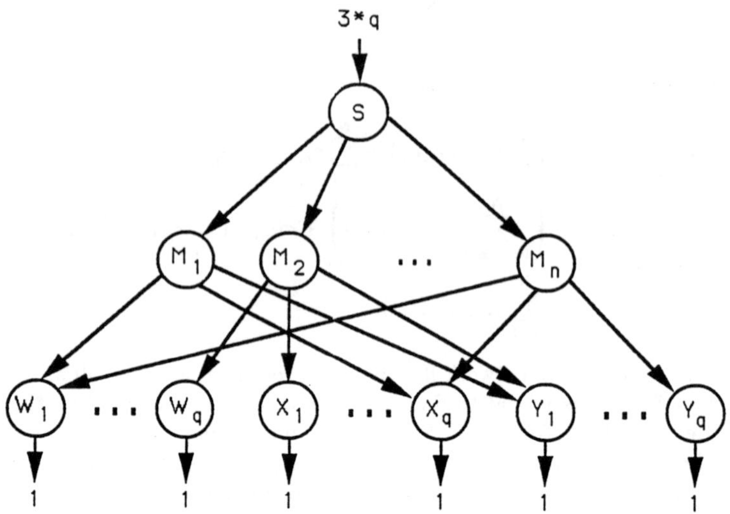

Figure 2: Network Resulting from 3DM Transformation

2. Create $n = |M|$ transshipment nodes M_i and arcs (S, M_i). These correspond to the elements (W_j, X_k, Y_l) for some j, k, and $l \ni 1 \leq j, k, l \leq q$.

3. Create $3 * q$ sinks $W_1, \ldots, W_q, X_1, \ldots, X_q, Y_1, \ldots, Y_q$. Each sink has flow requirement one. For each M_i, add arcs (M_i, W_j), (M_i, X_k), and (M_i, Y_l), where $M_i = (W_j, X_k, Y_l)$.

4. All arcs are uncapacitated.

5. All arcs have cost 0, except the arcs originating at the source.

The resulting network is pictured in Figure 2. Again, consider the cost functions:

1. Start-up costs (assume the nodes are labelled $1, 2, \ldots, N + 2$); this corresponds to the fixed-charge case: $c_{ij}(0) = 0$, $c_{ij}(x) = a + (b * x)$, $a > 0$.

2. Cost functions satisfying: $c_{ij}(0) = 0$, $c_{ij}(x + y) < c_{ij}(x) + c_{ij}(y)$ $x, y > 0$.

For case 1, consider when all arcs (S, M_i) have cost 1 for nonzero flow. Then the optimal cost of the network problem is q if and only if the 3DM instance has answer *yes*. In general, if the cost on each arc is $c_{ij}(x) = a + bx$, then the optimal cost of the network problem is $(q * a) + (b * 3 * q)$ if and only if the 3DM instance has answer *yes*. In both cases the result follows by noting that if flow is split across more than q of the outgoing source arcs, then we incur additional startup costs. In addition, each M_i has a directed path to exactly 3 sinks (one in each of W, X, and Y), implying at least q of the (S, M_i) have nonzero flow.

For case 2, the proof is similar to the result in Section 2.1. Consider where the cost of all (S, M_i) are identical and satisfy the specified constraint. Here we find that the optimal cost of the network problem is $\sum_{i=1}^{q} c_{S,M_i}(3)$ if and only if the 3DM instance has answer *yes*. This follows from the observation that if no split of the flow occurs, the cost is as specified above and the arcs of the flow establish the solution to the 3DM problem. If any split occurs, then the cost increases.

Noting that the above transformation from 3DM is a polynomial transformation indicates that the SSU MCNFP is NP-hard, even for cases with arc costs other than fixed-charge.

2.3   Local Search for the Uncapacitated Case

Each of the above results have addressed the complexity of locating a globally optimal solution for the SSU MCNFP. Here we investigate the complexity of checking if a solution is locally optimal, and of finding a local optimum for a SSU MCNFP. Before we can investigate these problems we must establish the criteria for a local optimum. For the MCNFP, the standard marginal definition of local optimality (i.e rerouting a small portion of flow [20]) is not satisfactory, as strictly concave cost functions result in all extreme flows being locally optimal. We use the definition of a local optimum as defined by Gallo and Sodini [5]. Here, a feasible solution is a local optimum if its objective value is less than or equal to all of it's neighboring vertices. Gallo and Sodini also demonstrate that the problem of checking if a feasible solution for SSU MCNFP is a local optimum is in P. This result indicates that the problem of checking if a solution for SSU MCNFP is a strictly local optimum is in P. The complexity of finding a local optimum for SSU MCNFP is an open problem. Using the following 3-SAT transformation, we can establish a result for the problem of finding a strictly local optimum.

The 3-Satisfiability problem (3-SAT) is stated formally as follows [6]:

Instance: Collection $C = \{c_1, c_2, \ldots, c_m\}$ of clauses on a finite set U of variables such that $|c_i| = 3$ for $1 \leq i \leq m$.

Question: Is there a truth assignment for U that satisfies all the clauses in C?

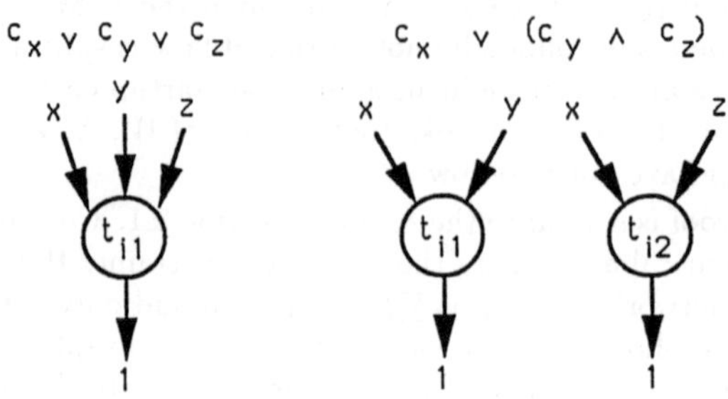

Figure 3: Transformation of the Clauses for 3-SAT

We construct the following flow problem:
Let S denote the source vertex.

1. Add $(S, V_i), \forall i = 1, \ldots, k$ where $k = |V|$ is the number of distinct variables in the 3-SAT instance.

2. Add $(V_i, T_i), (V_i, F_i), \forall i$. This corresponds to a TRUE or FALSE assignment for each variable.

3. Add $(T_i, FC_i), (F_i, FC_i), \forall i$. This forces a choice at each node.

4. The remaining arcs and nodes depend on the structure of the 3-SAT clauses. For example, if $C_i = c_x \wedge c_y \wedge c_z$ we add nodes t_{i1}, t_{i2}, t_{i3} and arcs $(T_x, t_{i1}), (T_y, t_{i2}), (T_z, t_{i3})$. If c_x were negated we would add arc (F_x, t_{i1}) in place of (T_x, t_{i1}). For the case where $C_i = c_x \wedge (c_y \vee c_z)$ we would add nodes t_{i1}, t_{i2} and arcs $(T_x, t_{i1}), (T_y, t_{i2}), (T_z, t_{i2})$. The addition of only two new nodes results from the choice of the conjunction in forcing a variables value of TRUE or FALSE. All other cases are handled in a similar manner. Figure 3 demonstrates instances of this subset of the transformation.

5. Let $|T|$ denote the number of t nodes added at step 4. Let $|V|$ denote the number of distinct variables occurring in clauses. Then set the source flow to $|T| + |V|$. $|V|$ units of flow are used to force the assignment of TRUE or FALSE to each variable. $|T|$ units of flow are used to force the satisfiabilty of each clause. This corresponds to each FC_i and each t_{ij} being a sink with requirement 1.

6. All arc flow costs are 0 except for (V_i, T_i) and (V_i, F_i) which have cost 0 if flow $= 0$ and 1 if flow > 0.

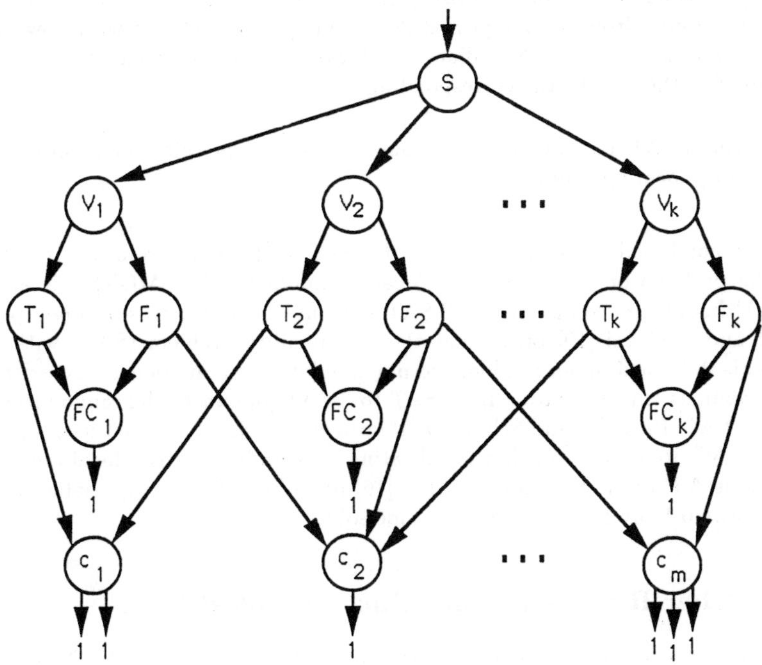

Figure 4: Network Resulting from 3-SAT Transformation

The resulting network, presented in Figure 4, has optimal flow cost $|V|$ if and only if the 3-SAT instance has a satisfying assignment. This can be seen by noting that any feasible flow has cost greater than, or equal to, $|V|$ due to the FC_i sinks forcing a unit of flow through each V_i. If the additional flow necessary to satisfy the sinks resulting from the clauses (t_{ij}) can be met without taking a path from V_i to T_i or F_i which currently has zero flow, then the cost of the flow remains at $|V|$. In this case, the assignment $v_i =$TRUE if the flow on (V_i, T_i) is greater than zero, else FALSE, results in a satisfying assignment for the 3-SAT instance. If no satisfying assignment exists, we see it is necessary to have some i such that both (V_i, T_i) and (V_i, F_i) have nonzero flows. This implies the optimal network flow has cost greater than $|V|$.

This result is similar to the result found in [13]. The general single-source, concave-cost, uncapacitated network flow problems with fixed costs are NP-hard.

To evaluate the complexity of local search we use the following properties of the polynomial time 3-SAT transformation:

1. If a feasible flow is not a global optimum, then it is not a strict local optimum. This results from a non-optimal flow having some variable with flow on both (V_i, T_i) and (V_i, F_i). An adjacent solution of equal cost can be obtained by altering the flow to the corresponding FC_i sink.

2. If the 3-SAT instance has a unique solution, then this global optimum is a strictly local optimum.

These combined facts indicate that if 3-SAT with unique solution is NP-hard, then the problem of finding a strict local optimum for SSU MCNFP is also NP-hard. Although the complexity of 3-SAT with unique solution is an open problem, Valient and Vazirani [19] prove that the Satisfiability problem (SAT) with unique solution is NP-hard under randomized polynomial-time reductions. The existence of a parsimonious transformation from SAT to 3-SAT [6] carries the randomized result over to 3-SAT. Pardalos and Schnitger [15] prove that checking strict local optimality for the indefinite case is NP-hard, indicating that finding a strict local optimum for this case is NP-hard. Pardalos and Jha [16] prove that finding the global minimum of quadratic 0-1 programming with unique solution is NP-hard.

2.4   Other Results for the Uncapacitated Case

In [10] it is proved that the single-source uncapacitated version of MCNFP in which a single arc has non-linear cost is solvable in polynomial time. This contrasts the NP-hard result for quadratic programming with one negative eigenvalue presented in [17]. In [3] we find that the concave version of the dynamic lot-size problem is NP-hard. In [11] the 3-SAT transformation is used to demonstrate that network problems in MCNFP exist that have an exponential number of non-global but locally optimal extreme solutions. This demonstrates that a heuristic based on random generation of extreme feasible solutions and local search, developed in [11], will be ineffective for computing a global optimum for the 3-SAT problem.

3   The Capacitated Case

The previous section addressed complexity issues for the uncapacitated MCNFP. The capacitated version of MCNFP is also NP-hard. This follows from the fact that every uncapacitated problem can be converted to an equivalent capacitated problem where the upper bounds on all arc flows are constrained to be the total sink requirement. In this section we present polynomial time transformations for the Subset Sum problem and the Traveling Salesman problem, demonstrating that the capacities, both upper and lower, add additional complexity.

3.1 Capacities on the Maximum Arc Flow

The uncapacitated network formulation of the Subset Sum problem, presented in Section 2.1, can be further transformed to a capacitated network. This is achieved by adding a sink and source node, and enforcing the original source and sink requirements by restricting flow into the original sources and sinks. The resulting network flow problem, as pictured in Figure 5, is a single-source single-sink, capacitated MCNFP.

This transformation extends the NP-hardness result to the single-source, single-sink capacitated case, for the arc cost functions described in the previous section. This result also has implications for local search in capacitated networks. Testing for a local optimum using the neighborhood defined by Gallo and Sodini [5] involves evaluating adjacent extreme flows. In the capacitated case, this corresponds to testing solutions obtained by rerouting a single subpath of flow in a given extreme feasible solution. The above result indicates that rerouting a single subpath of flow in an optimal fashion is NP-hard for the capacitated case. This can be seen by adding an additional arc (s, t) to the network in Figure 5, carrying all flow as the initial solution. Using the concept of an adjacent extreme flow of Gallo and Sodini for capacitated local search offers no improvement for the resulting Subset Sum network flow problem.

3.2 Capacities on the Minimum Arc Flow

The Traveling Salesman Problem can be reformulated as a capacitated MCNFP. In this case the flow into any source is limited to one, and the only arc capacities are lower bounds on the arc flow. The Traveling Salesman Problem is stated formally as follows:

Instance: A finite set $C = \{c_1, c_2, \ldots, c_m\}$ of "cities", a "distance" $d(c_i, c_j) \in Z^+$ (where Z^+ denotes the positive integers).

Question: Is there a "tour" of all the "cities" in C having total length no more than B, that is, an ordering $> c_{\Pi(1)}, c_{\Pi(2)}, \ldots, c_{\Pi(m)} >$ of C such that

$$\left[\sum_{i=1}^{m-1} d(c_{\Pi(i)}, c_{\Pi(i+1)})\right] + d(c_{\Pi(m)}, c_{\Pi(1)}) \leq B? \tag{3}$$

Construct the following flow problem:

1. Create $N = 2*m$ vertices $c_1, c_2, \ldots, c_m$ and $c_1', c_2', \ldots, c_m'$. Vertices $c_1', c_2', \ldots, c_m'$ are source nodes with source flow one. Vertex c_1 is the single sink with sink flow m.

2. For each $d(c_i, c_j) \leq B$, $i, j = 1, \ldots, m$, create an arc (c_i', c_j) with flow cost $d(c_i, c_j)$ if flow is nonzero, and 0 otherwise. Create arcs (c_i, c_i') for $i = 2, \ldots, m$, each arc having zero cost for any flow value.

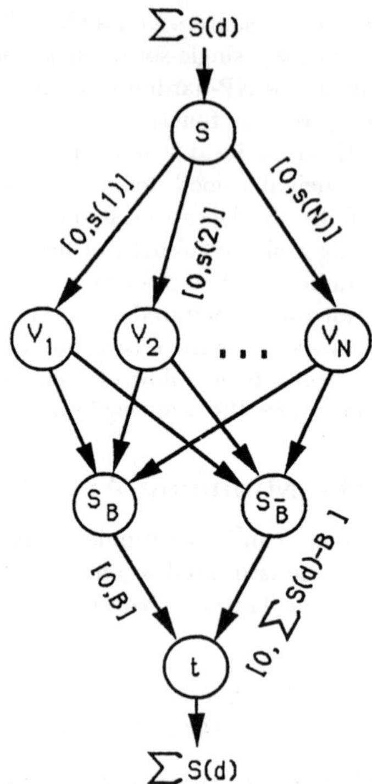

Figure 5: Capacitated Network for Subset Sum Problem Transformation

3. Arcs $(c_i, c_i'), i = 2, \ldots, m$ are required to have flow of magnitude of at least one.

An example network is pictured in Figure 6 . The source flows at each c_i' force a path from each "city" to the chosen "tour" start c_1. The arcs (c_i, c_i') with required minimum flow of one force a path to enter into each "city".

As stated the network formulation does not guarantee that the minimum cost flow is a "tour". For example, in Figure 6 a valid flow could be satisfied by the path $< c_1', c_4, c_4', c_1 >$, the cycle $< c_2, c_2', c_3, c_3' >$, and the path $< c_3', c_1 >$. In order to force the optimal flow to be the optimal tour we exploit the following:

a. A "tour" involves exactly $2m - 1$ arcs with nonzero flow.

b. A valid flow that is not a tour consists of more than $2m - 1$ arcs with nonzero flow.

Observation a results from a "tour" being equivalent to a simple path containing $2m$ nodes. Observation b follows from the fact that a path exists from each c_i' to the sink node, and a path enters into each node c_i. As a result, the undirected graph $G_U = \{N_U, E_U\}$ defined by $N_U = N$ and $E_U = \{(i,j)|(i,j) \in A$ and $x_{ij} > 0\}$ is connected. This implies that a spanning tree rooted at c_1 and containing all nodes in N_U exists. Any spanning tree with these properties contains $2m - 1$ edges. If G_U corresponds to a "tour" then the unique spanning tree for G_U is a simple path consisting of all edges in E_U. If G_U does not correspond to a "tour" then G_U is not a simple path. There are two possibilities for this case:

a. The spanning tree is a simple path (i.e., one leaf node), or

b. The spanning tree has more than one leaf.

For case a there must exist additional edges in E_U that do not occur in the spanning tree. This indicates $|E_U| > 2m - 1$ and that G_U contains at least one cycle.

For case b consider a leaf of the spanning tree other than c_1'. If the leaf is $c_j, j \neq 1$, there is an edge in E_U corresponding to arc (c_j', c_j) that is not in the spanning tree. If the leaf is $c_j', j \neq 1$, there is an edge in E_U that is necessary to satisfy the flow requirement on arc (c_j', c_j). Again, this indicates $|E_U| > 2m - 1$ and G_U contains a cycle.

Exploiting the above observations allows us to force the minimum cost flow to be a minimum cost "tour" by adding a large constant to all arc flows. It is sufficient to add a cost of $\sum_{(i,j) \in A} d(c_i, c_j)$ to the current costs of arcs having nonzero flow. With this modification any "tour" would now have cost

$$\left((2m - 1) \times \sum_{(i,j) \in A} d(c_i, c_j) \right) + \text{original "tour" cost.} \tag{4}$$

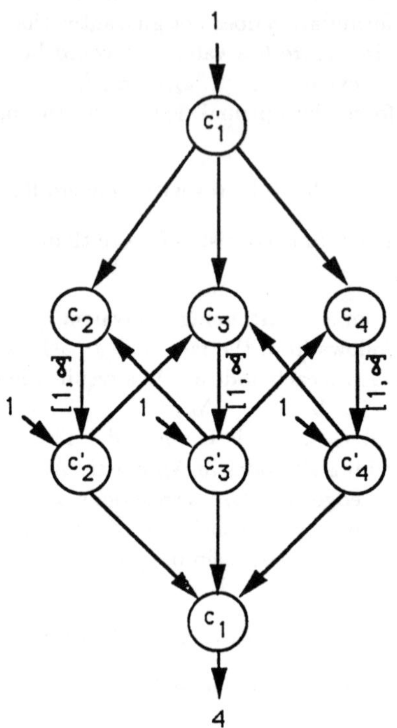

Figure 6: Network Resulting from Traveling Salesman Transformation

For the case of a "non-tour" the flow cost is greater than

$$2m \times \sum_{(i,j)\in A} d(c_i, c_j). \tag{5}$$

This is always larger than any "tour" cost.

The network formulation of the Traveling Salesman Problem also has implications in the usefulness of local search in generating approximate solutions to the Traveling Salesman Problem. Computing an acyclic flow for the network formulation of the Traveling Salesman Problem is equivalent to finding a Directed Hamiltonian Circuit in the underlying network $G = (A_G, E_G)$. This indicates finding an acyclic feasible flow for a capacitated network flow problem is NP-hard. Even if an acyclic feasible flow were provided, local search offers no benefit in terms of finding an improved solution. To achieve a change in the order that the nodes are visited in a "tour" a k-away version of local search would be required in the network formulation, with $k \geq 3$. In [18] it is demonstrated that various capacitated location-allocation problems on networks are also NP-hard.

4 Summary

We have presented network formulations of NP-complete problems. These formulations indicate that the complexity of nonconvex network flow problems arise from both network structure and constraints on the arc flows. Although the complexity of local search for this class of problems remains open, the results presented here indicate that local search offers little improvement in the generation of approximations to a global optimum for the networks resulting from NP-complete problems.

References

[1] Dantzig, G.B. (1963), *Linear Programming and Extensions*, Princeton University Press, Princeton, New Jersey.

[2] Eggleston, H.G. (1963), *Convexity, Cambridge Tracts in Mathematics and Mathematical Physics No. 47*, Cambridge University Press, Cambridge, Mass.

[3] Florian, M., Lenstra, J.K. and Rinnooy Kan, A.H.G. (1980), "Deterministic Production Planning: Algorithms and Complexity," *Management Science, Vol. 26, No. 7*, 669-679.

[4] Florian, M., Rossin-Arthiat, M., and de Werra, D. (1971), "A Property of Minimum Concave Cost Flows in Capacitated Networks," *Canadian Journal of Operations Research, Vol. 9*, 293-304.

[5]   Gallo, G. and Sodini, C. (1979), "Adjacent Extreme Flows and Application to Min Concave-Cost Flow Problems," *Networks 9*, 95-121.

[6]   Garey, M.R. and Johnson, D.S. (1979), *Computers and Intractability: A Guide to the Theory of NP-Completeness.* W.H. Freeman and Company, San Francisco, CA.

[7]   Guisewite, G.M. and Pardalos, P.M. (1990), "Minimum Concave-Cost Network Flow Problems: Applications, Complexity, and Algorithms," *Annals of Operations Research, Vol. 25*, 75-100.

[8]   Guisewite, G.M. and Pardalos, P.M. (1991), "Algorithms for the Single-Source Uncapacitated Minimum Concave-Cost Network Flow Problem," *Journal of Global Optimization 1*, 245-265.

[9]   Guisewite, G.M. and Pardalos, P.M. (1992), "Performance of Local Search in Minimum Concave-Cost Network Flow Problem," *In " Recent Advances in Global Optimization" (Eds: C.A. Floudas & P.M. Pardalos)*, Princeton University Press, 50-75.

[10] Guisewite, G.M. and Pardalos, P.M. (1993), "A Polynomial Time Solvable Concave Network Flow Problem," *Networks, Vol. 23*, 143-147.

[11] Guisewite, G.M. and Pardalos, P.M. (1991), "Global Search Algorithms For Minimum Concave-Cost Network Flow Problems," *Journal of Global Optimization 1*, 309-330.

[12] Hochbaum, D.S. and Segev, A. (1989), "Analysis of a Flow Problem with Fixed Charges," *Networks, Vol. 19*, 291-312.

[13] Lozovanu, D.D.(1983), "Properties of Optimal Solutions of a Grid Transport Problem with Concave Function of the Flows on the Arcs," *Engineering Cybernetics, Vol 20*, 34-38.

[14] Murty, K.G. and Kabadi, S.N. (1987), "Some NP-Complete Problems in Quadratic and Non-linear Programming," *Mathematical Programming, Vol. 39*, 117-129.

[15] Pardalos, P.M. and Schnitger, G. (1988), "Checking Local Optimallity in Constrained Quadratic Programming is NP-hard," *Operations Research Letters, Vol. 7, No. 1*, 33-35.

[16] Pardalos, P.M. and Jha, S. (1992), "Complexity of Uniqueness and Local Search in Quadratic 0-1 Programming," *Operations Research Letters 11*, 119-123.

[17] Pardalos, P.M. and Vavasis, S.A. (1991), "Quadratic Programming with One Negative Eigenvalue is NP-hard," *Journal of Global Optimization 1*, 15-23.

[18] Sherali, H.D. and Nordai, F.L. (1988), "NP-hard, Capacitated, Balanced p-median Problems on a Chain Graph With a Continuum of Link Demands", *Mathematics of Operations Research 13(1)*, 32-49.

[19] Valiant, L.G., and Vazirani, V.V. (1985), "NP Is As Easy As Detecting Unique Solutions," *17'th STOC of the ACM*, 458-463.

[20] Yaged, Jr. B.(1971), "Minimum Cost Routing for Static Network Models," *Networks, Vol. 1*, 139-172.

[21] Zangwill, W.I. (1968), "Minimum Concave-Cost Flows in Certain Networks," *Management Science, Vol. 14, No. 7*, 429-450.

Complexity in Numerical Optimization, pp.180-202
P.M. Pardalos, Editor
©1993 World Scientific Publishing Co.

Complexity of Smooth Convex Programming and its Applications

Osman Güler
Department of Mathematics and Statistics, University of Maryland Baltimore County, Baltimore, Maryland 21228 USA

Abstract

We describe the two methods of Y. E. Nesterov [18, 19, 20] for minimizing a convex function with a Lipschitz continuous derivative over a closed convex set. These methods are optimal in the oracle model of computational complexity. We extend the applicability of one of the methods and give a simple proof of the other. We present results of our numerical experiments with one of the methods in solving randomly generated quadratic problems with box constraints. These and other numerical experiments of the author in [7] indicate that Nesterov's methods are promising for solving sparse, large scale convex programs.

Keywords: Computational complexity, optimal methods, global convergence rates, convex programming, quadratic programming, conjugate gradient methods.

1   Introduction

The conjugate gradient method of Hestenes and Stiefel is well–known and often used for minimizing a convex quadratic function. It is especially popular for solving the large scale, sparse linear equation systems arising in the numerically solving partial differential equations. The popularity of the method is due to its iterative nature, simplicity, low memory requirement, and fast convergence rate, especially when it is properly preconditioned. The method possesses some well–known optimality properties, see for example Luenberger [12].

The success of the method has prompted researchers to extend the method to nonlinear problems. The conjugate gradient methods of Fletcher–Reeves and Polak–Ribière–Polyak are designed for unconstrained minimization of a nonlinear function. See [12] for a description of these methods. Although these methods have been successfully used in practice, it is shown in Chapter 8 of Nemirovsky and Yudin [17] that they can, in their worst cases, be as slow as the steepest descent method. Also, they cannot be used for constrained optimization problems. It is not clear, for example, how the Hestenes–Stiefel method can be modified to obtain an efficient algorithm for convex quadratic programs subject to box constraints.

Remarkable progress has been achieved during the last ten years in extending the Hestenes–Stiefel method to handle some classes of convex programs – both unconstrained and constrained. These methods are *optimal* for the classes of problems under consideration in the black box, or oracle model of complexity. Some versions of the methods can be used to solve convex quadratic programs with box constraints. Our numerical experiments with one such method indicate that it is promising for solving large scale, sparse problems.

A. S. Nemirovsky and D. B. Yudin investigate the complexity of smooth convex programs in Chapter 7 of their book [17]. They obtain the important result that the conjugate gradient method is optimal in the oracle model for minimizing a convex quadratic function. They determine the complexity of this method, and thus obtain a theoretical lower bound on the complexity of minimizing convex functions with Lipschitz continuous derivatives. They also propose an optimal method for this class of problems. Unfortunately, this method is not practical.

Subsequent research by Y. E. Nesterov [18, 19, 20], and Nemirovsky and Nesterov [16] have resulted in practical optimal methods for smooth convex programs. Nemirovsky [14] uses Nesterov's algorithm [18] with an appropriate scaling to develop a polynomial–time non–interior point method for linear programming. Nesterov and Nemirovsky [21] use Nesterov's algorithm [19] to accelerate polynomial–time interior point methods. The key ideas of Nesterov's methods have uses beyond smooth convex programming. The proximal point method is closely related to the augmented Lagrangian method and plays a central role in convex programming. Güler [7, 5] uses the main ideas of the methods in [18, 19] to accelerate the proximal point method for convex minimization. The new proximal point method is applied to linear programming in [6].

We now describe the classes of problems processed by Nesterov's algorithms. Let $f : \mathbb{R}^n \to \mathbb{R}$ be a differentiable convex function with Lipschitz continuous derivative f' satisfying

$$l\|x - y\| \leq \|f'(x) - f'(y)\| \leq L\|x - y\|,$$

where $l \geq 0$ and $L > 0$. Equivalently,

$$l\|x - y\|^2 \leq \langle f'(x) - f'(y), x - y \rangle \leq L\|x - y\|^2.$$

These functions also satisfy the inequalities

$$f(x) + \langle f'(x), y - x \rangle + \frac{l}{2}\|y - x\|^2 \leq f(y) \leq f(x) + \langle f'(x), y - x \rangle + \frac{L}{2}\|y - x\|^2. \quad (1)$$

We denote this class of functions by $C^{1,1}(l_f, L_f)$. The smallest constant L is called the Lipschitz constant of f' and is denoted by L_f. Similarly, the largest constant l is called the constant of strong convexity, and is denoted by l_f. If $l_f > 0$, then f is strongly convex, and the ratio L_f/l_f is called the modulus of strong convexity of f.

The simplest problem considered is the unconstrained problem,

$$\min f(x), \quad (2)$$

where $f \in C^{1,1}(l_f, L_f)$. Next, we have the constrained minimization problem,

$$\min_{x \in C} f(x), \quad (3)$$

where $C \subseteq \mathbb{R}^n$ is a closed convex set and $f \in C^{1,1}(l_f, L_f)$. We assume that the projection onto C can be performed inexpensively. The convex quadratic programming problem with box constraints satisfies this property. Nesterov's methods can be used to solve the composite convex minimization problem

$$\min_{x \in C} h(x), \quad (4)$$

where $h(x) = F(f(x))$ with $f(x) = (f_1(x), \ldots, f_m(x))$, $f_i \in C^{1,1}(l_f, L_f)$, and C a closed convex subset of $\mathbb{R}^n$. Here F is non–decreasing convex function and Lipschitz continuous with respect to its ith argument. Many convex programming problems can be cast in form (4), for example, the discrete minimax problem

$$min_{x \in C} \ \max_{1 \leq i \leq m} f_i(x).$$

See [18, 20, 16] for more details.

In order to explain the optimality properties of the methods, it is necessary to describe briefly the *oracle* model of computational complexity. A detailed description can be found in Chapter 1 of the seminal book [17]. This model is also called the *black box* model, see Vavasis [24]. In this model, we have under consideration a certain class of functions $\mathcal{F}$, and we have an *oracle* $\mathcal{O}$ which is our only means of obtaining information about a function $f \in \mathcal{F}$. For example, an oracle can be a FORTRAN subroutine which calculates the function values and the derivatives. Such an oracle is called a first order oracle.

A first order *algorithm* $\mathcal{A}$ for minimizing a function $f \in \mathcal{F}$ is an iterative procedure in which we try to obtain information about f by asking $\mathcal{O}$ questions about $f \in \mathcal{F}$. We start from an initial point x_0 and ask $\mathcal{O}$ to supply us with the values $f(x_0)$

and $f'(x_0)$. On the basis of this information, the algorithm $\mathcal{A}$ determines the next point x_1. We then ask $\mathcal{O}$ to supply us with $f(x_1)$ and $f'(x_1)$. On the basis of the accumulated information $\{f(x_0), f'(x_0), f(x_1), f'(x_1)\}$, $\mathcal{A}$ determines the next point x_2, and so on. In general, we cannot hope to obtain an exact optimal solution x^* of f, so we also need an error measure $\varepsilon(x)$ to gauge the closeness of x to optimality. For example, we can use $\varepsilon(x) = (f(x) - f(x^*))/(f(x_0) - f(x^*))$. For a given $\epsilon \in (0, 1)$, the *complexity* $N(\epsilon, \mathcal{A}, f)$ of $\mathcal{A}$ for a function $f \in \mathcal{F}$ is the minimum number of calls we have to make to the oracle in order to obtain a point $\tilde{x}$ for which $\varepsilon(\tilde{x}) \leq \epsilon$. The complexity $N(\epsilon, \mathcal{A})$ of $\mathcal{A}$ on the class $\mathcal{F}$ is $N(\epsilon, \mathcal{A}, f)$ for the worst $f \in \mathcal{F}$, i.e.,

$$N(\epsilon, \mathcal{A}) = \sup_{f \in \mathcal{F}} N(\epsilon, \mathcal{A}, f).$$

The complexity $N(\epsilon)$ of the class $\mathcal{F}$ is

$$N(\epsilon) = \inf_{\mathcal{A}} N(\epsilon, \mathcal{A}).$$

An algorithm $\mathcal{A}$ is *optimal* if $N(\epsilon) = N(\epsilon, \mathcal{A})$.

Most global optimization problems are intractable in the oracle model of complexity [17], [11]. If one considers convex programming, however, the situation is much more favorable. For example, the famous ellipsoid method of Nemirovsky and Yudin is a polynomial method for a general convex program in $\mathbb{R}^n$.

The main purpose of the paper is to present results of our numerical experiments with one version of Nesterov's algorithm [18] on box constrained convex quadratic programs. This is done in Section 4. Additional numerical results on the performance of the method on other problems can be found in Güler [7]. These experiments indicate that the methods show promise in solving sparse, large scale problems. Another purpose of this paper is to describe the basic results of Nesterov [18, 19], since these do not seem to be well-known in the West. In the process, we simplify and extend some of Nesterov's results. In section 2, we extend Nesterov's algorithm in [19, 20] for the constrained minimization problem (3) in which the function f may be strongly convex. See also [5]. Since (3) subsumes the unconstrained problem (2), the results in this section also apply to problem (2). In section 3, we give a proof of Nesterov's algorithm [18] for problem (3). This is different from Nesterov's original proof and gives a somewhat better convergence rate when f is strongly convex.

2   The Algorithm for Constrained Minimization

In this section we describe an optimal algorithm in the oracle model of complexity for the constrained minimization problem (3). It extends the algorithms of Nesterov [19, 20] to problems in which f may be strongly convex. (See also [5] for an application of the same ideas to the proximal point method.) It is assumed here that the constants l_f and L_f are known.

The key idea of the method is to generate recursively a sequence $\{\varphi_k\}_{k=0}^{\infty}$ of convex quadratic functions of the form

$$\varphi_k(x) = \varphi_k^* + \frac{A_k}{2}||x - v_k||^2$$

that approximate $f(x)$ in such a way that at step $k \geq 0$, the difference $\varphi_k(x) - f(x)$ is reduced by a fraction $1 - \alpha_k$. That is, for all $x \in C$

$$\varphi_{k+1}(x) - f(x) \leq (1 - \alpha_k)\left(\varphi_k(x) - f(x)\right), \tag{5}$$

where α_k is a number satisfying $0 < \alpha_k < 1$. If (5) is satisfied for each $k \geq 0$, then we obtain by induction

$$\varphi_k(x) - f(x) \leq \beta_k(\varphi_0(x) - f(x), \tag{6}$$

where

$$\beta_k = \prod_{j=0}^{k-1}(1 - \alpha_j).$$

If we have a point x_k such that

$$f(x_k) \leq \varphi_k^*, \tag{7}$$

then we obtain from (6) the inequality

$$f(x_k) - f(x) \leq \beta_k\left(\varphi_0(x) - f(x)\right). \tag{8}$$

If x^* minimizes f in C, then (8) yields the global convergence estimate

$$f(x_k) - f(x^*) \leq \beta_k\left(\varphi_0(x^*) - f(x^*)\right).$$

If $\beta_k \to 0$, then $\{x_k\}$ is a minimizing sequence for f. The magnitude of the constant β_k measures the rate of convergence of $f(x_k) - f(x^*)$ to zero.

The quadratic functions $\{\varphi_k\}$ are defined as follows. The initial quadratic function is given by

$$\varphi_0(x) = f(x_0) + \frac{A}{2}||x - x_0||^2,$$

where $x_0 \in C$ and $A \geq l_f$. The remaining quadratic functions $\{\varphi_k\}_{k \geq 1}$ will be defined recursively, so as to satisfy the inequality

$$\varphi_{k+1}(x) \leq \Phi_{k+1}(x) \equiv (1 - \alpha_k)\varphi_k(x) + \alpha_k\left(f(y_k) + \langle f'(y_k), x - y_k\rangle + \frac{l_f}{2}||x - y_k||^2\right), \tag{9}$$

where y_k is properly chosen. Note that if (9) is satisfied, then (5) holds true, since

$$\varphi_{k+1}(x) \leq \Phi_{k+1}(x) \leq (1 - \alpha_k)\varphi_k(x) + \alpha_k f(x) = f(x) + (1 - \alpha_k)(\varphi_k(x) - f(x)).$$

It is not obvious a priori how the points $x_k \in C$ and y_k can be chosen so as to satisfy the inequalities (7) and (9). Towards this end, we have the key technical Lemma 3 below. It extends a result of Nesterov [20] who proves it for the case $l_f = 0$. We need the following two lemmas in the proof of Lemma 3.

Lemma 1 *Suppose $h(x)$ is a closed, strongly convex function in $\mathbb{R}^n$ with constant $l = l_h > 0$, and $C \subseteq \mathbb{R}^n$ is a closed convex set. If x^* is the minimizer of h on C, then for any $x \in C$,*

$$h(x) \geq h(x^*) + \frac{l}{2}||x - x^*||^2.$$

Proof. The function $g(x)$ defined by $h(x) = g(x) + (l/2)||x||^2$ is convex. It is well–known that x^* minimizes h on C if and only if it satisfies the variational inequality

$$\langle w^* + lx^*, x - x^* \rangle \geq 0,$$

where $w^* \in \partial g(x^*)$. Thus, we have

$$g(x) \geq g(x^*) + \langle w^*, x - x^* \rangle \geq g(x^*) + l\langle x^*, x^* - x \rangle.$$

Since $2\langle a, b \rangle = ||a||^2 + ||b||^2 - ||a - b||^2$, we obtain,

$$g(x) \geq g(x^*) + \frac{l}{2}||x^*||^2 + \frac{l}{2}||x - x^*||^2 - \frac{l}{2}||x||^2.$$

The lemma is proved. $\qquad\qquad\square$

The proof of the following lemma is straightforward and we omit it.

Lemma 2 *Let $\gamma_1 > 0$ and $\gamma_2 > 0$. Then*

$$\gamma_1||x - a_1||^2 + \gamma_2||x - a_2||^2 = (\gamma_1 + \gamma_2)||x - \frac{\gamma_1 a_1 + \gamma_2 a_2}{\gamma_1 + \gamma_2}||^2 + \frac{\gamma_1 \gamma_2}{\gamma_1 + \gamma_2}||a_1 - a_2||^2.$$

We are ready for the key result of this section.

Lemma 3 *Suppose the point x_k satisfies the inequality $f(x_k) \leq \varphi_k^*$. If $x \in C$ and $0 < \alpha_k < 1$, then there exist φ_{k+1}^*, A_{k+1}, v_{k+1}, and y_k such that inequality (9) is satisfied, that is,*

$$\Phi_{k+1}(x) \geq \varphi_{k+1}(x) \equiv \varphi_{k+1}^* + \frac{A_{k+1}}{2}||x - v_{k+1}||^2.$$

We can choose

$$
\begin{aligned}
y_k &= (1 - \alpha_k)x_k + \alpha_k v_k, \\
A_{k+1} &= (1 - \alpha_k)A_k + \alpha_k l_f, \\
v_{k+1} &= \frac{(1 - \alpha_k)A_k/\alpha_k)(x_{k+1} - (1 - \alpha_k)x_k) + l_f \alpha_k y_k}{A_{k+1}} \\
&= v_k + \frac{(1 - \alpha_k)A_k}{A_{k+1}}\frac{(x_{k+1} - y_k)}{\alpha_k} + \frac{l_f \alpha_k}{A_{k+1}}(y_k - v_k), \\
\varphi_{k+1}^* &= f_k(x_{k+1}),
\end{aligned}
$$

where

$$x_{k+1} = \arg\min_{y \in C} f_k(y) = \pi_C\left(y_k - \frac{\alpha_k^2}{(1-\alpha_k)A_k}f'(y_k)\right),$$

$$f_k(y) = f(y_k) + \langle f'(y_k), y - y_k\rangle + \frac{(1-\alpha_k)A_k}{2\alpha_k^2}||y - y_k||^2.$$

Furthermore, if $(1-\alpha_k)A_k/\alpha_k^2 = L_f$, *then* $f(x_{k+1}) \le \varphi_{k+1}^*$.

Proof. Since x_k satisfies (7), we have

$$\begin{aligned}
\varphi_k(x) &= \varphi_k^* + \frac{A_k}{2}||x - v_k||^2 \\
&\ge f(x_k) + \frac{A_k}{2}||x - v_k||^2 \\
&\ge f(y_k) + \langle f'(y_k), x_k - y_k\rangle + \frac{l_f}{2}||x_k - y_k||^2 + \frac{A_k}{2}||x - v_k||^2.
\end{aligned}$$

It follows from this inequality and (9) that

$$\begin{aligned}
\Phi_{k+1}(x) &\ge f(y_k) + \langle f'(y_k), (1-\alpha_k)x_k + \alpha_k x - y_k\rangle \\
&\quad + \frac{(1-\alpha_k)A_k}{2}||x - v_k||^2 + \frac{\alpha_k l}{2}||x - y_k||^2 + \frac{(1-\alpha_k)l_f}{2}||x_k - y_k||^2.
\end{aligned}$$

Letting $y = (1-\alpha_k)x_k + \alpha_k x$, we have $x - v_k = (y - y_k)/\alpha_k$. Substituting this in the above inequality, we obtain

$$\begin{aligned}
\Phi_{k+1}(x) &\ge f(y_k) + \langle f'(y_k), y - y_k\rangle + \frac{(1-\alpha_k)A_k}{2\alpha_k^2}||y - y_k||^2 + \frac{\alpha_k l}{2}||x - y_k||^2 \\
&= f_k(y) + \frac{\alpha_k l_f}{2}||x - y_k||^2 \\
&\ge f_k(x_{k+1}) + \frac{(1-\alpha_k)A_k}{2\alpha_k^2}||y - x_{k+1}||^2 + \frac{\alpha_k l_f}{2}||x - y_k||^2,
\end{aligned}$$

where the last inequality follows from Lemma 1. Since $y = (1-\alpha_k)x_k + \alpha_k x$, we have we have

$$\begin{aligned}
\Phi_{k+1}(x) &\ge \varphi_{k+1}^* + \frac{(1-\alpha_k)A_k}{2}\left\|x - \frac{x_{k+1} - (1-\alpha_k)x_k}{\alpha_k}\right\|^2 + \frac{\alpha_k l_f}{2}||x - y_k||^2 \\
&\ge \varphi_{k+1}^* + \frac{A_{k+1}}{2}||x - v_{k+1}||^2,
\end{aligned}$$

where the last inequality follows from Lemma 2. If $(1-\alpha_k)A_k/\alpha_k^2 = L_f$, then the second inequality in (1) is satisfied, and it follows that $f(x_{k+1}) \le f_k(x_{k+1}) \equiv \varphi_{k+1}^*$. The lemma is proved. $\qquad\square$

Remark. In the case $l_f = 0$, we have $v_{k+1} = v_k + (x_{k+1} - y_k)/\alpha_k$.

Lemma 3 suggests the following algorithm.

The Constrained Minimization Algorithm I

Initialization. Choose a point $x_0 \in C$ and $A \geq l_f$. Define $y_0 = v_0 = x_0$, $A_0 = A$.

Iteration $k \geq 0$. Calculate α_k from the equation

$$L_f \alpha_k^2 = (1 - \alpha_k) A_k.$$

Define

$$
\begin{aligned}
y_k &= (1 - \alpha_k) x_k + \alpha_k v_k, \\
x_{k+1} &= \pi_C(y_k - L_f^{-1} f'(y_k)), \\
A_{k+1} &= (1 - \alpha_k) A_k + \alpha_k l_f, \\
v_{k+1} &= v_k + \frac{(1 - \alpha_k) A_k}{\alpha_k A_{k+1}} (x_{k+1} - y_k) + \frac{l_f \alpha_k}{A_{k+1}} (y_k - v_k).
\end{aligned}
$$

The following theorem summarizes the convergence properties of the above algorithm.

Theorem 1 *If $u \in C$, then the algorithm above satisfies*

$$f(x_k) - f(u) \leq \beta_k(f(x_0) - f(u) + \frac{A}{2}||u - x_0||^2), \tag{10}$$

where

$$
\beta_k \leq
\begin{cases}
\frac{2L_f/A}{k^2}, & \text{if } l_f = 0; \\
e^{-2\sqrt{q}k/3}, & \text{if } l_f > 0.
\end{cases}
$$

The method always produces a minimizing sequence $\{x_k\}$. If the problem (3) has optimal solutions and $f^* = \min_{x \in C} f(x)$, then the method is optimal in the oracle model and has the global convergence rate estimate

$$f(x_k) - f^* \leq \beta_k(f(x_0) - f^* + \frac{A}{2} d(x_0, C^*)^2),$$

where $d(x_0, C^*)$ is the distance from x_0 to the set of optimal solution C^*.

Proof. The inequality (10) follows from (8), and implies that $\{x_k\}$ is a minimizing sequence. The estimate for β_k is given in Nesterov [19, 20], see also [5]. The optimality of the method follows from the estimate on β_k, see [17, 20]. $\qquad\square$

One drawback of the algorithm above is that the constants l_f and L_f are assumed to be known. The algorithm described in the next section eliminates the necessity of knowing L_f, by performing an Armijo type line search. Such a line search is not possible here, since the term α_k appears in both y_k and x_{k+1} and makes these variables interdependent.

3   Another Algorithm for Constrained Minimization

In this section we describe another optimal algorithm in the oracle model of complexity for the constrained minimization problem (3). This algorithm is due to Nesterov [18]. We give a different proof of the method here, which can be modified to obtain a slightly better complexity estimate when f is strongly convex.

The idea of this algorithm is to generate a solution sequence $\{x_k\}_{k\geq 0}$ in C and an auxiliary sequence $\{y_k\}_{k\geq 1}$ which is appropriately chosen to satisfy some desirable properties.

The Constrained Minimization Algorithm II

Initialization. Choose a point $x_0 \in C$ and $L_0 > 0$. Define $y_1 = x_0$, $\gamma > 1$, and $\beta_1 = 1$.

Iteration $k \geq 1$. For $i = 0, \ldots, i_k$, let $L_k = \gamma^i L_{k-1}$,

$$x_k = \pi_C(y_k - L_k^{-1} f'(y_k))$$

and test until satisfied the inequality

$$f(x_k) \leq f(y_k) + \langle f'(y_k), x_k - y_k \rangle + \frac{L_k}{2}||x_k - y_k||^2. \tag{11}$$

Define

$$\beta_{k+1} = \frac{1 + \sqrt{4\beta_k^2 + 1}}{2} \tag{12}$$

$$y_{k+1} = x_k + \frac{\beta_k - 1}{\beta_{k+1}}(x_k - x_{k-1}). \tag{13}$$

Thus the point x_k is obtained from y_k using a gradient projection step together with an Armijo type line search procedure. This line search terminates since the inequality (11) is satisfied as soon as $L_k \geq L_f$. Its convergence properties are summarized in the following theorem.

Theorem 2 *The above algorithm has the global convergence rate estimate*

$$f(x_k) - f^* \leq \frac{2\max\{L_0, \gamma L_f\}}{(k+1)^2} d(x_0, X^*)^2.$$

Proof. From (11) we obtain

$$
\begin{aligned}
f(x_k) \;&\leq\; f(y_k) + \langle f'(y_k), x_k - y_k \rangle + \frac{L_k}{2}||x_k - y_k||^2 \\
&\leq\; f(y_k) + \langle f'(y_k), x - y_k \rangle + \frac{L_k}{2}||x - y_k||^2 - \frac{L_k}{2}||x - x_k||^2 \\
&\leq\; f(x) + \frac{L_k}{2}||x - y_k||^2 - \frac{L_k}{2}||x - x_k||^2,
\end{aligned}
$$

where the second inequality follows from Lemma 1 applied to the function $h(x) = f(y_k) + \langle f'(y_k), x - y_k \rangle + (L_k/2)||x - y_k||^2$, and the last one is due to the convexity of f. Rearranging the above inequalities, we have

$$
f(x) - f(x_k) \geq \frac{L_k}{2}||x - x_k||^2 + \frac{L_k}{2}||x - y_k||^2. \tag{14}
$$

Let $u \in C$ be arbitrary. From (14), we obtain in particular

$$
f(x_{k-1}) - f(x_k) \;\geq\; \frac{L_k}{2}||x_{k-1} - x_k||^2 - \frac{L_k}{2}||x_{k-1} - y_k||^2, \tag{15}
$$

$$
f(u) - f(x_k) \;\geq\; \frac{L_k}{2}||u - x_k||^2 - \frac{L_k}{2}||u - y_k||^2. \tag{16}
$$

Let $V_k = f(x_k) - f(u)$. Adding $\beta_k - 1$ times (15) to (16) gives

$$
\begin{aligned}
(\beta_k - 1)V_{k-1} - \beta_k V_k \;&\geq\; \frac{L_k}{2}\left((\beta_k - 1)||x_{k-1} - x_k||^2 + ||u - x_k||^2 \right) \\
&\quad - \frac{L_k}{2}\left((\beta_k - 1)||x_{k-1} - y_k||^2 + ||u - y_k||^2 \right) \\
&=\; \frac{L_k \beta_k}{2}||x_k - \frac{(\beta_k - 1)x_{k-1} + u}{\beta_k}||^2 \\
&\quad - \frac{L_k \beta_k}{2}||y_k - \frac{(\beta_k - 1)x_{k-1} + u}{\beta_k}||^2,
\end{aligned}
$$

where the equality follows from Lemma 2. The equation (12) implies $\beta_k(\beta_k - 1) = \beta_{k-1}^2$. Thus, multiplying the above inequality by β_k, we obtain

$$
\begin{aligned}
\beta_{k-1}^2 V_{k-1} - \beta_k^2 V_k \;&=\; \beta_k(\beta_k - 1)V_{k-1} - \beta_k^2 V_k \\
&\geq\; \frac{L_k}{2}||\beta_k x_k - (\beta_k - 1)x_{k-1} - u||^2 \\
&\quad - \frac{L_k}{2}||\beta_k y_k - (\beta_k - 1)x_{k-1} - u||^2.
\end{aligned}
$$

Note that (13) implies $\beta_k y_k - (\beta_k - 1)x_{k-1} = \beta_{k-1}x_{k-1} - (\beta_{k-1} - 1)x_{k-2}$. We also have $L_k \geq L_{k-1}$. Thus,

$$
\frac{\beta_{k-1}^2 V_{k-1}}{L_{k-1}} - \frac{\beta_k^2 V_k}{L_k} \geq \frac{1}{2}||\varphi_k||^2 - \frac{1}{2}||\varphi_{k-1}||^2,
$$

where

$$\varphi_k = \beta_k x_k - (\beta_k - 1)x_{k-1}. \tag{17}$$

Summing the above inequality for $j = 2, \ldots, k$, we obtain

$$\frac{\beta_1^2 V_1}{L_1} - \frac{\beta_k^2 V_k}{L_k} \geq \frac{1}{2}||\varphi_k||^2 - \frac{1}{2}||\varphi_1||^2,$$

which gives

$$\frac{\beta_k^2 V_k}{L_k} \leq \frac{\beta_1^2 V_1}{L_1} + \frac{1}{2}||\varphi_1||^2.$$

Note that $\beta_1 = 1$. Now, (16) and (17) imply

$$\begin{aligned}
\frac{\beta_k^2 V_k}{L_k} &\leq \frac{V_1}{L_1} + \frac{1}{2}||x_1 - u||^2 \\
&\leq \frac{1}{2}(||y_1 - u||^2 - ||x_1 - u||^2) + \frac{1}{2}||x_1 - u||^2 \\
&= \frac{1}{2}||y_1 - u||^2 = \frac{1}{2}||x_0 - u||^2.
\end{aligned}$$

Consequently, we have

$$V_k = f(x_k) - f(u) \leq \frac{L_k}{2\beta_k^2}||x_0 - u||^2. \tag{18}$$

The constant β_k can be estimated from (12). We have $\beta_{k+1} - \beta_k \geq \frac{1}{2}$, and thus $\beta_k \geq (k+1)/2$. It follows from (18) that

$$f(x_k) - f(u) \leq \frac{2L_k}{(k+1)^2}||x_0 - u||^2.$$

Let $x^* \in C$ be a minimizer of f on C. In particular, we have the global convergence rate estimate

$$f(x_k) - f^* \leq \frac{2L_k}{(k+1)^2}||x_0 - x^*||^2.$$

Due to the selection of L_k, we have $L_k \leq \max\{L_0, \gamma L_f\}$. The theorem is proved. $\square$

If f is strongly convex, then the algorithm above is not optimal. However, it is optimal when restarted every $O(\sqrt{L_f/l_f})$ steps. Lemma 1 implies

$$f(x_0) - f(x^*) \geq \frac{l_f}{2}d(x_0, X^*)^2.$$

Using this in Theorem 2 gives

$$f(x_k) - f^* \leq \frac{4 \max\{L_0, \gamma L_f/l_f\}}{(k+1)^2}(f(x_0) - f^*).$$

If $L_0 \leq \gamma L_f$ and $k = O(\sqrt{L_f/l_f})$, we obtain

$$\frac{f(x_k) - f^*}{f(x_0) - f^*} \leq \frac{1}{2}.$$

Thus, if we restart the algorithm every $O(\sqrt{L_f/l_f})$ steps, then the initial objective gap $f(x_0) - f^*$ is halved. It is shown in [17] that this is an optimal method for solving problem (3). The restart scheme above assumes that we know l_f, at least approximately. If such an approximation is not available, Nemirovsky [15] suggests adaptive restart schemes.

4   Applications to Convex Quadratic Programs

The method described in Section 3 is attractive for solving large scale, sparse convex quadratic programs with simple constraints, especially when the projection onto this set is easy to calculate. One such constraint set occurring in many practical engineering applications is the box constraint set $C = \prod_{i=1}^{n}[a_i, b_i]$, where $a_i = -\infty$ and $b_i = \infty$ are allowed. For example, the discretization of many obstacle problems, such as the elastic–plastic torsion problem [4], and the journal bearing problem [1] give rise to convex quadratic programs with box constraints.

Traditionally, such programs have been solved using the SOR method [3], [4]. Recently, Han, Pardalos, and Ye [8, 9, 10, 22] used interior point methods to solve such problems, and Moré [13] used a type of gradient projection method with an active–set identification scheme to solve such problems.

In [7], the author compares Nesterov's method described in Section 3 with the SOR method on a variety of problems including the elastic–plastic torsion problem, the journal bearing problem, and randomly generated convex quadratic programs. These numerical experiments indicate that Nesterov's method is competitive with the SOR method. It also has the advantage that it can be easily parallellized, although this is not tried in [7]. Nesterov's method is also used there to solve successfully a parametric minimal surface problem, which is a nonlinear elliptic partial differential equation. Discretization of this problem gives rise to an unconstrained convex minimization problem [2].

The interested reader can find the description of the SOR method in Cryer [3] and Glowinski [4]. We note that the SOR method can only be applied to convex quadratic programs with box constraints. A trivial example is given in [4] which shows that the method fails on a convex quadratic program over the two dimensional simplex. To our knowledge, the convergence rate of the SOR method is not known.

In this section we limit ourselves to reporting the results of our numerical experiments with Nesterov's method on minimizing randomly generated convex quadratic program subject to non–negativity constraints,

$$\min_{x \geq 0} \frac{1}{2}\langle Ax, x \rangle - \langle c, x \rangle + d, \tag{19}$$

where $A \in \mathbb{R}^{n \times n}$ is a random, symmetric positive–definite matrix, and $c \in \mathbb{R}^n$.

The matrix A is generated as follows. First, an $n \times n$ matrix B is generated with elements drawn from the standard normal distribution $N(0,1)$. The matrix B is factored by the QR decomposition as $B = QR$, where Q is an n by n orthogonal matrix and R is an upper triangular matrix. The matrices Q generated by this process are randomly distributed over the set of orthogonal matrices, see Stewart [23]. Next, we generate a diagonal matrix. We are interested in testing the performance of Nesterov's method for a given condition number since the condition number measures the rate of convergence of the method. For a fixed condition number κ, we generate the diagonal matrix $D = diag\{1, 1, \ldots, \kappa\}$. Finally, we form $A = Q^T D Q$.

We generate c indirectly as follows. First, we generate the optimal solution x^* to (19). We do this by randomly fixing a certain percentage of the coordinates of x^* at 0 and then generating the remaining coordinates of x^* from the uniform distribution in the interval $[0, 10]$. We then choose c in such a way to make x^* an optimal solution. This is done by satisfying the Karush–Kuhn–Tucker complementary slackness conditions for (19),

$$Ax - c \geq 0, \quad x \geq 0, \quad \langle Ax - c, x \rangle = 0.$$

Thus, if $x_i > 0$ we define $c_i = (Ax - c)_i$. If $x = 0$, we have some freedom in defining c, since we need only to satisfy the inequality $c_i \leq (Ax)_i$. We choose $c_i = (Ax)_i - r_i$, where r_i is uniformly distributed in the interval $[0, 10]$.

The methods were tested on a VAX 6410 machine. The codes were in double precision arithmetic and were compiled using the default options. Our convergence criterion was based on the Kuhn-Tucker conditions. The methods were stopped when these conditions were satisfied to within ε, that is, $\min\{|\partial f(x)/\partial x_i|, x_i\} \leq \varepsilon$. We chose $\varepsilon = 1.0e - 5$.

We used the subroutines URANDM and UNORMR of Linus Schrage to generate the random numbers from the uniform and the normal distributions, respectively. We used the IMSL subroutine DLQRRR and DLQERR to perform the QR decomposition.

We tested the methods for five different condition numbers, $\kappa=100, 200, 400, 800,$ and 1600. For each choice of the condition number, we solved QP problems of three different sizes, $n = 10, 20, 40$. For a given choice of κ and n, we generated 20 random problems.

We solved each random problem generated in this fashion with Nesterov's method with the restart values $1, 20, 40, \ldots, 400, 450, 500, \ldots, 1000$. In total we generated 300 random problems. This amounts to a total of 10200 runs of Nesterov's method. We also solved each random problem with the SOR method using the relaxation parameters $\omega = 1.95, 1.90, 1.80, \ldots, .1$. This adds up to total of 6000 runs of the SOR method.

The results of our experiments are illustrated in tables 1–6 and figures 1–8. In the tables, we give only the results for condition numbers $\kappa = 100, 400$, and 1600. In the figures, only the results for condition numbers $\kappa = 400$ and 1600 are given. The results for the other condition numbers tested are similar. The numbers in these tables and figures are the averages over 20 random problems. This means that in Nesterov's method, for a given κ, n, and restart value, 20 random problems were run and the average number of iterations and cpu seconds were recorded. The same was done in the SOR method for each κ, n, and ω.

The first conclusion to be drawn from these figures is the remarkable fact that for a given condition number, the number of steps required by Nesterov's method is independent of the dimension of the problem, as shown by the dramatic figures 1 and 5. This supports the theory nicely, which states that the convergence rate of Nesterov's method depends only on the condition number of the problem, not its dimension.

The same conclusion cannot be claimed for the SOR method. The tables 2, 4, 6 and figures 2, 4, 6, 8 show that for a given ω, the number of iterations changes with the dimension of the problem. The curves in these figures become more steep as the dimension of the problem increases, which means that the SOR method becomes very sensitive to the choice of the relaxation parameter as the dimension increases. Moreover, the number of iterations required at the optimal ω seems to increase with the problem size. This becomes more noticeable as the condition numbers of the problems increase.

The performance of both methods worsen as the condition number of the problems increase, but the SOR method worsens at a faster rate than Nesterov's method. In fact, we stopped testing the algorithms at condition number 1600 because the SOR method was taking unacceptable amounts of computing time. When we compare the methods at their best, we see that the SOR method takes more computing time already at condition number 200. This happens in spite of the fact that one iteration of Nesterov's method is about twice as costly as one iteration of the SOR method.

From these results it is obvious that the SOR method cannot compete with Nesterov's method for solving general quadratic programs (19) in large dimensions or ill conditioned problems. However, the numerical experiments in [4, 7] indicate that the

SOR method is fast for certain problems, such as the elastic–plastic torsion problem. This seems to indicate that the SOR method needs a certain structure in the matrix A in order to work well. Further research is needed to explain the fast and slow convergence of the SOR method.

References

[1] Cimatti, G. (1977), "On a problem of the theory of lubrication governed by a variational inequality," *Applied Mathematics and Optimization 3*, 227–242.

[2] Concus, P. (1967), "Numerical solution of the minimal surface problem," *Mathematics of Computation 21*, 340–350.

[3] Cryer, C. W. (1971), "The solution of a quadratic programming problem using systematic overrelaxation," *SIAM Journal on Control and Optimization 9*, 385–392.

[4] Glowinski, R. (1984), *Numerical Methods for Nonlinear Variational Problems*, Springer Verlag, New York, New York.

[5] Güler, O. (1992), "New proximal point algorithms for convex minimization," *SIAM Journal on Optimization 2*, 649–664.

[6] Güler, O. (1992), "Augmented Lagrangian algorithms for linear programming," *Journal of Optimization Theory and Applications 75*, 445–470.

[7] Güler, O, (1990), *Efficient Algorithms for Convex Programming*, Ph.D. Thesis, The University of Chicago.

[8] Han, C. G., Pardalos, P. M., and Ye, Y. (1990), "Computational aspects of an interior point algorithm for quadratic problems with box constraints," *Large–Scale Numerical Optimization*, 92–112 (Eds: T. Coleman and Y. Li), SIAM, Philadelphia, Pennsylvania.

[9] Han, C. G., Pardalos, P. M., and Ye, Y. (1990), "An interior point algorithm for large-scale quadratic problems with box constraints," *Springer–Verlag Lecture Notes in Control and Information, Vol. 144*, 413–422 (Eds: A. Bensoussan and J. L. Lions).

[10] Han, C. G., Pardalos, P. M., and Ye, Y. (1992), "On the solution of indefinite quadratic problems using an interior point algorithm," *Informatica 3*, 474–496.

[11] Ivanov, V. V. (1972), "Optimal algorithms of minimization of certain classes of functions," *Cybernetics 4*, 620–634.

[12] Luenberger, D. G. (1973), *Introduction to Linear and Nonlinear Programming*, Addison–Wesley, Reading, Massachusetts.

[13] Moré, J. J. and Toraldo, G. (1991), "On the solution of large quadratic programming problems with bound constraints," *SIAM Journal on Optimization 1*, 93–113.

[14] Nemirovsky, A. S. (1988), "A new polynomial algorithm for linear programming," *Soviet Mathematics Doklady 37*, 264–269.

[15] Nemirovsky, A. S. (1988), Private communication.

[16] Nemirovsky, A. S. and Nesterov, Y. E. (1985), "Optimal algorithms of smooth convex programming," *USSR Computational Mathematics and Mathematical Physics 25*, 21–30.

[17] Nemirovsky, A. S. and Yudin, D. B. (1983), *Problem Complexity and Method Efficiency in Optimization*, John Wiley and Sons, New York, New York.

[18] Nesterov, Y. E. (1983), "A method of solving a convex programming problem with convergence rate $O(1/k^2)$", *Soviet Mathematics Doklady 27*, 372–376.

[19] Nesterov, Y. E. (1988), "On an approach to the construction of optimal methods of minimization of smooth convex functions," (in Russian) *Ekonomika i Matematicheskie Metody 24*, 509–517.

[20] Nesterov, Y. E. (1989), *Efficient Methods of Nonlinear Programming*, Radio i Sviaz, Moscow.

[21] Nesterov, Y. E. and Nemirovsky, A. S. (1991), *Interior Point Polynomial Methods in Convex Programming: Theory and Applications*, to be published by SIAM, Philadelphia, Pennsylvania.

[22] Pardalos, P. M., Han, C. G., and Ye, Y. (1991), "Interior point algorithms for solving nonlinear optimization problems," *COAL Newsletter 19*, 45–54.

[23] Stewart, G. W. (1980), "The efficient generation of random orthogonal matrices with an application to condition estimators," *SIAM Journal on Numerical Analysis 17*, 403–409.

[24] Vavasis, S. A. (1991), *Nonlinear Optimization – Complexity Issues*, Oxford University Press, New York.

Dimension	Restart Value	Avg. Iter.	Avg. Secs.
10	1	276.4	0.333
10	20	115.2	0.157
10	1000	634.7	0.541
20	1	239.5	0.746
20	20	114.9	0.323
20	1000	700.4	1.698
40	1	265.1	2.553
40	20	105.8	0.895
40	40	144.0	1.181

Table 1: Nesterov's method on random QP with $\kappa = 100$

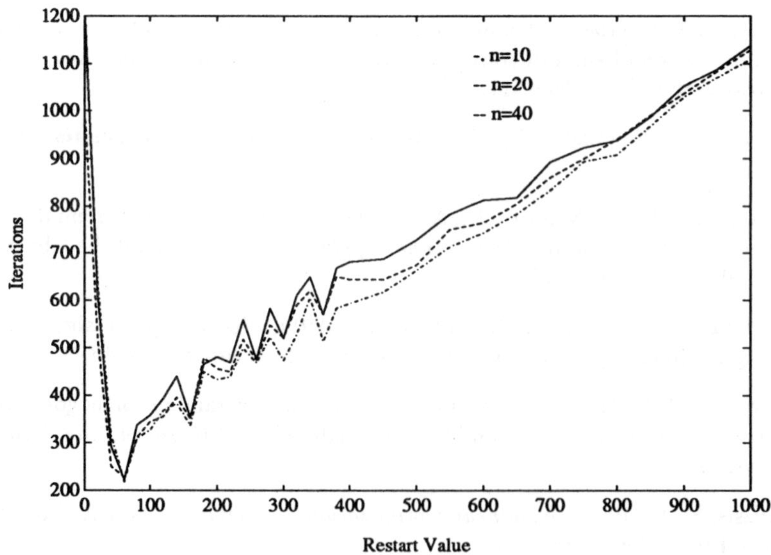

Figure 1: Iterations taken by Nesterov's method ($\kappa = 400$)

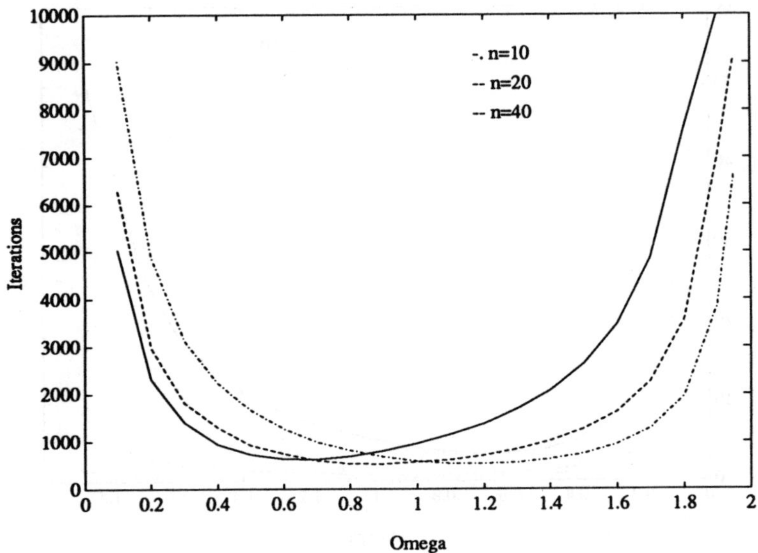

Figure 2: Iterations taken by SOR method ($\kappa = 400$)

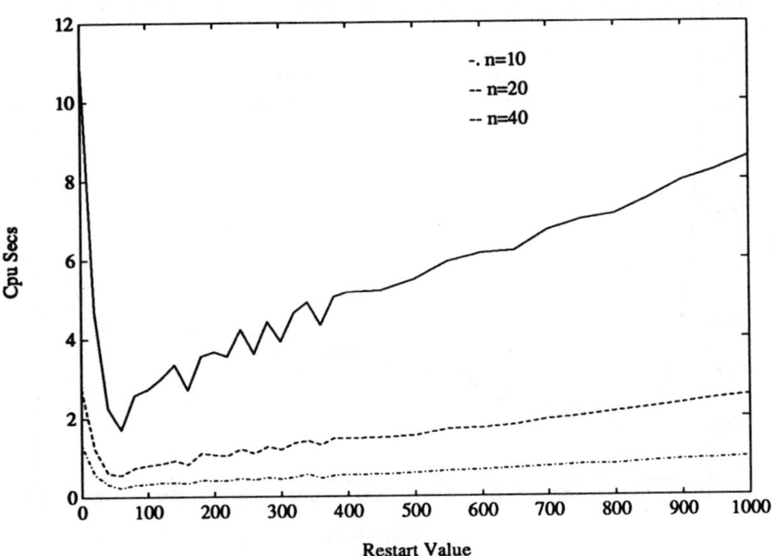

Figure 3: Cpu time taken by Nesterov's method ($\kappa = 400$)

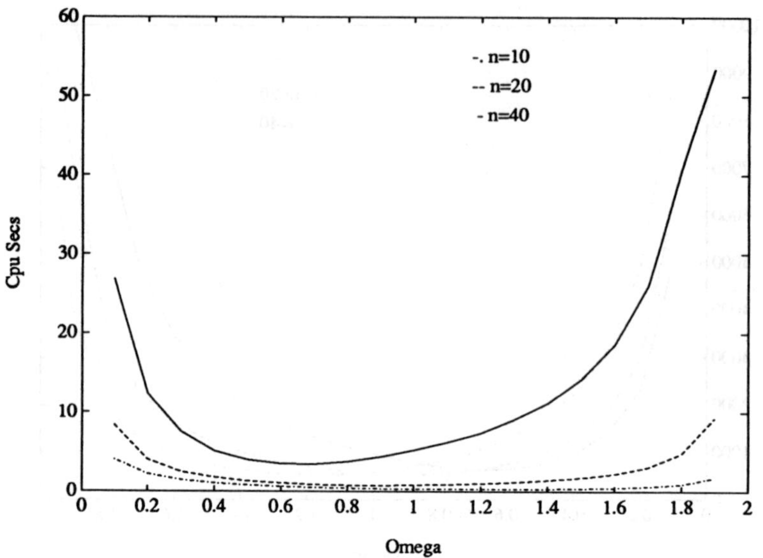

Figure 4: Cpu time taken by SOR method ($\kappa = 400$)

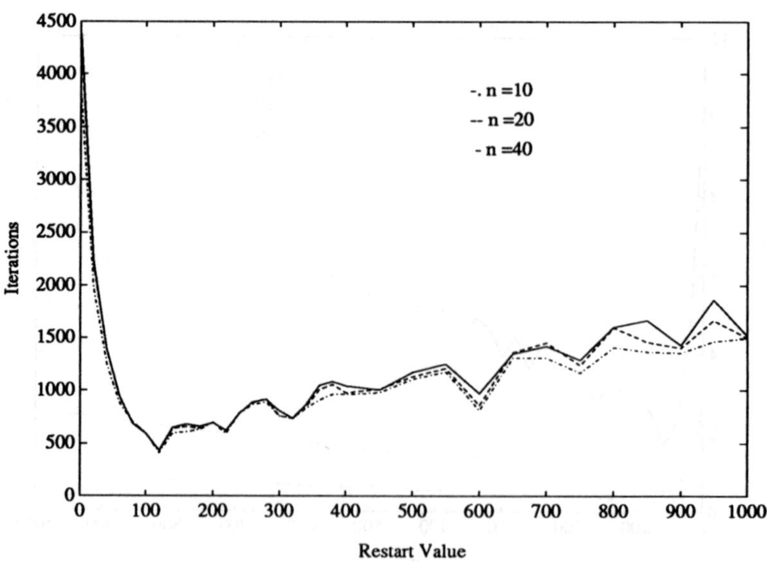

Figure 5: Iterations taken by Nesterov's method ($\kappa = 1600$)

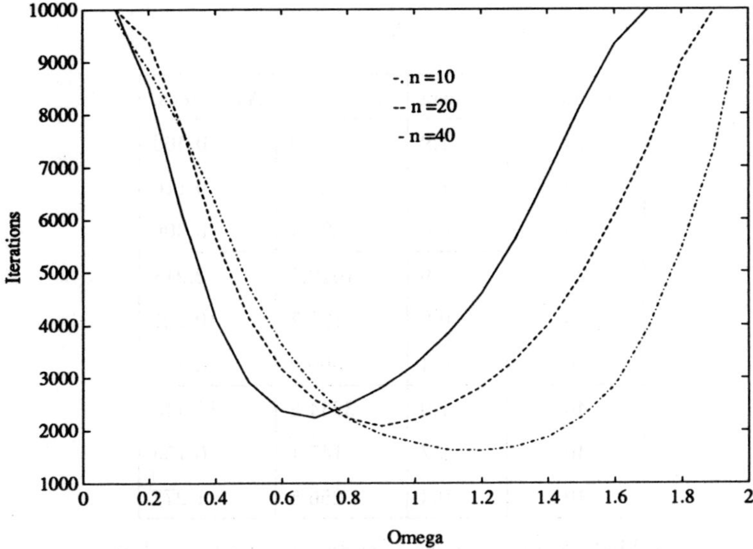

Figure 6: Iterations taken by SOR method ($\kappa = 1600$)

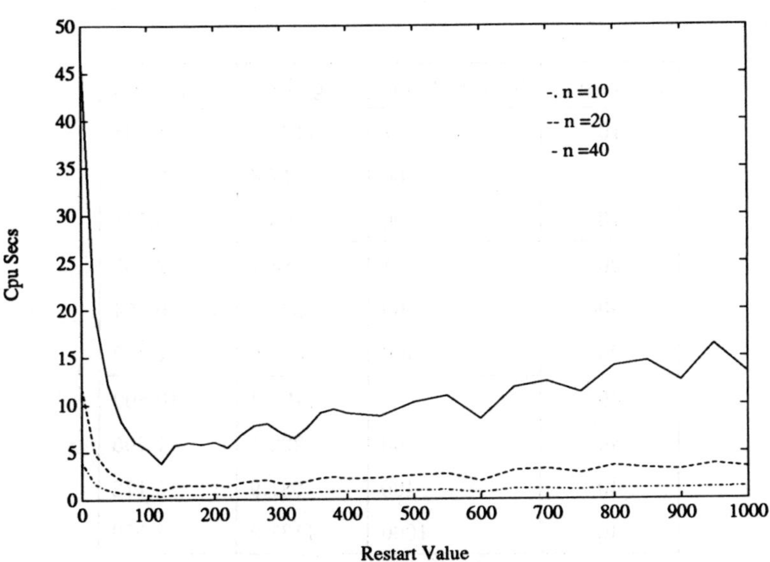

Figure 7: Cpu time taken by Nesterov's method ($\kappa = 1600$)

Dimension	Omega	Avg. Iter.	Avg. Secs.
10	1.9	860.0	0.468
10	1.0	125.7	0.116
10	0.4	520.4	0.306
20	1.9	1622.7	2.218
20	0.8	121.7	0.221
20	0.4	300.0	0.461
40	1.9	3234.5	15.413
40	0.7	147.1	0.753
40	0.4	256.5	1.277

Table 2: SOR method on random QP with $\kappa = 100$

Dimension	Restart Value	Avg. Iter.	Avg. Secs.
10	1	1182.8	1.293
10	60	217.8	0.228
10	400	593.4	0.551
20	1	988.1	2.705
20	60	227.4	0.554
20	400	644.6	1.460
40	1	1195.3	10.809
40	60	222.1	1.705
40	400	681.9	5.172
40	1000	1137.2	8.579

Table 3: Nesterov's method on random QP with $\kappa = 400$

Dimension	Omega	Avg. Iter.	Avg. Secs.
10	1.9	3861.9	1.736
10	1.0	592.3	0.311
10	0.4	2230.1	1.037
20	1.9	6976.9	9.275
20	0.9	530.9	0.751
20	0.4	1299.9	1.776
40	1.9	*	53.443
40	0.7	631.7	3.420
40	0.4	944.3	5.074

Table 4: SOR method on random QP with $\kappa = 400$
* The algorithm failed to converge in 10,000 iterations.

Dimension	Restart Value	Avg. Iter.	Avg. Secs.
10	1	3808.0	3.897
10	120	408.8	0.369
10	400	961.1	0.818
20	1	4282.0	11.557
20	120	432.4	0.985
20	400	976.8	2.158
40	1	4377.7	45.893
40	120	431.8	3.836
40	400	1041.7	9.103
40	10000	5278.7	45.970

Table 5: Nesterov's method on random QP with $\kappa = 1600$

Dimension	Omega	Avg. Iter.	Avg. Secs.
10	1.9	7374.4	3.184
10	1.2	1617.8	0.727
10	0.4	6319.5	2.735
20	1.9	*	13.666
20	0.9	2084.1	2.856
20	0.4	5641.4	7.700
40	1.9	*	55.040
40	0.7	2243.9	12.397
40	0.4	4124.3	22.750

Table 6: SOR method on random QP with $\kappa = 1600$
* The algorithm failed to converge in 10,000 iterations.

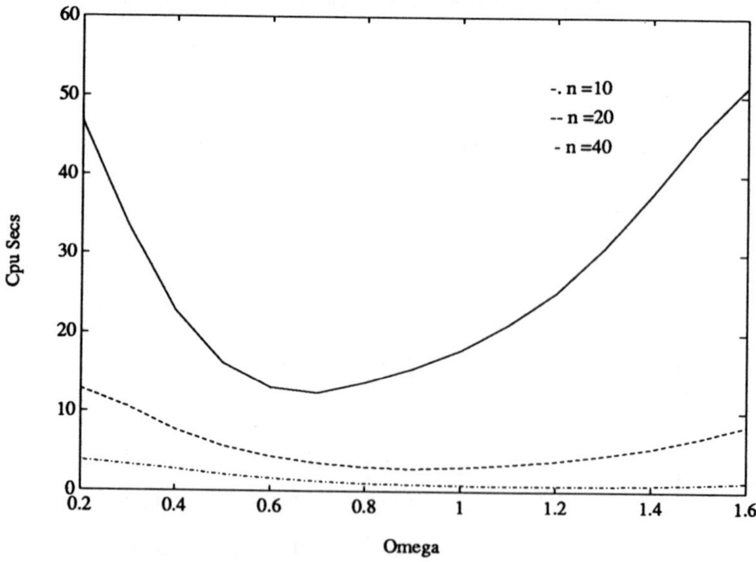

Figure 8: Cpu time taken by SOR method ($\kappa = 1600$)

Complexity in Numerical Optimization, 203-253
P.M. Pardalos, Editor

A Classification of Static Scheduling Problems

Jeffrey W. Herrmann
Chung-Yee Lee
*Department of Industrial and Systems Engineering, University of Florida,
Gainesville, FL 32611*

Jane L. Snowdon
IBM Corporation, CIM Technology Center, 051-1/2103, Boca Raton, FL 33432

Abstract

In the last four decades, scheduling problems have received much attention
by researchers. Recently, the Just-in-Time concept has inspired a renewed
interest in scheduling, especially among industry practitioners. Although a
number of papers have reviewed this field, this paper presents an easy-to-use
reference guide of static scheduling problems with complexity results for practitioners, students, and others. Every attempt has been made to be complete;
however, this survey is not exhaustive. This paper includes both regular and
non-regular measures of performance, separate sections on dual criteria and
multicriteria problems, and a discussion of recent developments in heuristic
search approaches to these problems.

Keywords: Complexity, production planning and scheduling, sequencing,
heuristic search.

1   Introduction

This paper presents a survey of static scheduling problems in an easy-to-use reference
guide. To the best of the authors' knowledge, the only paper to provide such a review
is Lageweg et al. (1982). This paper differs from Lageweg et al. in the following
way: instead of being organized by complexity results, this paper classifies problems

by their common characteristics. In addition, recent results and more sophisticated heuristics are included.

The paper has seven tables listing a number of deterministic machine scheduling problems and their algorithms and complexities. The paper divides the problems into three primary categories: one-machine, parallel-machine, and shop problems; it also covers four additional topics: dual-criteria problems, resource-constrained problems, stochastic scheduling problems, and heuristic searches, called in this paper smart-and-lucky searches.

Recently, some researchers have applied heuristic searches, such as simulated annealing, tabu search, and genetic algorithms, to scheduling problems. These searches are smart enough to escape most local optima; still, they must be lucky to find the global optimum. Some papers have shown promising results and research is continuing. This paper includes some of these studies in the tables.

Works on scheduling problems include the following books: Conway, Maxwell, and Miller (1967); Baker (1992); Rinnooy Kan (1976); Lenstra (1985); French (1982); Dempster, Lenstra, and Rinnooy Kan (1982); and Morton (1992). Surveys have been done by Graves (1981) and Lawler, Lenstra, Rinnooy Kan, and Shmoys (1989), which was a major source of information for some of these tables. Baker and Scudder (1990) also provided information on the earliness-tardiness problems. Fundamental papers on simulated annealing include Kirkpatrick et al. (1983), Cerny (1983), and Aarts and Van Laarhoven (1985); on tabu search, Glover (1989, 1990); and on genetic algorithms Holland (1975), Liepins and Hilliard (1989), Goldberg (1986), and Davis (1991).

2   Problem description

In deterministic machine scheduling, a set of m machines must process a set of n jobs, and all problem data is known in advance. The machine environment, job characteristics, objective function, and notation for the deterministic machine scheduling problem are defined in this section.

2.1   Machine environment

The first element of the problem description is the machine environment. A job may consist of one or more operations. If each job has only one operation, the environment is a single-machine problem or a parallel machine problem, where the job may be processed by any of the machines. Parallel machines may be identical, uniform, or unrelated machines.

Identical machines process jobs with the same speed.
Uniform machines have machine-dependent speeds.
Unrelated machines have machine-and-job-dependent speeds.

For shops, each job has a fixed sequence of operations requiring different machines. A shop may be a flow shop or a job shop. In a **job shop**, jobs may have different operation sequences. In a **flow shop**, all jobs have the same operation sequence.

2.2   Job characteristics

Each problem has a set of job characteristics, which may occur in any combination. **Preemption** (abbreviated **pmtn**) refers to environments in which a job's processing may be interrupted and later resumed (possibly on another machine). The jobs to be processed can have **precedence constraints** (abbreviated **prec**), that is, some jobs cannot be started until others are completed. The graph of these constraints may resemble a **tree** , where each job will have either a maximum of one successor (**intree**) or a maximum of one predecessor (**outtree**). The jobs may not be available until their individual **release dates** into the shop. Jobs can have **deadlines** which must be met or a **common due date** (all jobs are due at the same time). The jobs or operations can have unit or equal **processing requirements**.

2.3   Objective function

The objective function to be maximized or minimized is the third element of a problem description. This may be a sum of variables or the maximum of some variable or function. Typical objective functions include the following performance measures: **flowtime** is the sum of completion times; **makespan** is the maximum completion time. **Lateness** is the difference between the due date and completion time; this value can be positive (and thus is a measure of **tardiness**) or negative (**earliness**).

One section of the Table 1 focuses on single-machine **earliness-tardiness** problems. Earliness-tardiness problems have been the focus of much recent study, since they model some aspects of the Just-in-Time philosophy, in which it is desirable that jobs finish close to their due dates. This deviation from the due date is the sum of the earliness and tardiness, since each quantity is a positive deviation in one of two directions. In addition, researchers have studied different performance measures, such as the square of the deviation, the penalty cost of the deviation, and the cost of delivering the jobs.

2.4   Notation

The problem descriptions include a number of abbreviations and symbols that represent characteristics and functions or variables associated with deterministic scheduling problems. Included here is a list of these symbols and their meaning.

m	the number of machines.
n	the number of jobs.
J_j	job j, where $j = 1, \ldots, n$.
M_i	machine i, where $i = 1, \ldots, m$.
α_j	the unit earliness penalty of job j.
α	a common unit earliness penalty, all $\alpha_j = \alpha$.
β_j	the unit tardiness penalty of job j.
β	a common unit tardiness penalty, all $\beta_j = \beta$.
C_j	the completion time of job j.
C_{ij}	the completion time of operation i of job j.
C_{max}	the makespan, the maximum of C_j over all jobs j.
d_j	the due date of job j.
D_j	the deadline for the completion time of job j.
e_j	the start of a due date window for job j.
E_j	the earliness of job j; $E_j = max\ \{d_j - C_j, 0\}$.
E_{max}	the maximum earliness of a set of jobs.
f_j	a regular (nondecreasing) function of a job's completion time.
f_{max}	the maximum of the f_j over all jobs j.
I_{ij}	the idleness of operation i of job j; $I_{ij} = C_{i+1,j} - p_{i+1,j} - C_{ij}$.
L_j	the lateness of job j equals $C_j - d_j$.
L_{max}	the maximum lateness.
O_{ij}	the ith operation of job j.
p_j	the processing time of job j.
p_{ij}	the processing time of operation i of job j.
Q	the number of batch deliveries.
r_j	the release date of job j.
T_j	the tardiness of job j; $T_j = max\ \{C_j - d_j, 0\}$.
T_{max}	the maximum tardiness of a set of jobs; this also refers to the minimum possible T_{max}.
U_j	$U_j = 1$ if job j is tardy; otherwise $U_j = 0$.
U_{min}	the minimum possible number tardy.
w_j	the weight associated with job j.

The following definitions are used only in Table 6 for stochastic scheduling problems:

exp	exponentially distributed
LEPT	longest expected processing time first
SEPT	shortest expected processing time first
$E[C_{max}]$	expected makespan
$E[\Sigma\,C_j]$	expected flowtime
$E[\Sigma\,U_j]$	expected number of tardy jobs
$E[\Sigma\,f_j]$	expected sum of f_j
iid	identically, independently distributed

3   Table organization

Tables 1 through 8 list a number of machine scheduling problems in these areas: one-machine problems, dual-criteria problems, parallel-machine problems, shop problems, resource-constrained problems, stochastic scheduling problems, and heuristic searches. The problems in each table are grouped by some common characteristic or objective function. Each of these tables is composed of three columns, as described below.

Some of these tables deserve additional comment. Table 2 includes some previously-studied **dual-criteria** and **multicriteria** single-machine scheduling problems. In the dual-criteria problems, the primary objective is used as a constraint on the feasible schedules, and the secondary criteria is minimized over this more limited set. In multicriteria problems, the aim is to find efficient solutions that cannot be dominated by any other solution in all criteria simultaneously.

In Table 5, **resource-constrained project scheduling**, jobs may require the use of a part of some limited resource during job execution. In Table 6, **stochastic scheduling**, some problem data may be unknown at the beginning. Usually, this occurs in processing times that are random variables (traditionally exponentially distributed). Table 7 includes a number of papers that use **smart-and-lucky** searches to solve scheduling problems. Table 8 lists some problems that have a **class scheduling** structure; that is, the jobs to be scheduled are grouped into job classes, where a setup is performed when the machine switches from one job class to another.

These tables form an easy-to-use reference guide for practitioners, students, and others. Every attempt has been made to be complete; however, this survey is not exhaustive.

3.1   Column one: Problem description

The problem description is given in the first column. The standard description of a scheduling problem includes three elements: machine environment (denoted by x), job characteristics (denoted by y), and objective function (denoted by z); this description is expressed in a three- field classification: $x/y/z$.

The first field is the machine environment. The symbol may be the number 1, which denotes a **single-machine problem**. Parallel machine problems are denoted by the letters **P**, **Q**, or **R**, as follows:

P: identical machines
Q: uniform machines
R: unrelated machines

Two additional letters identify multi-operation (shop) problems:

J: job shop
F: flow shop

In any of these cases, a number appearing after the symbol denotes that the number of machines is fixed at this value (for example, **F3** denotes a three-machine flow shop).

The second field contains any special job characteristics, which may occur in any combination. If the second field is blank, the jobs are assumed to have individual due dates and be immediately available, non-preemptive, and without precedence constraints between jobs.

The objective function to be minimized is the third field. This can be a sum of variables or the maximum of some variable or function. The following examples may serve to clarify this notation:

$1/r_j/L_{max}$	One-machine problem with jobs that have unequal release dates; the objective function is the maximum lateness.
$1/r_j, p_j = p/L_{max}$	Same as above but processing times are identical.
$1//\Sigma C_J$	One-machine problem with all jobs available at time 0; the objective function is the sum of completion times, i.e., flowtime.
$P/pmtn/C_{max}$	Identical parallel machine problem with preemption; the objective function is the maximum completion time, i.e., makespan.
$F2//C_{max}$	Two-machine flowshop problem; the objective function is makespan.
$J//Cmax$	General jobshop problem; the objective function is makespan.

3.2   Column two: Complexity

The second column of the table contains information about the solution for the problem. For example, if the problem is **NP-hard** (or strongly NP-hard), then **NP** as well as an error bound for an approximate solution can be listed. For the polynomial problem, this column can include the complexity of the algorithm as well as a brief description of, or an abbreviated name for, the algorithm. Common algorithms are **SPT (shortest processing time)** and **EDD (earliest due date)**. The following examples may serve to clarify the notation in this column:

NP	The problem is NP-hard.
P	The problem is polynomially solvable.
$O(n^2)$	The complexity of the algorithm is proportional to the square of the number of jobs n.
$O(nlogn)$	The complexity of the algorithm is proportional to n times the log of n.
NP(2)	The problem is NP-hard, and the heuristic has a relative error bound of 2.
EDD	The algorithm used to solve the problem optimally is the earliest due date procedure.
ERD	The algorithm used to solve the problem optimally is the earliest release date procedure.
WSPT	Weighted shortest processing time: namely, sequence jobs by ratio of processing time to weight, smallest first.
DP	Problem solved with dynamic program.
LP	Problem solved with linear program.
LB	Algorithm finds lower bound on solution.

3.3   Column three: Reference

The last column of the table lists the papers that address the problem or establish the results listed in column two. Refer to the bibliography for a complete reference.

Table 1. The Single Machine Deterministic Scheduling Problem

Problem description	Complexity	Reference

MINIMAX CRITERIA

$1/\text{prec}/f_{max}$	$O(n^2)$	Lawler(1973)
$1/\text{pmtn}, r_j, \text{prec}$ $/f_{max}$	$O(n^2)$	Baker et al. (1983)

DELIVERY TIME MODEL: MINIMIZE MAXIMUM LATENESS; RELEASE DATES

$1//L_{max}$	$O(n \log n)$ EDD	Jackson (1955)
$1//L_{max} - L_{min}$	branch-and-bound	Gupta and Sen (1984b)
	DP $O(n \log n\ MS)$	Liao and Huang (1991)
$1/r_j/L_{max}$	NP(2) extended Jackson	Simons (1978)
$1/r_j/L_{max}$	NP(3/2)	Potts (1980b)
$1/r_j/L_{max}$	NP(4/3)	Hall and Shmoys (1988)
$1/r_j/L_{max}$	NP	Lenstra, Rinnooy Kan, Brucker (1977)
$1/r_j, d_j = d/L_{max}$	$O(n \log n)$ ERD	Jackson (1955)
$1/r_j, p_j = p/L_{max}$	P	Simons (1978)
$1/r_j, \text{prec}/L_{max}$	NP (elegant enumeration)	Simons (1978)

FLOWTIME PROBLEMS

$1//\Sigma\, w_j C_j$	WSPT	Smith (1956)
$1//\Sigma\, C_j$	SPT	Smith (1956)
$1/\text{prec}, p_j = 1/\Sigma\, C_j$	NP	Lawler (1978)
		Lenstra, Rinnooy Kan (1978)
$1/r_j/\Sigma\, C_j$	NP	Lenstra, Rinnooy Kan, Brucker (1977)
	asymptotic algorithms	Gazmuri (1985)
	heuristics	Liu and MacCarthy (1991)
$1/r_j, \text{pmtn}/\Sigma\, C_j$	P	Baker (1974)
$1/r_j, \text{pmtn}/\Sigma\, w_j C_j$	NP	Labetoulle et al. (1984)
$1/\text{prec}/\Sigma\, w_j C_j$	polynomial decomposition	Sidney and Steiner (1986)
$1/D_j/\Sigma\, C_j$	P extension of Smith (1956)	Lenstra, Rinnooy Kan, Brucker (1977)

FLOWTIME PROBLEMS, cont.

$1/D_j/\Sigma w_j C_j$	NP	Lenstra, Rinnooy Kan, Brucker (1977)
$1/r_j$, pmtn, $D_j/\Sigma C_j$	NP	Du and Leung (1988b)
$1/D_j/\Sigma w_j C_j$	NP	Potts and Van Wassenhove (1983)
$1/r_j/\Sigma w_j C_j$	NP	Hariri and Potts (1983)
$1/r_j$, clustered jobs $/\Sigma w_j C_j$	P	Posner (1986)
$1/\text{prec}/\Sigma w_j C_j$	branch-and-bound	Potts (1985b)
$1/\text{preventive}$ maintenance$/\Sigma C_j$	NP	Rinnooy Kan
	NP(2/7)	Lee and Liman

NUMBER OF TARDY JOBS

$1//\Sigma w_j U_j$	NP	Karp (1972)
$1//\Sigma w_j U_j$	NP	Lawler and Moore (1969)
$1//\Sigma U_j$	P	Moore (1968)
$1/D_j, D_j \geq d_j/\Sigma U_j$	NP	Lawler (1982b)
$1/p_j < p_k$ implies $w_j \geq w_k/\Sigma w_j U_j$	$O(n \log n)$	Lawler (1976b)
$1/r_j/\Sigma U_j$	NP	Lawler (1982b)
	dominance properties	Erschler et al. (1983)
$1/r_j$, pmtn$/\Sigma U_j$	$O(n^5)$	Lawler (1990)
$1/r_j$, pmtn$/\Sigma w_j U_j$	$O(n^3(\Sigma w_j)^2)$	Lawler (1990)
$1/r_j, r_j < r_k$ implies $d_j \leq d_k/\Sigma U_j$	$O(n^2)$	Kise, Ibaraki, Mine (1978)
	$O(n \log n)$	Lawler (1982b)
$1/\text{pmtn}, r_j, (r_j, d_j)$ nested$/\Sigma w_j U_j$	$O(n \log n)$	Lawler (1982b)
$1/\text{pmtn}, r_j, r_j < r_k$ implies $p_j \leq p_k$ and $w_j \geq w_k/\Sigma w_j U_j$	$O(n \log n)$	Lawler (1982b)
$1/p_j = 1/\Sigma U_j$	$O(n)$	Monma (1982)
$1/p_j = 1, \text{prec}/\Sigma U_j$	NP	Garey and Johnson (1976)
$1//\Sigma w_j U_j$	NP, branch and bound	Villarreal and Bulfin (1983)
	NP: LB in $O(n \log n)$	Potts and Van Wassenhove (1988)
	DP: $O(n \Sigma w_j)$	Sahni (1976)

OTHER SUMS

$1/p_j = 1/\Sigma f_j$	$O(n^3)$ weighted bipartite matching	Lawler et al. (1989)
$1//\Sigma f_j$	open: $O(n^3)$ for LB	Rinnooy Kan, Lageweg, Lenstra (1975)
$1/\text{prec}/\Sigma f_j$	DP	Steiner (1984)
$1//\Sigma T_j$	pseudopolynomial algorithm $O(n^4 \, MS)$	Lawler (1977)
	polynomial approximation	Lawler (1982c)
$1//\Sigma T_j$	NP	Du and Leung (1989b)
	adjacent pairwise interchange	Fry et al. (1989)
	branch-and-bound	Potts and Van Wassenhove (1985)
	branch-and-bound	Sen and Borah (1991)
$1/\text{chain}, p_j = 1/\Sigma T_j$	NP	Leung and Young (1990)
$1/\text{prec}, p_j = 1/\Sigma T_j$	NP	Lenstra and Rinnooy Kan (1978)
$1/\text{pmtn}$ $/\Sigma \, min \, \{T_j, p_j\}$	$O(n \log n)$	Potts and Van Wassenhove (1992)
$1//\Sigma \, min \, \{T_j, p_j\}$	NP	Potts and Van Wassenhove (1992)
$1//\Sigma \, w_j T_j$	NP	Lawler (1977)
		Lenstra, Rinnooy Kan, Brucker (1977)
	branch-and-bound	Potts and Van Wassenhove (1985)
	local precedence relationships	Rachamadugu (1987)
	analysis of local searches	Chang et al. (1990)
	heuristic decomposition	Chambers et al. (1991)
$1/r_j, p_j = 1/\Sigma \, w_j T_j$	$O(n^3)$ weighted bipartite matching	Lawler et al. (1989)
$1/\Sigma \, U_j = n/\Sigma \, w_j T_j$	WSPT	McNaughton (1959)
$1//\Sigma \, w_j C_j^2$	branch-and-bound	Townsend (1978)
	dominance properties	Gupta and Sen (1984a)
		Szwarc et al. (1988)
$1//\Sigma \, (C_j - \overline{C})^2$	heuristic	Vani and Ragchavachari (1987)
$1//\Sigma \, (C_j - \overline{C})^2$	DP $O(n^2 \, MS)$ polynomial approximation	De et al. (1992)

MULTIPLE RELEASE-DEADLINE INTERVALS

$1/p_j = p/\Sigma U_j$	NP	Simons and Sipser (1984)
$1/p_j = 1/\Sigma U_j$	$O(n^3)$	Simons and Sipser (1984)
$1/p_j = 1/\Sigma C_j$	$O(n^4)$	Simons and Sipser (1984)
$1/p_j = 1/C_{max}$	$O(n^4)$	Simons and Sipser (1984)

EARLINESS-TARDINESS PROBLEMS
SINGLE-MACHINE COMMON DUE DATE PROBLEMS

$1/d_j = d/\Sigma(E_j + T_j)$	unrestricted d, $O(n\log n)$	Kanet (1981)
	restricted, NP	Hall, Kubiak, Sethi (1991)
	heuristic (1.5)	Liman and Lee (1991)
	branch and bound	Szwarc (1989)
$1/d_j = d/\Sigma(\alpha E_j + \beta T_j)$	unrestricted d, $O(n\log n)$	Bagchi, Chang, Sullivan (1987)
$1/d_j = d/\Sigma(E_j^2 + T_j^2)$	interchange heuristic	Eilon and Chowdhury (1977)
$1/d_j = d/\Sigma(\alpha E_j^2 + \beta T_j^2)$	enumerative search	Bagchi, Chang, and Sullivan (1987)
$1/d_j = d, \alpha_j = \beta_j /\Sigma(\alpha_j E_j + \beta_j T_j)$	NP	Hall and Posner (1991)
$1/d_j = d, \alpha_j = \alpha p_j /\Sigma(\alpha_j E_j + \beta_j T_j)$		Bagchi (1985)
$1/d_j = d/\Sigma(F(E_j) + G(T_j))$	$O(n\ MS)$	Kahlbacher (1993)
$1/d_j = d/\Sigma(F(E_j) + G(T_j))$	$O(n^2 p_{max}^2)$	Federgruen and Mosheiov (1991)

DIFFERENT DUE DATES

$1//\Sigma(\alpha_j E_j + \beta_j T_j)$	NP	Garey, Tarjan, Wilfong (1988)
	Interchange heuristic	Fry, Armstrong, Blackstone (1987)
	Filtered beam search	Ow and Morton (1989)
$1//\Sigma(\alpha_j E_j^2 + \beta_j T_j^2)$	branch-and-bound	Gupta and Sen (1983)

DUE DATE WINDOW ($E_j = max\{0, e_j - C_j\}, T_j = max\{0, C_j - d_j\}$)

$1/C_j \le d_j/E_{max}$	allowable idle time, P no idle time, $O(n^2)$	Lee (1991)
$1//max$ $\{E_{max}, T_{max}\}$	bisection search	Lee (1991)
$1/d_j - e_j = K$, agreeable α_j, β_j $/\Sigma(\alpha_j E_j + \beta_j T_j)$	Pseudopolynomial	Kraemer and Lee (1992)
$1/d_j - e_j = K$, agreeable α_j, β_j $/\Sigma(\alpha E_j + \beta T_j)$	$O(n \log n)$	Kraemer and Lee (1992)

DELIVERY COSTS

$1/d_j = d$, agreeable α_j, β_j $/\Sigma(\alpha_j E_j + \beta_j T_j + \gamma_j U_j)$	$O(n\ MS + n^2 d)$	Lee, Danusaputro, Lin (1991)
$1/d_j = d/\Sigma(\alpha E_j + \beta T_j + \gamma U_j)$	NP, $O(n(d + p_{max}))$	Lee, Danusaputro, Lin (1991)

BATCH DELIVERIES

$1/\beta_j = 0/\Sigma \alpha_j E_j + f(Q)$	NP	Cheng and Kahlbacher
$1/\beta_j = 0/\Sigma \alpha E_j + f(Q)$	$O(n^2 \log n)$	Cheng and Kahlbacher
$1/d_j = 0$, two fixed deliveries$/\Sigma(\alpha E_j + \beta T_j)$	unrestricted, $O(n)$ restricted, NP	Chhajed (1991)
$1/d_j = d, \alpha \le \beta$ $/\Sigma(\alpha E_j + \beta T_j) + KQ$	$O(n^2 MS d)$	Herrmann and Lee (1991)
$1/d_j = d, \alpha \le \beta, p_j = p/\Sigma(\alpha E_j + \beta T_j) + KQ$	$O(n^2)$	Herrmann and Lee (1991)
$1/d_j = d, \alpha_j = 0$ $/\Sigma \beta T_j + KQ$	$O(n^2)$	Herrmann and Lee (1991)

ADDITIONAL PENALTIES, d A DECISION VARIABLE

$1/d_j = d/\Sigma(\alpha E_j + \beta T_j + \gamma(d - d_0)^+)$	enumeration	Panwalkar, Smith, Seidmann (1982)
$1//\Sigma(\alpha E_j + \beta T_j + \gamma(d - d_0)^+)$	SPT	Panwalkar, Smith, Seidmann (1982)
$1/d_j = d/\Sigma(\alpha E_j + \beta T_j + \gamma d)$	$O(n \log n)$	Panwalkar, Smith, Seidmann (1982)
$1/d_j = d/\Sigma(\alpha_j E_j + \beta_j T_j + \gamma d)$	V-shaped schedules optimal	Bector et al. (1989)
$1/d_j = d/\Sigma(\alpha E_j + \beta T_j + \theta C_j)$	$O(n \log n)$	Baker and Scudder (1990)

JOB DEADLINES

$1/D_j/\Sigma E_j$ $1/\text{order } j,$ $\text{job } ij, D_j$ $/\Sigma_j n_j \theta_j C_j + \Sigma_j \Sigma_i (\alpha_j E_{ij} + \beta_j T_{ij}$	NP, branch-and-bound	Ahmadi and Bagchi (1986) Bagchi (1989)

Table 2. Dual-Criteria Deterministic Scheduling Problems

Problem description	Complexity	Reference

DEADLINES
First, note that the following constraints are equivalent: $\Sigma U_j = 0$, $C_j \leq d_j$, D_j, $T_{max} = 0$.

$1/C_j \leq d_j/\Sigma C_j$	$O(n \log n)$	Smith (1956)
$1/C_j \leq d_j + T_{max}$ $/\Sigma C_j$	$O(n \log n)$	Heck and Roberts (1972)
$1/C_j \leq d_j/\Sigma w_j C_j$	counter-example to Smith's algorithm	Burns (1976)
	NP	Lenstra, Rinnooy Kan, Brucker (1977)
	branch-and-bound	Bansal (1980)
	pivoting heuristic $O(n^3)$	Miyazaki (1981)
	branch-and-bound, decomposition approach	Shanthikumar and Buzacott (1982)
	branch-and-bound	Potts and Van Wassenhove (1983)
	better branch-and-bound	Posner (1985)
	$p_i > p_j$ implies $w_i \leq w_j$ Smith's algorithm	Chand and Schneeberger (1986)
	w_j a convex (concave) function of p_j: Smith's algorithm	Chand and Schneeberger (1986)
	improved lower bound	Bagchi and Ahmadi (1987)
$1/C_j \leq d_j/\Sigma w_j E_j$	NP, dynamic goal programming, special cases	Chand and Schneeberger (1988)
	branch-and-bound	Ahmadi and Bagchi (1986)

MINIMAL NUMBER OF TARDY JOBS

$1/\Sigma U_j = U_{min}/\Sigma C_j$	branch-and-bound	Emmons (1975)
$1/\Sigma U_j = U_{min}/T_{max}$	branch-and-bound	Shanthikumar (1983)
$1/\Sigma U_j = U_{min}/\Sigma T_j$	branch-and-bound	Vairaktarakis and Lee (1992)

MULTIPLE MACHINE MODELS

$P2/p_j = 1$, prec, $C_j \leq d_j/C_{max}$	$O(n^2 \log n)$	Garey and Johnson (1976)
$F/r_j, C_j \leq d_j /\Sigma\, w_{ij} I_{ij}$	NP, branch-and-bound	Ahmadi and Bagchi (1992)
$F2/C_j \leq d_j/\Sigma\, I_{ij}$	NP, branch-and-bound	Ahmadi and Bagchi (1992)

MULTICRITERIA OBJECTIVE FUNCTIONS

$1//\Sigma\, C_j, T_{max}$	$O(MSn \log n)$	Van Wassenhove and Gelders (1980)
$1/a_j \leq p_j \leq b_j/T_{max}, \Sigma\, w_j(b_j - p_j)$	$O(n^2)$	Van Wassenhove and Baker (1982)
$1/a_j \leq p_j \leq b_j/max\ g_j(C_j), \Sigma\, w_j(b_j - p_j)$	$O(n^3)$	Van Wassenhove and Baker (1982)
$1//\Sigma\, C_j, T_{max}, \Sigma\, U_j$	branch-and-bound, heuristics	Nelson et al. (1986)
$1//\Sigma\, C_j, \Sigma\, U_j$	lower bound	Kiran and Unal (1991)
$1//\Sigma\, C_j, \Sigma\, T_j, \Sigma\, U_j$	branch-and-bound	Kao (1980)
$1//\Sigma\, C_j, \Sigma\Sigma |C_j - C_i|$	P	Bagchi (1989)
$1//\Sigma\,(C_j - \overline{C})^2, \overline{C}$	DP $O(n^2 MS)$	De et al. (1992)
$F//C_{max}, T_{max}$	branch-and-bound heuristics	Daniels and Chambers (1990)

 J.W. Herrmann et al.

Table 3. Parallel Machine Deterministic Scheduling Problems

Problem description	Complexity	Reference
MINIMUM SUM PROBLEMS		
FLOWTIME		
$R//\Sigma C_j$	$O(n^3)$	Horn (1973)
	Weighted bipartite matching	Bruno, Coffman, Sethi (1974)
$P//\Sigma C_j$	$O(n \log n)$	Conway, Maxwell, Miller (1967)
$Q//\Sigma C_j$	$O(n \log n)$	Horowitz and Sahni (1976)
UNIFORM MACHINES, UNIT PROCESSING TIMES		
$Q/p_j = 1/\Sigma f_j$	$O(n^3)$	Dessoukey et al. (1989)
$Q/p_j = 1/\Sigma w_j C_j$	$O(n \log n)$	Dessoukey et al. (1989)
$Q/p_j = 1/\Sigma T_j$	$O(n \log n)$	Dessoukey et al. (1989)
$Q/p_j = 1/\Sigma w_j U_j$	$O(n \log n)$	Dessoukey et al. (1989)
$P/p_j = 1/\Sigma U_j$	$O(n \log n)$	Lawler (1976a)
$Q/p_j = 1/f_{max}$	$O(n^2)$	Dessoukey et al. (1989)
$Q/p_j = 1/L_{max}$	$O(n \log n)$	Dessoukey et al. (1989)
$Q/p_j = 1/C_{max}$	$O(n \log n)$	Dessoukey et al. (1989)
IDENTICAL MACHINES		
$P2//\Sigma C_j$	$O(n)$	Assad (1985)
$P2/p_j = 1/\Sigma w_j C_j$	$O(n)$	Assad (1985)
$P/$increasing processing rate$/\Sigma C_j$	SPT	Meilijson and Tamir (1984)
P/r_j, increasing processing rate$/\Sigma C_j$	SPT	Huang (1986)
$P/$variable processing rate$/C_{max}$		Dror et al. (1987)
$P/$variable processing rate$/\Sigma C_j$	SPT on one machine	Dror et al. (1987)
$P2//\Sigma w_j C_j$	NP	Bruno, Coffman, Sethi (1974)
$P2/$tree$/\Sigma C_j$	NP	Sethi (1977)
$P2/$machine 2 has limited available time$/\Sigma C_j$	NP	Lee and Liman (1992)

IDENTICAL MACHINES, cont.

$P//\Sigma w_j C_j$	DP		Lawler and Moore (1969)
	DP		Lee and Uzsoy (1992)
	$NP((\sqrt{2}+1)/2)$		Kawaguchi and Kyan (1986)
$P//\Sigma T_j$	heuristic		Ho and Chang (1991)
$P/w_j = p_j/\Sigma w_j T_j$	NP		Arkin and Roundy (1991)
	$O(n^2 MS)$		

IDENTICAL MACHINES, PREEMPTION

$P/pmtn/\Sigma C_j$	$O(n \log n)$	McNaughton (1959)
$P/pmtn/\Sigma w_j C_j$	NP	Du, Leung, Young (1989)
$P2/pmtn, tree/\Sigma C_j$	NP	Du, Leung, Young (1989)
$P2/pmtn, r_j/\Sigma C_j$	NP	Du, Leung, Young (1988)
$P2/pmtn, p_j = p, r_j$ $/\Sigma C_j$	$O(n \log n)$	Herrbach (1990)
$P2/pmtn, r_j/\Sigma U_j$	NP	Du, Leung, Wong (1989)
$P/pmtn/\Sigma U_j$	NP	Lawler (1983)

OTHER PREEMPTIVE MACHINES

$Q/pmtn/\Sigma C_j$	$O(n \log n + nm)$	Gonzalez (1977)
$Q2/pmtn/\Sigma U_j$	$O(n^3)$	Lawler and Martel (1989)
$Q2/pmtn/\Sigma w_j U_j$	$O(n^2 \Sigma w_j)$ polynomial approximation	Lawler and Martel (1989)
$Q/r_j, pmtn/\Sigma U_j$	$O(mn^3)$	Federgruen and Groenvelt (1986)
$Q/r_j, pmtn/L_{max}$	$O(mn^3 \log(np_{max}s_{max}))$	Federgruen and Groenvelt (1986)
$Q/r_j, pmtn/\Sigma w_j C_j$	$O(mn^3 p_{max})$	Federgruen and Groenvelt (1986)
$R/r_j, pmtn/\Sigma U_j$	NP	Du and Leung (1991)

MINIMAX CRITERIA
NON-PREEMPTIVE PROBLEMS

$P//C_{max}$	NP	Garey and Johnson (1978)
	$NP(2 - 1/(2m))$	Graham (1966)
	$NP(4/3 - 1/(3m))$	Graham (1969)
	$NP(1.22)$	Coffman et al. (1978)
	$NP(6/5)$	Friesen and Langston (1986)
	$NP(10/9$ for 2 machines$)$	Lee and Massey(1988)
$P/|\{p_j\}| = k/C_{max}$	pseudopolynomial	Leung (1982)
P/communication delay$/C_{max}$	Error bound	Lee et al. (1988)
P/M_i not immediately available$/C_{max}$	NP (1.5 for LPT)	Lee (1991)
$P/r_j/L_{max}$	EDD $(((2m - 1)/m)p_{max})$	Gusfield (1984)
$P/r_j, p_j = p/L_{max}$	$O(mn^2)$	Simons and Warmuth (1989)
$R//C_{max}$	NP	Lenstra, Shmoys, Tardos (1989)
	List scheduling	Davis and Jaffe (1981)
	$NP(2)$	Potts (1985a)

PREEMPTION, MAKESPAN

$P/pmtn/C_{max}$	$O(n)$	McNaughton (1959)
$Q/pmtn/C_{max}$	$O(mn^2)$	Horvath, Lam, Sethi (1977)
	$O(n + m\log m)$	Gonzalez and Sahni (1978b)
$R/pmtn/C_{max}$	LP	Lawler and Labetoulle (1978)
$P/pmtn, r_j/C_{max}$	$O(n^2)$	Horn (1974)
	$O(mn)$	Gonzalez and Johnson (1980)
$Q/pmtn, r_j/C_{max}$	$O(n\log n + mn)$	Labetoulle et al. (1984)

PREEMPTION, MAXIMUM LATENESS

$P/pmtn/L_{max}$	$O(n^2)$	Horn (1974)
	$O(mn)$	Gonzalez and Johnson (1980)

PRECEDENCE CONSTRAINTS
IDENTICAL MACHINES, UNIT PROCESSING TIMES

P/prec, $p_j = 1/C_{max}$	NP	Ullman (1975)
	Critical path bound $2 - 1/(m - 1)$	Chen (1975)
	Coffman-Graham bound $2 - 2/m$	Lam and Sethi (1977)
P/tree, $p_j = 1/C_{max}$	$O(n)$	Hsu (1966)
		Sethi (1977)
P/opposing forest, $p_j = 1/C_{max}$	NP	Garey et al. (1983)
P/interval order/C_{max}	$O(m + n)$ list scheduling	Papadimitriou and Yannakakis (1979)
P/intree, $p_j = 1/L_{max}$	P	Brucker, Garey, Johnson (1977)
	$O(n)$	Monma (1982)
P/outtree, $p_j = 1$ /L_{max}	NP	Brucker, Garey, Johnson (1977)
P2/prec, $p_j = 1/C_{max}$	$O(n^2)$	Coffman and Graham (1972)
P2/prec, $p_j = 1, D_j$ /C_{max}	$O(n^2)$	Garey and Johnson (1976)
P2/prec, $p_j = 1, r_j, D_j$ /C_{max}	$O(n^3)$	Garey and Johnson (1977)

IDENTICAL MACHINES, SMALL PROCESSING TIMES

P2/prec, $p_j = 1$ or 2 /C_{max}	NP	Ullman (1975)
P2/prec, $p_j = 1$ or 2 /ΣC_j	NP	Lenstra and Rinnooy Kan (1978)
P2/tree, $p_j = 1$ or 2 /C_{max}	$O(n \log n)$	Nakajima, Leung, and Hakimi (1981)
	Heuristic, error of 1	Kaufman (1974)
P2/tree, $p_j = 1$ or 3 /C_{max}	$O(n^2 \log n)$	Du and Leung (1989a)
P2/tree, $p_j = 1$ or k /C_{max}	NP	Du and Leung (1988a)
	generalized Coffman-Graham, bound $3/2 - 1/2k$	Goyal (1977)

PRECEDENCE PROBLEMS WITHOUT PREEMPTION

$P/prec/C_{max}$	List scheduling bound $2 - 1/m$	Graham (1966)
$P2/tree/C_{max}$	NP	Du, Leung, and Young (1989)
$Q/prec/C_{max}$	List scheduling bound	Jaffe (1980a)
$Q2/prec,$ $p_j = 1/C_{max}$	approximation algorithms	Gabow (1988)

PRECEDENCE CONSTRAINTS AND PREEMPTION

$P/pmtn, prec, p_j = 1$ $/C_{max}$	NP	Ullman (1976)
$P/pmtn, tree/C_{max}$	P	Muntz and Coffman (1969)
$P2/pmtn, prec/C_{max}$	P	Muntz and Coffman (1970)
$P/pmtn, prec/C_{max}$	Muntz-Coffman bound $2 - 2/m$	Lam and Sethi (1977)
$P/pmtn, tree/C_{max}$	$O(n \log m)$	Gonzalez and Johnson (1980)
$Q2/pmtn, prec/C_{max}$	$O(mn^2)$	Horvath, Lam, and Sethi (1977)
$Q/pmtn, prec/C_{max}$	Muntz-Coffman bound $\sqrt{3m/2}$ heuristic, bound $\sqrt{m} + 1/2$	Horvath, Lam, and Sethi (1977) Jaffe (1980b)
$P/pmtn, intree/L_{max}$	$O(n^2)$	Lawler (1982a)
$Q2/pmtn, prec/L_{max}$	$O(n^2)$	Lawler (1982a)
$Q2/pmtn, intree, r_j$ $/L_{max}$	$O(n^6)$	Lawler (1982a)

Table 4. Deterministic Shop Scheduling Problems

Problem description	Complexity	Reference
FLOW SHOPS		
MAKESPAN		
$F2//C_{max}$	$O(n \log n)$	Johnson (1954)
$F3//C_{max}$ machine 2 dominated	$O(n \log n)$	Johnson (1954)
$F3//C_{max}$	NP	Garey, Johnson, Sethi (1976)
$F3//C_{max}$ machine 2 with infinite capacity	$O(n \log n)$	Conway, Maxwell, Miller (1967)
$F2//C_{max}$ non-bottleneck work between machines	$O(n \log n)$	Mitten (1958)
$F2/r_j/C_{max}$	NP	Lenstra, Rinnooy Kan, Brucker (1977)
$F2/prec/C_{max}$	branch-and-bound	McMahon and Lim (1993)
$F3/p_{2j} \leq min \{p_{1j}, p_{3j}\}/C_{max}$	$O(n \log n)$	Burns and Rooker (1978)
$F3/p_{1j} \leq p_{2j} \leq p_{3j}/C_{max}$	$O(n^7)$	Achugbue and Chin (1982)
$F3/p_{1j} \geq p_{2j} \geq p_{3j}/C_{max}$	$O(n^7)$	Achugbue and Chin (1982)
$F2/one\ transporter/C_{max}$	$O(n^2)$	Panwalker (1991)
OTHER CRITERIA		
$F2//L_{max}$	NP	Lenstra, Rinnooy Kan, Brucker (1977)
$F2/r_j/L_{max}$	branch-and-bound	Grabowski (1980)
$F2//\Sigma C_j$	NP	Garey et al. (1976)
	branch-and-bound	Ignall and Schrage (1965)
	Lagrangian relaxation	Van de Velde (1990)
$F2//\Sigma T_j$	branch-and-bound	Sen et al. (1989)
$F//\Sigma w_j T_j$	bottleneck scheduling	Ow (1985)
$F//\Sigma U_j$	bottleneck scheduling	Ow (1985)

FLOW SHOPS WITH PREEMPTION

$F2/pmtn/C_{max}$	Johnson's algorithm $O(n \log n)$	Johnson (1954)
$F3/pmtn/C_{max}$	NP	Lawler et al. (1989)
$F2/pmtn, r_j/C_{max}$	NP	Lawler et al. (1989)
$F2/pmtn/L_{max}$	NP	Lawler et al. (1989)
$F3/pmtn/\Sigma C_j$	NP	Lawler et al. (1989)

PRECEDENCE CONSTRAINTS

$F2/tree/C_{max}$	NP	Lenstra et al. (1977)

OPERATION PRECEDENCE

$F2/tree/C_{max}$	$O(n \log n)$	Sidney (1979)
$F2/prec/C_{max}$	NP	Monma (1980)
	Branch-and-bound	Hariri and Potts (1984)

PERMUTATION FLOW SHOP

$F//C_{max}$	Machine-based bound	Ignall and Schrage (1965)
	Job-based bound	McMahon (1971)
	LPT and active schedules relative error: m	Gonzalez and Sahni (1978a)
	$O(mn \log n)$, error: $m/2$	Gonzalez and Sahni (1978a)
	Front and back scheduling	Potts (1980a)
	$O(m^3n^2 + m^4n)$ error: $(m-1)(3m-1)p_{max}/2$	Barany (1981)
	Block approach	Grabowski (1982)
	Aggregation heuristic error proportional to m	Roeck and Schmidt (1982)
	Insertion method	Nawaz, Enscore, Ham (1983)
	integer programming	Frieze and Yadegar (1989)
$F/\text{time lags}/C_{max}$	LB, approximation	Szwarc (1983)
$F/r_j/L_{max}$	block approach	Grabowski et al. (1983)
$F//\Sigma C_j$	LB	Ahmadi and Bagchi (1990)
$F//\Sigma U_j$	branch-and-bound	Hariri and Potts (1989)

FLOW SHOPS WITH NO WAIT IN PROCESS
MAKESPAN

F2/no wait/C_{max}	$O(n^2)$	Gilmore and Gomory (1964)
F3/no wait/C_{max}	NP	Roeck (1984a)
F4/no wait/C_{max}	NP	Papadimitriou and Kanellakis (1980)
J2/no wait/C_{max}	NP	Sahni and Cho (1979)
$F/p_{2j} = b_j + a_j I_{2j}$ /C_{max}	NP	Sriskandarajah and Goyal (1989)

OTHER CRITERIA

F2/no wait/L_{max}	NP	Roeck (1984b)
F2/no wait/ΣC_j	NP	Roeck (1984b)

FLOW SHOPS WITH LOT STREAMING
The various forms of the problem can be classified according to three things (Trietsch and Baker, 1992):
whether the sublots are variable (V), consistent (C), or equal (E);
whether there is intermittent idling (I), or no idling (NI);
whether the continuous version (CV) or discrete version (DV) applies.

F2/CV/C_{max}	P	Trietsch (1987)
		Potts and Baker (1989)
F2/DV/C_{max}	P	Trietsch (1987)
		Trietsch and Baker (1992)
F3/C, II, CV/C_{max}	P	Potts et al. (1992)
F3/V, II, CV/C_{max}	P	Trietsch (1989)
		Trietsch and Baker (1992)
F3/V, II, DV/C_{max}	P	Trietsch and Baker (1992)
F/C, CV/C_{max}	P	Baker (1988)
		Trietsch and Baker (1992)
	2 sublots, $O(m^2)$	Williams and Tufekci (1992)
F/C, DV/C_{max}	NP	Trietsch and Baker (1992)
F/V, NI, CV/C_{max}	P	Trietsch and Baker (1992)
F/V, NI, DV/C_{max}	P	Trietsch (1987)
F/V, II, DV/C_{max}	NP	Trietsch and Baker (1992)

JOB SHOPS
MAKESPAN

$J2/m_j \leq 2/C_{max}$	Johnson's algorithm	Jackson (1956)
$J2/p_j = 1/C_{max}$	O(number of operations)	Hefetz and Adiri (1982)
$J2/m_j \leq 3/C_{max}$	NP	Lenstra et al. (1977)
$J3/m_j \leq 2/C_{max}$	NP	Gonzalez and Sahni (1978a)
$J2/p_{ij} = 1$ or $2/C_{max}$	NP	Lenstra and Rinnooy Kan (1979)
$J3/p_{ij} = 1/C_{max}$	NP	Lenstra and Rinnooy Kan (1979)
$J//C_{max}$	Enumerate active schedules	Giffler and Thompson (1960)
	LPT and active schedules	Gonzalez and Sahni (1978a)
	branch-and-bound	Barker and McMahon (1985)
	Shifting bottleneck	Adams, Balas, Zawack (1988)
	branch-and-bound	Carlier and Pinson (1989)
	Simulated annealing	Matsuo et al. (1988)
		Van Laarhoven et al. (1988)
$J2/$no wait, $p_{ij} = 1/C_{max}$	pseudopolynomial	Kubiak (1989)

OTHER CRITERIA

$J2/p_j = 1/L_{max}$	O(number of operations)	Brucker (1981, 1982)
$J//\Sigma T_j$	dispatching rules	Anderson and Nyirenda (1990)
$J/r_j/\Sigma w_j T_j$	priority rules	Vepsalainen and Morton (1987)

Table 5. Resource-constrained Project Scheduling

Problem description	Complexity	Reference

IDENTICAL MACHINES

$P2/p_j = 1/C_{max}$ with resources	maximum cardinality matching	Garey and Johnson (1975)
$P/$prec$/C_{max}$ with resources	integer programming	Talbot and Patterson (1978)
	graph theory approach	Bartusch et al. (1988)

FLOW SHOPS

$F/p_j = 1/C_{max}$ with resources	NP	Blazewicz et al. (1988)
$F2/p_j = 1/L_{max}$ with resources	$O(n(\log n)^2)$	Blazewicz et al. (1988)

Table 6. Stochastic Machine Scheduling

Problem description	Complexity	Reference

ONE MACHINE

$1/p_j \text{ exp}/E[\Sigma T_j]$	branch-and-bound	Cadambi (1991)
$1/p_j \text{ exp}, r_j \text{ random}, \text{pmtn}/E[\Sigma C_j]$	$O(n \log n)$	Pinedo (1983)
$1/p_j \text{ exp}/E[\Sigma w_j T_j]$	$O(n \log n)$	Pinedo (1983)
$1/p_j \text{ exp}/E[\Sigma w_j U_j]$	$O(n \log n)$	Pinedo (1983)
$1/p_j \text{ random}/\Sigma U_j$	NP	Kise and Ibaraki (1983)

LIST SCHEDULING

$P/p_j \text{ exp}/E[C_{max}]$	LEPT	Bruno and Downey (1977)
$P/p_j \text{ iid}/E[C_{max}]$	list scheduling heuristic	Coffman and Gilbert (1985)
$P/p_j \text{ exp}/E[\Sigma C_j]$	SEPT	Bruno, Downey, Frederickson (1981)
$P/p_j \text{ exp}, \lambda_i > \lambda_j$ implies $w_i \geq w_j/E[\Sigma w_j U_j]$	SEPT	Pinedo (1983)
$Q/\text{pmtn}, p_j \text{ exp}/E[C_{max}]$	LEPT	Weiss and Pinedo (1980)
$Q/\text{pmtn}, p_j \text{ exp}/E[\Sigma C_j]$	SEPT	Weiss and Pinedo (1980)

PARALLEL MACHINES

Problem description	Complexity	Reference
P/comparable processing times/$E[\Sigma\, C_j]$	SEPT	Weber et al. (1986)
P2/intree, p_j exp /$E[C_{max}]$	P	Pinedo and Weiss (1984)
P/p_j random /$E[\Sigma\, w_j C_j]$	performance of SEPT	Weiss (1990)

SHOPS

Problem description	Complexity	Reference
F2/p_j exp/$E[C_{max}]$	Johnson's algorithm	Brumelle and Sidney (1982)
F/p_j exp/$E[C_{max}]$		Weber (1979)
F//$E[C_{max}]$	nondecreasing, nonincreasing $E[p_{ij}]$	Pinedo (1982)
F/$p_{ij} = w_j$, s_i, w_j random, $s_1 \leq \ldots \leq s_m$/$E[C_{max}]$	$O(n \log n)$	Pinedo (1985)
F//$E[\Sigma\, U_j]$		Boxma and Forst (1986)
J2/$m_j \leq 2$, p_j exp /$E[C_{max}]$	Pinedo (1981)	
F/no wait, stochastically ordered p_j/$E[C_{max}]$	SEPT-LEPT	Wie and Pinedo (1986)
F/no wait, stochastically ordered p_j/$E[\Sigma\, C_j]$	SEPT	Wie and Pinedo (1986)

Table 7. Smart-and-Lucky Searches

Problem description	Complexity	Reference

SIMULATED ANNEALING

Problem description	Complexity	Reference
Traveling Salesman Problem		Cerny (1985)
		Kirkpatrick et al. (1983)
F//C_{max}		Osman and Potts (1989)
	probabilistic-exhaustive	Ogbu and Smith (1990)
	include setup times	Vakharia and Chang (1990)

SIMULATED ANNEALING, cont.

$J//C_{max}$	disjunctive graph	Van Laarhoven, Aarts, and Lenstra (1988)
	controlled search	Matsuo, Suh, and Sullivan (1988)
$1//\Sigma\, w_j T_j$	controlled search	Matsuo, Suh, and Sullivan (1989)

TABU SEARCH

$1//\Sigma\, w_j C_j$	include setup costs	Laguna, Barnes, and Glover (1991)
$P//\Sigma\, w_j C_j$	partition reduction	Barnes and Laguna (1992)
$J//C_{max}$	disjunctive graph	Barnes and Chambers (1991)
$F//C_{max}$	Traveling Salesman model	Widmer and Hertz (1989)
$F//C_{max}$		Taillard (1990)
Traveling Salesman Problem	parallel implementation	Malek et al. (1989)
$1/D_j/$profit maximization	class scheduling	Woodruff and Spearman (1992)

GENETIC ALGORITHMS

Traveling Salesman Problem	crossovers	Oliver et al. (1987)
		Starkweather et al. (1991)
		Whitley, Starkweather, and Shaner (1991)
		Fox and McMahon (1990)
$J//C_{max}$	special case	Davis (1985)
	problem and heuristic space	Storer, Wu, and Vaccari (1992)
		Nakano and Yamaha (1991)
$1/D_j/\Sigma\, C_j$	class scheduling problem space	Herrmann and Lee (1992)
$F//\Sigma\,(E_j + T_j^2)$		Cleveland and Smith (1989)
Resource-constrained scheduling		Syswerda (1991)
$1//\Sigma\,(C_j - \overline{C})^2$		Gupta et al. (1992)

Table 8. Class Scheduling Problems (Jobs are grouped into job classes where a setup is done when the machine switches from one job class to another.)

Problem description	Complexity	Reference

ONE-MACHINE PROBLEMS

Problem description	Complexity	Reference
$1//C_{max}$	NP	Monma and Potts (1989)
$1/\text{two job classes}/\Sigma\,C_j$	branch-and-bound	Sahney (1972)
		Gupta (1984)
		Potts (1991)
		Coffman, Nozari, and Yannakakis (1989)
$1//\Sigma\,C_j$	NP	Monma and Potts (1989)
$1//\Sigma\,C_j$	item-flow and batch-flow	Dobson, Karmarkar, and Rummel (1987)
		Gupta (1988)
		Ahn and Hyun (1990)
$1//\Sigma\,w_j C_j$	NP	Monma and Potts (1989)
$1//\Sigma\,w_j C_j$	branch-and-bound	Mason and Anderson (1991)
$1/D_j/\Sigma\,C_j$	problem space genetic algorithm	Herrmann and Lee (1992)
$1//\Sigma\,U_j$	NP	Bruno and Downey (1978)
		Monma and Potts (1989)
$1//L_{max}$	NP	Monma and Potts (1989)
$1/D_j/\text{profit maximization}$	tabu search	Woodruff and Spearman (1992)

IDENTICAL MACHINES

Problem description	Complexity	Reference
$P//\Sigma\,C_j$	heuristic	Tang (1990)
$P//\Sigma\,C_j$	item-flow and batch-flow	Dobson, Karmarkar, and Rummel (1989)
$P/d_j = d/\Sigma\,w_j U_j$	heuristics	So (1990)

References

Aarts, E. H. L., and P. J. M. van Laarhoven, "Statistical cooling: a general approach to combinatorial optimization problems," *Philips Journal of Research*, 40, p. 193, 1985.

Achugbue, J. O. and F. Y. Chin, "Complexity and solutions of some three-stage flow shop scheduling problems," *Mathematics of Operations Research*, Vol. 7, No. 4, pp. 532-544, 1982.

Adams, J., E. Balas, and D. Zawack, "The shifting bottleneck procedure for job shop scheduling," *Management Science*, 34, pp. 391-401, 1988.

Adiri, I., J. Bruno, E. Frostig, A.H.G. Rinnooy Kan, "Single machine flow-time scheduling with a single breakdown," *Acta Informatica*, 26, pp. 679-696, 1989.

Ahmadi, R., and U. Bagchi, "Single-machine scheduling to minimize earliness subject to deadlines," Working Paper 86-4-17, Dept. of Management, University of Texas, Austin, 1986.

Ahmadi, R. H. and U. Bagchi, "Improved lower bounds for minimizing the sum of completion times of n jobs over m machines in a flow shop," *European Journal of Operational Research*, Vol. 44, No. 3, pp. 331-336, 1990.

Ahmadi, R. H., and U. Bagchi, "Minimizing job idleness in deadline constrained environments," *Operations Research*, Vol. 40, No. 5, pp. 972-985, 1992.

Ahn, B.-H., and J.-H. Hyun, "Single facility multi-class job scheduling," *Computers and Operations Research*, 17, pp. 265-272, 1990.

Anderson, E. J. and J. C. Nyirenda, "Two new rules to minimize tardiness in a job shop," *International Journal of Production Research*, Vol. 28, No. 12, pp. 2277-2292, 1990.

Arkin, E.M., and R.O. Roundy, "Weighted-tardiness scheduling on parallel machines with proportional weights," *Operations Research*, Vol. 39, No. 1, pp. 64-81, 1991.

Assad, A. A., "Nested optimal policies for set functions with applications to scheduling," *Mathematics of Operations Research*, Vol. 10, No. 1, pp. 82-99, 1985.

Bagchi, U., "Scheduling to minimize earliness and tardiness penalties with a common due date," Dept. of Management, University of Texas, Austin, 1985.

Bagchi, U., "Due date or deadline assignment to multi-job customer orders," 1989.

Bagchi, U., "Simultaneous minimization of mean and variation of flow time and waiting time in single machine systems," *Operations Research*, Vol. 37, No. 1, pp. 118-125, 1989.

Bagchi, U. and R. H. Ahmadi, "An improved lower bound for minimizing weighted completion times with deadlines," *Operations Research*, Vol. 35, No. 2, pp. 311-312, 1987.

Bagchi, U., Y. Chang and R. Sullivan, "Minimizing absolute and squared deviation of completion times with different earliness and tardiness penalties and a common due date," *Management Science*, Vol. 33, pp. 894-906, 1987.

Baker, K. R., *Introduction to Sequencing and Scheduling*, 2nd ed., Wiley, New York, 1992.

Baker, K.R., "Lot streaming to reduce cycle time in a flow shop," Working paper 203, The Amos Tuck School of Business Administration, Dartmouth College, Hanover, N.H., 1988

Baker, K. R., E. L. Lawler, J. K. Lenstra, and A. H. G. Rinnooy Kan, "Preemptive scheduling of a single machine to minimize maximum cost subject to release dates and precedence constraints," *Operations Research*, 31, pp. 381-386, 1983.

Baker, K. R., and G. D. Scudder, "Sequencing with earliness and tardiness penalties: a review," *Operations Research*, 38:1, pp. 22-36, 1990.

Bansal, S. P., "Single-machine scheduling to minimize weighted sum of completion times with secondary criterion: a branch-and-bound approach," *European Journal of Operational Research*, 5, pp. 177-181, 1980.

Barany, I., "A vector-sum theorem and its application to improving flow shop guarantees," *Mathematics of Operations Research*, 6, pp. 445-452, 1981.

Barker, J. R. and G. B. McMahon, "Scheduling the general job-shop," *Management Science*, Vol. 31, No. 5, pp. 594-598, 1985.

Barnes, J. W., and J. B. Chambers, "Solving the job-shop scheduling problem using tabu search," Technical Report OPR91-06, Graduate Program in Operations Research, University of Texas at Austin, 1991.

Barnes, J. W., and M. Laguna, "Solving the multiple-machine weighted flow time problem using tabu search," to appear in *IIE Transactions*, 1992.

Bartusch, M., R. H. Moehring, and F. J. Radermacher, "M-machine unit-time scheduling: a report of ongoing research," in *Optimization Parallel Processing, and Applications, Lecture Notes in Economics and Mathematical Systems*, 304, Springer, Berlin, pp. 165-212, 1988.

Bector, C. R., et al., "V-shape property of optimal sequence of jobs about a common due date on a single machine," *Computers and Operations Research*, Vol. 16, No. 6, pp. 583-588, 1989.

Błazewicz, J., et al., "Scheduling unit-Time tasks on flow-Shops under resource constraints," *Annals of Operations Research*, Vol. 16, No. 3, pp. 255-266, 1988.

Boxma, O. J., and F. G. Forst, "Minimizing the expected weighted number of tardy jobs in stochastic flow shops," *Operation Research Letters*, 5, pp. 119-126, 1986.

Brucker, P., "Minimizing maximum lateness in a two-machine unit-time job shop," *Computing*, 27, pp. 367-370, 1981.

Brucker, P., "A linear time algorithm to minimize maximum lateness for the two-machine, unit-time, job-shop, scheduling problem," in *System Modeling and Optimization, Lecture Notes in Control and Information Sciences*, 38, Springer, Berlin, pp. 566-571, 1982.

Brucker, P., M. R. Garey, and D. S. Johnson, "Scheduling equal-length tasks under tree-like precedence constraints to minimize maximum lateness," *Mathematics of Operations Research*, 2, pp. 275-284, 1977.

Brumelle, S. L., and J. B. Sidney, "The two-machine makespan problem with stochastic flow times," Technical report, University of British Columbia, Vancouver, 1982.

Bruno, J. L., E. G. Coffman, and R. Sethi, "Scheduling independent tasks to reduce mean finishing time," *Communications ACM*, 17, pp. 382-387, 1974.

Bruno, J. L., and P. J. Downey, "Sequencing tasks with exponential service times on two machines," Technical report, Department of Electrical Engineering and Computer Science, University of California, Santa Barbara, 1977.

Bruno, J., and P. Downey, "Complexity of task sequencing with deadlines, set-up times and changeover costs," *SIAM Journal of Computing*, 7, pp. 393-404, 1978.

Bruno, J. L., P. J. Downey, and G. N. Frederickson, "Sequencing tasks with exponential service times to minimize the expected flowtime or makespan," *Journal of the Association of Computing Machinery*, 28, pp. 100-113, 1981.

Burns, R. N., "Scheduling to minimize the weighted sum of completion time with secondary criteria," *Naval Research Logistics Quarterly*, 23, pp. 125-129, 1976.

Burns, F., and J. Rooker, "Three-stage flow-shops with recessive second stage," *Operations Research*, 26, pp. 207-208, 1978.

Cadambi, B. V., "One machine scheduling to minimize expected mean tardiness-Part 1," *Computers and Operations Research*, Vol. 18, No. 8, pp. 787-796, 1991.

Carlier, J. and E. Pinson, "An algorithm for solving the job-shop problem," *Management Science*, Vol. 35, No. 2, pp. 164-176, 1989.

Cerny, V., "Thermodynamical approach to the traveling salesman problem: an efficient simulation algorithm," *Journal of Optimization Theory and Applications*, 45, p. 41, 1985.

Chambers, R.J., R.L. Carraway, T.J. Lowe, and T.L. Morris, "Dominance and decomposition heuristics for single machine scheduling," *Operations Research*, Vol. 39, No. 4, pp. 639-647, 1991.

Chand, S., and H. Schneeberger, "A note on the single-machine scheduling problem with minimum weighted completion time and maximum allowable tardiness," *Naval Research Logistics Quarterly*, 33, pp. 551-557, 1986.

Chand, S., and H. Schneeberger, "Single-machine scheduling to minimize weighted earliness subject to no tardy jobs," *European Journal of Operational Research*, 34, pp. 221-230, 1988.

Chang, S., et al., "Worst-case analysis of local search heuristics for the one-machine total tardiness problem," *Naval Research Logistics*, Vol. 37, No. 1, pp. 111-122, 1990.

Chen, N.-F., "An analysis of scheduling algorithms in multiprocessing computing systems," Technical Report UIUCDCS-R-75-724, Dept. of Computer Science, University of Illinois at Urbana-Champaign, 1975.

Cheng, T. C. E., and H. G. Kahlbacher, "Scheduling with delivery and earliness penalties," Dept. of Actuarial and Management Sciences, University of Manitoba, Winnipeg, Manitoba.

Chhajed, Dilip, "A fixed interval due-date scheduling problem with earliness and due-date costs," Dept. of Business Administration, University of Illinois, Champaign, IL, 1991.

Cleveland, G., and S. Smith, "Using genetic algorithms to schedule flow shop releases," in *Proceedings of the Third International Conference on Genetic Algorithms*, Morgan Kaufmann, 1989.

Coffman, E. G., Jr., M. R. Garey, and D. S. Johnson, "An application of bin-packing to multiprocessor scheduling," *SIAM Journal of Computing*, 7, pp. 1-17, 1978.

Coffman, E. G., Jr. and E. N. Gilbert, "On the expected relative performance of list scheduling," *Operations Research*, Vol. 33, No. 3, pp. 548-561, 1985.

Coffman, E.G., Jr., and R. L. Graham, "Optimal scheduling for two-processor systems," *Acta Informatica*, I, pp. 200-213, 1972.

Coffman, E.G., A. Nozari, and M. Yannakakis, "Optimal scheduling of products with two subassemblies on a single machine," *Operations Research*, 37, pp. 426-436, 1989.

Conway, R. W., W. L. Maxwell, and L. W. Miller, *Theory of Scheduling*, Addison-Wesley, Reading, Massachusetts, 1967.

Daniels, R. L. and R. J. Chambers, "Multiobjective flow-shop scheduling," *Naval Research Logistics*, Vol. 37, No. 6, pp. 981-995, 1990.

Davis, E., and J. M. Jaffe, "Algorithms for scheduling tasks on unrelated processors," *Journal of the Association for Computing Machinery*, 28, pp. 721-736, 1981.

Davis, L., "Job shop scheduling with genetic algorithms," in *Proceedings of an International Conference on Genetic Algorithms and their Applications*, 1985, J. Grefenstette, ed.

Davis, L., ed., *Genetic Algorithms and Simulated Annealing*, Pitman Publishing, London, 1987.

Davis, L., ed., *Handbook of Genetic Algorithms*, Van Nostrand Reinhold, New York, 1991.

De, P., J.B. Ghosh, and C.E. Wells, "On the minimization of completion time variance with a bicriteria extension," *Operations Research*, Vol. 40, No. 6, pp. 1148-1155, 1992.

Dempster, M. A. H., J. K. Lenstra, and A. H. G. Rinnooy Kan, *Deterministic and Stochastic Scheduling*, D. Reidel Publishing, Dordrecht, Holland, 1982.

Dessouky, M. I., B. J. Lageweg, and S. L. Van de Velde, "Scheduling identical jobs on uniform parallel machines," Report BS-89xx, Centre for Mathematics and Computer Science, Amsterdam, 1989.

Dobson, G., U.S. Karmarkar, and J.L. Rummel, "Batching to minimize flowtimes on one machine," *Management Science*, 33, pp. 784-799, 1987.

Dobson, G., U.S. Karmarkar, and J.L. Rummel, "Batching to minimize flowtimes on parallel heterogeneous machines," *Management Science*, 35, pp. 607-613, 1989.

Dror, M., et al., "Parallel machine scheduling: processing rates dependent on number of jobs in operation," *Management Science*, Vol. 33, No. 8, pp. 1001-1009, 1987.

Du, J., and J. Y.-T. Leung, "Scheduling tree-structured tasks with restricted execution times," *Inform. Process. Lett.*, 28, pp. 183-188, 1988a.

Du, J., and J. Y.-T. Leung, "Minimizing mean flow time with release time and deadline constraints," Technical report, Computer Science Program, University of Texas, Dallas, 1988b.

Du, J., and J. Y.-T. Leung, "Scheduling tree-structured tasks on two processors to minimize schedule length," *SIAM Journal of Discrete Mathematics*, 2, pp. 176-196, 1989a.

Du, J., and J. Y.-T. Leung, "Minimizing total tardiness on one machine is NP-hard," *Mathematics of Operations Research*, 15, pp. 483-495, 1989b.

Du, J. and J. Y.-T. Leung, "Minimizing the number of late jobs on unrelated machines," *Operations Research Letters*, Vol. 10, No. 3, pp. 153-158, 1991.

Du, J., J. Y.-T. Leung, and C. S. Wong, "Minimizing the number of late jobs with release time constraints," Technical report, Computer Science Program, University of Texas, Dallas, 1989.

Du, J., J. Y.-T. Leung, and G. H. Young, "Minimizing mean flow time with release time constraint," Technical report, Computer Science Program, University of Texas, Dallas, 1988.

Du, J., J. Y.-T. Leung, and G. H. Young, "Scheduling chain-structured tasks to minimize makespan and mean flow time," *Information and Computers*, 1989.

Eilon, S., and I. Chowdhury, "Minimizing waiting time variance in the single machine problem," *Management Science*, 23, pp. 567-575, 1977.

Emmons, H., "One machine sequencing to minimize mean flow time with minimum number tardy," *Naval Research Logistics Quarterly*, 22, pp. 585-592, 1975a.

Erschler, J., et al., "A new dominance concept in scheduling n jobs on a single machine with ready times and due dates," *Operations Research*, Vol. 31, No. 1, pp. 114-127, 1983.

Federgruen, A. and H. Groenevelt, "Preemptive scheduling of uniform machines by ordinary network flow techniques," *Management Science*, Vol. 32, No. 3, pp. 341-349, 1986.

Federgruen, A., and G. Mosheiov, "Efficient algorithms for scheduling problems with general earliness and tardiness cost structures," Graduate School of Business, Columbia University, 1991.

Fox, B.R., and M.B. McMahon, "Genetic operators for sequencing problems," Planning and Scheduling Group, McDonnell Douglas Space Systems, Houston, Texas, 1990.

French, S., *Sequencing and Scheduling: An Introduction to the Mathematics of the Job-Shop*, Ellis Horwood Limited, Chichester, England, 1982.

Friesen, D. K., and M. A. Langston, "Evaluation of a MULTIFIT-based scheduling algorithm," *Journal of Algorithms*, 7, pp. 35-59, 1986.

Frieze, A. M. and J. Yadegar, "A new integer programming formulation for the permutation flowshop problem," *European Journal of Operational Research*, Vol. 40, No. 1, pp. 90-98, 1989.

Fry, T., R. Armstrong, and J. Blackstone, "Minimizing weighted absolute deviation in single machine scheduling," *IIE Transactions*, 19, pp. 445-450, 1987.

Fry, T. D., et al., "A heuristic solution procedure to minimize $\bar{T}$ on a single machine," *Journal of the Operational Research Society*, Vol. 40, No. 3, pp. 293-297, 1989.

Gabow, H. N., "Scheduling UET systems on two uniform processors and length-two pipelines," *SIAM Journal of Computing*, 17, pp. 810-829, 1988.

Garey, M. R., and D. S. Johnson, "Complexity results for multiprocessor scheduling under resource constraints," *SIAM Journal of Computing*, 4, pp. 397-411, 1975.

Garey, M. R., and D. S. Johnson, "Scheduling tasks with nonuniform deadlines on two processors," *Journal of the Association for Computing Machinery*, 23, pp. 461-

467, 1976.

Garey, M. R., and D. S. Johnson, "Two-processor scheduling with start-times and deadlines," *SIAM Journal of Computing*, 6, pp. 416-426, 1977.

Garey, M. R., and D. S. Johnson, "Strong NP-completeness results- motivation, examples and implications," *Journal of the Association for Computing Machinery*, 25, pp. 499-508, 1978.

Garey, M. R., D. S. Johnson, and R. Sethi, "The complexity of flowshop and jobshop scheduling," *Mathematics of Operations Research*, I, pp. 117-129, 1976.

Garey, M. R., D. S. Johnson, R. E. Tarjan, and M. Yannakakis, "Scheduling opposing forests," *SIAM Journal of Algebraic Discrete Methods*, 4, pp. 72-93, 1983.

Garey, M., R. Tarjan, and G. Wilfong, "One-processor scheduling with symmetric earliness and tardiness penalties," *Mathematics of Operations Research*, 13, pp. 330-348, 1988.

Gazmuri, P. G., "Probabilistic analysis of a machine scheduling problem," *Mathematics of Operations Research*, Vol. 10, No. 2, pp. 328-339, 1985.

Giffler, B., and G. L. Thompson, "Algorithms for solving production-scheduling problems," *Operations Research*, 8, pp. 487-503, 1960.

Gilmore, P. C., and R. E. Gomory, "Sequencing a one-state variable machine: a solvable case of the traveling salesman problem," *Operations Research*, 12, pp. 655-679, 1964.

Glover, F., "Tabu Search - Part I," *ORSA Journal on Computing*, 1, p. 190, 1989.

Glover, F., "Tabu Search - Part II," *ORSA Journal on Computing*, 2, p. 4, 1990.

Goldberg, D.E., *Genetic Algorithms in Search, Optimization, and Machine Learning*, Addison-Wesley, Reading, Massachusetts, 1989.

Gonzalez, T., "Optimal mean finish time preemptive schedules," Technical report 220, Computer Science Department, Pennsylvania State University, 1977.

Gonzalez, T., and D. B. Johnson, "A new algorithm for preemptive scheduling of trees," *Journal of the Association for Computing Machinery*, 27, pp. 287-312, 1980.

Gonzalez, T., and S. Sahni, "Flowshop and jobshop schedules: complexity and approximations," *Operations Research*, 26, pp. 36-52, 1978a.

Gonzalez, T., and S. Sahni, "Preemptive scheduling of uniform processor systems," *Journal of the Association for Computing Machinery*, 25, pp. 92-101, 1978b.

Goyal, D. K., "Non-preemptive scheduling of unequal execution time tasks on two identical processors," Technical report CS 77-039, Computer Science Department, Washington State University, Pullman, 1977.

Grabowski, J., "On two-machine scheduling with release dates to minimize maximum lateness," *Opsearch*, 17, pp. 133-154, 1980.

Grabowski, J., "A new algorithm for solving the flow-shop problem," in *Operations Research in Progress*, Feichtinger, G., and P. Kall, (eds.), Reidel, Dordrecht, pp. 57-75, 1982.

Grabowski, J., E. Skubalska, and C. Smutnicki, "On flow-shop scheduling with release and due dates to minimize maximum lateness," *Journal of the Operations Research Society*, 34, pp. 615-620, 1983.

Graham, R. L., "Bounds for certain multiprocessing anomalies," *Bell System Technical Journal*, 45, pp. 1563-1581, 1966.

Graham, R. L., "Bounds on multiprocessing timing anomalies," *SIAM Journal of Applied Mathematics*, 17, pp. 263-269, 1969.

Graves, S. C., "A review of production scheduling," *Operations Research*, 29, pp. 646-675, 1981.

Gupta, J.N.D., "Optimal schedules for a single facility with classes," *Computers and Operations Research*, 11, pp. 409-413, 1984.

Gupta, J.N.D., "Single facility scheduling with multiple job classes," *European Journal of Operational Research*, 8, pp. 42-45, 1988.

Gupta, M.C., Y.P. Gupta, and A. Kumar, "Minimizing flow time variance in a single machine system using genetic algorithms," 1992.

Gupta, S., and T. Sen, "Minimizing a quadratic function of job lateness on a single machine," *Engn. Costs Prod. Econ.*, 7, pp. 181-194, 1983.

Gupta, S. K. and T. Sen, "On the single machine scheduling problem with quadratic penalty function of completion times: an improved branching procedure," *Management Science*, Vol. 30, No. 5, pp. 644-647, 1984a.

Gupta, S., and T. Sen, "Minimizing the range of lateness on a single machine," *Journal of the Operational Research Society*, Vol. 35, pp. 853-857, 1984b.

Gusfield, D., "Bounds for naive multiple machine scheduling with release times and deadlines," *Journal of Algorithms*, 5, pp. 1-6, 1984.

Hall, L. A., and D. B. Shmoys, "Jackson's rule for one-machine scheduling: Making a good heuristic better," Department of Mathematics, Massachusetts Institute of Technology, Cambridge, 1988.

Hall, N. G., and M. E. Posner, "Earliness-tardiness scheduling problems, I: Weighted deviation of completion times about a common due date," *Operations Research*, 39:5, pp. 836-846, 1991.

Hall, N. G., W. Kubiak, and S. P. Sethi, "Earliness-tardiness scheduling problems, II: Deviation of completion times about a restrictive common due date," *Operations Research*, 39:5, pp. 847-856, 1991.

Hariri, A. M. A., and C. N. Potts, "Algorithms for two-machine flow-shop sequencing with precedence constraints," *European Journal of Operational Research*, 17, pp. 238-248, 1984.

Hariri, A. M. A. and C. N. Potts, "A branch and bound algorithm to minimize the number of late jobs in a permutation flow-shop," *European Journal of Operational Research*, Vol. 38, No. 2, pp. 228-237, 1989.

Heck, N., and S. Roberts, "A note on the extension of a result on scheduling with secondary criteria," *Naval Research Logistics Quarterly*, 19, pp. 403-405, 1972.

Hefetz, N., and I. Adiri, "An efficient optimal algorithm for the two-machines unit-time jobshop schedule-length problem," *Mathematics of Operations Research*, 7, pp. 354-360, 1982.

Herrbach, L. A. and J. Y.-T. Leung, "Preemptive scheduling of equal length jobs on two machines to minimize mean flow time," *Operations Research*, Vol. 38, No. 3, pp. 487-494, 1990.

Herrmann, J. W., and C.-Y. Lee, "On scheduling to minimize earliness-tardiness and batch delivery costs with a common due date," Research Report 91-16, Department of Industrial and Systems Engineering, University of Florida, Gainesville, Florida, 1991.

Herrmann, J.W., and C.-Y. Lee, "Solving a Class Scheduling Problem with a Genetic Algorithm," Research Report, Department of Industrial and Systems Engineering, University of Florida, Gainesville, Florida, 1992.

Ho, J. C. and Y.-L. Chang, "Heuristics for minimizing mean tardiness for m parallel machines," *Naval Research Logistics*, Vol. 38, No. 3, pp. 367-381, 1991.

Holland, J. H., *Adaption in Natural and Artificial Systems*, University of Michigan Press, 1975.

Horn, W. A., "Minimizing average flow time with parallel machines," *Operations Research*, 21, pp. 846-847, 1973.

Horn, W. A., "Some simple scheduling algorithms," *Naval Research Logistics Quarterly*, 21, pp. 177-185, 1974.

Horowitz, E., and S. Sahni, "Exact and approximate algorithms for scheduling nonidentical processors," *Journal of the Association for Computing Machinery*, 23, pp. 317-327, 1976.

Horvath, E. C., S. Lam, and R. Sethi, "A level algorithm for preemptive scheduling," *Journal of the Association for Computing Machinery*, 24, pp. 32-43, 1977.

Hsu, N. C., "Elementary proof of Hu's theorem on isotone mappings," *Proceedings of the American Mathematical Society*, 17, pp. 111-114, 1966.

Huang, H. C., "On minimizing flow time on processors with variable unit processing time," *Operations Research*, Vol. 34, No. 5, pp. 801-802, 1986.

Ignall, E., and L. Schrage, "Application of the branch and bound technique to some flow-shop scheduling problems," *Operations Research*, 13, pp. 400-412, 1965.

Jackson, J. R., "Scheduling a production line to minimize maximum tardiness," Research Report 43, Management Science Research Project, University of California, Los Angeles, 1955.

Jackson, J. R., "An extension of Johnson's results on job lot scheduling," *Naval Research Logistics Quarterly*, 3, pp. 201-203, 1956.

Jaffe, J. M., "Efficient scheduling of tasks without full use of processor resources," *Theoretical Computer Science*, 12, pp.1-17, 1980a.

Jaffe, J. M., "An analysis of preemptive multiprocessor job scheduling," *Mathematics of Operations Research*, 5, pp. 415-421, 1980b.

Johnson, S. M., "Optimal two- and three-stage production schedules with setup times included," *Naval Research Logistics Quarterly*, I, pp. 61-68, 1954.

Kahlbacher, H.G., "Scheduling with monotonous earliness and tardiness penalties," *European Journal of Operational Research*, Vol. 64, No. 2, 1993.

Kanet, J.J., "Minimizing the average deviation of job completion times about a common due date," *Naval Research Logistics Quarterly*, 28, pp. 643-651, 1981.

Kao, E. P. C., "A multiple objective decision theoretic approach to one-machine scheduling problems," *Computers and Operations Research*, 7, pp. 251-259, 1980.

Karp, R. M., "Reducibility among combinatorial problems," in *Complexity of Computer Computations*, Miller, R.E., Thatcher, J.W., (eds.), Plenum Press, New York, pp. 85-103, 1972.

Kaufman, M. T., "An almost-optimal algorithm for the assembly line scheduling problem," *IEEE Transactions on Computing*, C-23, pp. 1169-1174, 1974.

Kawaguchi, T., and S. Kyan, "Worst case bound of an LRF schedule for the mean weighted flow-time problem," *SIAM Journal of Computing*, 15, pp. 1119-1129, 1986.

Kiran, A. S. and A. T. Unal, "A single-machine problem with multiple criteria," *Naval Research Logistics*, Vol. 38, No. 5, pp. 721-727, 1991.

Kirkpatrick, S., C. D. Gelatt, and M. P. Vecchi, "Optimization by simulated annealing," *Science*, 220, p. 671, 1983.

Kise, H. and T. Ibaraki, "On Balut's algorithm and NP-completeness for a chance-constrained scheduling problem," *Management Science*, Vol. 29, No. 3, pp. 384-388, 1983.

Kise, H., T. Ibaraki, and H. Mine, "A solvable case of the one-machine scheduling problem with ready and due times," *Operations Research*, 26, pp. 121-126, 1978.

Kraemer, F.-J., and C.-Y. Lee, "Common due window scheduling," Research Report 92-5, Department of Industrial and Systems Engineering, University of Florida, Gainesville, 1992.

Kubiak, W., "A pseudo-polynomial algorithm for a two-machine no-wait job-shop scheduling problem," *European Journal of Operational Research*, Vol. 43, No. 3, pp. 267-270, 1989.

Labetoulle, J., E. L. Lawler, J. K. Lenstra, and A. H. G. Rinnooy Kan, "Preemptive scheduling of uniform machines subject to release dates," Pulleyblank, pp. 245-261, 1984.

Lageweg, B. J., J. K. Lenstra, E. L. Lawler, and A. H. G. Rinnooy Kan, "Computer-aided complexity classification of combinatorial problems," *Communications of the Association for Computing Machinery*, 25, pp. 817-822, 1982.

Laguna, M., J. W. Barnes, and F. Glover, "Scheduling jobs with linear delay penalties and sequence dependent setup costs and times using tabu search," Graduate School of Business and Administration, University of Colorado at Boulder, 1990.

Lam, S., and R. Sethi, "Worst-case analysis of two scheduling algorithms," *SIAM Journal of Computing*, 6, pp. 518-536, 1977.

Lawler, E. L., "Optimal sequencing of a single machine subject to precedence constraints," *Management Science*, 19, pp. 544-546, 1973.

Lawler, E. L., "Sequencing to minimize the weighted number of tardy jobs," *RAIRO Rech. Oper.*, 10(5), Suppl. 27-33, 1976a.

Lawler, E. L., *Combinatorial Optimization: Networks and Matroids*, Holt, Rinehart and Winston, New York, 1976b.

Lawler, E. L., "A 'pseudopolynomial' algorithm for sequencing jobs to minimize total tardiness," *Annals of Discrete Mathematics*, 1, pp. 331-342, 1977.

Lawler, E. L., "Sequencing jobs to minimize total weighted completion time subject to precedence constraints," *Annals of Discrete Mathematics*, 2, pp. 75-90, 1978.

Lawler, E. L., "Preemptive scheduling of precedence-constrained jobs on parallel machines," in Dempster, Lenstra and Rinnooy Kan, 1982a.

Lawler, E. L., "Scheduling a single machine to minimize the number of late jobs," Preprint, Computer Science Division, University of California, Berkeley, 1982b.

Lawler, E.L., "A fully polynomial approximation scheme for the total tardiness problem," *Operations Research Letters*, 1, pp. 207-208, 1982c.

Lawler, E. L., "Recent results in the theory of machine scheduling," in *Mathematical Programming: the State of the Art*, Bonn 1982, A. Bachem, M. Groetschel, B. Korte (eds.), Springer, Berlin, pp. 202-234, 1983.

Lawler, E. L., "A dynamic programming algorithm for preemptive scheduling of a single machine to minimize the number of late jobs," *Annals of Operations Research*, Vol. 26, No. 5, pp. 125-133, 1990.

Lawler, E. L., and J. Labetoulle, "On preemptive scheduling of unrelated parallel processors by linear programming," *Journal of the Association for Computing Machinery*, 25, pp. 612-619, 1978.

Lawler, E. L., J. K. Lenstra, and A. H. G. Rinnooy Kan, "Recent developments in deterministic sequencing and scheduling: A survey," in Dempster, Lenstra and Rinnooy Kan, pp. 35-73, 1982.

Lawler, E. L., J. K. Lenstra, A. H. G. Rinnooy Kan, and D. B. Shmoys, "Sequencing and scheduling: Algorithms and complexity," to appear in *Handbooks in Operations Research and Management Science, Volume 4: Logistics of Production and Inventory*, 1989.

Lawler, E. L. and C. U. Martel, "Preemptive scheduling of two uniform machines to minimize the number of late jobs," *Operations Research*, Vol. 37, No. 2, pp. 314-318, 1989.

Lawler, E. L., and J. M. Moore, "A functional equation and its application to resource allocation and sequencing problems," *Management Science*, 16, pp. 77-84, 1969.

Lee, C.-Y., "Single-machine scheduling with constant due-date window," Dept. of Industrial and Systems Engineering, University of Florida, Gainesville, FL, 1991.

Lee, C.-Y., S. L. Danusaputro, and C. S. Lin, "Minimizing weighted number of tardy jobs and weighted earliness-tardiness penalties about a common due date," *Computers and Operations Research*, 18, pp. 379-389, 1991.

Lee, C.-Y., J. J. Hwang, Y. C. Chow, and F. D. Anger, "Multiprocessor scheduling with interprocessor communication delays," *Operations Research Letters*, 7, pp. 141-147, 1988.

Lee, C.-Y., and S.D. Liman, "Single machine flow-time scheduling with scheduled maintenance," *Acta Informatica*, 29, pp. 375-382, 1992.

Lee, C.-Y., and J. D. Massey, "Multiprocessor scheduling: Combining LPT and MULTIFIT," *Discrete Applied Mathematics*, 20, pp. 233-242, 1988.

Lee, C.-Y., and R. Uzsoy, "A new dynamic programing algorithm for the parallel machines total weighted completion time problem," *Operations Research Letters*, Vol. 11, pp. 73-75, 1991.

Lenstra, J. K., *Sequencing by Enumerative Methods*, Mathematisch Centrum, Amsterdam, 1985.

Lenstra, J. K., and A. H. G. Rinnooy Kan, "Complexity of scheduling under precedence constraints," *Operations Research*, 26, pp. 22-35, 1978.

Lenstra, J. K., and A. H. G. Rinnooy Kan, "Computational complexity of discrete optimization problems," *Annals of Discrete Mathematics*, 4, pp. 121-140, 1979.

Lenstra, J. K., and A. H. G. Rinnooy Kan, P. Brucker, "Complexity of machine-scheduling problems," *Annals of Discrete Mathematics*, 1, pp. 343-362, 1977.

Lenstra, J. K., D. B. Shmoys, and E. Tardos, "Approximation algorithms for scheduling unrelated parallel machines," *Mathematical Programming*, 46, pp. 259-271, 1990.

Leung, J. Y.-T., "On scheduling independent tasks with restricted execution times," *Operations Research*, Vol. 30, No. 1, pp. 163-171, 1982.

Leung, J. Y.-T. and G. H. Young, "Minimizing Total Tardiness on a Single Machine with Precedence Constraints," *ORSA Journal on Computing*, Vol. 2, No. 4, pp. 346-352, 1990.

Liao, C.-J. and R.-H. Huang, "An algorithm for minimizing the range of lateness on a single machine," *Journal of the Operational Research Society*, Vol. 42, No. 2, pp. 183-186, 1991.

Liepins, G. E., and M. R. Hilliard, "Genetic algorithms: foundations and applications," *Annals of Operations Research*, 21, pp. 31-58, 1989.

Liman, S. D., and C.-Y. Lee, "Error bound of a heuristic for the common due date scheduling problem," Research Report 91-1, Department of Industrial and Systems Engineering, University of Florida, Gainesville, Florida, 1991.

Liu, J. and B. L. MacCarthy, "Effective heuristics for the single machine sequencing problem with ready times," *International Journal of Production Research*, Vol. 29, No. 8, pp. 1521-1533, 1991.

Malek, M., M. Guruswamy, M. Pandya, and H. Owens, "Serial and parallel simulated annealing and tabu search algorithms for the traveling salesman problem," *Annals of Operations Research*, 21, pp. 59-84, 1989.

Mason, A.J., and E.J. Anderson, "Minimizing flow time on a single machine with job classes and setup times," *Naval Research Logistics*, 38, pp. 333-350, 1991.

Matsuo, H., C. J. Suh, and R. S. Sullivan, "Controlled search simulated annealing for the general job shop scheduling problem," Working Paper 03-04-88.

Matsuo, H., C. J. Suh, and R. S. Sullivan, "A controlled search simulated annealing method for the single-machine weighted tardiness problem," *Annals of Operations Research*, 21, pp. 85-108, 1989.

McMahon, G. B., "A study of algorithms for industrial scheduling problems," Ph.D. dissertation, University of New South Wales, Kensington, 1971.

McMahon, G.B., and C.-J. Lim, "The two-machine flow shop problem with arbitrary precedence relations," *European Journal of Operational Research*, Vol. 64, No. 2, 1993.

McNaughton, R., "Scheduling with deadlines and loss functions," *Management Science*, 6, pp. 1-12, 1959.

Meilijson, I. and A. Tamir, "Minimizing flow time on parallel identical processors with variable unit processing time," *Operations Research*, Vol. 32, No. 2, pp. 440-448, 1984.

Mitten, L. G., "Sequencing n jobs on two machines with arbitrary time lags," *Management Science*, 5, pp. 293-298, 1958.

Miyazaki, S., "One machine scheduling problem with dual criteria," *Journal of the Operations Society of Japan*, 24, pp. 37-50, 1981.

Monma, C. L., "Sequencing to minimize the maximum job cost," *Operations Research*, 28, pp. 942-951, 1980.

Monma, C. L., "Linear-time algorithms for scheduling on parallel processors," *Operations Research*, 30, pp. 116-124, 1982.

Monma, C.L., and C.N. Potts, "On the complexity of scheduling with batch setup times," *Operations Research*, 37, pp. 798-804, 1989.

Moore, J. M., "An n-job, one-machine sequencing algorithm for minimizing the number of late jobs," *Management Science*, 15, pp. 102-109, 1968.

Morton, T.E., *Heuristic Scheduling Systems*, 1992.

Muntz, R. R., and E. G. Coffman, Jr., "Optimal preemptive scheduling on two-processor systems," *IEEE Transactions on Computing*, C-18, pp. 1014-1020, 1969.

Muntz, R. R., and E. G. Coffman, Jr., "Preemptive scheduling of real-time tasks on multiprocessor systems," *Journal of the Association for Computing Machinery*, 17, pp. 324-338, 1970.

Nakajima, K., J. Y.-T. Leung, and S. L. Hakimi, "Optimal two-processor scheduling of tree precedence constrained tasks with two execution times," *Performance Evaluation*, 1, pp. 320-330, 1981.

Nakano, R., and T. Yamada, "Conventional genetic algorithms for job shop problems," in *Proceedings of the Fourth International Conference on Genetic Algorithms*, Morgan Kaufmann Publishers, Inc., 1991.

Nawaz, M., E. E. Enscore, Jr., and I. Ham, "A heuristic algorithm for the m-machine, n-job flowshop sequencing problem," *Omega*, 11, pp. 91-95, 1983.

Nelson, R. T., R. K. Sarin and R. L. Daniel, "Scheduling with multiple performance measures: the one-machine case," *Management Science*, 32:4, pp. 464-479, 1986.

Ogbu, F. A., and D. K. Smith, "The application of the simulated annealing algorithm to the solution of the $n/m/C_{max}$ flowshop problem," *Computers and Operations Research*, 17, p.243, 1990.

Oliver, I. M., D. J. Smith, and J. R. C. Holland, "A study of permutation crossover operators on the traveling salesman problem," in *Genetic Algorithms and their Applications: Proceedings of the Second International Conference on Genetic Algorithms*, J. Grefenstette, ed., 1987.

Osman, I. H., and C. N. Potts, "Simulated annealing for permutation flow-shop scheduling," *OMEGA*, 17, p. 551, 1989.

Ow, P. S., "Focused scheduling in proportionate flowshops," *Management Science*, Vol. 31, No. 7, pp. 852-869, 1985.

Ow, P. S., and T. E. Morton, "The single-machine early/tardy problem," *Management Science*, 35, 177-191, 1989.

Panwalkar, S.S., "Scheduling of a two-machine flowshop with travel time between machines," *Journal of the Operational Research Society*, 42, pp. 609-613, 1991.

Panwalker, S. S., M. L. Smith, and A. Seidmann, "Common due-date assignment to minimize total penalty for the one-machine scheduling problem," *Operations Research*, 30, pp. 391-399, 1982.

Papadimitriou, C. H., and P. C. Kannelakis, "Flowshop scheduling with limited temporary storage," *Journal of the Association for Computing Machinery*, 27, pp. 533-549, 1980.

Papadimitriou, C. H., and M. Yannakakis, "Scheduling interval-ordered tasks," *SIAM Journal on Computing*, 8, pp. 405-409, 1979.

Pinedo, M. L., "A note on the two-machine job shop with exponential processing times," *Naval Research Logistics Quarterly*, 28, pp. 693-696, 1981.

Pinedo, M. L., "Minimizing the expected makespan in stochastic flow shops," *Operations Research*, 30, pp. 148-162, 1982.

Pinedo, M., "Stochastic scheduling with release dates and due dates," *Operations Research*, Vol. 31, No. 3, pp. 559-572, 1983.

Pinedo, M., "A note on stochastic shop models in which jobs have the same processing requirements on each machine," *Management Science*, Vol. 31, No. 7, pp. 840-846, 1985.

Pinedo, M. L., and G. Weiss, "Scheduling jobs with exponentially distributed processing times and intree precedence constraints on two parallel machines," *Operations Research*, 33, pp. 1381-1388, 1985.

Posner, M. E. "Minimizing weighted completion times with deadlines," *Operations Research*, 33, pp. 562-574, 1985.

Posner, M. E., "A sequencing problem with release dates and clustered jobs," *Management Science*, Vol. 32, No. 6, pp. 731-738, 1986.

Potts, C. N., "An adaptive branching rule for the permutation flow-shop problem," *European Journal of Operations Research*, 5, pp. 19-25, 1980a.

Potts, C. N., "Analysis of a heuristic for one-machine sequencing with release dates and delivery times," *Operations Research*, 28, pp. 1436-1441, 1980b.

Potts, C. N., "Analysis of heuristics for two-machine flow-shop sequencing subject to release dates," *Mathematics of Operations Research*, 10, pp. 576-584, 1985a.

Potts, C. N., "A Lagrangean based branch and bound algorithm for single machine sequencing with precedence constraints to minimize total weighted completion time," *Management Science*, Vol. 31, No. 10, pp. 1300-1311, 1985b.

Potts, C.N., "Scheduling two job classes on a single machine," *Computers and Operations Research*, 18, pp. 411-415, 1991.

Potts, C. N., and K. R. Baker, "Flow shop scheduling with lot streaming," *Operations Research Letters*, Vol. 8, pp. 297-303, 1989.

Potts, C. N., J. Gupta, and C. Glass, "Lot streaming in three-stage production processes," Working paper, University of Southampton, 1992.

Potts, C. N., and L. N. Van Wassenhove, "An algorithm for single-machine sequencing with deadlines to minimize total weighted completion time," *European Journal of Operations Research*, 12, pp. 379-387, 1983.

Potts, C. N. and L. N. Van Wassenhove, "A branch and bound algorithm for the total weighted tardiness problem," *Operations Research*, Vol. 33, No. 2, pp. 363-377, 1985.

Potts, C. N., and L. N. Van Wassenhove, "Algorithms for scheduling a single machine to minimize the weighted number of late jobs," *Management Science*, 34, pp. 843-858, 1988.

Potts, C.N., and L.N. Van Wassenhove, "Single machine scheduling to minimize total late work," *Operations Research*, Vol. 40, No. 3, pp. 586-595, 1992.

Rachamadugu, R. M. V., "A note on the weighted tardiness problem," *Operations Research*, Vol. 35, No. 3, pp. 450-452, 1987.

Rinnooy Kan, A. H. G., *Machine Scheduling Problems: Classification, Complexity and Computations*, Nijhoff, The Hague, Holland, 1976.

Rinnooy Kan, A. H. G., B. J. Lageweg, and J. K. Lenstra, "Minimizing total costs in one-machine scheduling," *Operations Research*, 23, pp. 908-927, 1975.

Roeck, H., "The three-machine no-wait flow-shop problem is NP-complete," *Journal of the Association for Computing Machinery*, 31, pp. 336-345, 1984a.

Roeck, H., "Some new results in flow-shop scheduling," *Z. Oper. Res.*, 28, pp. 1-16, 1984b.

Roeck, H., and G. Schmidt, "Machine aggregation heuristics in shop scheduling," Bericht 82-11, Fachberech 20 Mathematik, Technische Unversitaet Berlin, 1982.

Sahney, V.K., "Single-server, two-machine sequencing with switching time," *Operations Research*, 20, pp. 24-36, 1972.

Sahni, S., "Algorithms for scheduling independent tasks," *Journal of the Association for Computing Machinery*, 23, pp. 116-127, 1976.

Sahni, S., and Y. Cho, "Complexity of scheduling jobs with no wait in process," *Mathematics of Operations Research*, 4, pp. 448-457, 1979.

Sen, T., et al., "The two-machine flowshop scheduling problems with total tardiness," *Computers and Operations Research*, Vol. 16, No. 4, pp. 333-340, 1989.

Sen, T. and B. N. Borah, "On the single-machine scheduling problem with tardiness penalties," *Journal of the Operational Research Society*, Vol. 42, No. 8, pp. 695-702, 1991.

Sethi, R., "On the complexity of mean flowtime scheduling," *Mathematics of Operations Research*, 2, pp. 320-330, 1977.

Shanthikumar, J. G., "Scheduling n jobs on one machine to minimize the maximum tardiness with minimum number tardy," *Computers and Operations Research*, Vol. 10, No. 3, pp. 255-266, 1983.

Shanthikumar, J. G., and J. A. Buzacott, "On the use of decomposition approaches in a single machine scheduling problem," *Journal of the Operations Society of Japan*, 25:1, pp. 29-47, 1982.

Sidney, J. B., "The two-machine maximum flow-time problem with series parallel precedence relations," *Operations Research*, 27 pp. 782-791, 1979.

Sidney, J. B. and G. Steiner, "Optimal sequencing by modular decomposition: polynomial algorithms," *Operations Research*, Vol. 34, No. 4, pp. 606-612, 1986.

Simons, B., "A fast algorithm for single processor scheduling," *Proceedings of the 19th Annual Symposium Foundations of Computer Science*, pp. 246-252, 1978.

Simons, B., and M. Sipser, "On scheduling unit-length jobs with multiple criteria release time/deadline intervals," *Operations Research*, 32, pp. 80-88, 1984.

Simons, B., and M. Warmuth, "A fast algorithm for multiprocessor scheduling of unit-length jobs," *SIAM Journal of Computing*, to appear in 1989.

Smith, W. E., "Various optimizers for single-stage production," *Naval Research Logistics Quarterly*, 3, pp. 59-66, 1956.

So, K. C., "Some heuristics for scheduling jobs on parallel machines with setups," *Management Science*, Vol. 36, No. 4, pp. 467-475, 1990.

Sriskandarajah, C., and S.K. Goyal, "Scheduling of a two-machine flowshop with processing time linearly dependent on job waiting-time," *Journal of the Operational Research Society*, Vol. 40, No. 10, pp. 907-921, 1989.

Starkweather, T., S. McDaniels, K. Mathias, C. Whitley, and D. Whitley, "A comparison of genetic sequencing operators," in *Proceedings of the Fourth International Conference on Genetic Algorithms*, Morgan Kaufmann Publishers, Inc., 1991.

Steiner, G., "Single machine scheduling with precedence constraints of dimension 2," *Mathematics of Operations Research*, Vol. 9, No. 2, pp. 248-259, 1984.

Storer, R. H., S. Y. D. Wu, and R. Vaccari, "New search spaces for sequencing problems with application to job shop scheduling," *Management Science*, Vol. 38, No. 10, pp. 1495-1509, 1992.

Syswerda, G., "Schedule optimization using genetic algorithms," in *Handbook of Genetic Algorithms*, L. Davis, ed., Van Nostrand Reinhold, 1991.

Szwarc, W., "Flow shop problems with time lags," *Management Science*, Vol. 29, No. 4, pp. 477-481, 1983.

Szwarc, W., "Single-machine scheduling to minimize absolute deviation of completion times from a common due date," *Naval Research Logistics*, Vol. 36, No. 5, pp. 663-673, 1989.

Szwarc, W., et al., "The single machine problem with a quadratic cost function of completion times," *Management Science*, Vol. 34, No. 12, pp. 1480-1488, 1988.

Taillard, E., "Some efficient heuristic methods for the flow shop sequencing problem," *European Journal of Operational Research*, Vol. 47, No. 1, pp. 65-74, 1990.

Talbot, F. B., and J. H. Patterson, "An efficient integer programming algorithm with network cuts for solving resource-constrained scheduling problems," *Management Science*, 24, pp. 1163-1174, 1978.

Tang, C. S., "Scheduling batches on parallel machines with major and minor set-ups," *European Journal of Operational Research*, Vol. 46, No. 1, pp. 28-37, 1990.

Townsend, W., "Single machine problems with quadratic penalty function of completion times: a branch and bound solution," *Management Science*, Vol. 24, No. 5, pp. 530-534, 1978.

Trietsch, D., "Optimal transfer lots for batch manufacturing: a basic case and extensions," Technical report NPS-54-87-010, Naval Postgraduate School, Monterey, Calif., 1987.

Trietsch, D., "Polynomial transfer lot sizing techniques for batch processing on consecutive machines," Technical report NPS-54-89-011, Naval Postgraduate School,

Monterey, Calif., 1989.

Trietsch, D., and K. R. Baker, "Basic techniques for lot streaming," to appear in *Operations Research*, 1992.

Ullman, J. D., "NP-Complete scheduling problems," *J. Comput. System Sci.*, 10, pp. 384-393, 1975.

Ullman, J. D., "Complexity of sequencing problems," in *Computer and Job Shop Scheduling Theory*, E.G. Coffman, ed., Wiley, New York, pp. 139-164, 1976.

Vairaktarakis, G.L., and C.-Y. Lee, "The single machine problem to minimize total tardiness subject to minimum number of tardy jobs," Research Report 92-4, Department of Industrial and Systems Engineering, University of Florida, Gainesville, 1992.

Vakharia, A. J., and Y. L. Chang, "A simulated annealing approach to scheduling a manufacturing cell," *Naval Research Logistics*, 37, p. 559, 1990.

Van de Velde, S. L., "Minimizing the sum of the job completion times in the two-machine flow shop by Lagrangian relaxation," *Annals of Operations Research*, Vol. 26, No. 12, pp. 257-268, 1990.

Van Laarhoven, P. J. M., E. H. L. Aarts, and J. K. Lenstra, "Job-shop scheduling by simulated annealing," report OS-R8809, Centre for Mathematics and Computer Science, Amsterdam, 1988.

Van Wassenhove, L. N., and K. R. Baker, "A bicriterion approach to time/cost trade-offs in sequencing," *European Journal of Operational Research*, 11, pp. 48-54, 1982.

Van Wassenhove, L. N., and L. F. Gelders, "Solving a Bicriterion Scheduling Problem," *European Journal of Operational Research*, 4, pp. 42-48, 1980.

Vani, V. and M. Raghavachari, "Deterministic and random single machine sequencing with variance minimization," *Operations Research*, Vol. 35, No. 1, pp. 111-120, 1987.

Vepsalainen, A. P. J. and T. E. Morton, "Priority rules for job shops with weighted tardiness costs," *Management Science*, Vol. 33, No. 8, pp. 1035-1047, 1987.

Villarreal, F. J., and R. L. Bulfin, "Scheduling a single machine to minimize the weighted number of tardy jobs," *AIIE Transactions*, 15, pp. 337-343, 1983.

Weber, R. R., "The interchangeability of */M/1 queues in series," *Journal of Applied Probability*, 16, pp. 690-695, 1979.

Weber, R. R., P. Varaiya, and J. Walrand, "Scheduling jobs with stochastically ordered processing times on parallel machines to minimize expected flowtime," *Journal of Applied Probability*, 23, pp. 841-847, 1986.

Weiss, G., "Approximation results in parallel machines stochastic scheduling," *Annals of Operations Research*, Vol. 26, No. 12, pp. 195-242, 1990.

Weiss, G., and M. L. Pinedo, "Scheduling tasks with exponential service times on non-identical processors to minimize various cost functions," *Journal of Applied Probability*, 17, pp. 187-202, 1980.

Whitley, D., T. Starkweather, and D. Shaner, "The traveling salesman and sequence scheduling: quality solutions using genetic edge recombination," in *Handbook of Genetic Algorithms*, L. Davis, ed., Van Nostrand Reinhold, 1991.

Widmer, M., and A. Hertz, "A new heuristic method for the flow-shop sequencing problem," *European Journal of Operational Research*, 41, pp. 186-193, 1989.

Wie, S.-H. and M. Pinedo, "On minimizing the expected makespan and flow time in stochastic flow shops with blocking," *Mathematics of Operations Research*, Vol. 11, No. 2, pp. 336-342, 1986.

Williams, E.F., and S. Tufekci, "Polynomial time algorithms for the m x 2 lot streaming problem," Research Report 92-10, Department of Industrial and Systems Engineering, University of Florida, Gainesville, 1992.

Woodruff, D.L., and M.L. Spearman, "Sequencing and batching for two classes of jobs with deadlines and setup times," *Production and Operations Management*, 1, pp. 87-102, 1992.

Complexity in Numerical Optimization, pp. 254-268
P.M. Pardalos, Editor
©1993 World Scientific Publishing Co.

An $O(nL)$ Iteration Algorithm for Computing Bounds in Quadratic Optimization Problems

Anil P. Kamath [1]
Department of Computer Science, Stanford University, Stanford, CA 94305 USA

Narendra K. Karmarkar
AT&T Bell Laboratories, Murray Hill, NJ 07974 USA

Abstract

We consider the problem of optimizing a quadratic function subject to integer constraints. This problem is NP-hard in the general case. We present a new polynomial time algorithm for computing bounds on the solutions to such optimization problems. We transform the problem into a problem for minimizing the trace of a matrix subject to positive definiteness condition. We then propose an interior-point method to solve this problem. We show that the algorithm takes no more than $O(nL)$ iterations (where L is the the number of bits required to represent the input). The algorithm does two matrix inversions in each iteration .

Keywords: Bounds, complexity, quadratic optimization, interior point methods.

1   Outline

The second section of the paper shall introduce the problem of computing upper bounds on a quadratic optimization problem. We shall also motivate an interior point approach to solving the problem. The third section gives an interior point method for solving the problem. The algorithm described in the third section assumes knowledge

---

[1]Research partially supported by US Army Office Research Grant DAAL-03091-G-0102

of the optimal value of the objective function. In the fourth section, we analyse the complexity of this algorithm. Then in the fifth section, we shall generalize the algorithm to the case where we do not have knowledge about the optimal value. Finally, the conclusions are presented in the sixth section.

2 Introduction

Consider the quadratic optimization problem

$$\min f(x) = x^T Q x$$

$$s.t. \ x \in S = \{-1, 1\}^n \tag{1}$$

where $Q \in R^{n \times n}$ is a real symmetric matrix. Let f_{min} be the optimal solution to problem (1).

In this paper we shall be addressing the problem of finding good lower bounds on f_{min}. Note that if problem (1) were solvable in polynomial time we could obtain a lower bound by simply computing the value of f_{min}. However problem (1) is NP-hard [2] in the general case and hence computing f_{min} may be difficult. The problem of computing the lower bound needs to be contrasted with that of finding the upper bound which is comparatively easy since we simply need to compute $f(x)$ at some x in S.

To apply an interior point method to this problem we need to embed the discrete set S in a continuous set $T \supseteq S$. Notice that the minimum of $f(x)$ over T is a lower bound on f_{min}.

A commonly used approach is to choose the continuous set to be the box B defined as follows

$$B = \{x \in R^n | -1 \le x_i \le 1, i = 1, \dots, n\}.$$

The quadratic optimization problem on a box can be solved in polynomial time if Q is a positive definite matrix [3] [9] [10] [11]. But if Q doesn't satisfy positive definiteness then the problem is again shown to be NP-hard [13] [12]. We are interested in solving the difficult problem in which Q has at least one negative eigenvalue [1].

We observed that optimizing over a box can be hard and so instead we choose to enclose the box in an ellipsoid E that contains the box B. Let

$$U = \{w = (w_1, \dots, w_n) \in R^n | \sum_{i=1}^{n} w_i = 1 \ and \ w_i > 0\}.$$

Consider the ellipsoid

$$E(w) = \{x \in R^n | x^T W x \le 1\}$$

where $w \in U$ and $W = diag(w)$.

The set S is contained in $E(w)$. Let $\lambda_{min}(w)$ be the minimum eigenvalue of $W^{-1/2}QW^{-1/2}$ then we have

$$\min \frac{x^T Q x}{x^T W x} = \min \frac{x^T W^{-1/2} Q W^{-1/2} x}{x^T x} = \lambda_{min}(w),$$

and so we may conclude that

$$x^T Q x \geq \lambda_{min}(w) , \, \forall x \in E(w).$$

Hence the minimum value of $f(x)$ over $E(w)$ can be obtained by simply computing the minimum eigenvalue of $W^{-1/2}QW^{-1/2}$. To obtain a better bound on f_{min} we need to optimize $\lambda_{min}(w)$ over the set U.

3　Interior Point Approach to the Problem

We have transformed the problem of finding an upper bound into the following optimization problem

$$\max \mu$$

subject to

$$\frac{x^T Q x}{x^T W x} \geq \mu , \, \forall x \in R^n - \{0\} \text{ and } w \in U.$$

We can further simplify the problem by defining new variables $d = (d_1, \ldots, d_n) \in R^n$ where $\sum_{i=1}^n d_i = 0$. Let $D = diag(d)$ then we notice that if

$$\frac{x^T (Q - D) x}{x^T W x} \geq \mu$$

then since $\sum_{i=1}^n w_i = 1$ and $\sum_{i=1}^n d_i = 0$ we may conclude that for $x \in S$ we have

$$x^T Q x \geq \mu x^T W x + x^T D x = \mu.$$

Let $z = \mu w + d$ and $Z = diag(z)$ then for all $x \in S$ we get

$$x^T Z x = e^T z = \mu$$

and so the problem transforms into

$$\max e^T z$$

$$s.t. \ x^T (Q - Z) x \geq 0.$$

Let $M(z) = Q - Z$. We observe that solving above problem amounts to minimizing the trace of $M(z)$ while keeping $M(z)$ positive semi-definite. Let $\lambda_i(M(z))$ denote the ith eigenvalue of $M(z)$ [14]. Since $M(z)$ is a real and symmetric matrix, it has n real eigenvalues $\lambda_i(M(z)), i = 1, \ldots, n$. To ensure positive definiteness, the eigenvalues of $M(z)$ must be nonnegative. Hence the above problem may be reformulated as follows

$$\min tr(M(z))$$

$$s.t. \ \lambda_i(M(z)) \geq 0 , \ i = 1, \ldots, n \tag{2}$$

3.1   Centering Transformation

We shall be using an iterative interior point approach to solve problem (2). As in linear programming we would like to define a centering transform that centers the current iterate with respect to the region over which we are optimizing. Hence we examine the following family of transformations.

Let $P = \{A \in R^{n \times n} | A = A^T$ and $x^T A x > 0, \forall x \in R^n - \{0\}\}$ be a set of positive definite matrices. We define a family G of transformation $L_F : P \to P$ where F is an $n \times n$ real and invertible matrix. The transformations are defined as

$$L_F(A) = FAF^T.$$

The transformations are one-one and onto and they form a Lie group under the composition operation.

3.2   Properties of Transformation

In this subsection, we shall prove some lemmas about the transformations in G and about optimization over the transformed space.

Lemma 3.1 Centering Property
Given any positive definite matrix M there exists $L_F \in G$ such that $L_F(M) = I$.

Proof:
Since M is a positive definite matrix, we know that $M^{-1/2}$ exists.
Let $F = M^{-1/2}$ then note that $L_F(M) = I$.   ∎

If we are at a point z^o then $L_{M(z^o)^{-1/2}}$ maps the matrix $M(z^o)$ to the identity matrix which has all eigenvalues equal to 1. Hence it can be used as a centering transform in our interior point method for solving problem (2).

To measure distances in the transformed space, we define a Riemannian metric [6][5]. Let $F = M(z^o)^{-1/2}$. The Riemannian distance between the points z^o and z that is invariant under the transformation L_F is denoted by $d^2(M(z), M(z^o))$.

Lemma 3.2 Riemannian metric in transformed space
Let $z, z^o \in R^n$ be any two points in the interior of the region of optimization and let the matrix $H(z^o)$ be defined as

$$[H(z^o)]_{ij} = (e_i^T M(z^o)^{-1} e_j)^2$$

then the distance in the Riemannian metric may be expressed as

$$d^2(M(z), M(z^o)) = (z - z^o)^T H(z^o)(z - z^o).$$

Proof:

Let $\Delta z = z - z^o$ and $\Delta Z = diag(\Delta z)$ then we have

$$L_F(M(z)) = I + \Delta M'$$

where $\Delta M' = F \Delta Z F$.

Since the Riemannian distances are invariant under transform L_F we have

$$
\begin{aligned}
d^2(M(z), M(z^o)) &= d^2(L_F(M(z)), L_F(M(z^o))) \\
&= d^2(I, I + \Delta M') \\
&= tr(\Delta M' \Delta M'^T) \\
&= tr(F \Delta Z F \Delta Z F) \\
&= \sum_i \sum_j (e_i^T M(z^o)^{-1} e_j)^2 \Delta z_i \Delta z_j \\
&= (z - z^o)^T H(z^o)(z - z^o).
\end{aligned}
$$

$\blacksquare$

The following lemma will describe an inscribed ball in the feasible region for problem (2).

Lemma 3.3 Inscribed Open Ball in the transformed space

An open ball of radius 1 centered at z^o in the transformed space is contained in the interior of the feasible region for problem (2).

Proof:

If z is within an open ball of radius 1 centered at z^o in the transformed space then

$$d^2(M(z), M(z^o)) < 1$$

and hence we get

$$tr(M(z^o)^{-1/2} \Delta Z M(z^o)^{-1} \Delta Z M(z^o)^{-1/2}) < 1.$$

In other words we can say that

$$\sum_{i=1}^n \lambda_i^2(M(z^o)^{-1/2} \Delta Z M(z^o)^{-1/2}) < 1.$$

But we know that

$$\lambda_i(L_F(M(z))) = 1 + \lambda_i(M(z^o)^{-1/2} \Delta Z M(z^o)^{-1/2})$$

and hence we can conclude that

$$\lambda_i(L_F(M(z))) > 0.$$

Consequently, we get

$$\lambda_i(M(z)) > 0.$$

Hence z is an interior point of (2).

In the next lemma, we show how we can optimize a linear function on an open ball in the transformed space.

Lemma 3.4 Optimizing Linear function over a ball
Consider the following optimization problem

$$\min c^T z$$

$$s.t.\ d^2(M(z), M(z^o)) \le \alpha^2. \tag{3}$$

The optimal solution to problem (3) is given by

$$z = z^o - \beta H(z^o)^{-1} c$$

where $\beta = \dfrac{\alpha}{\sqrt{c^T H(z^o)^{-1} c}}$.

Proof:
We notice that problem (3) corresponds to minimizing $c^T z$ over a ball of radius α in the transformed space. Hence by lemma (2) the problem may be rewritten as

$$\min c^T z$$

$$s.t.(z - z^o)^T H(z^o)(z - z^o) \le \alpha^2$$

where $[H(z^o)]_{ij} = (e_i^T M(z^o)^{-1} e_j)^2$.
The matrix $H(z^o)$ can be shown to be positive definite. Hence by the Kuhn-Tucker conditions [8], the optimal solution to the above problem (3) can be obtained by solving the linear system given by

$$H(z^o)(z - z^o) = -\beta c$$

where $\beta = \dfrac{\alpha}{\sqrt{c^T H(z^o)^{-1} c}}$.

We now consider a potential function corresponding to the constraints in (2). The potential function is defined as

$$\phi(z) = -\sum_{i=1}^{n} \ln \lambda_i(M(z)) = -\ln \det(M(z)).$$

The potential function is defined only over the interior of the region of optimization in problem 2.

Lemma 3.5 Invariance of Potential Function

The potential function $\phi(z)$ is invariant under the transformations in G.

Proof:

We notice that

$$
\begin{aligned}
\phi(L_F(A)) - \phi(L_F(B)) &= -\ln\det(FAF^T) - (-\ln\det(FBF^T)) \\
&= -\ln\frac{\det(F)\det(A)\det(F)}{\det(F)\det(B)\det(F)} \\
&= \phi(A) - \phi(B).
\end{aligned}
$$

$\blacksquare$

In the next subsection, we shall use the properties that we proved in the above lemmas, to construct an interior point method to solve the optimization problem (2).

3.3   Description of an Algorithm

Our interior point method to solve problem (2) is an iterative method. We shall first describe a simple method that assumes knowledge of the optimal value v^* of the objective function in (2). Later we shall modify the algorithm to work in the case where the value of v^* is not known.

The method uses the potential function defined as

$$
f(z) = q\ln(tr((M(z)) - v^*)) - \sum_{i=1}^{n}\ln\lambda_i(M(z)) \tag{4}
$$

where $q = 2n$.

We start with some value $z^{(0)}$ in the interior of the region of optimization. At iteration K, let $z^{(K)}$ denote the current value of the iterate. The algorithm may then be described by the following sequence of steps:

1. start with some feasible interior point $z^{(0)}$.

2. Apply the centering transform $L_{M(z^{(K)})^{-1/2}}$.

3. Construct a linear approximation to the potential function $f(z)$ at $z^{(K)}$.

4. find a point $z^{(K+1)}$ that optimizes this linear approximation over a ball of radius $\alpha < 1$ centered at $z^{(K)}$ in the transformed space.

5. Move to the new point $z^{(K+1)}$.

6. Repeat steps 2-6 until we are reasonably close to the solution.

Note that a ball of radius α will be completely contained in the feasible region by lemma 3 and hence the point $z^{(K+1)}$ is strictly in the interior of the feasible region. Also, the optimization over the ball may be done by simply solving a linear system of equations as shown in lemma 4.

4 Complexity Analysis

We shall measure the progress made by our algorithm in terms of the reduction we obtain in the potential function defined in (4). Let L_F be the centering transform and let $\tilde{M}(z) = FM(z)F$ then the transformed potential function is given by

$$\tilde{f}(z) = q\ln(tr(M(z)) - v^*) - \sum_{i=1}^{n} \ln \lambda_i(\tilde{M}(z)). \tag{5}$$

Note that changes in the potential function are invariant with respect to the transformation and we can just as well work with the transformed potential function (5). The linear approximation to (5) at the point $z^{(K)}$ is given by

$$\tilde{f}_1(z) = \tilde{f}(z^{(K)}) + q\frac{tr(M(z)) - tr(M(z^{(K)}))}{tr(M(z^{(K)})) - v^*} - (tr(\tilde{M}(z)) - tr(\tilde{M}(z^{(K)}))). \tag{6}$$

At each iteration we shall seek to reduce the potential function by minimizing its linear approximation.

4.1 Reduction of Potential function

We shall first show that the linear approximation (6) to the potential function can be reduced by a constant.

Lemma 4.1 :
Let $z^{(K+1)}$ be the point that minimizes $\tilde{f}_1(z)$ on a ball of radius α centered at $z^{(K)}$ in the transformed space at iteration K. Let $\Delta \tilde{f}_1^{(K)} = \tilde{f}_1(z^{(K+1)}) - \tilde{f}_1(z^{(K)})$ then

$$\Delta \tilde{f}_1^{(K)} \leq -\alpha.$$

Proof:
Let z^* be an optimal solution to problem (2). Let the Riemannian distance between z^* and $z^{(K)}$ be R i.e. we have

$$(z^* - z^{(K)})^T H(z^{(K)})(z^* - z^{(K)}) = \sum_{i=1}^{n} (\lambda_i(\tilde{M}(z^*)) - 1)^2 = R^2.$$

If we join the current iterate $z^{(K)}$ to the point z^* , it will intersect the ball of radius α in the transformed space at a point $z^\alpha = (1 - \lambda)z^{(K)} + \lambda z^*$ where $\lambda = \alpha/R$. The change in $\tilde{f}_1(z)$ at the point z^α is given by

$$\begin{aligned}
\Delta \tilde{f}_1^\alpha &= \tilde{f}_1(z^\alpha) - \tilde{f}_1(z^{(K)}) \\
&= q\frac{tr(M((1 - \lambda)z^{(K)} + \lambda z^*)) - tr(M(z^{(K)}))}{tr(M(z^{(K)})) - tr(M(z^*))} \\
&\quad -(tr(\tilde{M}((1 - \lambda)z^{(K)} + \lambda z^*)) - tr(\tilde{M}(z^{(K)}))) \\
&= -q\lambda - \lambda tr(\tilde{M}(z^*) - I).
\end{aligned}$$

Let $\Gamma = \{i | \lambda_i(\tilde{M}(z^*)) < 1\}$. We know that $\lambda_i(\tilde{M}(z^*)) \geq 0$ and hence we get

$$tr(\tilde{M}(z^*) - I) = \sum_{i \in \Gamma}(\lambda_i(\tilde{M}(z^*)) - 1) + \sum_{i \notin \Gamma}(\lambda_i(\tilde{M}(z^*)) - 1)$$

$$\geq -n + \begin{cases} 0 & if \ R^2 < n \\ \sqrt{R^2 - n} & if \ R^2 \geq n \end{cases}.$$

Hence we conclude that the change in the linear approximation to the potential function is bounded by

$$\Delta \tilde{f}_1^\alpha \leq -n\lambda - \lambda \begin{cases} 0 & if \ R^2 < n \\ \sqrt{R^2 - n} & if \ R^2 \geq n \end{cases}.$$

If $\lambda \geq \alpha/n$ then clearly $\Delta \tilde{f}_1^\alpha \leq -\alpha$. On the other hand, if $\lambda < \alpha/n$ then since $\lambda R = \alpha$, we can again show that $\Delta \tilde{f}_1^\alpha \leq -\alpha$. Since $z^{(K+1)}$ minimizes $\tilde{f}_1(z)$ on the ball of radius α

$$\Delta \tilde{f}_1^{(K)} \leq \Delta \tilde{f}_1^\alpha \leq -\alpha.$$

∎

So we have shown that the linear approximation of $\tilde{f}(z)$ can be reduced by a constant. To determine the corresponding reduction in the potential function we need to determine how close the potential function value is to its linear approximation.

Lemma 4.2 :
Let $z^{(K+1)}$ be the point that minimizes $\tilde{f}_1(z)$ on the ball of radius α centered at $z^{(K)}$ in the transformed space at iteration K. Then the reduction in the potential function at iteration K is bounded by

$$\tilde{f}(z^{(K+1)}) - \tilde{f}(z^{(K)}) \leq -\alpha + \frac{\alpha^2}{2(1-\alpha)}.$$

Proof:
Consider the two terms in the potential function given by $\Psi(z) = q \ln(tr(M(z)) - v^*)$ and $\Phi(z) = -\sum_{i=1}^{n} \ln(\lambda_i(\tilde{M}(z)))$. Let $\Psi_1(z)$ and $\Phi_1(z)$ denote their respective linear approximations. Now since $\Psi(z)$ is a concave function we have $\Psi(z) \leq \Psi_1(z)$. On the other hand, since $\ln(1 - \lambda) \geq \lambda - \frac{\lambda^2}{2(1-|\lambda|)}$ for $|\lambda| < 1$ we get the relation that

$$\Phi(z^{(K+1)}) \leq \Phi_1(z^{(K+1)}) + \frac{1}{2(1-\alpha)}\sum_{i=1}^{n}(\lambda_i(\tilde{M}(z)) - 1)^2 \leq \Phi_1(z^{(K+1)}) + \frac{\alpha^2}{2(1-\alpha)}.$$

Hence the reduction in the potential function is bounded by

$$\tilde{f}(z^{(K+1)}) - \tilde{f}(z^{(K)}) \leq \Delta \tilde{f}_1^{(K)} + \frac{\alpha^2}{2(1-\alpha)} \leq -\alpha + \frac{\alpha^2}{2(1-\alpha)}.$$

∎

Let $\delta = -\alpha + \frac{\alpha^2}{2(1-\alpha)}$ then for $\alpha = 1/2$ we get $\delta = -1/4$. Thus we can reduce the potential function by a constant at each iteration.

4.2   Convergence to the optimal solution

Let λ_{min} be the minimum eigenvalue of Q then if we initialize $z^{(0)} = (\lambda_{min} - 1)e$ where $e = (1, \ldots, 1) \in R^n$ then we know that $Q - Z^{(0)}$ is a positive definite matrix and hence $z^{(0)}$ is in the interior of the region of optimization.

Lemma 4.3 :

If λ_{max} is the maximum absolute eigenvalue of Q and we need the optimal value of the objective to be within a precision of ϵ then the algorithm terminates in $O(n \ln(\frac{2n\lambda_{max}+n}{\epsilon}))$ iterations.

Proof:

Since $Q - Z^*$ is a positive definite matrix we know that $v^* \geq 0$. Hence we have

$$f^{(0)}(z^{(0)}) \leq q \ln tr(M(z^{(0)})) - \ln det(M(z^{(0)})).$$

Let N be the number of iterations required to attain the optimal solution within a precision of ϵ then

$$f(z^{(N)}) \leq f(z^{(0)}) - N\delta.$$

This gives us the following inequalities.

$$
\begin{aligned}
N\delta &\leq q \ln(\frac{tr(M(z^{(0)}))}{\epsilon}) + \ln(\frac{det(M(z^{(N)}))}{det(M(z^{(0)}))}) \\
&\leq q \ln(\frac{2n\lambda_{max} + n}{\epsilon}) + n \ln(\frac{2n\lambda_{max} + n}{n}).
\end{aligned}
$$

Hence we have

$$N \leq \frac{3n}{\delta} \ln(\frac{2n\lambda_{max} + n}{\epsilon}).$$

$\blacksquare$

If the matrix Q is over integers then $\epsilon = 1$ should be sufficient. If the number of bits in the input is L then we can in fact show that $O(nL)$ iterations are sufficient to obtain the solution. Each iteration requires two matrix inversions . If we use fast matrix multiplication to do matrix inversion then the worst case complexity of the algorithm is $O(n^{3.38}L)$.

5   Generalizing the Algorithm

The algorithm in the previous section, assumed knowledge about the optimal value v^* of the objective in problem (2). But in practice this value may not be known. Hence the potential function $f(z)$ is not known. In this section we show how the previous algorithm can be adapted to make do without the value of v^*.

Let us consider the parametric family of potential functions given by

$$g(z, v) = q \ln(tr(M(z)) - v) - \ln det(M(z)).$$

where $v \in R$ is a parameter.

Our new algorithm will generate a monotonically increasing sequence of parameters $v^{(K)}$ that converges to the optimal value v^*. We shall in the process get a sequence of functions $g(z, v^{(K)})$ that converges to the desired potential function $g(z, v^*) = f(z)$. Note that since $v^{(K)} \le v^*$ at each iteration $g(z, v^{(K)}) \ge f(z)$ for all z in the domain of f.

The sequence $v^{(K)}$ is constructed in conjunction with the sequence $z^{(K)}$ of interior points. The algorithm to do this outlined in Figure 1. Since $Q - Z^*$ is a positive definite matrix we may use $v^{(0)} = 0 \le v^*$ as the starting point in the sequence. For this method to terminate in a polynomial number of iterations, we must give a technique for updating $v^{(K)}$ at each iteration, so that the potential function is reduced substantially at each iteration.

Let $g_1^{(K)}(z, v)$ denote the linear approximation of $g(z, v)$ at $z^{(K)}$. Let $\sigma(v) = \frac{q}{tr(M(z^{(K)}) - v}$. Then we can write

$$g_1^{(K)}(z, v) = -\sigma(v)e^T z + \nabla \Phi(z^{(K)})^T z + const.$$

We shall show how $g_1^{(K)}(z, v)$ can be reduced by a constant amount at each iteration.

Lemma 5.1 :
We can find $v^{(K+1)} \in R$ and a point $z^{(K+1)}$ in a closed ball of radius α centered at $z^{(K)}$ such that $v^{(K)} \le v^{(K+1)} \le v^*$ and

$$g_1^{(K)}(z^{(K+1)}, v^{(K+1)}) - g_1^{(K)}(z^{(K)}, v^{(K+1)}) \le -\alpha.$$

Proof:
Let $\Delta g_1^\alpha(v)$ be a function in v that denotes the amount by which $g_1^{(K)}(z, v)$ can be decreased over a ball of radius α centered at $z^{(K)}$. Then by Lemma 3.4 we may conclude that

$$\Delta g_1^\alpha(v) = -\alpha\sqrt{(-\sigma(v)e + \nabla\Phi(z^{(K)}))^T H(z^{(K)})(-\sigma(v)e + \nabla\Phi(z^{(K)}))}. \tag{7}$$

Now if can find $v^{(K+1)} \le v^*$ such that $\Delta g_1^\alpha(v^{(K+1)}) \le -\alpha$ then we can find the corresponding $z^{(K+1)}$ by simply minimizing $g_1^{(K)}(z, v^{(K+1)})$ over a ball of radius α centered at $z^{(K)}$.

We need to consider the following 2 cases:
Case 1: $\Delta g_1^\alpha(v^{(K)}) \le -\alpha$
In this case we simply set $v^{(K+1)} = v^{(K)}$.
Case 2: $\Delta g_1^\alpha(v^{(K)}) > -\alpha$
In this case we need to increase the value of v. We know from Lemma 4.1 that

$g_1^\alpha(v^*) = \Delta \tilde{f}_1^{(K)} \leq -\alpha$. Moreover, we know that $\Delta g_1^\alpha(v^{(K)}) > -\alpha$. But $\Delta g_1^\alpha(v)$ is a continuous function in v over the closed interval $[v^{(K)}, v^*]$. Hence by the Intermediate Value Theorem there exists a point $v^{(K+1)} \in [v^{(K)}, v^*]$ such that $\Delta g_1^\alpha(v^{(K+1)}) = -\alpha$. To compute the exact value of $v^{(K+1)}$ we need to simply solve a quadratic equation in v arising from (7) to get a value of $v \in [v^{(K)}, v^*]$ that gives the necessary reduction. This proves the lemma for the second case.

∎

The above lemma gives us a technique for updating $v^{(K)}$ and $z^{(K)}$ at each iteration. To prove convergence of this method, we prove the following lemma.

Lemma 5.2 :
Let $z^{(K)}$ be the current interior point and $v^{(K)} \leq v^*$ be the current estimate of the optimal value. Let $z^{(K+1)}$ and $v^{(K+1)}$ be the new interior point and new estimate respectively, that are obtained by applying Lemma 5.1 then

$$g(z^{K+1}, v^{(K+1)}) - g(z^K, v^{(K)}) \leq -\alpha + \frac{\alpha^2}{2(1-\alpha)}.$$

Proof:
We know from Lemma 5.1 that

$$g_1^{(K)}(z^{(K+1)}, v^{(K+1)}) - g_1^{(K)}(z^{(K)}, v^{(K+1)}) \leq -\alpha.$$

Now using an argument similar to that in Lemma 4.2 we can show that

$$g(z^{(K+1)}, v^{(K+1)}) - g(z^{(K)}, v^{(K+1)}) \leq -\alpha + \frac{\alpha^2}{2(1-\alpha)}.$$

But since $v^{(K)} \leq v^{(K+1)}$ we know that

$$g(z^{(K)}, v^{(K+1)}) \leq g(z^{(K)}, v^{(K)}).$$

and hence the result follows.

∎

Now we can prove the convergence result for our generalized algorithm.

Lemma 5.3 :
The algorithm described in Figure 1 terminates in $O(nL)$ iterations.

Proof:
From Lemma 5.2, we know that at each iteration, we reduce the parametrized potential function $g(z, v)$ by at least a constant amount $\delta = -\alpha + \frac{\alpha^2}{2(1-\alpha)}$. If the algorithm takes N iterations to converge then we have

$$f(z^{(N)}) \leq g(z^{(N)}, v^{(N)}) \leq g(z^{(0)}, v^{(0)}) - N\delta.$$

We can then use an analysis similar to that in Lemma 4.3 to prove a similar bound on the number of iterations.

∎

Lower bound *(in: Q, ϵ, out: z, opt)*

begin

1. $z^{(0)} = (\lambda_{min}(Q) - 1)e \; ; \; v^{(o)} = 0$

2. $M(z^{(0)}) = Q - Z^{(0)} \, ; \, K = 0$

   While $tr(M(z^{(K)})) - v^{(K)}) \geq \epsilon$ do

3. Construct $H^{(K)}$ where $H_{ij}^{(K)} = (e_i^T M(z^{(K)})^{-1} e_j)^2$

4. Define $f^{(K)}(z) = q \ln(tr(M(z^{(K)})) - v^{(K)}) - \ln \det M(z^{(K)})$

5. Find $g^{(K)} = \nabla f^{(K)}(z^{(K)})$

6. Solve linear system
$$H^{(K)}\Delta z = -\beta g^{(K)}$$
   where $\beta = \dfrac{0.5}{\sqrt{g^{(K)T}H^{(K)-1}g^{(K)}}}$

7. If $g^{(K)T}\Delta z < 0.5$ then increase $v^{(K)}$ until $g^{(K)T}\Delta z = 0.5$

8. $z^{(K+1)} = z^{(K)} + \Delta z \; ; \; K = K + 1$

   endwhile

9. $z = z^{(K)}$, $opt = tr(Q) - v^{(K)}$

end

Figure 1: Algorithm for computing Lower Bounds

6   Conclusions

We have described an $O(n^{3.38}L)$ algorithm for finding a lower bound on quadratic minimization problems subject to integer constraints. We have analyzed the complexity of the algorithm in this paper and are working on further reducing this complexity. We also hope to extend the techniques developed here to solve other optimization problems. A more generalized version of our problem can be found in [7]. The analysis of the quality of the lower bound that can be obtained by this method is an interesting open problem.

References

[1] Gantmakher, F. (1959), *Theory of Matrices*, Chelsea.

[2] Garey, M. and Johnson D. (1979), *Computers and Intractability: A guide to the theory of NP-completeness*, W. H. Freeman, San Francisco.

[3] Han, C. Pardalos, P. and Ye, Y. (1990), "Computational aspects of an interior point algorithm for quadratic problems with box constraints," *Large-Scale Numerical Optimization*, SIAM , 92-112.

[4] Kamath, A. and Karmarkar, N. (1992), "A continuous method to Compute Upper Bounds in Quadratic Maximization Problems with Integer Constraints," *Recent Advances in Global Optimization* (Edit. C.A. Floudas and P.M.Pardalos), Princeton University Press, 125-140.

[5] Kamath, A. and Karmarkar, N. (1992), "A continuous method for computing bounds in integer quadratic optimization problems," *Journal of Global Optimization 2*, 229-241.

[6] Karmarkar, N. (1988), "Riemannian Geometry underlying Interior-Point Methods for Linear Programming," *Contemporary Mathematics 114*, 51-75.

[7] Karmarkar, N. and Thakur, S. (1992), " An interior-point approach to tensor optimization with application to upper bounds in integer quadratic optimization problems," *Proceedings of IPCO 1992*, 406-420.

[8] Luenberger, D. (1973), *Linear and Nonlinear Programming*, Addison-Wesley.

[9] Mehrotra, S. and Jie Sun (1990), "An algorithm for convex quadratic programming that requires $O(n^{3.5}L)$ arithmetic operations," *Mathematics of Operations Research vol.15, no.2*, 342-363.

[10] Monteiro, R. and Adler, G. (1988), "Interior Path Following Primal-Dual Algorithms, Part II; Convex Quadratic Programming," *Mathematical Programming* *44*, 43-66.

[11] Nesterov N., Ye Y. (1988), "Polynomial methods in the linear and quadratic programming," *Soviet Journal of Computer and Systems Sciences vol.26, no.5*, 98-101.

[12] Pardalos, P.M. and Schnitger, G. (1988), "Checking Local Optimality in Constrained Quadratic Programming is NP-hard," *Operations Research Letters, Vol. 7, No. 1*, 33-35.

[13] Pardalos, P.M. and Vavasis, S.A. (1991), "Quadratic Programming with One Negative Eigenvalue is NP-hard," *Journal of Global Optimization 1*, 15-23.

[14] Wilkinson, J. (1965), *The Algebraic Eigenvalue Problem* , Clarendon Press, Oxford.

Complexity in Numerical Optimization, pp. 269-298
P.M. Pardalos, Editor
©1993 World Scientific Publishing Co.

Complexity of Single Machine Hierarchical Scheduling : A Survey [1]

Chung-Yee Lee and George L. Vairaktarakis

Department of Industrial and Systems Engineering
University of Florida, Gainesville FL, 32611

Abstract

In this paper we explore the complexity status of hierarchical scheduling on a single machine. We consider the case of two criteria where the second criterion is optimized subject to the constraint that the first meets its minimum value. Pairwise combinations of all performance measures traditionally treated in scheduling theory are considered. For every problem we either provide an $\mathcal{NP}$-completeness proof or else a polynomial time algorithm. We survey most existing results in the topics involved and develop new results for problems not considered previously. Convenient tables summarizing the results are provided. Also, we discuss the managerial issues relating to each of the problems considered.

1   Introduction.

Scheduling theory has resolved many problems found in industry since 1950. Most of the research in this area involves a single criterion. However, in reality operational effectiveness has many attributes including customer satisfaction, on time delivery, work-in-process inventory, etc. In order for scheduling to be in touch with reality, multicriteria problems must be studied.

The simplest multicriteria problems consider only two criteria and try to minimize the less important criterion subject to the constraint that the most important

---

[1]This research was supported in part by NSF grant DDM-9201627

criterion meets some requirement. Literature refers to this subclass as *dual criteria* optimization. As we will see, hierarchical optimization is a special case of dual criteria optimization, where the most important criterion is at its optimal value.

Dual criteria problems are very important because they can give solutions where the different operational costs are balanced in an acceptable and profitable way. The interplaying issues involved with each problem are discussed in the subsection where the problem is considered. Typically, one tries to balance two kinds of costs; *internal* and *external.* Internal costs are the ones that are absorbed by the producer without customer interference. Such costs include work-in-process and inventories. External costs are the costs relating to customer satisfaction. Such costs are usually measured by late delivery penalties such as total tardiness and number of tardy jobs.

We denote by $F(B|A)$ a typical dual criteria problem involving the performance measures A, B. This notation will mean that we want to find a schedule that minimizes criterion B subject to the constraint that $A \leq Y$ for a specified value of Y. In case that the value Y equals the optimum value $Y^\star$ of the criterion A, we have a special class of dual criteria problems called *hierarchical* problems in the literature. Here we want to minimize the second and less important criterion subject to the constraint that the first and most important criterion is at its optimal level. We will denote such a problem by $F_h(B|A)$. The managerial significance of this class of problems is apparent. Moreover, these problems are interesting from the theoretical stand point because in case they are $\mathcal{NP}$-complete, the corresponding dual criteria problems are $\mathcal{NP}$-complete as well. In addition, a polynomial time algorithm for $F(B|A)$ also solves $F_h(B|A)$.

Hierarchical optimization problems are scattered in literature and only recently researchers started to study these problems intensively. In this paper we discuss the complexity status of hierarchical minimization of all pairwise combinations of total flowtime, total weighted flowtime, maximum lateness, maximum cost, maximum earliness, total tardiness, total weighted tardiness, total earliness, number of tardy jobs and weighted number of tardy jobs. For every problem we either provide an $\mathcal{NP}$-completeness proof or else a polynomial time algorithm. We survey most existing results in the topics involved and provide new results for problems not considered previously. Convenient tables summarizing the results are provided. Also, we discuss the managerial issues relating to each of the problems considered.

We start out by reviewing some basic results in Section 2. At the end of Section 2 we give the layout of the rest of the paper.

2   Notation and basic results.

The following notation is used throughout the paper:

n = number of jobs

$$p_i = \text{processing time of job } i$$

$$MS = \sum_{i=1}^{n} p_i$$

$$d_i = \text{due date of job } i$$

$$r_i = \text{release time of job } i$$

$$w_i = \text{nonnegative weight associated with job } i$$

$$C_i = \text{completion time of job } i$$

$$\sum_{i=1}^{n} C_i = \text{total flowtime}$$

$$\sum_{i=1}^{n} w_i C_i = \text{total weighted flowtime}$$

$$E_i = \max(d_i - C_i, 0); \text{ the earliness of job } i$$

$$E_{\max} = \max_{1 \leq i \leq n} E_i; \text{ the maximum earliness}$$

$$\sum_{i=1}^{n} E_i = \text{the total earliness}$$

$$L_{\max} = \max_{1 \leq i \leq n}(C_i - d_i); \text{ the maximum lateness}$$

$f_{\max} = \max_{1 \leq i \leq n}(f_i(C_i))$; the maximum cost. (It is assumed that all penalty functions f_i $i = 1, 2, \ldots, n$ are nondecreasing in the job completion times.) Note that if $f_i = C_i - d_i$ then $f_{\max} = L_{\max}$.

$$T_i = \max(C_i - d_i, 0); \text{ the tardiness of job } i$$

$$T_{\max} = \max_{1 \leq i \leq n} T_i; \text{ the maximum tardiness}$$

$$\sum_{i=1}^{n} T_i = \text{the total tardiness}$$

$$\sum_{i=1}^{n} w_i T_i = \text{the total weighted tardiness}$$

$$U_i = 1 \text{ if job } i \text{ is tardy, } 0 \text{ otherwise}$$

$$\sum_{i=1}^{n} U_i = \text{the number of tardy jobs}$$

$$\sum_{i=1}^{n} w_i U_i = \text{the total weighted number of tardy jobs}$$

$$\tau = \text{the minimum number of tardy jobs}$$

$$n_T = \text{any integer in } \{\tau, \ldots, n\}.$$

We will use the following scheduling disciplines in the paper:

SPT: jobs are sequenced in nondecreasing order of processing times. This rule is known to minimize total flowtime, see Smith (1956).

WSPT: jobs are sequenced in nondecreasing order of the ratio $\frac{p_i}{w_i}$. This rule is known to minimize total weighted flowtime, see Smith (1956).

EDD: jobs are sequenced in nondecreasing order of due dates. This rule is known to minimize maximum lateness, see Jackson (1955).

MST: jobs are sequenced according to minimum slack time order, i.e. in nondecreasing order of $d_i - p_i$. This rule is known to minimize maximum earliness subject to no machine idle time.

From the above rules we can describe others depending on how ties are broken. For example **SPT/EDD** is the **SPT** sequence where between two jobs having equal processing times, the one with the smallest due date is scheduled first.

The Moore-Hodgson algorithm, denoted by **MH**, (see [27]) to minimize the number of tardy jobs in a single machine follows in the obvious way from the following algorithm. The time requirement of this algorithm is $\mathcal{O}(n \log n)$. By running **MH** we can determine τ.

Moore-Hodgson Algorithm[27]

Schedule job k last, where:

1. Job j is the first tardy job in EDD

2. $J = \{1, 2, \ldots, j\}$

3. $p_k = \max_{i \in J} p_i$ (break ties with larger due-date).

Lawler's algorithm, see Lawler (1973), to minimize the maximum cost $f_{\max}$ subject to precedence constraints on a single machine, is described next. The algorithm has complexity $\mathcal{O}(n^2)$.

Lawler's Algorithm[21]

1. $T := MS;\ J := \{J_1, J_2, \ldots, J_n\}$.

2. Determine the set L that contains the jobs that have no successors in J.

3. Choose from L a job J_j that has minimal $f_j(T)$ value; break ties arbitrarily. Process J_j from time $T - p_j$ to time T.

4. $T := T - p_j;\ J := J \setminus \{J_j\}$.

5. If $J \neq \emptyset$ then go to Step 1; otherwise stop.

In what follows we use the three field notation of Graham et al. [12], $\alpha\backslash\beta\backslash\gamma$, where α denotes the number of machines, β includes all constraints of the problem and γ is the objective function. In this paper $\alpha = 1$.

The notation $F(B|A)$ introduced earlier, will be used to denote the problem $1\backslash A \leq Y\backslash B$ for some specified value Y. Thus, $1\backslash A \leq Y\backslash B$ and $1\backslash\backslash F(B|A)$ will mean the same thing when the value of Y is specified. Since in our treatment Y takes on the optimum value Y^* that A can obtain, the problem $1\backslash A \leq Y^*\backslash B$ will be denoted by $1\backslash\backslash F_h(B|A)$ where the subscript h denotes hierarchical optimization.

To ascertain the difficulty of a problem, we either have to provide a polynomial time algorithm or else we have to provide $\mathcal{NP}$-completeness proof. In the former case we say that the problem is polynomially solvable. In the latter, we have to reduce a well known $\mathcal{NP}$-complete problem to our problem, in such a way that our problem has a solution if and only if the well known $\mathcal{NP}$-complete problem does. In scheduling theory, many reductions are made from the Even-Odd Partition and 3-Partition problems which are well known to be ordinary and strongly $\mathcal{NP}$-complete, respectively (see [9], [19]). These two problems, are stated as follows:

Even-Odd Partition: Given a set of positive integers $A = \{a_i\}_{i=1}^{2n}$ such that $a_1 < a_2 < \ldots < a_{2n-1} < a_{2n}$, is there a set A_1 such that $\sum\{a_i : a_1 \in A_1\} = \sum\{a_i : a_1 \in A \setminus A_1\}$ and $\forall\ 1 \leq i \leq n$ precisely one of $\{a_{2i-1}, a_{2i}\}$ belongs to A_1 ?

3-Partition: Given a set of positive integers $A = \{a_i\}_{i=1}^{3n}$ such that $\sum_{i=1}^{3n} a_i = nB$ and $\frac{B}{4} < a_i < \frac{B}{2}$, is there a partition of A into $A_1, A_2, \ldots, A_n$ such that $\sum\{a_i : a_i \in A_k\} = B$ for all $1 \leq k \leq n$?

At this point we will show a property relating the complexity of hierarchical problems to single criterion problems.

Property 1 *If for some criterion f the unconstrained problem $1\backslash\backslash f$ is $\mathcal{NP}$-complete then the hierarchical problem $F_h(g|f)$ is $\mathcal{NP}$-complete for any criterion g.*

Proof: Indeed, let I be an $\mathcal{NP}$-complete instance for $1\backslash\backslash f \leq K$ for some threshold value K. Then the problem $1\backslash f \leq K\backslash g$ is $\mathcal{NP}$-complete because a solution to the latter problem is a solution to the former. Thus, the problem $F_h(g|f)$ is $\mathcal{NP}$-complete. $\square$

The complexity of all problems encountered in this research are summarized at Table 4 in Section 7. In each of the Sections 3-6 we discuss the results corresponding to 5 rows of Table 4. In Section 3 we review the set of problems where A is any criterion among $\sum_i C_i, \sum_i w_i C_i, L_{max}, f_{max}, E_{max}$ and B is any criterion among $\sum_i C_i, \sum_i \hat{w}_i C_i, L_{max}, \hat{f}_{max}, \hat{E}_{max}$. In Section 4 we consider the next 25 combinations where $A \in \{\sum_i E_i, \sum_i w_i T_i, \sum_i T_i, \sum_i w_i U_i, \sum_i U_i\}$ while $B \in \{\sum_i C_i, \sum_i \hat{w}_i C_i, L_{max},$

$\hat{f}_{\max}, \hat{E}_{\max}\}$.

Section 5 considers the problems with $A \in \{\sum_i C_i, \sum_i w_i C_i, L_{\max}, f_{\max}, E_{\max}\}$ and $B \in \{\sum_i E_i, \sum_i w_i T_i, \sum_i T_i, \sum_i w_i U_i, \sum_i U_i\}$. Finally, we consider the problems with $A, B \in \{\sum_i E_i, \sum_i w_i T_i, \sum_i T_i, \sum_i w_i U_i, \sum_i U_i\}$ in Section 6. In Section 7 we summarize the complexity of all the problems considered, in convenient tables where we include the section where each problem is treated. Closing comments are given in Section 8.

3 $A \in \{\Sigma_i\, C_i, \Sigma_i\, w_i C_i, L_{\max}, f_{\max}, E_{\max}\}$
    $B \in \{\Sigma_i\, C_i, \Sigma_i\, \hat{w}_i C_i, L_{\max}, \hat{f}_{\max}, \hat{E}_{\max}\}$

This set of 25 problems has been surveyed by Han Hoogeveen in his thesis, [16]. Some of the results are reported in [15], [17]. A summary is given in the following table.

	$\sum_{i=1}^n C_i$	$\sum_{i=1}^n \hat{w}_i C_i$	$L_{\max}$	$\hat{f}_{\max}$	$\hat{E}_{\max}$
$\sum_{i=1}^n C_i$	$\mathcal{O}(n \log n)$	$\mathcal{O}(n \log n)$	$\mathcal{O}(n \log n)$	$\mathcal{O}(n^2)$	$\mathcal{O}(n \log n)$
$\sum_{i=1}^n w_i C_i$	$\mathcal{O}(n \log n)$	$\mathcal{O}(n \log n)$	$\mathcal{O}(n \log n)$	$\mathcal{O}(n^2)$	$\mathcal{O}(n \log n)$
$L_{\max}$	$\mathcal{O}(n \log n)$	!!	$\mathcal{O}(n \log n)$	$\mathcal{O}(n^2)$	$\mathcal{O}(n^2)$
$f_{\max}$	$\mathcal{O}(n^2)$	!!	$\mathcal{O}(n^2)$	$\mathcal{O}(n^2)$	!!
$E_{\max}$	!!	!!	$\mathcal{O}(n^2)$	!!	$\mathcal{O}(n \log n)$

Table 1:

We will mention some of these problems starting with the classical ones i.e. the first multicriteria problems ever considered. The primary criterion of these problems is one that measures lateness. Such criteria include $L_{\max}$ and $f_{\max}$. These criteria take into account customer satisfaction by measuring how late the orders are delivered to the customer. A permissible upper bound on lateness expresses the idea that lateness not exceeding this upper bound is tolerable and does not induce serious disturbance in customer-producer relation.

The dual criteria problem $F(\sum w_i C_i | L_{\max})$ is the most studied dual criteria problem in the literature. The special case $1 \backslash L_{\max} \leq 0 \backslash \sum C_i$ is solved by Smith's algorithm (Smith 1956). Smith's algorithm is a backward algorithm that chooses as last job $J_{[n]}$, the biggest job that does not become tardy when it completes at time MS. The next job is the biggest among the unscheduled jobs that does not become tardy even if scheduled to finish at time $MS - p_{[n]}$ and so on.

Smith indicated that this algorithm can be extended for the problem $1\backslash L_{\max} \le 0\backslash \sum w_i C_i$. Heck and Roberts (1972) later relaxed the constraint to the case where job tardiness had value no greater than the one obtained by the **EDD** rule, $L^\star$. They provided an extension of Smith's approach for the problem $F_h(\sum C_i | L_{\max})$ that takes $\mathcal{O}(n \log n)$ time that can be used to solve the dual criteria problem $F(\sum C_i | L_{\max})$. They conjectured that it also provides the optimal solution for $F_h(\sum w_i C_i | L_{\max})$. This claim, however, was disproved by Burns (1976) and Emmons (1975) independently. Burns (1976) provided an algorithm to generate a local optimum that cannot be improved by any pairwise interchange of the jobs. Miyazaki (1981) indicated that Smith's algorithm is optimal only if the completion time of the job in the $(i+1)$-st position is no greater than the due date of the job in the i-th position. Lenstra *et al.* (1977) proved that $1\backslash L_{\max} \le 0\backslash \sum w_i C_i$ is $\mathcal{NP}$-complete by a reduction from the Knapsack problem. Branch and bound algorithms have been proposed for this problem by Bansal (1980), Shanthikumar and Buzacott (1982), Potts and Van Wassenhove (1983) and Posner (1985). The general problem $F(\sum w_i C_i | L_{\max})$ has been studied by Chand and Schneeberger (1986). The dual criteria problems $F(\sum C_i | L_{\max})$, $F(\sum C_i | f_{\max})$ have been considered by Van Wassenhove and Gelders (1980), Nelson *et al.* (1986), John (1989).

Emmons (1975A) proposed an $\mathcal{O}(n^2)$ algorithm for $F_h(\sum C_i | f_{\max})$. The following hierarchical problems have been indicated as easy by several researchers (see [1]). The problem $F_h(L_{\max} | \sum w_i C_i)$ is solved in $\mathcal{O}(n \log n)$ time by the **WSPT/EDD** rule i.e. **WSPT** rule with ties broken by **EDD** rule. Similarly, $F_h(f_{\max} | \sum w_i C_i)$ is solved in $\mathcal{O}(n^2)$ time by **WSPT** where ties are settled so that the maximum cost is minimized. All hierarchical problems of the form $F_h(\cdot | \sum C_i)$ are special cases of the hierarchical problems $F_h(\cdot | \sum w_i C_i)$ and hence the solution algorithms described still apply.

Recently, Hoogeveen and Van de Velde (1990) presented an $\mathcal{O}(n^3)$ algorithm that solves $F(L_{\max} | \sum C_i)$. Also, they presented an $\mathcal{O}(n^4)$ algorithm that solves both $F(f_{\max} | \sum C_i)$ and $F(\sum C_i | f_{\max})$.

In case that one of A, B is the criterion $E_{\max}$, important results include the following: If preemption is allowed, Hoogeveen and Van de Velde (1990) proposed an $\mathcal{O}(n^4)$ algorithm that solves both $1\backslash pmtn, nmit \backslash F(E_{\max} | \sum C_i)$ and $1\backslash pmtn, nmit \backslash F(\sum C_i | E_{\max})$.

Hoogeveen (1990) proposed an algorithm that can solve $F(E_{\max} | L_{\max})$, in $\mathcal{O}(n^2 \log n)$ time when no idle time ($nmit$) is allowed. Also the special case $1\backslash nmit \backslash F_h(E_{\max} | L_{\max})$ can be solved in $\mathcal{O}(n^2)$ time. In case that idle time is allowed, the problem is solved in $\mathcal{O}(n^2 \log n)$ time and therefore $F_h(E_{\max} | L_{\max})$, and $F_h(L_{\max} | E_{\max})$ are polynomially solvable as well.

Also, Hoogeveen (1991) proposed an $\mathcal{O}(n^2)$ algorithm that can be used to solve the dual criteria problem $1\backslash nmit\backslash F(E_{\max}|\hat{E}_{\max})$, where $\hat{E}_{\max}$ is a nondecreasing penalty function of the job completion times.

Tuzikov (1991) proposed an ϵ-approximation method for the dual criteria problem involving two different cost functions $f_{\max}$, $g_{\max}$, $F(f_{\max}|g_{\max})$. Hoogeveen (1991) proposed an $\mathcal{O}(n^2)$ algorithm that can solve optimally the dual criteria problem $F(f_{\max}|g_{\max})$. The algorithm also allows for precedence constraints. Since $L_{\max}$ is a special cost function, a variant of the previous algorithm solves both dual criteria problems combining $L_{\max}$, $f_{\max}$ in $\mathcal{O}(n^2)$ time. Thus, the corresponding hierarchical problems are polynomially solvable. In the same paper, Hoogeveen (1991) provides algorithms for any combination of K nondecreasing (nonincreasing) cost functions in the job completion times. The algorithms run in time polynomial on K; namely $\mathcal{O}(n^{K(K+1)-6})$ for $K \geq 4$ and $\mathcal{O}(n^{2K})$ for $K = 2, 3$.

The hierarchical problem $1\backslash nmit\backslash F_h(E_{\max}|f_{\max})$ is shown to be $\mathcal{NP}$-complete in the strong sense by a reduction from 3-Partition (see Hoogeveen 1992).

4 $A \in \{\Sigma_i\, E_i, \Sigma_i\, w_iT_i, \Sigma_i\, T_i, \Sigma_i\, w_iU_i, \Sigma_i\, U_i\}$
         $B \in \{\Sigma_i\, C_i, \Sigma_i\, \hat{w}_iC_i, L_{\max}, \hat{f}_{\max}, \hat{E}_{\max}\}$

In this section we consider problems where the primary criterion A could be any of $\Sigma_i\, E_i, \Sigma_i\, w_iT_i, \Sigma_i\, T_i, \Sigma_i\, w_iU_i, \Sigma_i\, U_i$ while the secondary criterion B is one of $\Sigma_i\, C_i$, $\Sigma_i\, \hat{w}_iC_i, L_{\max}, \hat{f}_{\max}, \hat{E}_{\max}$.

The unconstrained total earliness and total tardiness problems are ordinary $\mathcal{NP}$-complete, see Du and Leung (1990). The unconstrained total weighted tardiness problem is strongly $\mathcal{NP}$-complete, see Lawler (1977). The unconstrained total weighted number of tardy jobs is ordinary $\mathcal{NP}$-complete, see Karp (1973). Then by Property 1 the corresponding hierarchical optimization problems are $\mathcal{NP}$-complete. In Table 2 we denote ordinary $\mathcal{NP}$-completeness by ! and strong $\mathcal{NP}$-completeness by !!. Note that all the hierarchical problems with primary criterion $\Sigma_i\, w_iT_i$ are strongly $\mathcal{NP}$-complete because $1\backslash\backslash \Sigma_i\, w_iT_i$ is such.

We will consider the complexity of the problems unaccounted by this table. First we show that the problems $F_h(\Sigma\, \hat{w}_iC_i|\Sigma\, U_i)$ and $F_h(E_{\max}|\Sigma\, U_i)$ are strongly $\mathcal{NP}$-complete.

The former problem is useful in situations where tardiness costs outweigh flowtime costs. Under this assumption a manager would try to provide the best possible customer service by carrying the least work in process inventory costs that can make this service possible.

	$\sum_{i=1}^{n} C_i$	$\sum_{i=1}^{n} \hat{w}_i C_i$	$L_{\max}$	$\hat{f}_{\max}$	$\hat{E}_{\max}$
$\sum_{i=1}^{n} E_i$	!	!	!	!	!
$\sum_{i=1}^{n} w_i T_i$	!!	!!	!!	!!	!!
$\sum_{i=1}^{n} T_i$	!	!	!	!	!
$\sum_{i=1}^{n} w_i U_i$	!	!	!	!	!
$\sum_{i=1}^{n} U_i$					

Table 2:

Theorem 1 *The 3-Partition problem reduces to* $1 \backslash \sum U_i = \tau \backslash \sum w_i C_i$.

Proof: Let $A = \{a_i\}_{i=1}^{3n}$ be an instance of 3-Partition such that $\sum_{i=1}^{3n} a_i = nB$. We construct an instance for $1 \backslash \sum U_i = \tau \backslash \sum w_i C_i$ as follows:

$$\begin{aligned} p_i &= a_i \ \ for \ \ 1 \le i \le 3n \\ d_i &= 0 \ \ for \ \ 1 \le i \le 3n \\ w_i &= a_i \ \ for \ \ 1 \le i \le 3n \end{aligned}$$

Let $W_1, W_2, \ldots, W_n$ be n additional jobs defined by

$$\begin{aligned} p_{W_j} &= n^2 B^2 + 1 = L \ \ for \ \ 1 \le j \le n \\ d_{W_j} &= jL + (j-1)B \ \ for \ \ 1 \le j \le n \\ w_{W_j} &= 0 \ \ for \ \ 1 \le j \le n \end{aligned}$$

Clearly $\tau = 3n$ since all of the $3n$ regular jobs must be tardy while by scheduling the jobs $W_1, W_2, \ldots, W_n$ in this order followed by the regular jobs in any order results to a schedule with precisely $3n$ tardy jobs.

Let $Y = \frac{1}{2} Bn(n+1)(B+L)$ be a threshold value. We will show that there exists a solution to $1 \backslash \sum U_i = \tau \backslash \sum w_i C_i$ such that $\sum U_i = \tau$ and $\sum w_i C_i \le Y$ if and only if there exists a solution to 3-Partition.

First, assume $\{A_i\}_{i=1}^{n}$ is a solution to 3-Partition. Then, $\sum_{j:a_j \in A_i} p_j = B \ \ for \ \ 1 \le i \le n$. Consider the schedule $S = (W_1, \{j : a_j \in A_1\}, W_2, \{j : a_j \in A_2\}, \ldots, W_n, \{j :$

$a_j \in A_n\}$). Then all W-jobs are scheduled just in time and therefore $\sum_{i=1}^{n} U_i = 3n$. Since the W-jobs have zero weights, the total weighted flowtime is:

$$
\begin{aligned}
\sum_{i=1}^{3n} w_i C_i &= \sum_{i=1}^{3n} a_i C_i \\
&\leq \sum_{\{j:a_j \in A_1\}} p_j(B+L) + \ldots + \sum_{\{j:a_j \in A_n\}} p_j n(B+L) \\
&= \frac{1}{2} Bn(n+1)(B+L) = Y
\end{aligned}
$$

On the other hand, let σ be a solution to the proposed instance such that the total weighted flowtime is $\leq Y$. For this solution to be feasible the W-jobs must be the only nontardy jobs. Assume that the **EDD** order of the W-jobs is $\{W_1, W_2, \ldots, W_n\}$. A simple interchange argument shows that in an optimal schedule we can assume that job W_i precedes W_{i+1} for $1 \leq i < n$. Now, let P_i be the total processing time of regular jobs following the job W_i, and define $P_{n+1} = 0$. Since W_i is non-tardy, we have $P_i \geq (n-i+1)B$ *for* $1 \leq i \leq n$. Then, the total weighted flowtime in σ is:

$$
\begin{aligned}
\sum_{i=1}^{3n} w_i C_i &= \sum_{i=1}^{3n} a_i C_i \\
&\geq \sum_{i=1}^{n} (P_i - P_{i+1}) i L \\
&= L \sum_{i=1}^{n} P_i
\end{aligned}
$$

If there exists $1 \leq i \leq n$ such that $P_i \geq (n-i+1)B + 1 \Rightarrow$ $\sum_{i=1}^{n} P_i \geq B \sum_{i=1}^{n} (n-i+1) + 1 = \frac{1}{2} n(n+1)B + 1$. Then,

$$
\begin{aligned}
\sum_{i=1}^{3n} w_i C_i &\geq \frac{1}{2} n(n+1)BL + L \\
&> \frac{1}{2} n(n+1)B(B+L) = Y \\
&\quad since \quad L > \frac{1}{2} n(n+1)B^2.
\end{aligned}
$$

Hence, we have $P_i = (n-i+1)B$ *for* $1 \leq i \leq n$. This means that 3-Partition has a solution the i-th part of which consists of the 3 jobs immediately following W_i. This completes the proof.$\square$

Now we consider the problem $F_h(E_{\max} | \sum U_i)$. This problem not only considers the number of tardy jobs, but also internal considerations such as inventory costs. This problem allows the manager to schedule effectively in customer satisfaction intensive environments with significant inventory costs.

Theorem 2 *The 3-Partition problem reduces to* $1\backslash \sum_{i=1}^{n} U_i = \tau, \ nmit\backslash E_{\max}$.

Proof: Let $A = \{a_i\}_{i=1}^{3n}$ be an instance of 3-Partition and let $\sum_{i=1}^{3n} a_i = nB$. Construct the instance:

$$p_i = a_i \ \ for \ \ 1 \leq i \leq 3n$$
$$d_i = 0 \ \ for \ \ 1 \leq i \leq 3n$$

Let $W_1, W_2, \ldots, W_n$ be n additional jobs defined by

$$p_{W_j} = 1 \ \ for \ \ 1 \leq j \leq n$$
$$d_{W_j} = j(B+1) \ \ for \ \ 1 \leq j \leq n$$

Let the threshold value be $\bar{E} = 0$. In the given instance the jobs $1 \leq i \leq n$ have to be tardy while scheduling all the W-jobs in the head of the schedule followed by the remaining $3n$ jobs results to a schedule with presicely $3n$ tardy jobs. Thus, $\tau = 3n$.

If $A_1, A_2, \ldots, A_n$ is a solution to 3-Partition, then $\sum_{j \in A_k} p_j = B$, *for every* $1 \leq k \leq n$ and the schedule $\{j : a_j \in A_1\}, W_1, \{j : a_j \in A_2\}, W_2, \ldots, \{j : a_j \in A_n\}, W_n$ has precisely $3n$ tardy jobs while the earliness of each W-job is zero. Thus, $E_{\max} = \bar{E} = 0$.

On the other hand, if $1\backslash \sum_{i=1}^{n} = \tau\backslash E_{\max}$ has a solution such that $E_{\max} = \bar{E} = 0$ then since the jobs $1 \leq i \leq 3n$ have to be tardy and $\tau = 3n$, all of the W-jobs must be early. Since the W-jobs have equal processing times, we can assume that W_k precedes W_{k+1} for $1 \leq k \leq n - 1$.

Since $E_{\max} = 0 \Rightarrow E_{W_1} = E_{W_2} = \ldots = E_{W_n} = 0$. But then, W_k must finish at time $j(B+1)$ for every $1 \leq j \leq n$. Then the three jobs immediately preceding W_j form the A_j set of the 3-Partition i.e. $A_1, A_2, \ldots, A_n$ is a solution to 3-Partition. This completes the proof.$\square$

The last two theorems establish that the problems $F_h(\sum \hat{w}_i C_i | \sum w_i U_i)$ and $F_h(E_{\max} | \sum w_i U_i)$ are strongly $\mathcal{NP}$-complete because $\sum U_i$ is a special case of $\sum w_i U_i$ where every job has the same weight $w_i = 1$.

When the criteria of interest are $\sum C_i$ and $\sum U_i$, the hierarchical problem $F_h(\sum C_i | \sum U_i)$ has been considered by Emmons (1975) who proposed a number of optimal properties and used them in a fast branch and bound algorithm. In particular, he proposed an $\mathcal{O}(n \log n)$ algorithm that finds an optimal solution with respect to a given set of tardy jobs. However, the complexity of this problem remains open.

Another enumerative algorithm has been proposed for $F_h(T_{\max} | \sum U_i)$ by Shanthikumar (1983) who proposed a number of optimal properties that lead to a branch and bound algorithm. Also, he proposed an $\mathcal{O}(n \log n)$ algorithm that finds an optimal solution with respect to a given tardy set. The complexity of this problem is still open. The problem $F_h(f_{\max}|\sum U_i)$ has not been considered in the literature and therefore the complexity of this problem remains open as well.

In addition, no pseudopolynomial time algorithm exists for any of the problems having a !* in Table 4. To fully determine the complexity of these problems, one has to find such an algorithm or provide a reduction from a strongly $\mathcal{NP}$-complete problem. A summary of all the results considered in this section are included in rows 6-10 of Table 4.

5 $A \in \{\Sigma_i\, C_i, \Sigma_i\, w_i C_i, L_{\max}, f_{\max}, E_{\max}\}$ $B \in \{\Sigma_i\, E_i, \Sigma_i\, \hat{w}_i T_i, \Sigma_i\, T_i, \Sigma_i\, \hat{w}_i U_i, \Sigma_i\, U_i\}$

In this section we consider problems where the primary criterion A could be any of $\sum_i C_i, \sum_i w_i C_i, L_{\max}, f_{\max}, E_{\max}$ while the secondary criterion B is one of $\sum_i E_i$, $\sum_i \hat{w}_i T_i, \sum_i T_i, \sum_i \hat{w}_i U_i, \sum_i U_i$. No complexity result has been reported in the literature for this set of problems.

We start out by providing polynomial time algorithms for four hierarchical problems with primary objective to minimize total flowtime; namely $F_h(\sum w_i T_i|\sum C_i)$, $F_h(\sum T_i|\sum C_i)$, $F_h(\sum E_i|\sum C_i)$ and $F_h(\sum w_i U_i|\sum C_i)$. The importance of these problems is realized in enviroments where raw materials are very expensive and therefore the value of work-in-process is very high. We treat these problems in the order introduced, in the next subsection.

5.1 $\sum C_i$ and one of $\sum_i E_i, \sum_i w_i T_i, \sum_i T_i, \sum_i w_i U_i, \sum_i U_i$

For all five problems we have to schedule jobs in **SPT** in order to minimize total flowtime. Let $p_1 < p_2 < \ldots < p_k$ be the distinct values of processing time of the n jobs of J. Let also

$$V_i := \{j : p_j = p_i\} \quad for \quad 1 \le i \le k.$$

To minimize the total flowtime, all jobs in V_i must precede jobs in V_{i+1} for $1 \le i \le k-1$ (see [34]). Hence, to solve any of the four problems we must sequence the jobs within each V_i in such a way that the second criterion is minimized. To do this for the $F_h(\sum w_i T_i|\sum C_i)$ problem, we use the following weighted bipartite matching construction **WBM**.

Let C be the starting time of the first job of V_i. Assume that $|V_i| = r$ and that all jobs in V_i have processing time p. Then, the completion times of the r jobs of V_i will be $C + p, C + 2p, \ldots, C + rp$. Let $d_1, d_2, \ldots, d_r$ be the due-dates of jobs in V_i. Construct a complete bipartite graph $B = (X, Y)$ where X is a set of vertices labeled by $C + p, C + 2p, \ldots, C + rp$ and Y is a set of vertices labeled by $d_1, d_2, \ldots, d_r$. Define the weight w_{jl} of the edge $(C + jp, d_l)$ by $w_{jl} = w_l(C + jp - d_l)^+$, for every $1 \le j, l \le k$. Clearly, if $(C + jp, d_l)$ is an edge of an optimal assignment then assigning a job with due-date d_l to finish at time $C + jp$ results in a sequence that minimizes the total weighted flowtime within V_i.

By applying **WBM** for all V_i batches we get an optimal sequence for $F_h(\sum w_i T_i | \sum C_i)$. Since the number of batches cannot exceed n and every iteration of **WBM** takes time $\mathcal{O}(n^3)$ (see Hungarian algorithm in [29]), the total time to solve the problem is $\mathcal{O}(n^4)$. We denote the algorithm just descibed by **SPT/WBM** because we first find the **SPT** sequence and then we apply **WBM** on the resulting batches.

A similar approach solves the other four problems as well. This establishes that all four problems under consideration are polynomially solvable. However, we will present algorithms for the remaining four problems that have complexity better than $\mathcal{O}(n^4)$.

Theorem 3 *The $F_h(\sum T_i | \sum C_i)$ problem is solved by the* **SPT/EDD** *rule.*

Proof: As seen before, the minimum flowtime constraint induces batches $V_1, \ldots, V_k$. Effectively, we want to solve the problem $1 \backslash p_i = p \backslash \sum T_i$ within each batch. Consider the batch V_1. Let S be an optimal sequence for $F_h(\sum T_i | \sum C_i)$ and suppose there are two jobs i, j in V_1 where job j immediately precedes i in V_1 but $d_i < d_j$. Note that jobs i, j have equal processing times. In S, either both i, j are tardy, or both are nontardy or i is tardy and j is nontardy. A simple interchange argument shows that in every one of these three cases, we can interchange the positions of i, j without increasing total tardiness in S.

By repeating the argument for all jobs that are not in their **EDD** order within a batch, we conclude that the **SPT/EDD** sequencing rule solves $F_h(\sum C_i, \sum T_i)$ to optimality.$\square$

Clearly the **SPT/EDD** rule can be implemented in $\mathcal{O}(n \log n)$ time and hence this is the complexity of $F_h(\sum T_i | \sum C_i)$.

Theorem 4 *When no machine idle time is allowed, the $F_h(\sum E_i | \sum C_i)$ problem is solved by the* **SPT/EDD** *rule.*

Proof: Given an instance for $F_h(\sum E_i | \sum C_i)$, construct an instance for $F_h(\sum T_i | \sum C_i)$ where each job has the same processing time and new due-date defined by $d'_i = MS - d_i + p_i$. We have seen in previous subsections that if S is an optimal sequence for $F_h(\sum T_i | \sum C_i)$, then the reverse sequence S' is optimal for $F_h(\sum E_i | \sum C_i)$ and vice versa. Moreover, the total earliness incurred by S equals the total tardiness incurred by S'.

By the previous theorem we know that the **SPT/EDD** rule is optimal for $F_h(\sum T_i | \sum C_i)$. Therefore, within a batch of jobs with equal processing times, the sequence S' has jobs ordered in earliest due date with respect to the due-dates d'_i. Thus, S has jobs ordered in largest due date with respect to d'_i and therefore in earliest due date with respect to the due-dates d_i. Therefore, S is the sequence obtained by the **SPT/EDD** rule.$\square$

Clearly the **SPT/EDD** rule can be implemented in $\mathcal{O}(n \log n)$ time and hence this is the complexity of $F_h(\sum E_i | \sum C_i)$.

The next theorem considers the problem $F_h(\sum w_i U_i | \sum C_i)$. For this, we will need a modification of **MH** algorithm described in Section 2, that solves the $1 \backslash p_i = p \backslash \sum w_i U_i$ problem. We will denote this modification by **MMH** and it is as follows: Replace Step 3 of **MH** by:

3a. $p_k = p_i$ where $w_i \leq w_k$ for $k \in J$ (break ties with larger due-date)

The fact that **MMH** solves $1 \backslash p_i = p \backslash \sum w_i U_i$ has been demonstrated by Lawler (1976).

Theorem 5 *The $F_h(\sum w_i U_i | \sum C_i)$ problem is solved by the* **SPT/MMH** *rule.*

Proof: To attain the minimum flowtime constraint, we have to schedule the jobs in **SPT** order. As seen earlier, this results to blocks V_i where all jobs within a block have equal processing times. Then, according to the previous observation, in order to minimize the total weighted number of tardy jobs, we apply **MMH** within each block. Thus, the **SPT/MMH** rule solves $F_h(\sum w_i U_i | \sum C_i)$ optimally.$\square$

Clearly the **SPT/MMH** rule can be implemented in $\mathcal{O}(n \log n)$ time and hence this is the complexity of $F_h(\sum w_i U_i | \sum C_i)$.

In the following subsection we consider all five problems with $\sum w_i C_i$ as primary criterion.

5.2 $\sum w_i C_i$ and one of $\sum_i E_i, \sum_i \hat{w}_i T_i, \sum_i T_i, \sum_i \hat{w}_i U_i, \sum_i U_i$

The weighted flowtime represents work inventory costs absorbed by the operation of a machine. A balance of these costs with earliness and tardiness costs are very often the tradeoffs in managerial decisions. It is important to know the difficulty of these problems. The following theorems address this problem.

Theorem 6 *The problems $F_h(\sum \hat{w}_i U_i | \sum w_i C_i)$, $F_h(\sum E_i | \sum w_i C_i)$, $F_h(\sum T_i | \sum w_i C_i)$ are $\mathcal{NP}$-complete. The problem $F_h(\sum \hat{w}_i T_i | \sum w_i C_i)$ is strongly $\mathcal{NP}$-complete.*

Proof: Consider the problem $F_h(\sum \hat{w}_i U_i | \sum w_i C_i)$. We will reduce the problem $1 \backslash\backslash \sum \hat{w}_i U_i$ to it. Indeed, let I be an instance of the latter problem. Assume that each job J_i has been given the characteristics p_i, d_i, $\hat{w}_i$. Construct an instance I' for $F_h(\sum \hat{w}_i U_i | \sum w_i C_i)$ so that the characteristics of each job are p_i, d_i, $w_i = p_i$, $\hat{w}_i$. Observe that every sequence achieves minimum total flowtime for the instance I'

because the ratio w_i/p_i equals 1 for all jobs. Thus, the hierarchical optimization problem $F_h(\sum \hat{w}_i U_i | \sum w_i C_i)$ for the instance I' is equivalent to solving optimally the problem $1\backslash\backslash \sum \hat{w}_i U_i$ for the instance I. As seen in [19], this problem is ordinary $\mathcal{NP}$-complete and therefore so does $F_h(\sum \hat{w}_i U_i | \sum w_i C_i)$. A similar argument works for the remaining problems. In particular, the problem $F_h(\sum \hat{w}_i T_i | \sum w_i C_i)$ is strongly $\mathcal{NP}$-complete because $1\backslash\backslash \sum \hat{w}_i T_i$ is.
This completes the proof.$\Box$

The following problem couples the number of tardy jobs with an internal criterion such as the total weighted flowtime, which is known to minimize work in process inventory.

The hierarchical problem $F_h(\sum U_i | \sum w_i C_i)$ is solved by the **WSPT/MH** rule in $\mathcal{O}(n \log n)$ time. To see this, assume that C^* is the total weighted flowtime obtained by the **WSPT** order. Order the ratios $\frac{p_i}{w_i}$ $i = 1, \ldots, n$ in nondecreasing order. From this order, delete repeated values. Suppose that there remain $k \leq n$ values. Let

$$\frac{p_1}{w_1} < \frac{p_2}{w_2} < \ldots < \frac{p_k}{w_k} \quad k \leq n$$

be the k distinct values of the ratios $\frac{p_i}{w_i}$. Let also

$$V_i := \{j : \frac{p_j}{w_j} = \frac{p_i}{w_i}\} \quad for \quad 1 \leq i \leq k.$$

To minimize the total weighted flowtime, all jobs in V_i must precede jobs in V_{i+1} for $1 \leq i \leq k - 1$ (see [34]).

Hence, to minimize the number of tardy jobs, it is enough to apply Moore-Hodgson's algorithm inside each batch V_i. Then, Moore-Hodgson's algorithm will be applied at most $k \leq n$ times which results to $\mathcal{O}(n \log n)$ complexity.

In the next subsection we consider the case where $A = L_{\max}$.

5.3 $L_{\max}$ with one of $\sum T_i, \sum w_i T_i, \sum E_i, \sum w_i U_i, \sum U_i$

Theorem 7 *The problem $F_h(\sum T_i | L_{\max})$ is $\mathcal{NP}$-complete. The problem $F_h(\sum w_i T_i | L_{\max})$ is strongly $\mathcal{NP}$-complete.*

Proof: First, we will give a detailed proof for $F_h(\sum T_i | L_{\max})$. For this problem we will use a reduction from the ordinary $\mathcal{NP}$-complete problem $1\backslash\backslash \sum T_i$ (see Du and Leung, 1990). Let I be an instance of this problem. Assume that $d_i \leq MS$ otherwise disregard the jobs not satisfying this condition beacause they are not tardy even if scheduled last. Construct an instance I' for $F_h(\sum T_i | L_{\max})$ as follows:

$$d_i' = d_i \quad \forall J_i \in J$$

$$
\begin{aligned}
p'_i &= p_i \ \ \forall J_i \in J \\
p'_{n+1} &= MS \\
d'_{n+1} &= MS
\end{aligned}
$$

There exists an optimal solution to the instance I' of $F_h(\sum T_i | L_{\max})$ where the completion time of the last job is $2MS$, otherwise we can exclude the inserted idle time without worsening $\sum T_i$ nor $L_{\max}$. Therefore, for the set of due dates given in I', the minimum value for $L_{\max}$ equals MS. If S' is an optimal solution to the instance I' of $F_h(\sum T_i | L_{\max})$, then J_{n+1} must be the last job in S'. Indeed, if J_{n+1} is followed by a job say J_i, then J_i is tardy and thus an interchange in the positions of J_{n+1} and J_i does not worsen total tardiness nor it increases $L_{\max}$. Thus, we can assume that the last job in S' is J_{n+1} and $T_{n+1} = MS$. Then, the problem $1\backslash L_{\max} \le MS\backslash \sum T_i \le Y + MS$ is equivalent to the problem $1\backslash\backslash \sum T_i \le Y$ for the instance I. This means that the instance I' of $F_h(\sum T_i | L_{\max})$ is equivalent to the problem $1\backslash\backslash \sum T_i$ and therefore it is $\mathcal{NP}$-complete.

This completes the proof for the problem $F_h(\sum T_i | L_{\max})$. The same argument along with a reduction from the strongly $\mathcal{NP}$-complete problem $1\backslash\backslash \sum w_i T_i$ (see Lawler, 1977) with $w'_i = w_i \ \ \forall J_i \in J$ and $w'_{n+1} = 0$ shows that the $F_h(\sum w_i T_i | L_{\max})$ problem is strongly $\mathcal{NP}$-complete.

This completes the proof.$\square$

For the next theorem we assume that no machine idle time is allowed.

Theorem 8 *The problems $F_h(\sum E_i | L_{\max})$ and $F_h(\sum w_i U_i | L_{\max})$ are $\mathcal{NP}$-complete.*

Proof: Consider an instance with $L_{\max} = 0$. Then it is true that $\tau = 0$. Hence the problem $1\backslash L_{\max} = 0\backslash \sum E_i$ is equivalent to $1\backslash \sum U_i = 0\backslash \sum E_i$. An instance of this problem is proved to be $\mathcal{NP}$-complete in the proof of Theorem 10 in Section 6. Thus, $F_h(\sum E_i | L_{\max})$ is $\mathcal{NP}$-complete.

The problem $F_h(\sum w_i U_i | L_{\max})$ is solved by the **EDD** sequence when $L_{\max} \le 0$ and hence $U_i = 0 \ \forall i$. When $L_{\max} > 0$ we can show that the problem is $\mathcal{NP}$-complete with a reduction from the known $\mathcal{NP}$-complete problem $1\backslash\backslash \sum w_i U_i$, see Karp (1972).

Let I be an instance of $1\backslash\backslash \sum w_i U_i$. For each job J_i we are given the processing time p_i and due date d_i. Then, construct an instance for $F_h(\sum w_i U_i | L_{\max})$ as follows:

$$
\begin{aligned}
p'_{n+1} &= MS \\
d'_{n+1} &= 0 \\
d'_i &= d_i + MS \ \ \forall J_i \in J \\
p'_i &= p_i \ \ \forall J_i \in J \\
w'_i &= w_i \ \ \forall J_i \in J \\
w'_{n+1} &= 0
\end{aligned}
$$

It is easy to verify that $L_{\max} = L_{n+1} = MS$ and that every feasible schedule must process job J_{n+1} first. Then no job $J_i \in J$ violates the lateness constraint even if it is scheduled last. Thus, (since $w'_{n+1} = 0$) the problem reduces to schedule jobs $J_i \in J$ so that $\sum w'_i U_i$ is minimized. Clearly, for every sequence S for I, the value $\sum w_i U_i$ is equal to the value $\sum w'_i U_i$ obtained by the sequence J_{n+1}, S and vice versa. Hence, $F_h(\sum w_i U_i | L_{\max})$ is $\mathcal{NP}$-complete. $\square$

For the problem $F_h(\sum U_i | L_{\max})$ the **EDD** sequence is optimal when $\tau = 0$, see Smith (1956). The complexity of the case $\tau > 0$ has not been considered in the literature.

Since $L_{\max}$ is a special case of $f_{\max}$, the last two theorems tell us that the problems $F_h(\sum T_i | f_{\max})$, $F_h(\sum E_i | f_{\max})$ and $F_h(\sum w_i U_i | f_{\max})$ are $\mathcal{NP}$-complete. Also, the problem $F_h(\sum w_i T_i | f_{\max})$ is strongly $\mathcal{NP}$-complete.

Lastly, we consider the case where $A = E_{\max}$.

5.4 $E_{\max}$ with one of $\sum T_i, \sum w_i T_i, \sum E_i, \sum w_i U_i, \sum U_i$

Usually $E_{\max}$ takes into account inventory costs. In the area of manufacturing of perishable products however, a product cannot be completed too early. In this case the primary concern is to to limit earliness to an acceptable level. The first two secondary criteria treated in this subsection, account for customer satisfaction measured by the total tardiness absorbed by the customers. A solution to this class of problems tries to reduce the production cost by using the tolerance of the customers on late delivery as a tradeoff. Although such a policy is risky, there are instances that this is the only way for an organization to keep prices at competitive levels. Throughout this subsection we assume that no machine idle time is allowed.

Theorem 9 *The problems $F_h(\sum T_i | E_{\max})$, $F_h(\sum E_i | E_{\max})$ are $\mathcal{NP}$-complete. The problem $F_h(\sum w_i T_i | E_{\max})$ is strongly $\mathcal{NP}$-complete.*

Proof: First, we will give a detailed proof for $F_h(\sum T_i | E_{\max})$. Let I be any instance of the $\mathcal{NP}$-complete problem $1 \backslash\backslash \sum T_i$ (see Du and Leung, 1990). As usual, MS denotes the sum of processing times of the n jobs in J. We can assume that for every job J_i in I, $MS \geq d_i > p_i$, otherwise we can exclude the jobs not satisfying this condition. Let $p_{\max} := \max_{J_i \in J} p_i$. Then we construct an instance I' for $F_h(\sum T_i | E_{\max})$ as follows:

$$
\begin{aligned}
d'_i &= d_i + MS + p_{\max} \quad \forall J_i \in J \\
p'_i &= p_i \quad \forall J_i \in J \\
p'_{n+1} &= MS + p_{\max} \\
d'_{n+1} &= 2MS + p_{\max}
\end{aligned}
$$

Since we assume no machine idle time, given a sequence S' for I' the maximum earliness is realized by the job sequenced first in S'. To minimize maximum earliness, this job must be J_{n+1} and $E_{max} = MS$. We will show that for a given threshold value Y there exists a solution for I such that $1\backslash\backslash\sum T_i \leq Y$ if and only if there exists a sequence for the instance I' that solves the problem $1\backslash E_{max} = MS\backslash \sum T_i \leq Y$.

Suppose that I has a solution S, such that $\sum T_i \leq Y$. Then the sequence S' formed by J_{n+1} followed by the sequence S, is a solution for $1\backslash E_{max} = MS\backslash \sum T_i \leq Y$. Indeed, the earliness of J_{n+1} in S' equals MS. For every $1 \leq i \leq n$ the completion time of J_i is greater or equal to $p'_{n+1} = MS + p_{max}$ and therefore the earliness of J_i is less than or equal to $d'_i - p'_{n+1} = d_i \leq MS$. Thus in S' $E_{max} = MS$. The tardiness of job J_i in S' equals its tardiness in S for every $1 \leq i \leq n$. Therefore, the total tardiness in S' is $\sum T_i \leq Y$ and therefore S' solves I'.

On the other hand, assume S' is a solution for $1\backslash E_{max} = MS\backslash \sum T_i \leq Y$. We saw above that if J_{n+1} is the first job in S', then $E_{max} = MS$. Note that job J_i $1 \leq i \leq n$ cannot be the first job in S' because then $E_i = d'_i - p'_i = d_i + MS + p_{max} - p_i > MS$ since $p_{max} - p_i \geq 0$ and $d_i > p_i > 0$. Therefore, the first job in S' must be J_{n+1}. Then, as we saw above $E_{max} = E_{n+1} = MS$. Hence, the total tardiness of the jobs J_i for $1 \leq i \leq n$ is $\sum T_i \leq Y$.

Let S be the sequence formed by S' excluding J_{n+1} from S'. We will show that S is a solution to $1\backslash\backslash \sum T_i \leq Y$ for the instance I. Indeed, the tardiness of each job in S for the instance I equals its tardiness in S' for the instance I' and hence S solves $1\backslash\backslash \sum T_i \leq Y$. This completes the proof for the problem $F_h(\sum T_i | E_{max})$.

The same proof works for the problem $F_h(\sum E_i | E_{max})$ where we use a reduction from the ordinary $\mathcal{NP}$-complete problem $1\backslash\backslash \sum E_i$ (see Du and Leung, 1990). Finally, a reduction from the strongly $\mathcal{NP}$-complete problem $1\backslash\backslash \sum w_i T_i$ (see Lawler, 1977) with $w'_i = w_i$ $\forall J_i \in J$, $w'_{n+1} = 0$ and a similar argument, shows that the $F_h(\sum w_i T_i | E_{max})$ problem is strongly $\mathcal{NP}$-complete.
This completes the proof.$\square$

For the problem $F_h(\sum U_i | E_{max})$ we observe the following. Note that in the proof of Theorem 2, there exists a solution for 3-Partition if and only if there exists a schedule for which $1\backslash \sum U_i = \tau\backslash E_{max} = 0$. However, if such a schedule exists it would have to satisfy $1\backslash E_{max} = 0\backslash \sum U_i = \tau$. This shows that the problem $F_h(\sum U_i | E_{max})$ is strongly $\mathcal{NP}$-complete which in turn proves that the more general problem $F_h(\sum \hat{w}_i U_i | E_{max})$ is such.

All the results considered in this section are summarized in rows 11-15 of Table 4. If the complexity of a problem is not completely determined, a * appears in the corresponding box of Table 4.

6 $A \in \{\Sigma_i E_i, \Sigma_i w_i T_i, \Sigma_i T_i, \Sigma_i w_i U_i, \Sigma_i U_i\}$
$B \in \{\Sigma_i E_i, \Sigma_i \hat{w}_i T_i, \Sigma_i T_i, \Sigma_i \hat{w}_i U_i, \Sigma_i U_i\}$

In this section we consider problems where both criteria A, B could be any of $\sum_i E_i$, $\sum_i w_i T_i$, $\sum_i T_i$, $\sum_i w_i U_i$, $\sum_i U_i$. As we mentioned at Section 4 the problems $1\backslash\backslash \sum_i E_i$, $1\backslash\backslash \sum_i T_i$, $1\backslash\backslash \sum_i w_i T_i$ and $1\backslash\backslash \sum_i w_i U_i$ are $\mathcal{NP}$-complete.

Then, Property 1 implies the complexity of hierarchical problems as shown in the next table. Again ordinary $\mathcal{NP}$-completeness is denoted by ! and strong $\mathcal{NP}$-completeness by !!. Note that all the hierarchical problems with primary criterion $\sum_i w_i T_i$ are strongly $\mathcal{NP}$-complete because $1\backslash\backslash \sum_i w_i T_i$ is such.

	$\sum_{i=1}^n E_i$	$\sum_{i=1}^n \hat{w}_i T_i$	$\sum_{i=1}^n T_i$	$\sum_{i=1}^n \hat{w}_i U_i$	$\sum_{i=1}^n U_i$
$\sum_{i=1}^n E_i$	!	!	!	!	!
$\sum_{i=1}^n w_i T_i$	!!	!!	!!	!!	!!
$\sum_{i=1}^n T_i$	!	!	!	!	!
$\sum_{i=1}^n w_i U_i$	!	!	!	!	!
$\sum_{i=1}^n U_i$					$\mathcal{O}(n \log n), 2$

Table 3:

Of course, the problem $F_h(\sum U_i | \sum U_i)$ is solved by Moore-Hodgson algorithm in $\mathcal{O}(n \log n)$ time.

Theorem 2 established that the problem $F_h(E_{\max} | \sum U_i)$ is strongly $\mathcal{NP}$-complete. Note that in the proof of Theorem 2, there exists a solution for 3-Partition if and only if there exists a schedule for which $1\backslash \sum U_i = \tau \backslash \sum E_i = 0$ (because $E_{\max} = \bar{E} = 0 \Leftrightarrow \sum E_i = 0$). However, if such a schedule exists it would have to satisfy $1\backslash \sum E_i = 0\backslash \sum U_i = \tau$. This observation and an argument similar to Theorem 2 shows that the problem $F_h(\sum U_i | \sum E_i)$ is strongly $\mathcal{NP}$-complete which in turn proves that the more general problem $F_h(\sum \hat{w}_i U_i | \sum E_i)$ is such.

The next theorem shows that the problem $F_h(\sum E_i | \sum U_i)$ is strongly $\mathcal{NP}$-complete. We will use a reduction from $1\backslash r_i, nmit\backslash \sum_{i=1}^n C_i$ which is proved to be strongly $\mathcal{NP}$-complete, see [11], [25].

Theorem 10 $1\backslash r_i, nmit\backslash \sum_{i=1}^{n} C_i$ *reduces to* $1\backslash \sum_{i=1}^{n} U_i = \tau, \; nmit\backslash \sum_{i=1}^{n} E_i$.

Proof: Let I_1 be any instance of $1\backslash r_i, nmit\backslash \sum_{i=1}^{n} C_i$ and Y a given threshold value. We will construct an instance I_2 of $1\backslash \sum_{i=1}^{n} U_i = \tau, \; nmit\backslash \sum_{i=1}^{n} E_i'$ such that $\sum_{i=1}^{n} C_i \leq Y$ iff $\sum_{i=1}^{n} E_i' \leq Y + MS - \sum_{i=1}^{n} r_i$. Let p_i, r_i, d_i be the processing time, release time and the due-date of job i in I_1 for $1 \leq i \leq n$ and p_i', r_i', d_i' the corresponding times for I_2. Define the instance I_2 by:

$$\begin{aligned}
p_i' &= p_i \; for \; 1 \leq i \leq n \\
r_i' &= 0 \; for \; 1 \leq i \leq n \\
d_i' &= MS - r_i \; for \; 1 \leq i \leq n.
\end{aligned}$$

We can show that if $S_1 = (j_1, j_2, \ldots, j_n)$ is a solution to I_1 such that $\sum_{i=1}^{n} C_i \leq Y$ then $S_2 = (j_n, j_{n-1}, \ldots, j_1)$ is a solution to I_2 such that $\sum_{i=1}^{n} E_i' \leq Y + MS - \sum_{i=1}^{n} r_i$ and vice versa. Before showing this, we will see that in S_2 no job is tardy and therefore $\tau = 0$. Indeed, in S_1, job i starts no earlier than r_i and completes no later than MS, since no machine idle time is allowed. Thus, in S_2, job i starts no earlier than time zero and finishes no later than d_i'. Therefore, $\tau = 0$ and $E_i' \geq 0, \; for \; 1 \leq i \leq n$.

If C_i, C_i' are the completion times of job i in S_1, S_2 respectively, then $E_i' = d_i' - C_i' = (MS - r_i) - (MS - C_i - p_i) = C_i + p_i - r_i$. Thus, $\sum_{i=1}^{n} E_i' = \sum_{i=1}^{n} C_i + \sum_{i=1}^{n} p_i - \sum_{i=1}^{n} r_i = \sum_{i=1}^{n} C_i + MS - \sum_{i=1}^{n} r_i$.

Clearly, if $\sum_{i=1}^{n} C_i \leq Y \Leftrightarrow \sum_{i=1}^{n} E_i' \leq Y + MS - \sum_{i=1}^{n} r_i$. This completes the proof. $\square$

It is easy to see that we can extend the instance I_2 that has 0 as the minimum number of tardy jobs, to an instance with $\tau > 0$. We only need to add τ more jobs $W_1, W_2, \ldots, W_\tau$ such that

$$\begin{aligned}
p_{W_k}' &= MS \; for \; 1 \leq k \leq \tau \\
r_{W_k}' &= 0 \; for \; 1 \leq k \leq \tau \\
d_{W_k}' &= 0 \; for \; 1 \leq k \leq \tau,
\end{aligned}$$

where $MS = \sum_{i=1}^{n} p_i$. By construction, the W-jobs are tardy in every schedule. Thus, in order to minimize the number of tardy jobs, they must be scheduled at the last τ positions. We can assume that job W_k starts at time $k\,MS$ and completes at time $(k+1)MS$. Then, to minimize total earliness while none of the remaining jobs is tardy reduces to the argument in the previous proof.

The problem $F_h(\sum T_i | \sum U_i)$ is unaccounted in the previous table. Although its complexity is unknown, Vairaktarakis and Lee (1992) found a number of optimal properties and polynomially solvable cases and presented a fast branch and bound algorithm. Also, they proposed an $\mathcal{O}(n \log n)$ algorithm that finds an optimal solution with respect to a given set of tardy jobs.

Now we consider the problem $F_h(\sum w_i T_i | \sum U_i)$. We can see that the reduction used by Lawler (1977) to show strong $\mathcal{NP}$-completeness of the unconstrained problem

$1\backslash\backslash\sum w_i T_i$ can be used to prove that $F_h(\sum w_i T_i|\sum U_i)$ is strongly $\mathcal{NP}$-complete as well.

Indeed, if $\{a_i\}_{i=1}^{3n}$ is an instance of 3-Partition, construct the following instance:

$$
\begin{aligned}
V - jobs: \quad & V_i, \ 1 \le i \le 3n \\
W - jobs: \quad & W_i, \ 1 \le i \le n \\
processing\ times: \quad & p_{V_i} = B + a_i, \ 1 \le i \le 3n \\
& p_{W_i} = L = 16B^2\frac{n(n+1)}{2} + 1, \ 1 \le i \le n \\
weights: \quad & w_{V_i} = p_{V_i} = B + a_i, \ 1 \le i \le 3n \\
& w_{W_i} = W = 4B(L+4B)\frac{n(n+1)}{2} + 1, \ 1 \le i \le n \\
due\ dates: \quad & d_{V_i} = 0, \ 1 \le i \le 3n \\
& d_{W_i} = iL + 4(i-1)B, \ 1 \le i \le n \\
threshold\ value: \quad & W - 1.
\end{aligned}
$$

It is enough to observe that the minimum number of tardy jobs in the above instance is $3n$. Furthermore, the schedule used in the proof in [23] has precisely $3n$ tardy jobs. This observation establishes that the total weighted tardiness problem subject to minimum number of tardy jobs is strongly $\mathcal{NP}$-complete.

Since the problem $F_h(\sum w_i T_i|\sum U_i)$ is strongly $\mathcal{NP}$-complete, the more general problem $F_h(\sum \hat{w}_i T_i|\sum w_i U_i)$ is at least as difficult. To account all the problems of the previous table, we are left with $F_h(\sum \hat{w}_i U_i|\sum U_i)$. For this problem not only we will prove ordinary $\mathcal{NP}$-completeness but also we will provide a pseudopolynomial time algorithm to solve it, thus determining completely the complexity of the problem. This is done in the next subsection.

6.1 The problem $F_h(\sum \hat{w}_i U_i|\sum U_i)$

Assume that a manager wants to control the number of tardy jobs, but each job has a fixed cost associated with its being tardy; for example job dependent rush delivery costs. Then, the weighted number of tardy jobs is interpreted as an internal cost to the organization. Therefore, the problems with this pair of criteria consider an acceptable balance of the internal and external costs induced by late deliveries.

We show that $F_h(\sum w_i U_i|\sum U_i)$ is $\mathcal{NP}$-complete in the ordinary sense by a reduction from Modified Even-Odd Partition accompanied with a dynamic program that solves the problem optimally.

Modified Even-Odd Partition:
Given a set of $2n$ positive integers $V = \{v_1, v_2, \ldots, v_{2n}\}$ such that $v_i < v_{i+1}$ for all $1 \le i < 2n$. Define the quantities:

$$
B_1 \ = \ 0
$$

$$\begin{aligned}
B_2 &= v_2 \\
B_3 &= v_2 + v_4 \\
&\vdots \\
B_n &= v_2 + v_4 + \ldots + v_{2n-2}.
\end{aligned}$$

Assume that for $1 \leq i \leq n$, $v_{2i} > B_i$, and $v_{2i-1} > B_i$.

Is there a partition of V into V_1, V_2 such that $\sum_{v_i \in V_1} v_i = \sum_{v_i \in V_2} v_i$, and such that for each $1 \leq i \leq n$, V_1 contains exactly one of $\{v_{2i-1}, v_{2i}\}$?

Proposition 1 *The Modified Even-Odd Partition problem is $\mathcal{NP}$-complete.*

Proof: The Modified Even-Odd Partition problem is in $\mathcal{NP}$ because it is a sub-problem of the Even-Odd Partition. Next, we construct a reduction of the Even-Odd partition to the Modified Even-Odd Partition problem.

Let $B = \{u_1, u_2, \ldots, u_{2n}\}$ be ordered so that $u_i < u_{i+1}$ for all $1 \leq i < 2n$, be an arbitrary instance of the Even-Odd Partition problem. Construct an instance V of the Modified Even-Odd Partition problem as follows:

Define $v_1, v_2, \ldots, v_{2n}$ recursively as follows: Let $v_1 := u_1 + B_1$ and $v_2 := u_2 + B_1$, where $B_1 := 0$. Then let $v_3 := u_3 + B_2$ and $v_4 := u_4 + B_2$, where $B_2 := v_2$. In general, having found v_{2i-1}, v_{2i}, define v_{2i+1}, v_{2i+2} as $v_{2i+1} := u_{2i+1} + B_{i+1}$ and $v_{2i+2} := u_{2i+2} + B_{i+1}$, where $B_{i+1} := v_2 + v_4 + \ldots + v_{2i}$. Assume that the values $v_1, v_2, \ldots, v_{2n}$ are computed as described. Then, $v_{2i-1} < v_{2i}$ because $u_{2i-1} < u_{2i}$ and $v_{2i-1} > B_i$, $v_{2i} > B_i$. Thus, the instance V satisfies all the constraints of the Modified Even-Odd Partition problem.

Clearly the above construction can be done in polynomial time. Thus, we need only show that instance B has a solution if and only if V does. Suppose $\bar{U}_1$, $\bar{U}_2$ is a solution for B. Let I_1, I_2 be the set of indices of the elements of $\bar{U}_1$, $\bar{U}_2$ respectively, ordered in increasing order. Then,

$$\begin{aligned}
\sum_{i \in I_1} u_i &= \sum_{j \in I_2} u_j \quad \Leftrightarrow \\
\sum_{i \in I_1} (u_i + B_i) &= \sum_{j \in I_2} u_j \quad \Leftrightarrow \\
\sum_{i \in I_1} v_i &= \sum_{i \in I_2} v_i
\end{aligned}$$

Define $V_1 = \{v_i : i \in I_1\}$ and $V_2 = \{v_i : i \in I_2\}$. Then, $\sum_{v_i \in V_1} v_i = \sum_{v_i \in V_2} v_i$. Also V_1 contains precisely one of $\{v_{2i-1}, v_{2i}\}$ because $\bar{U}_1$ contains precisely one of $\{u_{2i-1}, u_{2i}\}$. Hence, V_1, V_2 constitutes a solution for V.

Conversely, if V_1, V_2 is a solution for V, the above equivalences mean that the sets $\bar{U}_1 = \{u_i : v_i \in V_1\}$ and $\bar{U}_2 = \{u_i : v_i \in V_2\}$ constitute a solution of B. Therefore, V has a solution if and only if B does. $\square$

Let $V = \{v_i\}_{i=1}^{2n}$ be an instance of the Modified Even-Odd Partition problem and let $\sum_{i=1}^{2n} v_i = MS$. Construct an instance I for $1\backslash \sum U_i \leq Y^*\backslash \sum w_i U_i$ as follows:

$$
\begin{aligned}
p_i &= v_i \ \ for \ \ 1 \leq i \leq 2n \\
w_i &= v_i \ \ for \ \ 1 \leq i \leq 2n \\
d_{2i-1} = d_{2i} &= \sum_{j=1}^{i} v_{2j} \ \ for \ \ 1 \leq i \leq n \\
d_{2n-1} &= \frac{MS}{2} \\
d_{2n} &= \frac{MS}{2}
\end{aligned}
$$

First, we find the minimum number of tardy jobs τ, for I.

Lemma 1 *The minimum number of tardy jobs for the instance I is n.*

Proof: First, observe that the sequence $j_1, j_3, \ldots, j_{2n-1}, j_2, j_4, \ldots, j_{2n}$ has its first n jobs nontardy since for every $1 \leq i \leq n$,

$$
\sum_{j=1}^{i} p_{2j-1} = \sum_{j=1}^{i} v_{2j-1} < \sum_{j=1}^{i} v_{2j} = d_{2i-1}.
$$

Thus, $\tau \leq n$.

Second, to show that τ cannot be less than n it is enough to observe that for every $i = 1, 2, \ldots, n$, at most one job between $2i - 1$, $2i$ can be nontardy. Indeed, note that

$$
\begin{aligned}
v_{2i-1} + v_{2i} &> B_i + v_{2i} \ \ since \ \ v_{2i-1} > B_i \\
&= \sum_{j=1}^{i-1} v_{2j} + v_{2i} \\
&= d_{2i}.
\end{aligned}
$$

This means that one of the two jobs must be tardy. Therefore, for every $1 \leq i \leq n$, at least one of the jobs $2i - 1$, $2i$ must be tardy which means that $\tau \geq n$. The inequalities $\tau \geq n$ and $\tau \leq n$ combined, show that $\tau = n$. $\square$

The following lemma gives a property of the dual criteria problem to minimize $\sum w_i U_i$ subject to n_T tardy jobs, which can be deduced by a simple interchange argument. This lemma will be useful in the construction of the dynamic program that will be developed later.

Lemma 2 *There is an optimal schedule to the dual criteria problem $F(\sum w_i U_i | \sum U_i)$ in which all the nontardy jobs precede all tardy jobs and the nontardy jobs are in* **EDD** *order.*

In the next theorem we show that the problem to minimize the weighted number of tardy jobs subject to minimum number of tardy jobs is $\mathcal{NP}$-complete.

Theorem 11 *The Modified Even-Odd Partition problem reduces to* $1\backslash \sum U_i = \tau \backslash \sum w_i U_i$.

Proof: Consider the instance I constructed earlier. We know that for this instance $\tau = n$ and for every $1 \leq i \leq n$, precisely one of the jobs $2i - 1$, $2i$ will be tardy.

Let the threshold value be $\bar{W} = \frac{MS}{2}$. In the given instance the total processing time of tardy jobs is at least $\frac{MS}{2}$ and thus, by construction, the weighted number of tardy jobs is no less than $\frac{MS}{2}$ i.e. $\sum_{i=1}^{2n} w_i U_i \geq \frac{MS}{2}$.

We will show that there is a solution to $1\backslash \sum_{i=1}^{2n} U_i = \tau \backslash \sum_{i=1}^{2n} w_i U_i = \frac{MS}{2}$ if and only if there is a solution for the instance V of the Modified Even-Odd Partition problem.

Indeed, suppose S is a solution to $1\backslash \sum_{i=1}^{2n} U_i = \tau \backslash \sum_{i=1}^{2n} w_i U_i = \frac{MS}{2}$ and E is the set that contains the nontardy jobs in S, while T is the set that contains the tardy jobs in S. Since the number of tardy jobs in S is $\tau = n$ we have $|E| = |L| = n$. Also, in S, precisely one of the jobs $2i - 1$, $2i$ is nontardy and therefore precisely one of the numbers v_{2i-1}, v_{2i} belongs to E, for all $1 \leq i \leq n$. Further, since

$$\sum_{i=1}^{2n} w_i U_i = \frac{MS}{2} \Leftrightarrow \sum_{i=1}^{2n} v_i U_i = \frac{MS}{2} \Leftrightarrow \sum_{v_i \in T} v_i = \frac{MS}{2}$$

and therefore $\sum_{v_i \in E} v_i = \frac{MS}{2}$. Hence, E, T forms a solution for V.

On the other hand assume E, T is a solution for V. Then for $1 \leq i \leq n$, precisely one of v_{2i-1}, v_{2i} lies in E, say $v_{i,1}$. The other, say $v_{i,2}$ lies in T. Let the job corresponding to $v_{i,1}$, $v_{i,2}$ be $j_{i,1}$, $j_{i,2}$ respectively. Consider the sequence $S = (j_{1,1}, j_{2,1}, \ldots, j_{n,1}, j_{1,2}, j_{2,2}, \ldots, j_{n,2})$. We will show that S has precisely n tardy jobs and that the total weighted number of tardy jobs equals $\frac{MS}{2}$.

Indeed, the jobs $j_{1,1}, j_{2,1}, \ldots, j_{n-1,1}$ are nontardy because for every $1 \leq i \leq n-1$,

$$C_{j_{i,1}} = \sum_{k=1}^{i} p_{j_{k,1}} = \sum_{k=1}^{i} v_{k,1} \leq \sum_{k=1}^{i} v_{2k} = d_{j_{i,1}}.$$

Also, job $j_{n,1}$ is nontardy since $C_{j_{n,1}} = d_{j_{n,1}} = \frac{MS}{2}$.

Then, by Lemma 2, the jobs in the last n positions of S are tardy and hence $\sum_{i=1}^{2n} w_i U_i = \sum_{i=1}^{n} v_{i,2} = \frac{MS}{2}$. This means that S solves $1\backslash \sum_{i=1}^{2n} U_i = \tau \backslash \sum_{i=1}^{2n} w_i U_i = \frac{MS}{2}$, and the theorem is proved. $\square$

Next, we will present a dynamic program that solves the dual criteria problems $F(\sum U_i | \sum w_i U_i)$ and $F(\sum w_i U_i | \sum U_i)$ in pseudopolynomial time. As we saw the hierarchical problems $F_h(\sum U_i | \sum w_i U_i)$ and $F_h(\sum w_i U_i | \sum U_i)$ are $\mathcal{NP}$-complete. Then, the algorithm that follows determines completely the complexity of these problems as ordinary $\mathcal{NP}$-complete.

Let $\{j_1, j_2, \ldots, j_n\}$ be the **EDD** order of the jobs. Then, define

$f(i, t, n_T) :=$ Minimum total weighted number of tardy jobs if we have scheduled jobs $\{j_1, j_2, \ldots, j_i\}$ given that the total processing time of nontardy jobs is t and the number of tardy jobs is n_T.

Boundary Conditions:

$$
\begin{aligned}
f(0, 0, 0) &= 0 \\
f(0, t, n_T) &= +\infty, \quad t > 0, \ n_T > 0 \\
f(i, t, n_T) &= +\infty, \quad t < 0, \ n_T < 0.
\end{aligned}
$$

For $i = 1, 2, \ldots, n, \quad t = 1, 2, \ldots, \min\{MS, d_n\}, \quad n_T = 1, \ldots, i.$
Recursive Relation:

$$
f(i, t, n_T) := \min \begin{cases} f(i - 1, t - p_i, n_T) & \text{if} \quad 0 \le t \le d_i; \\ f(i - 1, t, n_T - 1) + w_i & \text{if} \quad t > d_i \end{cases}
$$

Optimal Points for $n_T = \tau, \ldots, n,$

$$
\min_{1 \le t \le \min\{MS, d_n\}} f(n, t, n_T)
$$

i.e. the optimal solution to $1 \backslash \sum U_i = n_T \backslash \sum w_i U_i$ is given by $\min\{f(n, t, n_T) : t = 1, 2, \ldots, \min\{MS, d_n\}\}$ for $n_T \in \{\tau, \ldots, n\}$.

Clearly, the complexity of the above algorithm is $\mathcal{O}(n^2 \min\{MS, d_n\})$.

Justification:
By Lemma 2, the nontardy jobs are scheduled in **EDD** order. Therefore, if we have scheduled jobs $\{j_1, j_2, \ldots, j_{i-1}\}$ and the total processing time t does not exceed d_i, then job j_i either will immediately follow the nontardy jobs in $\{j_1, j_2, \ldots, j_{i-1}\}$ (first quantity of first branch of recurrence relation), or it will be tardy and thus increase the weighted number of tardy jobs by w_i (second branch of recurrence relation). If the total processing time t of nontardy jobs exceeds d_i, then j_i must become tardy and thus increase the weighted number of tardy jobs by w_i (second branch of recurence relation).

All the results considered in this section are summarized in the last 5 rows of Table 4. Note that the problem $F_h(\sum T_i | \sum T_i)$ is solved in pseudopolynomial time by Lawler (1977) and hence its complexity status is completely determined. If the complexity of a problem is not completely determined, a * appears in the corresponding box of Table 4.

7   Complexity Tables

In the following tables we summarize the complexity status of all hierarchical problems with two criteria. For each problem we give its complexity status and the section at which it is discussed. One can check at the appropriate section to find the relevant references. The symbol ! means that the corresponding problem is ordinary $\mathcal{NP}$-complete, !! means that the corresponding problem is strongly $\mathcal{NP}$-complete while ? means that the corresponding problem is open. A $\star$ next to a ! means that the corresponding problem is proved $\mathcal{NP}$-complete by a reduction from an ordinary $\mathcal{NP}$-complete problem but no pseudopolynomial time algorithm exists in the literature. If a problem is polynomial, then we give its complexity.

	$\sum_{i=1}^{n} C_i$	$\sum_{i=1}^{n} \hat{w}_i C_i$	$L_{\max}$	$\hat{f}_{\max}$	$\hat{E}_{\max}$
$\sum_{i=1}^{n} C_i$	$\mathcal{O}(n \log n),3$	$\mathcal{O}(n \log n),3$	$\mathcal{O}(n \log n),3$	$\mathcal{O}(n^2),3$	$\mathcal{O}(n \log n),3$
$\sum_{i=1}^{n} w_i C_i$	$\mathcal{O}(n \log n),3$	$\mathcal{O}(n \log n),3$	$\mathcal{O}(n \log n),3$	$\mathcal{O}(n^2),3$	$\mathcal{O}(n \log n),3$
$L_{\max}$	$\mathcal{O}(n \log n),3$	!!,3	$\mathcal{O}(n \log n),3$	$\mathcal{O}(n^2),3$	$\mathcal{O}(n^2),3$
$f_{\max}$	$\mathcal{O}(n^2),3$	!!,3	$\mathcal{O}(n^2),3$	$\mathcal{O}(n^2),3$	!!,3
$E_{\max}$	!!,3	!!,3	$\mathcal{O}(n^2),3$	!!,3	$\mathcal{O}(n \log n),3$
$\sum_{i=1}^{n} E_i$	!*,4	!*,4	!*,4	!*,4	!*,4
$\sum_{i=1}^{n} w_i T_i$	!!,4	!!,4	!!,4	!!,4	!!,4
$\sum_{i=1}^{n} T_i$	!*,4	!*,4	!*,4	!*,4	!*,4
$\sum_{i=1}^{n} w_i U_i$	!*,4	!!,4	!*,4	!*,4	!!,4
$\sum_{i=1}^{n} U_i$	?,4	!!,4	?,4	?,4	!!,4

8   Conclusion

In this paper we examined the complexity status of two criteria hierarchical scheduling problems on a single machine. All the performance measures often encountered in scheduling theory are considered. In this effort we surveyed most of the existing

	$\sum_{i=1}^n E_i$	$\sum_{i=1}^n \hat{w}_i T_i$	$\sum_{i=1}^n T_i$	$\sum_{i=1}^n \hat{w}_i U_i$	$\sum_{i=1}^n U_i$
$\sum_{i=1}^n C_i$	$\mathcal{O}(n\log n)$,5.1	$\mathcal{O}(n^4)$,5.1	$\mathcal{O}(n\log n)$,5.1	$\mathcal{O}(n\log n)$,5.1	$\mathcal{O}(n\log n)$,5.1
$\sum_{i=1}^n w_i C_i$	!*,5.2	!!,5.2	!*,5.2	!*,5.2	$\mathcal{O}(n\log n)$,5.2
$L_{\max}$	!*,5.3	!!,5.3	!*,5.3	!*,5.3	?,5.3
$f_{\max}$	!*,5.3	!!,5.3	!*,5.3	!*,5.3	?,5.3
$E_{\max}$	!*,5.4	!!,5.4	!*,5.4	!!,5.4	!!,5.4
$\sum_{i=1}^n E_i$	!,6	!*,6	!*,6	!!,6	!!,6
$\sum_{i=1}^n w_i T_i$	!!,6	!!,6	!!,6	!!,6	!!,6
$\sum_{i=1}^n T_i$	!*,6	!*,6	!,6	!*,6	!*,6
$\sum_{i=1}^n w_i U_i$	!!,6	!!,6	!*,6	!*,6	!,6.1
$\sum_{i=1}^n U_i$	!!,6	!!,6	?,6	!,6.1	$\mathcal{O}(n\log n)$,6

Table 4: Complexity tables for Hierarchical Optimization

results along with several new results. We presented the material in an organized way leading to a comprehensive complexity table. From this table, one can see that most hierarchical problems are $\mathcal{NP}$-complete although the exact complexity status of most of these problems is open. Also, we identify a few problems the status of which is unknown. Since hierarchical problems are special cases of the more general dual criteria problems described in the introduction, this research shows that most of the multicriteria decisions faced by managers today correspond to $\mathcal{NP}$-complete problems. This fact motivates the need to create the problem solving tools that will help the managers to improve their decision making ability.

References

[1] Baker K.R. (1974), Introduction to Sequencing and Scheduling. *Wiley, New York.*

[2] Burns R. N. (1976). Scheduling to Minimize the Weighted Sum of Copletion Times with Secondary Criteria. *Naval Research Logistics Quarterly* 23:125-129.

[3] Chand S. and H. Schneeberger (1986). A note on the Single Machine Scheduling Problem to minimize Weighted Completion Time and Maximum Allowable Tardiness. *Naval Research Logistics Quarterly* 33:551-557.

[4] Chand S. and H. Schneeberger (1988). Single Machine Scheduling to Minimize Weighted Earliness Subject to No Tardy Jobs. *European Journal of Operational Research* 34:221-230.

[5] Du J. and J. Y.-T. Leung (1990). Minimizing Total Tardiness on One Machine is NP-hard, *Mathematics of Operations Research* 15:483-495.

[6] Emmons H. (1969). One machine sequencing to minimize certain functions of job tardiness. *Operations Research* 17:701-715.

[7] Emmons H. (1975A). A note on a scheduling problem with dual criteria. *Naval Research Logistics Quarterly* 22:615-616.

[8] Emmons H. (1975B). One Machine Sequencing to Minimize Mean Flow Time With Minimum Number Tardy. *Naval Research Logistics Quarterly* 22:585-592.

[9] Garey, M.R. and D.S. Johnson (1975). Complexity Results for Multiprocessor Scheduling Under Resource Constraints. *SIAM J. Comput.* 4:397-411.

[10] Garey, M.R. and D.S. Johnson (1979). Computers and Intractability. *W.H. Freeman*, San Francisco, CA.

[11] Garey, M.R., D.S. Johnson and R. Sethi (1976). The Complexity of Flowshop and Jobshop Scheduling. *Mathematics of Operations Research* 1(2):117-129.

[12] Graham R.L., Lawler E.L., Lenstra J.K. and Rinnooy Kan A.H.G. (1979). Optimization and approximation in deterministic sequencing and scheduling: A survey. *Annals of Discrete Mathematics* 5:287-326.

[13] Heck H. and S. Roberts (1972). A note on the extension of a result on scheduling with secondary criteria. *Naval Research Logistics Quarterly* 19:59-66.

[14] Herrmann J.W., C.-Y. Lee, J.L. Snowdon 1992. A Classification of Static Scheduling Problems. Research Report No.92-3, Department of Industrial and Systems Engineering, University of Florida.

[15] Hoogeveen J.A. (1991). Single Machine Scheduling to Minimize a function of K Maximum Cost Criteria. *Report BS-R9113*, CWI, Amsterdam.

[16] Hoogeveen J.A. (1992). Single-Machine Bicriteria Scheduling. Thesis, Amsterdam.

[17] Hoogeveen J.A. and Van De Velde (1990). Polynomial-time Algorithms for Single-machine Multicriteria Scheduling. *Report BS-R9008*, CWI, Amsterdam.

[18] John T.C. (1989). Tradeoff Solutions in Single Machine Production Scheduling for Minimizing Flow Time and Maximum Penalty. *Computers and Operations Research* 16:471-479.

[19] Karp R.M. (1972). Reducibility Among Combinatorial Problems. R.E. Miller and J.W. Thatcher, eds. *Complexity of Computer Computations* (Plenum Press, New York) 85-103.

[20] Kise H., T. Ibaraki, H. Mine (1978). A solvable case of the one-machine scheduling problem with ready and due times. *Operations Research* 26:121-126.

[21] Lawler E.L. (1973). Optimal sequencing of a single machine subject to precedence constraints. *Management Science* 19:544-546.

[22] Lawler E.L. (1976). Sequencing to Minimize the Weighted Number of Tardy Jobs. *RAIRO Rech. Oper.* 10.5 Suppl. 27-33.

[23] Lawler E.L. (1977). A Pseudopolynomial Algorithm for Sequencing Jobs to Minimize Total Tardiness. *Annals of Discrete Mathematics* 1:331-342.

[24] Lawler E.L., J.M. Moore (1969). A functional equation and its application to resource allocation and sequencing problems. *Management Science* 16:77-84.

[25] Lenstra J.K., A.H.G. Rinnooy Kan, and P. Brucker (1977). Complexity of Machine Sequencing Problems. *Annals of Discrete Mathematics* 1:343-362.

[26] Miyazaki S. (1981). One Machine Scheduling Problem With Dual Criteria. *Journal of the Operations Research Society of Japan* 24:37-50.

[27] Moore J. M. (1968). An n Job, One Machine Sequencing Algorithm For Minimizing The Number Of Late Jobs. *Management Science* 15:102-109.

[28] Nelson R. T., R. K. Sarin and R. L Daniels (1986). Scheduling With Multiple Performance Measures: The One Machine Case. *Management Science* 32:464-479.

[29] Papadimitriou C. H. and K. Steiglitz (1982). Combinatorial Optimization. Algorithms and Complexity. *Prentice Hall.*

[30] Sahni S. (1976). Algorithms for Scheduling Independent tasks. *J. Assoc. Comput. Mach.* 23:116-127.

[31] Shanthikumar J.G. (1983). Scheduling Jobs On One Machine To Minimize The Maximum Tardiness With Minimum Number Tardy., *Journal of Computers and Operations Research* 10:255-266.

[32] Shanthikumar J.G. and J. A. Buzacott (1982). On the Use of Decomposition Approaches in a Single Machine Scheduling Problem. *Journal of Operations Society of Japan* 25:29-47.

[33] Sidney J. B. (1973). An Extension Of Moore's Due Date Algorithm. Symposium on the Theory of Scheduling and its Application. S. E. Elmaghraby (Ed.) *Springer-Verlag, New York* 393-398.

[34] Smith W. E. (1956). Various Optimizers for Single Stage Production, *Naval Research Logistics Quarterly* 3:59-66.

[35] Tuzikov A. V. (1990). One Approach to Solving Bicriterion Scheduling Problems. *Report 33*, Academy of Sciences of Byelorussian SSR, Minsk.

[36] Vairaktarakis G. and C.-Y. Lee (1992). The Single Machine Problem to Minimize Total Tardiness Subject to Minimum Number of Tardy Jobs. Research Report No.92-4, Department of Industrial and Systems Engineering, University of Florida.

[37] Van Wassenhove L. N. and F. Gelders (1980). Solving a bicriterion Scheduling Problem. *European Journal of Operational Research* 4:42-48.

Complexity in Numerical Optimization, pp. 299-322
P.M. Pardalos, Editor
©1993 World Scientific Publishing Co.

Performance Driven Graph Enhancement Problems[1]

Doowon Paik
AT&T Bell Laboratories, Murray Hill, NJ 07974

Sartaj Sahni
Computer and Information Sciences Department, University of Florida, Gainesville, FL 32611

Abstract

Graphs may be used to model systems in which performance issues are crucial. Cost effective performance enahancement of these systems can be accomplished by solving a graph enhancement problem on the associated graph. We define several graph enhancement problems. Some are shown to be NP-hard while others are polynomially solvable.

1   Introduction

When designing systems such as VLSI circuits or communication networks, one needs to make decisions that affect the performance of the resulting design. Often, the system is designed making one set of choices. The performance of the resulting design is determined. If this is found to be unsatisfactory, then one proceeds to change some of the design decisions so as to bring the system performance into the desired range. For example, we may design a circuit using certain circuit modules. Associated with each module is a delay. The circuit can be modeled as a directed acyclic graph (dag) with vertex weights. The vertices correspond to the circuit modules and the weights to the module delays. The sum of the vertex weights on any path gives the path length. The length of the longest path in the dag gives the circuit delay. If this delay exceeds the maximum allowable delay, then one can reduce the delay by choosing a different (and faster) implementation. However, choosing the faster implementation has a cost or weight associated with it. This results in a dag optimization problem: find a least weight vertex set whose upgrading results in a dag in which no path has length more than δ. In a simplified version of this problem, there is a factor x, $0 \le x < 1$ such that the upgraded module has a delay that is x times that of the original module. Let $\mathrm{DVUP}(x,\delta)$ denote the dag vertex upgrade problem in which $d(i)$ is the delay of vertex i and $w(i)$ its weight.

In an alternate modeling of signal flow in electronic circuits by dags [CHAN90, GHAN87, MCGE90], vertices represent circuit modules and directed edges represent signal flow. In a simplistic model, each edge has a delay of one. A module can be upgraded by replacing it with a functionally equivalent one using a superior technology. This reduces the delay of all edges incident to/from the module by a multiplicative factor x, $0 \le x < 1$. In a simplistic model, this reduction factor is the same for all circuit modules. The cost of the upgrade is reflected in the weight associated with the vertex. Again, in a simplistic model, each vertex has unit weight (i.e., all vertices cost the same to upgrade). Since signals can travel along any of the paths of the dag, the performance of the circuit is governed by the length of the longest path in the dag. We wish to meet certain performance requirements by upgrading the fewest possible number of vertices. This is stated formally below [PAIK91d]:

LongestPath(x,δ)

Given a dag $G = (V, E)$ with positive edge delays upgrade the smallest number of vertices so that the longest path in the upgraded graph has delay δ. When a vertex is upgraded, all edges incident to/from it have their delay changed by the multiplicative factor x. So, if edge $< v, w >$ has delay d before the upgrade, its delay is $x * d$ following the upgrade. If both v and w are upgraded, its delay becomes $x^2 * d$.

As another example, consider a communication network. This can be modeled as an undirected connected graph in which the edge delays (≥ 0) represent the time taken to communicate between a pair of vertices that are directly connected. Two vertices that are not directly connected can communicate by using a series of edges that form a path from one vertex to the other. The total delay along the communication path is the sum of the delays on each of the edges on the path. With respect to this undirected graph model, the following problems may be defined [PAIK91d]:

1. *LinkDelay(x,δ)*

In this problem, it is possible to upgrade each of the vertices in the undirected graph. If vertex v is upgraded, then the delay of each edge incident to v reduces by a factor x, $0 \le x < 1$. The problem is to upgrade the smallest number of vertices so that following the upgrades, no edge has delay $> \delta$.

2. *ShortestPath(x,δ)*

Upgrading a vertex has the same effect on edge delays as in LinkDelay(x,δ). This time, however, we seek to upgrade the smallest number of vertices so that following the upgrade there is no pair of vertices u and v for which the shortest path between them has delay $> \delta$.

3. *Satellite(δ)*

When a vertex is upgraded, a satellite up link and down link are placed there. Two vertices with satellite links can communicate in zero time. Let $dist(x, y)$ be the length of the shortest communication path between vertices x and y. Let $\mathrm{CommTime}(G)$ be $\max_{x, y \in V(G)} \{dist(x, y)\}$ where $V(G)$ is the set of vertices in G. The objective is to upgrade the smallest number of vertices so that $\mathrm{CommTime}(G) \leq \delta$. Note that there is always a shortest communication path between two vertices that uses either 0 or 2 satellite vertices (to use a satellite link there must be a send and a receive vertex; further there is no advantage to using more than one satellite link in any communication).

Each of the problems stated above is a simplified version of a more realistic problem. The more realistic problem has different costs associated with the upgrade of different vertices and the upgrade factor also varies from vertex to vertex.

Paik, Reddy, and Sahni [PAIK90, 93] model the optimal placement of scan registers in a partial scan design as well as the placement of signal boosters in lossy circuits as a vertex splitting problem in a dag. The input dag (which represents the circuit) has edge delays and the objective is to split the fewest number of vertices so that the resulting dag has no path of length $> \delta$. When a vertex is split, it is replaced by two copies; one retains the incoming edges and the other the outgoing edges. The dag vertex splitting problem is denoted DVSP(δ).

The dag vertex deletion problem, DVDP(δ), is concerned with deleting the fewest number of vertices from an edge weighted dag so that the resulting dag has no path whose length exceeds δ. In [PAIK91a], Paik, Reddy, and Sahni used this problem to model the problem of upgrading circuit modules so as to control signal loss.

Krishnamoorthy and Deo [KRIS79] have shown that for many properties, the vertex deletion problem isNP-hard. These properties include: resulting graph has no edges, resulting graph is a clique, each component of the resulting graph is a tree, each component of the remaining graph is planar, etc. We shall not discuss any of these results here as none of the properties considered in [KRIS79] apply to graphs with vertex and/oredge weights.

In subsequent sections, we summarize the known results regarding the problems stated above. We shall make use of the following known NP-hard problems [GARE79].

1. Partition

Input: A set of n positive integers $a_i, 1 \leq i \leq n$.

output: "Yes" iff there is a subset, I, of $\{1, 2, \cdots, n\}$ such that
$$\sum_{i \in I} a_i = \sum_{i=1}^{n} a_i / 2$$

2. Vertex Cover

Input: An undirected graph $G = (V, E)$ and a positive integer $k \leq |V|$

output: "Yes" iff there is a subset $V' \subseteq V$ with $|V'| \leq k$ such that for each edge $(u, v) \in E$ at least one of u and v belongs to V'.

3. Dominating Set

Input: An undirected graph $G = (V, E)$ and a positive integer $k \leq |V|$

output: "Yes" iff there is a subset $V' \subseteq V$ with $|V'| \leq k$ such that for $u \in V - V'$ there is a $v \in V'$ for which $(u, v) \in E$.

4. Maximum Clique

Input: A connected undirected graph $G = (V, E)$ and a positive integer $k \leq |V|$

output: "Yes" iff there is a subset $V' \subseteq V$ with $|V'| \geq k$ such that two vertices in V' are joined by an edge in E.

5. Exact Cover By 3-Sets (X3C)

Input: Set X with $|X| = 3q$ and a collection $C = \{C_1, C_2, \cdots, C_m\}$ of three element subsets of X such that $\bigcup_{i=1}^{n} C_i = X$.

output: "Yes" iff C contains an exact cover for X, i.e., a subcollection $C' \subseteq C$ such that every element of X appears in exactly one member of C'.

6. 3SAT Problem

Input: A boolean function $F = C_1, C_2, \cdots, C_m$ in n variables $x_1, x_2, \cdots, x_n$. Each clause C_i is the disjunction of exactly three literals.

output: "Yes" iff there is a binary assignment for the n variables such that $F = 1$. "No" otherwise.

2   DVUP

Let $G = (V, E, w)$ be a weighted *directed acyclic graph* (wdag) with vertex set V, edge set E, and edge weighting funtion w. $w(i, j)$ is the weight of the edge $< i, j > \in E$. $w(i, j)$ is a positive integer for $< i, j > \in E$ and $w(i, j)$ is undefined if $< i, j > \notin E$. A *source vertex* is a vertex with zero in-degree while a *sink vertex* is a vertex with zero out-degree. The *delay*, $d(P)$, of a path P is the sum of the weights of the edges on that path. The delay, $d(G)$, of the graph G is the maximum path delay in the graph, i.e.,

$$d(G) = \max_{P \ in \ G} \{ \ d(P) \ \}.$$

Let $G|X$ be the wdag that results when the vertices of X are deleted from the wdag G. Note that the deletion of a vertex also requires the deletion of all incident edges.

It is easy to see that DVUP$(0, \delta)$ is NP-hard. In fact, the problem is NP-hard for dags that are chains. To prove this, we use the partition problem. Construct a chain of n vertices with $d(i) = w(i) = a_i$, $1 \leq i \leq n$. It is easy to see that there is an I such that $\sum_{i \in I} a_i = \sum_{i=1}^{n} a_i/2$ iff the minimum cost X such that $d(G|X) \leq \sum_{i=1}^{n} a_i/2$ has cost $\sum_{i=1}^{n} a_i/2$. Note that when $x = 0$, a vertex upgrade is equivalent to a vertex deletion. In the remainder of this section, we assume $x = 0$.

2.1　General Dags

2.1.1　Unit Delay Unit Weight DVUP

A dag $G = (V, E)$ is a *unit delay unit weight dag* iff $d(v) = w(v) = 1$ for every vertex $v \in V$. A subset X of V is *k-colorable* iff we can label the vertices of X using at most k labels and such that no two adjacent vertices have the same label. A *maximum k-coloring* of a dag G is a maximum subset $X \subseteq V$ that is k-colorable. A dag G is *transitive* iff for every $u, v, w \in G$ such that $< u, v > \in E$ and $< v, w > \in E$, the edge $< u, w >$ is also in E. $G^+ = (V, E^+)$ is the transitive closure of (V, E) iff $< u, v > \in E^+$ if there is a path (with at least one edge on it) in G form u to v. Note that if G is a dag then G^+ is a transitive dag.

The unit delay unit weight DVUP for any δ, $\delta \geq 1$ can be solved in $O(n^3 logn)$ time by using the $O(n^3 logn)$ maximum k-coloring algorithm of [GAVR87] for transitive dags. Before showing how this can be done, we establish a relationship between the number of vertices of a given set X that can be on any path of G and colorability of X in G^+. Intuitively, X represents the set of vertices that are not upgraded. So, the number of X vertices on a path is the delay for that path.

Theorem 1 *Let $G = (V, E)$ be a dag and let $G^+ = (V, E^+)$ be its transitive closure. Let X be a subset of the vertices in V. G has no path with δ vertices of X iff X is δ colorable in G^+.*

Proof: [PAIK91b]. □

From the preceding theorem, it follows that if X is a maximum δ-coloring of G^+, $V - X$ is the smallest set such that $d(G|(V - X)) \leq \delta$. This implies the correctness of the following three step algorithm.

> **step 1:** Compute G^+ from G
> **step 2:** Compute X, a maximum δ-coloring of G^+
> **step 3:** $B = V - X$ is the solution for the DVUP instance (G, δ)

The complexity of this is governed by that of step 2 which is $O(n^3 logn)$. Note that when $\delta = 1$, a maximum δ-coloring is just a maximum independent set and such a set can be found in $O(ne)$ time for transitive closure graphs with n vertices and e edges [GAVR87]. So, the case $\delta \leq 1$ can be solved in $O(n^3)$ time as the graph G^+ computed in step 1 may have $O(n^2)$ edges even though G may not.

2.1.2   Nonunit Delay Unit Weight DVUP

A dag $G = (V, E)$ is a unit weight dag if $w(v) = 1$ for every $v \in V$. The case when $d(v)$ is also 1 for all v was considered in the previous section. So, in this section we are only concerned with the case of unit weight dags that have at least one vertex with delay > 1. In this section we show that the nonunit delay unit weight DVUP can be solved in $O(n^3)$ time when $\delta = 1$. The problem is NP-hard for $\delta \geq 2$(see [PAIK91b] for a proof).

Let X be a minimum set of vertices such that $d(G|X) \leq 1$. Clearly, every $v \in V$ with $d(v) > 1$ must be in X. For each $v \in V$ with $d(v) > 1$, let $a_1^v, a_2^v, \cdots, a_q^v$ be the vertices such that $< a_i^v, v > \in E$ and let $b_1^v, b_2^v, \cdots, b_r^v$ be such that $< v, b_i^v > \in E$ and let G' be the dag that results when each such v (together with all edges incident to/from v) are deleted from G and all edges of the form $< a_i^v, b_j^v >$ are added. To get G', this transformation is applied serially to all v with $d(v) > 1$.

Let $B = \{v | d(v) > 1 \text{ and } v \in V\}$. Let $G' = (V', E')$ and let $C \subseteq V'$ be a minimum vertex set such that $d(G'|C) \leq 1$. It is easy to that $A = B \cup C$ is a minimum vertex set such that $d(G|X) \leq 1$. C can be obtained in $O(n^3)$ time using the unit delay unit weight algorithm (note that G' is a unit delay unit weight dag), B can be obtained in $O(n)$ time, and G' can be constructed in $O(n^3)$ time. So, the overall complexity of our algorithm to compute X is $O(n^3)$.

Theorem 2 *Non unit delay unit weight DVUP is NP-hard for every δ, $\delta \geq 2$.*

Proof: See [PAIK91b]. Since the construction of [PAIK91b] generates a multistage graph, DVUP is NP-hard even when the dags are restricted to be multistage graphs. $\square$

2.2   Trees

2.2.1   Trees With Unit Weight And Unit Delay

When the dag is a rooted tree T such that $w(v) = d(v) = 1$ for every vertex, the minimum weight vertex subset X such that $d(T|X) \leq \delta$ can be found in $O(n)$ time by computing the height, h, of each vertex as defined by:

$$h(v) = \begin{cases} 1 & v \text{ is a leaf} \\ 1 + \max\{h(u) | u \text{ is a child of } v\}, & \text{otherwise} \end{cases}$$

X is selected to be the set $X = \{v | h(v) > \delta\}$. The vertex heights can be computed in $O(n)$ time by a simple postorder traversal of the tree T [HORO90]. The correctness of the procedure outlined above is established in Theorem 3. Note that when all vertices have unit weight, the weight of a set, Y, of vertices is simply its cardinality $|Y|$.

Theorem 3 *For any tree T let $h(v)$ be the height of vertex v. The set $X = \{v \mid h(v) > \delta\}$ is a minimum cardinality vertex set such that $d(T|X) \leq \delta$.*

Proof: The fact that $d(T|X) \leq \delta$ is easily seen. The minimality of X is by induction on the number, n, of vertices in T. For the induction base, we see that when $n = 0$, $|X| = 0$ and the theorem is true. Assume the theorem is true for all trees with $\leq m$ vertices where m is an arbitrary natural number. We shall see that the theorem is true when $n = m + 1$. Consider any tree with $n = m + 1$ vertices. Let X be the set of all vertices with height $> \delta$. If $|X| = 0$, then X is clearly a minimal set with $d(T|X) \leq \delta$. Assume $|X| = 0$. In this case the root, r, of T has $h(r) > \delta$ and so is in X. First, we show that there is a minimal vertex set W that contains r and for which $d(T|W) \leq \delta$. Let Z be a minimal vertex set for which $d(T|Z) \leq \delta$. If $r \notin Z$, then let $r, u_1, u_2, \cdots, u_{h(r)-1}$ be a longest root to leaf path in T. Since $h(r) > \delta$, at least one of the u_i's is in Z. Let u_j be any one of the u_i's in Z. Let $W = Z + \{r\} - \{u_j\}$. Clearly, $|W| = |Z|$. Since all root to leaf paths that include u_j also include r, the length of these paths in $T|W$ is the same as in $T|Z$. The length of the remaining paths in $T|W$ is no more than in $T|Z$. So, W is a minimal cardinality vertex set such that $d(T|W) \leq \delta$ and furthermore W contains the root r.

Let $A(v)$, $A \in \{X, W\}$, denote the subset of A that consists only of vertices in the subtree, $T(v)$, rooted at v. Since $d(T(v)|X(v)) \leq \delta$, $d(T(v)|W(v)) \leq \delta$, and $|T(v)| \leq m$ for each v that is a child of r, it follows from the induction hypothesis that $|X(v)| = |W(v)|$ for each v that is a child of r. Hence, $|X| = 1 + \sum_{v \text{ is a child of } r} |X(v)| = |W|$. $\square$

2.2.2 General Trees

Since a chain is a special case of a tree and since DVUP for chains with arbitrary weights and delays has been shown to be NP-hard, we do not expect to find a polynomial time algorithm for general trees. In this section we develop a pseudo polynomial time algorithm (i.e., one whose complexity is polynomial in the number of vertices and the actual values of the vertex delays and weights). We modify the definition of height used in the preceding section to account for the vertex delays. We use H to denote this modified height.

$$H(v) = \begin{cases} d(v) & v \text{ is a leaf} \\ d(v) + \max\{H(u) \mid u \text{ is a child of } v\}, & \text{otherwise} \end{cases}$$

For each vertex v, let $L(v)$ be a set of pairs (l, c) such that there is a subset $X \subseteq T(v)$ such that $d(T(v)|X) = l \leq \delta$ and $\sum_{u \in X} w(u) = c$. Let (l_1, c_1) and (l_2, c_2) be two different pairs such that $l_1 \leq l_2$ and $c_1 \leq c_2$. In this case, pair (l_1, c_1) *dominates* (l_2, c_2). Let $S(v)$ be the subset of $L(v)$ that results from the deletion of all dominated pairs. Let $S(r)$ be this set of dominating pairs for the root r of T. Let $(l', c') \in S(r)$ be the pair with least cost c'. It is easy to see that the least weight vertex set W such that $d(T|W) \leq \delta$ has weight c'. We shall describe how to compute $S(r)$. Using the backtrace strategy of [HORO78, cf. chapter on dynamic programming] we can compute the W that results in $d(T|W) \leq \delta$ and $\sum_{u \in W} w(u) = c'$ in less time than needed to compute $S(r)$ (however, $S(r)$ and some of the other S's computed while computing $S(r)$ are needed for this).

For a leaf vertex v, $S(v)$ is $\{(0, w(v))\}$ when $d(v) > \delta$ and $\{(0, w(v)), (d(v), 0)\}$ otherwise. For a non leaf vertex v, $S(v)$ may be computed from the $S(u)$'s of its children $u_1, \cdots, u_{k_v}$. First, we compute $U(v)$ as the set of nondominated pairs of the form (l, c) where $l = \max \{l_1, l_2, \cdots, l_{k_v}\}$ and $c = \sum_{i=1}^{k_v} c_i$ for some set of pairs $(l_i, c_i) \in S(u_i)$, $1 \leq i \leq k_v$. Let $V(v)$ and $Y(v)$ be as below:

$$V(v) = \{ (l, c + w(v)) \mid (l, c) \in U(v) \}$$
$$Y(v) = \{ (l + d(v), c) \mid l + d(v) \leq \delta \text{ and } (l, c) \in U(v) \}$$

Now, $S(v)$ is the set of nondominated pairs in $V(v) \cup Y(v)$. Since $S(v)$ contains only nondominated pairs, all pairs in $S(v)$ have different l and c values. So, $|S(v)| \leq \min \{\delta, \omega\}$ where $\omega = \sum_{u=1}^{n} w(u)$. Using the technique of [HORO78], $S(v)$ can be computed from the $S(u)$'s of its children in time $O(\min\{\delta, \omega\} * k_v)$. To compute $S(r)$ we need to compute $S(v)$ for all vertices v. The time needed for this is $O(\min\{\delta, \omega\} * \sum k_v) = O(\min\{\delta, \omega\} * n)$.

Note that for unit delay trees, $\delta \leq n$ and for unit weight trees $\omega = n$. So in both of these cases the procedure described above has complexity $O(n^2)$.

2.3 Series-Parallel Dags

A *series-parallel digraph*, SPDAG, may be defined recursively as:

1. A directed chain is an SPDAG.

2. Let s_1 and t_1 respectively, be the source and sink vertices of one SPDAG G_1 and let s_2 and t_2 be these vertices for the SPDAG G_2. The parallel combination of G_1 and G_2, $G_1//G_2$, is obtained by identifying vertex s_1 with s_2 and vertex t_1 with t_2 (Figure 1(c)). $G_1//G_2$ is an SPDAG. We restrict G_1 and G_2 so that at most one of the edges $< s_1, t_1 >$, $< s_2, t_2 >$ is present. Otherwise, $G_1//G_2$ contains two copies of the same edge.

3. The series combination of G_1 and G_2, $G_1 G_2$, is obtained by identifying vertex t_1 with s_2 (Figure 1(d)). $G_1 G_2$ is an SPDAG.

The strategy we employ for SPDAGs is a generalization of that used for trees with general delays and weights. Let s and t, respectively, be the source and sink vertices of the SPDAG G. Let $D(l, Y, G)$ be a minimum weight vertex set that contains the vertices in Y, $Y \subseteq \{s, t\}$, and such that $d(G|D(l, Y, G)) \leq l$ and let $f(G)$ be as below:

$$f(G) = \{ (l, c, Y) \mid 0 \leq l \leq \delta, \ c = \sum_{u \in D(l, Y, G)} w(u) \}.$$

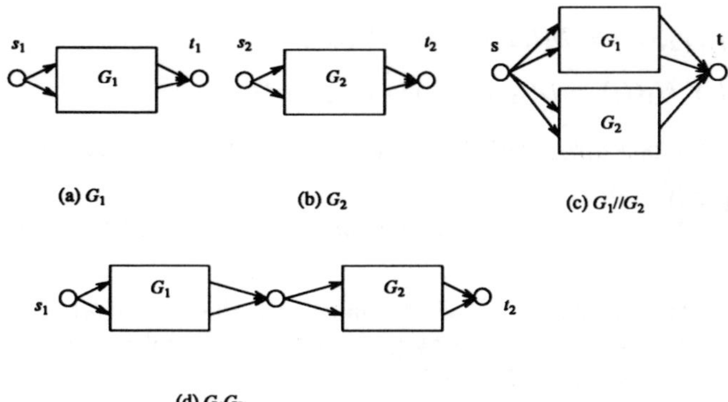

Figure 1: Series-parallel digraph.

Let (l_1, c_1, Y_1) and (l_2, c_2, Y_2) be two different triples in $f(G)$. (l_1, c_1, Y_1) *dominates* (l_2, c_2, Y_2) iff $l_1 \leq l_2$, $c_1 \leq c_2$ and $Y_1 = Y_2$. Let $F(G)$ be the set of triples obtained by deleting all dominated triples of $f(G)$. If (l', c', Y') is the least weight triple (i.e., the one with least c) in $F(G)$ then the least weight W such that $d(G|W) \leq \delta$ has weight c'. We shall show how to compute $F(G)$ and hence (l', c', Y'). The actual W may be obtained using a backtrace step as described in [HORO78].

2.3.1 G Is A Chain

Consider the case when G has only two vertices s and t. $F(G)$ is constructed using the code:

$$F(G) := \{ (0, w(s) + w(t), \{s, t\}) \}$$
if $d(s) \leq \delta$ **then** $F(G) := F(G) \cup \{ (d(s), w(t), \{t\}) \}$
if $d(t) \leq \delta$ **then** $F(G) := F(G) \cup \{ (d(t), w(s), \{s\}) \}$
if $d(s) + d(t) \leq \delta$ **then** $F(G) := F(G) \cup \{ (d(s) + d(t), 0, \emptyset) \}$

When G is a chain with more than two vertices, it may be regarded as the series composition of two smaller chains G_1 and G_2. In this case $F(G)$ may be constructed from $F(G_1)$ and $F(G_2)$ using the algorithm to construct $F(G_1G_2)$ described in the next section.

2.3.2 G is of the form G_1G_2

The following lemma enables us to construct $F(G_1G_2)$ from $F(G_1)$ and $F(G_2)$.

Lemma 1 *If* $(l, c, Y) \in F(G_1G_2)$, *then there is an* $(l_1, c_1, Y_1) \in F(G_1)$ *and an* $(l_2, c_2, Y_2) \in F(G_2)$ *such that*

(a) $D(l_1, Y_1, G_1) = D(l, Y, G_1 G_2) \cap V(G_1)$

(b) $D(l_2, Y_2, G_2) = D(l, Y, G_1 G_2) \cap V(G_2)$

(c) $D(l, Y, G_1 G_2) = D(l_1, Y_1, G_1) \cup D(l_2, Y_2, G_2)$

(d) $c = \sum_{u \in D(l,Y,G_1 G_2)} w(u), \quad c_1 = \sum_{u \in D(l_1,Y_1,G_1)} w(u), \quad c_2 = \sum_{u \in D(l_2,Y_2,G_2)} w(u)$

Proof: [PAIK91c]. $\square$

Lemma 1 suggests the following approach to obtain $F(G_1 G_2)$ from $F(G_1)$ and $F(G_2)$:

> **step 1:** Construct a set Z of triples such that $F(G_1 G_2) \subseteq Z$. This is obtained by combining together all pairs of triples $(l_1, c_1, Y_1) \in F(G_1)$ and $(l_2, c_2, Y_2) \in F(G_2)$.

> **step 2:** Eliminate Z all triples that are dominated by at least one other triple of Z.

The triples (l_1, c_1, Y_1) and (l_2, c_2, Y_2) are compatible iff ($t_1 \in Y_1$ and $s_2 \in Y_2$) or ($t_1 \notin Y_1$ and $s_2 \notin Y_2$). Only compatible triples may be combined. Assume that we are dealing with two compatible triples. We first obtain the triple (l, c, Y) as below:

> **if** $t_1 \in Y_1$
> **then** $(l, c, Y) := (l_1 + l_2, c_1 + c_2 - w(t_1), Y_1 \cup Y_2 - \{t_1\})$
> **else** $(l, c, Y) := (l_1 + l_2 - d(t_1), c_1 + c_2, Y_1 \cup Y_2 - \{t_1\})$

Next, (l, c, Y) is added to Z provided $l \leq \delta$.

2.3.3 $G = G_1 // G_2$

When $G = G_1 // G_2$ we use Lemma 2 which is the analogue of Lemma 1.

Lemma 2 f $(l, c, Y) \in F(G_1 // G_2)$, *then there is an* $(l_1, c_1, Y_1) \in F(G_1)$ *and an* $(l_2, c_2, Y_2) \in F(G_2)$ *such that*

(a) $D(l_1, Y_1, G_1) = D(l, Y, G_1 // G_2) \cap V(G_1)$

(b) $D(l_2, Y_2, G_2) = D(l, Y, G_1 // G_2) \cap V(G_2)$

(c) $D(l, Y, G_1 // G_2) = D(l_1, Y_1, G_1) \cup D(l_2, Y_2, G_2)$

(d) $c = \sum_{u \in D(l,Y,G_1 // G_2)} w(u), \quad c_1 = \sum_{u \in D(l_1,Y_1,G_1)} w(u), \quad c_2 = \sum_{u \in D(l_2,Y_2,G_2)} w(u)$

Proof: Similar to that of Lemma 1. $\square$

To obtain $F(G_1 // G_2)$ from $F(G_1)$ and $F(G_2)$ we use the two step approach used to compute $F(G_1 G_2)$. For step 1, we compute the triple (l, c, Y) obtained by combining $(l_1, c_1, Y_1) \in F(G_1)$ and $(l_2, c_2, Y_2) \in F(G_2)$. The triples are compatible iff $Y_1 = Y_2$. Again, only compatible triples may be combined. For compatible triples, (l, c, Y) is obtained as below:

$$l := \max \{l_1, l_2\}$$
$$l := c_1 + c_2 - \sum_{u \in Y_1} w(u)$$
$$Y := Y_1$$

Next, (l, c, Y) is added to Z.

2.3.4 Complexity

The series-parallel decomposition of an SPDAG can be determined in $O(n)$ time
[VALD79]. By keeping each $F(G_i)$ as four separate lists of triples, one for each of the
four possible values for the third coordinate of the triples, $F(G_1 G_2)$ and $F(G_1//G_2)$
can be obtained in $O(|F(G_1)| * |F(G_2)|)$ time from $F(G_1)$ and $F(G_2)$. Since $F(G_1)$
$(F(G_2))$ contains only non dominated triples, it can contain at most four triples for
each distinct value of the first coordinate and at most four for each distinct value
of the second coordinate (these four must differ in their third coordinate). Hence,
$|F(G_1)| \leq 4^* \min \{ \delta + 1, \sum_{u \in V(G_1)} w(u) \}$ and $|F(G_2)| \leq 4^* \min \{ \delta + 1, \sum_{u \in V(G_2)} w(u)$
$\}$. So, we can obtain $F(G)$ for any SPDAG in time $O(n * \min \{\delta^2, (\sum_{u \in V(G)} w(u))^2 \}$
$)$. For an SPDAG, G, with unit delay or unit weight, the complexity, $t(G)$, is $O(n^2)$.
To see this, note that for unit weight or unit delay SPDAGs, $|F(G_1)| \leq 4|V(G_1)|$ and
$|F(G_2)| \leq 4|V(G_2)|$. So, $t(G) \leq t(G_1) + t(G_2) + 16|V(G_1)| * |V(G_2)|$.

2.3.5 Extension To General Series Parallel Dags

The algorithm for series parallel dags may be extended to obtain an algorithm of
the same asymptotic complexity for general series parallel dags (GSPDAG). These
were introduced in [LAWL78, MONM77, SIDN76]. The extension may be found in
[PAIK91c].

3 LinkDelay(x, δ)

When $\delta = 0$ and $x > 0$, LinkDelay(x, δ) can be solved in linear time. In case G has an
edge with delay > 0, then the link costs cannot be made 0 by upgrading any subset of
the vertices. If G has no edge with delay > 0, then no vertex needs to be upgraded.
For all other combinations of δ and x, LinkDelay(x, δ) is NP-hard.

Theorem 4 [PAIK91d] *LinkDelay(x, δ) is NP-hard whenever $\delta \neq 0$ or $x = 0$.*

Proof: Let $G = (V, E)$ be an instance of the vertex cover problem. We obtain from
G, an instance G' of LinkDelay(x, δ) by associating a delay with each edge of G. If
$\delta = 0$, this delay is one and if $\delta > 0$, this delay is any number in the range $(\delta, \delta/x]$
(in case $x = 0$, the range is $(0, \infty)$). Since $0 \leq x < 1$, upgrading a set, A, of vertices
results in all links having a delay $\leq \delta$ iff A is a vertex cover of links in G' and hence

of the edges in G. So, G' has an upgrading vertex set of size $\leq k$ iff G has a vertex cover of size $\leq k$. $\square$

[PADH92] considers the link delay problem for trees and series-parallel graphs. For both cases, she develops a linear time algorithm. The case for trees is easily solved using a tree traversal algorithm. First, note that when $\delta = 0$ and $x > 0$, the problem has a solution iff all edges have zero delay. In this case, no vertex is to be upgraded. A simple examination of the edge delays suffices to solve this case. When $\delta \geq 0$ and $x = 0$, we begin by removing all edges with delay less than δ. This leaves behind a forest of trees. Trees with a single node are removed from consideration. From each of the remaining trees, the parents of all leaves are marked for upgrading and edges incident to these parents deleted (as their delay is now zero). The result is a new forest from which single node trees are removed and the parents of leaves upgraded. This process continues until the forest becomes empty. When, $\delta > 0$ and $x > 0$, we first verify that there is no edge with delay greater than δ/x^2. This is a necessary and sufficient condition for the existence of a solution. Next, for each edge (u, v) with delay d such that $x * d > \delta \geq x^2 * d$, both u and v are upgraded. Following this, we remove all edges with delay $\leq \delta$ from the tree and upgrade vertices in the resulting forest using the forest upgrading scheme just described for the case $\delta \geq 0$ and $x = 0$. The correctness of the algorithm is easily established and using appropriate data structures, it can be implemented to have complexity $O(n)$ where n is the number of vertices in the tree.

For the case of series-parallel graphs, [PADH92] proposes a dynamic programming algorithm which uses the series-parallel decomposition of the graph. For each (series-parallel) graph in the decomposition, she keeps track of the best solution that (a) necessarily upgrades the source vertex but not the sink, (b) necessarily upgrades the sink vertex but not the source, (c) necessarily upgrades both the source and the sink, and (d) upgrades neither the source nor the sink. Since only four solutions are recorded in each stage of the decomposition, the resulting dynamic programming algorithm has complexity $O(n)$ where n is the number of vertices in the input series-parallel graph.

4   ShortestPath(x,δ)

We note that while at first glance ShortestPath(0,0) may appear to be identical to either the vertex cover or the dominating set problem, this is not so.

Theorem 5 [PAIK91d] *ShortestPath(x,δ) is NP-hard whenever $x = 0$ or $\delta > 0$.*

Proof: We shall prove this here only for the case $x = \delta = 0$. Let X, q, C, and m be as in the definition of X3C. Construct an instance $G = (V, E)$ of ShortestPath(0,0) as below:

a) G is a three level graph with a root vertex r. This is the only vertex on level 1 of the graph.

b) The root has $m + q + 2$ children labeled $C_1, C_2, \cdots, C_m, Z_1, Z_2, \cdots, Z_{q+2}$. These are the level 2 nodes of the graph. Child C_i represents set C_i, $1 \leq i \leq m$ while child Z_i is just a dummy node in G.

c) The graph G has $3q$ nodes on level 3. These are labeled 1, 2, $\cdots$, $3q$. Node i represents element i of X, $1 \leq i \leq 3q$.

d) Each node C_i on level 2 has edges to exactly three nodes on level 3. These are to the nodes that represent the members of C_i, $1 \leq i \leq m$.

We shall show that the input X3C instance has answer "yes" iff the Shortest-Path(0,0) instance G has an upgrade sets of size $\leq q + 1$. First, suppose that the answer to the X3C instance is "yes". Then there is a $C' \subseteq C$ such that C' is an exact cover of X. Since $|X| = 3q$ and $|C_i| = 3$, $1 \leq i \leq m$, $|C'| = q$. Let $S = \{r\} \cup C'$. One may verify that S is an upgrade set for G and $|S| = q + 1$.

Next, suppose that G has an upgrade set S of size $\leq q + 1$. If $r \notin S$, then the shortest path from r to at least one of the Z_i's has length > 0 as at least one of the $q + 2$ Z_i's is not in S and every r to Z_i path must use the edge (r, Z_i). So, $r \in S$. When the vertices in S are upgraded, every vertex in G must have at least one zero length edge incident to it as otherwise the shortest paths to it have length ≥ 1. In particular, this must be the case for all $3q$ level three vertices. Upgrading the root r does not result in any of these $3q$ vertices having a zero length edge incident to it. So, this is accomplished by the remaining $\leq q$ vertices in S. The only way this can be accomplished by an upgrade of $\leq q$ vertices is if these remaining vertices are a subset of $\{C_1, C_2, \cdots, C_m\}$ and this subset is an exact cover of X (this would, of course, require $|S| = q + 1$). So, $S - \{r\}$ is an exact cover of the input X3C instance.

Hence, the X3C instance has output "yes" iff G has an upgrade set of size $\leq q + 1$. $\square$

5 Satellite(δ)

Satellite(0) is trivially solved. First, zero length edges are eliminated by combining their endpoints u and v into a single vertex uv. Remaining edges previously incident to u or v are now incident to uv. Duplicate edges are replaced by a single edge with the lower cost. If the resulting communication graph has at most one vertex, no satellite links are needed. When the number of vertices exceeds one, each vertex must be a satellite vertex. We shall show that Satellite(δ) is NP-hard for every δ, $\delta > 0$.

Lemma 3 *Let G be a communication graph. Let S be the subset of vertices of G that are satellite vertices and let N be the remaining vertices (i.e., the non-satellite vertices). CommTime(G) $\leq$ 1 iff the following are true:*

  (a) The vertices of N define a clique of G.

  (b) Each vertex of N is adjacent to a vertex of S (i.e., S is a dominating set of G) unless $S = \emptyset$.

Proof: If N contains two vertices that are not adjacent in G, then the shortest path between them is of length at least two regardless of whether or not this shortest path utilizes satellites. So, the subgraph of G induced by the vertices of N is a clique. For (b), assume that $S \neq \emptyset$. If N contains a vertex v that is not adjacent to at least one satellite vertex then v is at least distance two from every satellite vertex $s \in S$. Hence, CommTime$(G) > 1$. $\square$

Theorem 6 [PAIK91d] *Satellite(δ) is NP-hard for $\delta \geq 1$.*

Proof: We shall only present the proof for the case $\delta = 1$. The complete proof can be found in [PAIK91d]. The proof uses the max clique problem. Let G be a connected undirected graph with n vertices. Let G' be the graph obtained by adding to G n edges of the type (i, i') where i is a vertex of G and i' is a new vertex (Figure 2). The number of vertices in G' is $2n$. All edges of G' have unit cost. We claim that G has a clique of size $\geq k$, for any $k \geq 3$ iff G' has a vertex subset S, $|S| \leq 2n - k \leq 2n - 3$, such that by making the vertices of S satellite vertices, CommTime$(G')=1$. To see this, let A, $|A| \geq 3$, be any subset of vertices in G. If A forms a clique in G, then by making all vertices of G' except those in A satellites, the communication time in G' becomes at most one as the vertices of A satisfy the conditions of Lemma 3 on non-satellite vertices. So, if G has a clique of size $\geq k$, then G' has a satellite subset S of size $\leq 2n - k$. Next, suppose that G' has a satellite subset S of size $\leq 2n - k \leq 2n - 3$. Let N be the remaining vertices of G'. If N contains a vertex i' that is not in G, then from Lemma 3 it follows that $|N| \leq 2$ as the largest clique in G' that includes vertex i' has only two vertices in it (the other vertex being vertex i). In this case, $|S| \geq 2n - 2$. So, N contains no vertex that is not in G. Since N forms a clique in G' (Lemma 3), it forms a clique in G. Hence, if $|S| \leq 2n - k \leq 2n - 3$, G contains a clique of size $\geq k \geq 3$. While the NP-hard formulation of the max clique problem does not restrict k to be ≥ 3, it is easy to see that the problem remains NP-hard under this restriction. So, Satellite(1) is NP-hard. $\square$

6   LongestPath(x,δ)

Lemma 4 *LongestPath (0,0) is NP-hard.*

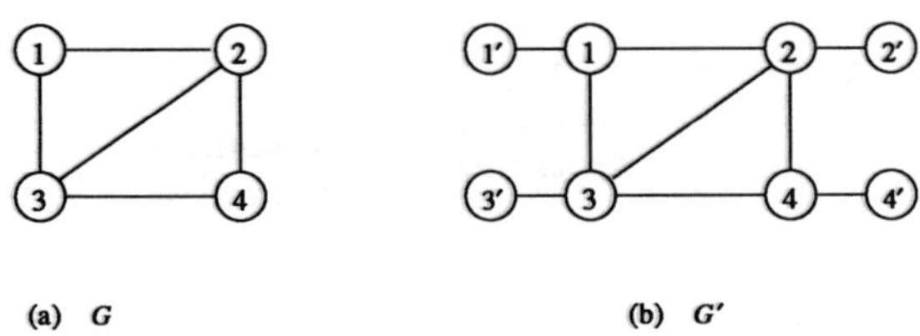

(a) *G*   (b) *G'*

Figure 2: Construction of Theorem 6.

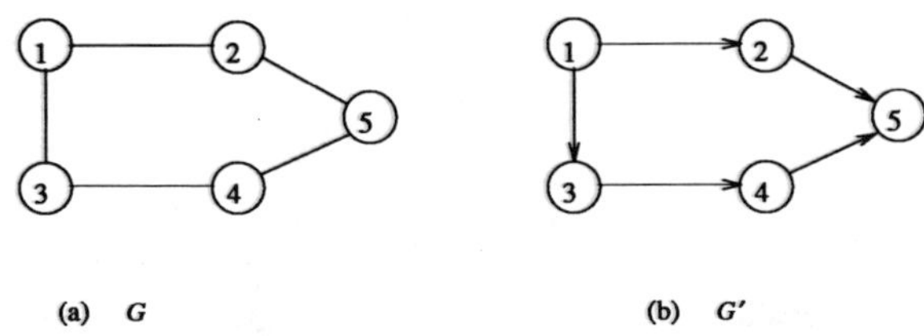

(a) *G*   (b) *G'*

Figure 3: Construction for Lemma 4.

Proof: Let G be a connected undirected graph. We shall construct an instance G' of LongestPath(0,0) such that G has a vertex cover of size $\leq k < n$ iff G' has a vertex upgrade set A' of size $\leq k < n$. To get G', orient the edges of G to begin at a higher index vertex. I.e., if (i, j), $i < j$, is an edge of G, then $< i, j >$ is a directed edge of G' (Figure 3). All edges of G' have unit delay. It is easy to see that G' is a dag and that A is a vertex cover of G iff $A' = A$ is a vertex upgrade set of the LongestPath(0,0) instance G'. $\square$

To show that LongestPath(x,δ) is NP-hard for $x = 0$ and $\delta > 0$, we use the subgraph $J_{n,q}$ (Figure 4) which is comprised of n directed chains of size q that share a common vertex r. This set of n chains is connected to a two vertex directed chain $< r, s >$. Each edge of $J_{n,q}$ has unit delay. We see that the longest path in $J_{n,q}$ has length q and that by upgrading the single vertex r, we can make the delay of this subgraph $q - 2$. However, to reduce the delay to $q - 3$, we need to upgrade at least $n + 1$ vertices.

Lemma 5 *LongestPath(x,δ) is NP-hard for $\delta > 0$.*

Proof: Let G be any instance of LongestPath(0,0) in which all edge delays are one. A corresponding instance G' of LongestPath(x,δ) is obtained by attaching a copy of $J_{n,\lfloor \delta \rfloor +2}$ in case $x = 0$ and Q_n (Figure 5) in case $x > 0$ to each vertex v of G that has in-degree zero. This is done by identifying the s vertex of $J_{n,\lfloor \delta \rfloor +2}$ with vertex v (i.e., these two vertices are the same). Note that n is the number of vertices in G. Let m

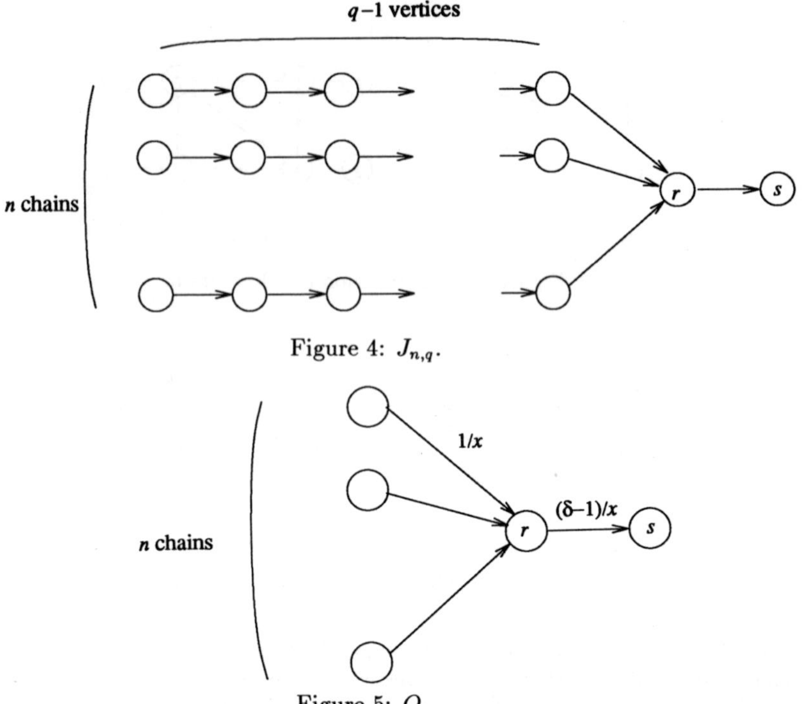

Figure 4: $J_{n,q}$.

Figure 5: Q_n

be the number of vertices in G that have zero in-degree. One may verify that for any k, $k \leq n$, G has an upgrade set of size $\leq k$ iff G' has one of size $\leq m + k$. Hence, LongestPath(x, δ) is NP-hard for $\delta > 0$. $\square$

Theorem 7 [PAIK91d]

   (a) Longest Path (x, δ) is NP-hard when $x = 0$ and $\delta = 0$ and also when $x \geq 0$ and $\delta > 0$.

   (b) Longest Path (x, δ) is polynomially solvable when $x > 0$ and $\delta = 0$.

Proof: (a) has been proved in Lemmas 4 and 5. For (b), if the dag has an edge with delay > 0, then it has no vertex upgrade set. If there is no such edge, then no vertex needs to be upgraded to ensure that the longest path has zero length. $\square$

7 DVSP

Let G/X be the wdag that results when each vertex v in X is split into two v^i and v^o such that all edges $< v, j > \in E$ are replaced by edges of the form $< v^o, j >$ and all

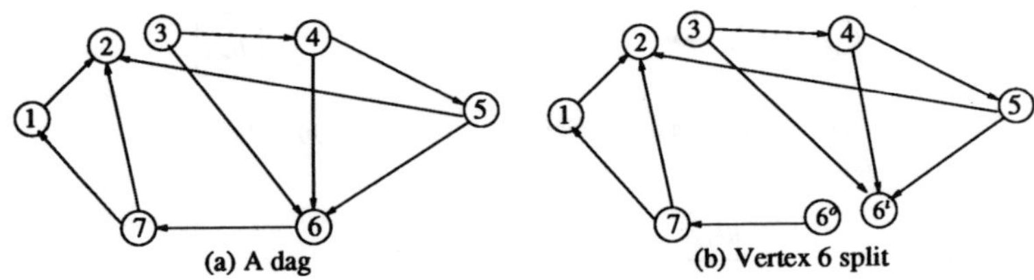

(a) A dag   (b) Vertex 6 split

Figure 6: Example of vertex splitting

edges $< i, v >\in E$ are replaced by edges of the form $< i, v^i >$. I.e., outbound edges of v now leave vertex v^o while the inbound edges of v now enter vertex v^i. Figure 6(b) shows the result, G/X, of splitting the vertex 6 of the dag of Figure 6(a). The *dag vertex splitting problem* (DVSP) is to find a least cardinality vertex set X such that $d(G/X) \leq \delta$, where δ is a prespecified delay. For the dag of Figure 6(a) and $\delta = 3$, $X = \{6\}$ is a solution to the DVSP problem.

7.1 Complexity Results

If $w(i,j) = 1$ for every edge in the wdag, then the edge weighting function w is said to be a *unit weighting function* and we say that G has unit weights. When $\delta = 1$, the unit weight DVSP can be solved in linear time as every vertex that is not a source or sink has to be split. However, for every $\delta \geq 2$, the problem is NP-hard [PAIK90]. To show this, for each instance F of 3SAT, we construct an instance G_F of the unit weight DVSP such that from the size of the solution to G_F we can determine, in polynomial time, the answer to the 3SAT problem for F. This construction employs two unit weight dag subassemblies: variable subassembly and clause subassembly.

Variable Subassembly

Figure 7(a) shows a chain with $\delta - 1$ vertices. This chain is represented by the schematic of Figure 7(b). The variable subassembly, $VS(i)$, for variable x_i is given in Figure 7(c). This is obtained by combining together three copies of the chain $H_{\delta-1}$ with another chain that has three vertices. Thus, the total number of vertices in the variable subassembly $VS(i)$ is 3δ. Note that $d(VS(i)) = \delta + 1$. Also, note that if $d(VS(i)/X) \leq \delta$, then $|X| \geq 1$. The only X for which $|X| = 1$ and $d(VS(i)/X) \leq \delta$ are $X = \{x_i\}$ and $X = \{\bar{x}_i\}$. Figure 7(d) shows the schematic for $VS(i)$.

Clause Subassembly

The clause subassembly $CS(j)$ is obtained by connecting together four $\delta - 1$ vertex chains with another three vertex subgraph as shown in Figure 8(a). The schematic for $CS(j)$ is given in Figure 8(b). The number of vertices in $CS(j)$ is $4\delta - 1$ and $d(CS(j)) = 2\delta$. One may easily verify that if $|X| = 1$, then $d(CS(j)/X) > \delta$. So, if $d(CS(j)/X) \leq \delta$,then $|X| > 1$. Since $\delta \geq 2$, the only X with $|X| = 2$ for which $d(CS(j)/X) \leq \delta$ are such that $X \subseteq \{l_{j1}, l_{j2}, l_{j3}\}$. Furthermore, every

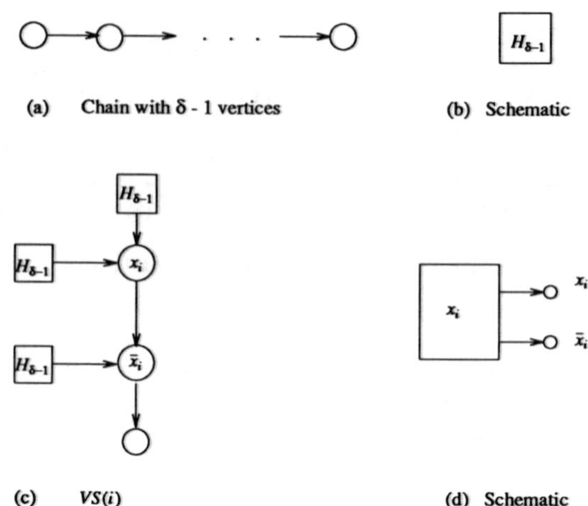

(a)   Chain with δ - 1 vertices   (b)   Schematic

(c)   *VS(i)*   (d)   Schematic

Figure 7: Variable subassembly for DVSP.

$X \subseteq \{l_{j1}, l_{j2}, l_{j3}\}$ with $|X| = 2$ results in $d(CS(j)/X) \leq \delta$.

To construct G_F from F, we use n $VS(i)$'s, one for each variable x_i in F and m $CS(j)$'s, one for each clause C_j in F. There is a directed edge from vertex x_i ($\overline{x}_i$) of $VS(i)$ to vertex l_{jk} of $CS(j)$ iff x_i ($\overline{x}_i$) is the k'th literal of C_j (we assume the three literals in C_j are ordered). For the case $F = (x_1 + \overline{x}_2 + \overline{x}_4)(\overline{x}_1 + \overline{x}_3 + \overline{x}_4)(x_1 + x_2 + x_3)$, the G_F of Figure 9 is obtained.

Since the total number of vertices in G_F is $3\delta n + (4\delta - 1)m$, the construction of G_F can be done in polynomial time for any fixed δ.

Theorem 8 *Let F be an instance of 3SAT and let G_F be the instance of unit weight DVSP obtained using the above construction. For $\delta \geq 2$, F is satisfiable iff there is a vertex set X such that $d(G_F/X) \leq \delta$ and $|X| = n + 2m$.*

Proof: If F is satisfiable then there is a binary assignment to the x_i's such that F has value 1. Let $b_1, b_2, \cdots, b_n$ be this ssignment. Construct a vertex set X in the following way:

1. x_i is in X if $b_i = 1$. If $b_i = 0$, then $\overline{x}_i$ is in X.

2. From each $CS(j)$ add exactly two of the vertices l_{j1}, l_{j2}, l_{j3} to X. These are chosen such that the literal corresponding to the vertex not chosen has value 1. Each clause has at least one literal with value 1.

We readily see that $|X| = n + 2m$ and that $d(G_F/X) \leq \delta$.

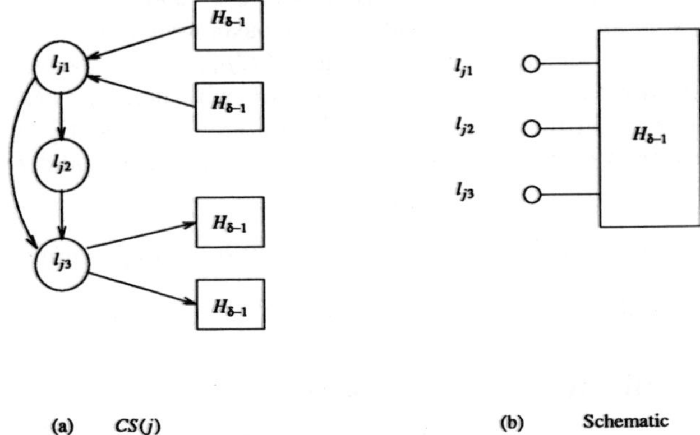

Figure 8: Clause assembly for DVSP.

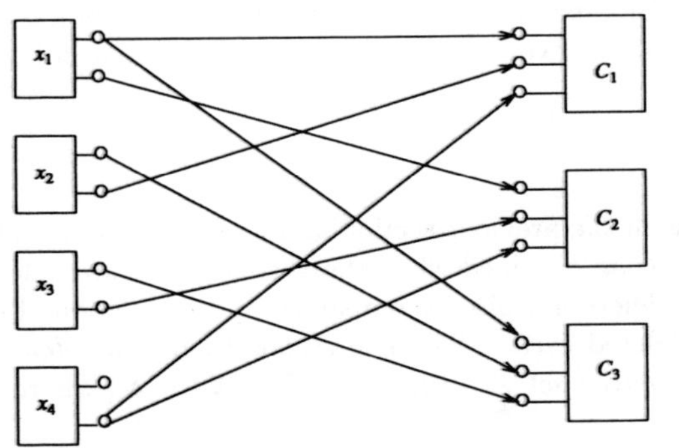

Figure 9: $F = (x_1 + \overline{x}_2 + \overline{x}_4)\,(\overline{x}_1 + \overline{x}_3 + \overline{x}_4)\,(x_1 + x_2 + x_3).$

Next, suppose that there is an X such that $|X| = n + 2m$ and $d(G_F/X) \leq \delta$. From the construction of the variable and clause assemblies and from the fact that $|X| = n + 2m$, it follows that X must contain exactly one vertex from each of the sets $\{x_i, \overline{x}_i\}$, $1 \leq i \leq n$ and exactly 2 from each of the sets $\{l_{j1}, l_{j2}, l_{j3}\}$, $1 \leq j \leq m$. Hence there is no i such that both $x_i \in X$ and $\overline{x}_i \in X$ and there is no j for which $l_{j1} \in X$ and $l_{j2} \in X$ and $l_{j3} \in X$. Consider the Boolean assignment $b_i = 1$ iff $x_i \in X$. Suppose that $l_{jk} \notin X$ and $l_{jk} = x_i$ $(\overline{x}_i)$. Since $d(G_F/X) \leq \delta$, vertex x_i $(\overline{x}_i)$ must be split as otherwise there is a source to sink path with delay greater than δ. So, x_i $(\overline{x}_i) \in X$ and $b_i = 1$ (0). As a result, the k'th literal of clause C_j is true. Hence, $b_1, b_2, \cdots, b_n$ results in each clause having at least one true literal and F has value 1. $\square$

Theorem 9 *DVSP is NP-hard for unit weight multistage graphs when $\delta \geq 4$.*

Proof: See [PAIK90]. $\square$

7.2 Polynomially Solvable Cases

When the wdag G is a rooted tree the DVSP problem can be solved in linear time by performing a postorder [HORO90] traversal of the tree. During this traversal we compute, for each node x, the maximum delay, $D(x)$, from x to any other node in its subtree. If x has a parent z and $D(x) + w(z, x)$ exceeds δ, then the node x is split and $D(z)$ is set to 0. Note that $D(x)$ satisfies:

$$D(x) = \max_{v \ is \ a \ child \ of \ r}\{D(y) + w(x, y)\}.$$

Another polynomially sovlable case is when the dag is a series-parallel graph. For such graphs dynamic programming can be used to obtain a quadratic time algorithm [PAIK91a]. For general dags, a backtracking algorithm has been formulated in [PAIK90] and heuristics have been developed and evaluated in [PAIK90, 93].

8 DVDP

Let G be a wdag as in the previous section and let X be a subset of the vertices of G. Let $G - X$ be the wdag obtained when the vertices in X are deleted from the wdag G. This vertex set deletion is also accompanied by the deletion of all edges in G that are incident to a deleted vertex. The *dag vertex deletion problem* (DVDP) is to find a least cardinality vertex set X such that $d(G - X) \leq \delta$, where δ is a prespecified graph delay.

Lemma 6 *Let $G = (V, E, w)$ be a weighted wdag and let δ be a prespecified delay value. Let $MaxEdgeDelay = max_{<i,j> \in E}\{w(i, j)\}$.*

(a) The DVDP has a solution iff $\delta \geq 0$.

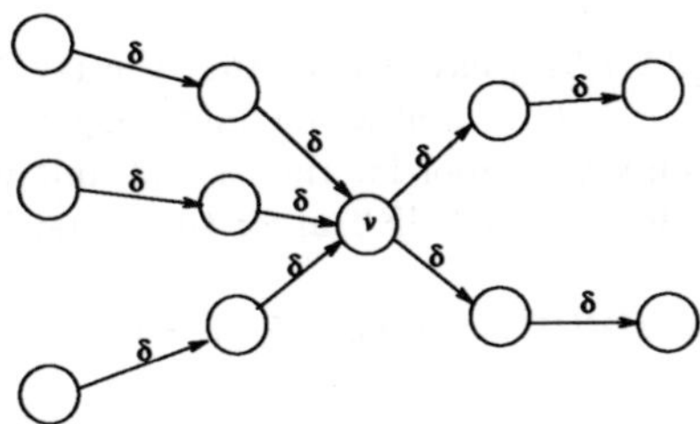

Figure 10: Construction for Lemma 7.

(b) The DVSP has a solution iff $\delta \geq$ MaxEdgeDelay.

(c) For every $\delta \geq$ MaxEdgeDelay, the size of the DVDP solution is less than or equal to that of the DVSP solution.

Proof:

(a) Since $d(G-V) = 0$, there must be a least cardinality set X such that $d(G-X) \leq \delta$.

(b) Vertex splitting does not eliminate any edges. So, there is no X such that $d(G/X) <$ MaxEdgeDelay. Further, $d(G/V) =$ MaxEdgeDelay. So, for every $\delta \geq$ MaxEdgeDelay, there is a least cardinality set X such that $d(G/X) \leq \delta$.

(c) Let X be a solution to the DVSP. Since $d(G/X) \leq \delta$, $d(G-X) \leq \delta$. Hence the cardinality of the DVDP solution is $\leq |X|$. $\square$

Let $|DVSP|$ $(|DVDP|)$ be the size of solution to the DVSP (DVDP).

Lemma 7 *For every δ, $\delta \geq 0$, there is a wdag $G = (V, E, w)$ with MaxEdgeDelay $\leq \delta$ such that $|DVSP|\bigwedge DVDP| =$ number of nodes that are neither source nor sink.*

Proof: Consider the wdag of Figure 10. $d(G - \{v\}) = \delta$. However, since every edge has weight δ, it is necessary to split every vertex that is not a source or sink to get the delay down to δ.

Corollary 1 *For every $\delta \geq$ MaxEdgeDelay and every wdag G such that $d(G) > \delta$, $1 \leq |DVSP|\bigwedge DVDP| =$ number of nodes that are neither source nor sink.*

Proof: The lower bound follows from Lemma 6 part (c) and the upper bound follows from the observation that $|DVSP| \leq$ number of nodes that are neither source nor sink and $|DVDP| \geq 1$. Note that the source and sink vertices of a wdag never need to be split. $\square$

8.1   Complexity Results

Paik, Reddy, and Sahni [PAIK91a] have shown that the DVDP problem is NP-hard for unit weight dags with $\delta \geq 0$ as well as for unit weight multistage graphs with $\delta \geq 2$. We shall present only the proof for the case of unit weight dags and $\delta = 0$. The interested reader is referred to [PAIK91a] for the remaining proofs.

Theorem 10 *Unit weight DVDP is NP-hard for $\delta = 0$.*

Proof: Let G be an instance of unit weight DVDP and let X be such that $d(G-X) = 0$. So, X must contain at least one of the two end-points of each edge of G. Hence, X is a vertex cover of the undirected graph obtained from G by removing directions from the edges. Actually, every vertex cover problem can be transformed into an equivalent DVDP with $\delta = 0$. Let U be an arbitrary undirected graph. Replace each undirected edge (u, v) of U by the directed edge $<\min\{u, v\}, \max\{u, v\}>$ to get the directed graph V. V is a wdag as one cannot form a cycle solely from edges of the form $<i, j>$ where $i < j$. Furthermore the DVDP instance V with $\delta = 0$ has a solution of size $\leq k$ iff the corresponding vertex cover instance U does. Hence, unit weight DVDP with $\delta = 0$ is NP-hard. $\square$

8.2   Polynomially Solvable Cases

As in the case of the DVSP problem, the DVDP problem can be solved in linear time when the wdag is a tree and in quadratic time when the wdag is a series-parallel graph. The algorithms are similar to those for the corresponding DVSP cases and can be found in [PAIK91a].

References

[CHAN90] P. K. Chan, "Algorithms For Library-Specific Sizing Of Combinational Logic", *Proc. 27th DAC Conf.*, 1990 pp. 353-356.

[GARE79] M. R. Garey, and D. S. Johnson, "Computers and Intractability", W. H. Freeman and Company, San Francisco, 1979.

[GAVR87] F. Gavril, "Algorithms For Maximum k-colorings And k-coverings Of Transitive Graphs", *Networks*, Vol. 17, pp. 465-470, 1987.

[GHAN87] S. Ghanta, H. C. Yen, and H. C. Du, "Timing Analysis Algorithms For Large Designs", University of Minnesota, Technical Report, 87-57,1987.

[HORO78] E. Horowitz, and S. Sahni, "Fundamentals of Computer Algorithms", Computer Science Press, Maryland, 1978.

[KRIS79] M. Krishnamoorthy and N. Deo, "Node deletion NP-complete problems", SIAM Jr on Computing, Vol 8, No 4, 1979, pp 619-625.

[LAWL78] E. L. Lawler, "Sequencing Jobs To Minimize Total Weighted Completion Time subject to precedence constraints", *Annals of Discrete Math. 2*, 1978, 75-90.

[LEE90] D. H. Lee and S. M. Reddy, "On Determining Scan Flip-flops In Partial-scan Designs", *Proc. of International Conference on Computer Aided Design*, November 1990.

[MCGE90] P. McGeer, R. Brayton, R. Rudell, and A. Sangiovanni-Vincentelli, "Extended Stuck-fault Testability For Combinational Networks", *Proc. of the 6th MIT Conference on Advanced Research in VLSI*, MIT Press, April 1990.

[MONM77] C. L. Monma and J. B. Sidney, "A General Algorithm For Optimal Job Sequencing With Series-Parallel Constraints", Technical Report No. 347, School of Operations Research and Industrial Engineering, Cornell University, Ithaca, N.Y., July 1977.

[PADH92] V. Padhye, "Upgrading vertices in trees and series-parallel graphs to bound link delays," University of Florida, Sept. 1992.

[PAIK90] D. Paik, S. Reddy, and S. Sahni, "Vertex Splitting In Dags And Applications To Partial Scan Designs And Lossy Circuits", University of Florida, Technical Report, 90-34,1990.

[PAIK91a] D. Paik, S. Reddy, and S. Sahni, "Deleting Verticies To Bound Path Lengths", University of Florida, Technical Report, 91-4, 1991.

[PAIK91b] D. Paik, and S. Sahni, "Upgrading Circuit Modules To Improve Performance", University of Florida, Technical Report, 1991.

[PAIK91c] D. Paik, and S. Sahni, "Upgrading Vertices In Trees, Series-Parallel Digraphs And General Series Parallel Digraphs", University of Florida, Technical Report, 1991.

[PAIK91d] D. Paik, and S. Sahni, "NP-hard Network Upgrading Problems", University of Florida, Technical Report, 1991.

[PAIK93] D. Paik, S. Reddy, and S. Sahni, "Heuristics for the placement of flip-flops in partial scan designs and the placement of signal boosters in lossy circuits", *Proc. VLSI Design '93*, IEEE, 1993.

[SIDN76]   J. B. Sidney, "The Two Machine Flow Line Problem With Series-Parallel Precedence Relations", Working paper 76-19, Faculty of Management Science, University of Ottawa, November 1976.

[VALD79]   J. Valders, R. E. Tarjan, and E. L. Lawler, "The recognition of Series Parallel digraphs", *SIAM J. Comput.*, 11 (1982), pp. 298-313.

Complexity in Numerical Optimization, 323-350
P.M. Pardalos, Editor
©1993 World Scientific Publishing Co.

Efficient Algorithms for δ-Near-Planar Graph and Algebraic Problems [1]

Venkatesh Radhakrishnan   Harry B. Hunt III   Richard E. Stearns
*Department of Computer Science, University at Albany – SUNY, Albany,
NY 12222, USA*

Abstract

For each $\delta > 0$, we introduce a natural generalization of planar graphs called
δ-near-planar graphs. We compare and contrast the δ-near-planar graphs with
the class of graphs of genus $\leq g$ (for any integer $g \geq 0$). We observe that a
number of NP-complete problems, that are polynomial time solvable for planar graphs(more generally for the graphs of genus g), remain NP-complete for
δ-near-planar graphs. We also show that δ-near-planar graphs do not have an
efficient recursively applicable $O(n^r)$ separator theorem for any $r < 1$. However, we show that a number of problems for δ-near-planar graphs, including
polynomial time solvable and NP-hard problems, are solvable in time bounded
by linear functions of the best known bounds on the times of the corresponding
problems for planar graphs. Examples include the following:

1. The problems of solving a δ-near-planar system of linear equations and for
solving the single source shortest path problem, for δ-near-planar graphs, are
solvable using $O(n^{3/2})$ operations.

2. The all-pairs shortest path problem, for δ-near-planar graphs, is solvable
using $O(n^2 \log n)$ operations.

3. The problems 3SAT, max-3SAT and #-SAT, etc. restricted to formulas f
whose associated interaction graphs are δ-near-planar are solvable using only
$|f|2^{O(\sqrt{n})}$ operations and low level polynomial space, where n is the number of
variables of f.

4. The Hamiltonian Circuit problem is solvable and the chromatic polynomial
is computable in $n^{O(\sqrt{n})}$ operations and low level polynomial space for n-vertex
δ-near-planar graphs.

---

[1]Supported in part by NSF Grant CCR 89-03319. A preliminary version of some of the results
of this paper was presented at the 9th STACS, February 1992[20].

Key words: Algorithms and data structures, Systems of Linear Equations, Path Problems, NP-complete, 3SAT, Chromatic Polynomial, Nonserial optimization.

1   Introduction

Throughout this paper, δ is a positive rational; and a "δ-near-planar graph" is a graph with vertex set V presented together with one of its planar layouts with $\leq \delta \cdot |V|$ crossovers of edges. We show that the δ-near-planar graphs are a robust extension of the planar graphs, that have no forbidden subgraphs but are sufficiently structured that various efficient algorithms for planar graphs can be extended to apply to them within the same complexity bounds.

Our motivation for considering δ-near-planar graphs is the following:

1.Planarity of graphs is unstable, in the sense that the addition of a single edge can destroy planarity. The addition of a single edge only causes a δ-near-planar graph to become at most a $(4\delta + 1)$-near-planar graph.

2. Except for path problems for planar graphs and certain grid graphs, planar systems of linear equations seem unnatural. In particular, the planarity of G_A (the graph corresponding to the matrix A, formally defined later) imposes tight constraints on the matrix A. For example, the complete graph on five vertices is not a subgraph of any planar graph. However, for all $\delta > 0$, δ-near-planar graphs can have arbitrarily large cliques as subgraphs (and hence have no forbidden subgraphs).

3.There are several different application areas which naturally yield δ-near-planar graphs that are not planar. These include wide-area communication networks (eg. the ARPA network in [23] is 0.2-near-planar, and the NSFNET backbone in [7] is 0.4-near-planar) and digital circuits laid out on chips. These graphs naturally come with δ-near-planar layouts.

4. The ideas in this paper extend the range of applicability of separator-based techniques in two ways. First for $\delta > 0$, δ-near-planar graphs are not closed under subgraph because a subgraph of size m can inherit more than $\delta \cdot m$ crossovers. Thus, the various recursively applications of separator theorems in [15, 16, 18] do not directly apply. Second by Theorem 12 in [16], the ideas and techniques of [9, 15, 16, 18] are not applicable to any class of graphs with more than linearly bounded numbers of edges. For functions $f(n)$ growing faster than linear in n, n-vertex graphs laid out in the plane with $\leq f(n)$ crossovers of edges can have more than a linear number of edges in n. The proofs of all of our results, for δ-near-planar graphs, are actually uniform in the sum $n + c$ for n vertex graphs laid out in the plane with c crossovers of edges. Thus in contrast, analogues of each of our results for δ-near-planar graphs hold for n vertex graphs laid out in the plane with $\leq f(n)$ crossovers of edges, for all functions $f(n)$.

5. The interaction graphs for a nonserial optimization problem Π [15, 22] applied to planar graphs are **not** usually planar. When this is so, the ideas in [15] using the planar separator theorem to solve such problems are **not** directly applicable. For many problems Π, the interaction graphs of Π applied to planar graphs are easily seen to be δ-near-planar.

6. A number of NP-hard graphs problems remain NP-hard when restricted to δ-near-planar graphs, even when these problems restricted to planar graphs are polynomially solvable. For example, the clique problem, and for $k \geq 4$, the k-coloring problem remain NP-hard, for δ-near-planar graphs.

7. Many reductions between combinatorial problems in the literature [10, 24], especially those involving "local replacement" do not preserve planarity but do preserve near-planarity.

Here, we consider two general types of problems, namely, path problems[11] and generalized satisfiability problems including nonserial optimization problems[3, 22] and their associated counting problems.

We formalize path problems as in [11]. Thus, let F be a field; and let R be a closed semiring [4, 13, 11]. Let A be the $n \times n$ matrix $(a_{i,j})$ over F or R; and let G_A be the directed graph (V_A, E_A), where $V_A = \{x_i | 1 \leq i \leq n\}$ and $E_A = \{(x_i, x_j) | a_{i,j} \neq 0\}$. Let $\mathcal{L}_{A,b}$ be a system of linear equations on F or on R of the form[2]

$$x_i = a_{i,1}x_1 + a_{i,2}x_2 + \cdots + a_{i,n}x_n + b_i \ (1 \leq i \leq n)$$

We study the complexity of the problems **P1**, **P2** and **P3** below for n vertex graphs laid-out in the plane with $c \geq 0$ crossovers of edges.

    P1 : Solve $\mathcal{L}_{A,b}$.

    P2 : Compute the matrix A^{-1}, if A^{-1} exists.

    P3 : Compute the matrix A^*, when A is a matrix over a closed semiring R.

A number of researchers [4, 1, 21, 6, 16, 25, 13, 11, 18] have studied the complexity of the problems **P1**, **P2** and **P3** and the use of these problems in solving path problems for graphs, including both the **single-source** and the **all-pairs shortest path problems** (denoted by sssp and apsp respectively) [1, 11, 18]. In particular, Lipton et al. [16] and Pan et al. [18] have shown how the ideas of [21] and [6] and the planar separator theorem of [15] can be used to solve problem **P1** in $O(n^{3/2})$ operations and to solve problems **P2** and **P3** in $O(n^2 \log n)$ operations. Here, we use the ideas of [21] and [6] to solve problems **P1**, **P2** and **P3**, in $O((n + c)^{3/2})$, $O(n(n+c) \log (n + c))$ and $O(n(n+c) \log (n + c))$ respectively for n vertex graphs G_A laid out on the plane with c crossovers of edges. For δ-near-planar, our results imply that the problems **P1**, **P2** and **P3** are solvable using only $O(n^{3/2})$, $O(n^2 \log n)$ and

---

[2]For fields, the equations can be of the equivalent form

$$a_{i,1}x_1 + a_{i,2}x_2 + \cdots + a_{i,n}x_n = b_i \ (1 \leq i \leq n)$$

$O(n^2 \log n)$ operations, respectively, the same bounds that hold for planar graphs and systems of equations. Exactly analogous results hold for the shortest path problems in [11] and the path expression problems in [25].

We consider the complexity of a wide variety of NP- and #P-hard problems, when restricted to δ-near-planar instances. These problems include 3SAT, max 3SAT, each of the generalized satisfiability problems in [24], many nonserial optimization problems[3, 22], many of the NP-hard problems in [10], the counting versions of these problems, and the problem of computing the chromatic polynomial of a graph. Several of these results were previously unknown, even for planar graphs. To obtain these results, we use a data structure called a "structure tree" which can be used to display variable independence. When given as input, the structure tree controls the order of computation in a way which exploits the displayed independence and reduces the computation time. We identify a measure of subproblem independence called "weighted depth" and describe how a structure tree of weighted depth $O(\sqrt{n})$ can be obtained in $O(n \log n)$ time for a problem whose "interaction graph" is presented along with one of its planar layout with $\leq \delta \cdot |V|$ crossovers of edges. Using this structure tree, we obtain algorithms for solving the problems listed above in time exponential only in the weighted depth and using only polynomial space. As corollaries, we obtain $2^{O(\sqrt{n})}$ or $n^{O(\sqrt{n})}$ time and low-level polynomial space algorithms for solving the problems 3SAT,#3SAT, max-3SAT, Hamiltonian circuit, and computing the Chromatic polynomial for δ-near-planar problem instances or graphs. We also show that identical simultaneous time and space upper bounds hold, for solving a wide collection of nonserial optimization problems with δ-near-planar interaction graphs together with their associated counting problems.

The rest of the paper is organized as follows. Section 2 provides definitions of some important concepts which are used in the rest of the paper. Section 3 contains comparisons of the δ-near-planar graphs with the graphs of genus $g \geq 0$, emphasizing the planar graphs. Section 4 contains algorithms for solving systems of linear equations over a field, for evaluating determinants; and for computing the inverse of a matrix. It also contains algorithms for path and path expression problems, for computing A^* and for solving systems of linear equations over a closed semiring. Section 5 gives efficient algorithms for finding $O(\sqrt{n})$ weighted depth structure trees for δ-near-planar graphs; and it also gives simultaneous $2^{O(\sqrt{n})}$ or $n^{O(\sqrt{n})}$ time and low-level polynomial space algorithms for solving several algebraic and graph problems including their counting versions. Finally in section 6, we present a summary of our results and list some open problems.

2   Preliminaries

We present the definitions of the following concepts needed here: planar layout with crossovers, generalized satisfiability problem, R-formula, interaction graph of an R-formula, structure tree and genus of a graph.

Definition 2.1 *Let $G = (V, E)$ be a graph. A **planar layout with crossovers** for G is a planar graph $G' = (V', E')$ together with a set C of **crossover nodes** and a function $f : E' \to E$ such that:*

1. $C \cap V = \phi$.

2. $V' = V \cup C$.

3. Each node of C has degree 4.

4. For all (a, b) in E, $\{e \in E' | f(e) = (a, b)\}$ is the set of edges on a simple path from a to b in G' involving no other nodes of V.

*A **δ-near-planar graph** is a graph with vertex set V which is presented together with one of its planar layouts with $\leq \delta \cdot |V|$ crossovers of edges.*

*We say that crossover node c is **associated** with edges e_1 and e_2 of E if and only if there are edges e_1' and e_2' in E' with endpoint c such that $e_1 = f(e_1')$ and $e_2 = f(e_2')$.*

Definition 2.2 *Given a set of variables V, an **assignment** on V is a pairing in which each variable v from V is paired with a value in the domain of v. The set of assignments to V will be designated by $\Gamma(V)$. $\Gamma(\phi)$ contains one assignment, namely the empty set of pairs. For any assignment γ, we denote the variables in γ by $VAR(\gamma)$ (i.e. $VAR(\gamma) = V$ if and only if $\gamma \in \Gamma(V)$). If γ_1 and γ_2 are assignments such that $\Gamma(VAR(\gamma_1) \cap VAR(\gamma_2)) = \phi$, we let $\gamma_1 + \gamma_2$ be the assignment in $\Gamma(VAR(\gamma_1) \cup VAR(\gamma_2))$ formed by taking the union of the two assignments.*

Definition 2.3 *Let $R = (S, +, \cdot, 0, 1)$ be a commutative semiring. An **R-term** t is a string of the form $\mathbf{f}(v_1, \ldots, v_k)$, where $\mathbf{f}$ is a k-ary function symbol denoting a k-ary function $f : D_1 \times \ldots \times D_k \to S$ $k \geq 1$ and D_i with $1 \leq i \leq k$ are finite sets and each v_i is a variable or a constant symbol whose type matches the corresponding argument D_i of f. We write $VAR(t)$ for the set of variable symbols occurring in $\{v_1, \ldots, v_k\}$, and we define the **size** of t, denoted by $|t|$, to be $k + 1$. If γ is an assignment such that $VAR(\gamma) \supset VAR(t)$, we define $t[\gamma]$ to be the semiring element $f(d_1, \ldots, d_k)$ where d_i is the value assigned to v_i by γ. If P is a set of R terms, we define $VAR(P) = \bigcup_{p \in P} VAR(p)$.*

*Given R, an **R-formula** F equals (V, P), where V is a finite set of variables, P is a finite set of R-terms, and $V \supset VAR(P)$. The **size** of F denoted by $\|F\|$, equals $|V| + \sum_{p \in P} |p|$. The **generalized satisfiability problem** for R is the problem of computing, given an R-formula $F = (V, P)$, the value of the sum*

$$\sum_{\gamma \in \Gamma(V)} \Pi_{p \in P}\, p[\gamma].$$

By varying the commutative semiring R, many different problems are modellable as generalized satisfiability problems. Such problems include 3SAT, #-SAT, max 3SAT, the generalized satisfiability problems in [24], many nonserial optimization problems[3, 22] and many of the NP-hard problems in [10]. These problems can be solved more efficiently by exploiting subproblem independence, which is displayed by the data structure called structure tree. One measure of subproblem independence in an R-formula or in a structure tree is weighted depth. The concepts of structure trees and weighted depth are defined as follows:

Definition 2.4 *Let* $F = (V, P)$ *be a formula. A* **structure tree** S *for* F *is an ordered triple* (T, α, β) *where*
(1) T *is a rooted tree with node set* N,
(2) $\alpha : V \to N$ *gives the variable association,*
(3) $\beta : P \to N$ *gives the predicate association,*
(4) for all y *in* V *and* p *in* P, y *in* $VAR(p)$ *implies* $\alpha(y)$ *is an ancestor of* $\beta(p)$ *in* T.
For a structure tree S, *we also define the following:*
(5) $A(n) = \{y \in V | \alpha(y) = n\}$, *the variables associated with* n,
(6) $B(n) = \{p \in P | \beta(p) = n\}$, *the predicates associated with* n,
(7) $AD(n)$ *is the union of the* $A(n')$ *such that* n' *is a descendant of* n.
(8) $BD(n)$ *is the union of the* $B(n')$ *such that* n' *is a descendant of* n.
(9) y *in* V *is a* **branch variable** *at node* n *if and only if* $\alpha(y)$ *is an ancestor of* n. *Let* $BV(n)$ *be the set of branch variables at* n.
(10) The **weighted depth of a structure tree**, *denoted by* $WD(S)$ *equals* $\max\{|BV(n)| \mid n \in N\}$.
(11) The **weighted depth of a formula** F *is the minimum of the weighted depth of all of its structure trees.*

Definition 2.5 *Let* $F = (V, P)$ *be a formula. The* **interaction graph** *of* F *is a graph* $G = (V, E)$ *such that* $\{u, v\} \in E$ *if and only if* u *and* v *appear together in some* R-*term of* F.

The structure tree of a graph is analogously defined as follows:

Definition 2.6 *Let* $G = (V, E)$ *be a graph. A* **structure tree** S *for* G *is an ordered triple* (T, α, β) *where*
(1) T *is a rooted tree with node set* N,
(2) $\alpha : V \to N$,
(3) $\beta : E \to N$,
(4) for all $e = \{u, v\} \in E$, $\alpha(v)$ *and* $\alpha(u)$ *are ancestors of* $\beta(e)$ *in* T.

Definition 2.7 *The* **genus** *of a graph* G *is the minimum number of handles which must be added to a sphere so that* G *can be embedded on the resulting surface.*

3   Comparisons with other classes of graphs

We compare the classes of δ-near-planar graphs and of graphs of genus g $(g \geq 0)$. Our first result and its corollary show that these classes are incomparable.

Proposition 3.1 *The problem of determining, given a δ-near-planar graph and an integer $k \geq 1$, if G is of genus k is NP-complete.*

Proof. This problem is known to be NP-complete, for arbitrary graphs[26]. Let $G = (V, E)$ be a graph of genus $g + 1$. Then G can be laid out in the plane in polynomial time with $\leq |E|^2$ crossovers of edges. By adding at most $\lceil |E|^2/\delta \rceil$ isolated vertices to G, we obtain a δ-near-planar graph G' such that the genuses of G and G' are equal. $\square$

Corollary 3.2 *For all $g \geq 0$, the class of δ-near-planar graphs contain graphs of genus g.*

Proof. Immediate from the proof of Proposition 3.1. $\square$

Simple padding techniques also imply that a number of NP-complete problems remain NP-complete, when restricted to δ-near-planar instances. The next theorem gives several examples, each of which is polynomial time solvable when restricted to planar instances.

Theorem 3.3 *The problems Clique, Graph k-coloring($k \geq 4$), Crossing Number, and Not-All-Equal 3SAT[3] are NP-complete, when restricted to δ-near-planar instances.*

Since δ-near-planar graphs allow arbitrary size cliques, we observe that they do not have any forbidden subgraphs. Finally, we recall from [15, 9] that, for all $g \geq 0$, the class of graphs of genus g has a recursively applicable $O(\sqrt{n})$ separator theorem. For example for planar graphs, Lipton and Tarjan[15] have shown the following:

Theorem 3.4 *Let G be any n-vertex planar graph. The vertices of G can be partitioned into three sets A, B, C, such that no edge joins a vertex in A with a vertex in B, neither A nor B contains $2n/3$ vertices, and C contains no more than $2\sqrt{2}\sqrt{n}$ vertices.*

Theorem 3.4 is applicable recursively to planar graphs, since the class of planar graphs is closed under subgraph. In contrast, a subgraph of a δ-near-planar graph need not be δ-near-planar, and in addition, δ-near-planar graphs can contain arbitrarily large cliques. Thus, the class of δ-near-planar graphs does not have a recursively applicable n^r separator theorem, for any $r < 1$.

Planar graphs have the property that it is possible to destroy planarity by adding just one edge, eg. the clique on 5 vertices itself is not planar, but removing one edge makes it planar. A δ-near-planar graph $G = (V, E)$ has a layout such that there are $\leq \delta|V|$ crossovers of edges. The layout itself is a planar graph. Hence, it has $\leq 3(\delta + 1)|V| - 6$ edges. Since, each edge of G consists of some collection of edges in the layout, the number of edges in G is also $\leq 3(\delta + 1)|V| - 6$. Thus any new edge

---

[3]Moret[17] has shown that Not-All-Equal 3SAT is polynomial time solvable for planar instances.

added can cross each of these edges and introduce $\leq 3(\delta + 1)|V| - 6$ crossovers, so that the resulting graph is at most $4(\delta + 1)$-near-planar.

Thus, we observe that δ-near-planar graphs are vastly different from planar graphs and from graphs of fixed genus and are closer to arbitrary graphs as far as hardness results are concerned. However, in sections 4 and 5, we show that their structure can be exploited to obtain many algorithms that are as efficient as the best known algorithms for the corresponding problems on planar graphs.

4 Path problems and systems of linear equations

In this section, we discuss systems of linear equations over fields and closed semirings and their applications to path problems, when these problems are restricted to instances which are δ-near-planar.

4.1 Solution of a system of linear equations over a field

As in [12], we use LDU decomposition to solve by Gaussian elimination the system of linear equations $Ax = b$. Here A is an $n \times n$ matrix, x is an $n \times 1$ vector of variables and b is an $n \times 1$ vector of constants. The solution process consists of two steps. First, we factor A by means of row operations into $A = LDU$ where L is lower triangular, D is diagonal and U is upper triangular. Second, we solve the simplified systems $Lz = b$, $Dy = z$ and $Ux = y$.

The LDU decomposition is found as follows: Let $A_0 = A = \begin{bmatrix} a_{1,1} & r_1 \\ c_1 & B_1 \end{bmatrix}$,

where r_1 is a $1 \times (n-1)$ vector and c_1 is an $(n-1) \times 1$ vector and B_1 is an $(n-1) \times (n-1)$ matrix.

Then $A_0 = \begin{bmatrix} 1 & 0 \\ c_1/a_{1,1} & I \end{bmatrix} \begin{bmatrix} a_{1,1} & 0 \\ 0 & A_1 \end{bmatrix} \begin{bmatrix} 1 & r_1/a_{1,1} \\ 0 & I \end{bmatrix}$ where $A_1 = B_1 - c_1 r_1/a_{1,1}$.

This is the system obtained by eliminating x_1; and its graph is the graph obtained by removing vertex v_1 and joining every pair of vertices adjacent to v_1. The LDU decomposition is obtained one column at a time of L, one diagonal element at a time of D and a row at a time of U. The elimination of the variable x_i proceeds as follows

Let $A_{i-1} = \begin{bmatrix} a_{i,i}^{i-1} & r_i \\ c_i & B_i \end{bmatrix}$. Then $A_{i-1} = \begin{bmatrix} 1 & 0 \\ c_i/a_{i,i}^{i-1} & I \end{bmatrix} \begin{bmatrix} a_{i,i}^{i-1} & 0 \\ 0 & A_i \end{bmatrix} \begin{bmatrix} 1 & r_i/a_{i,i}^{i-1} \\ 0 & I \end{bmatrix}$

where $A_i = B_i - c_i r_i/a_{i,i}^{i-1}$.

Corresponding to the factorization $A = LDU$ is the graph $G^* = (V, E^*)$ such that $\{v_i, v_j\} \in E^*$ iff $i > j$ and $l_{i,j}$ is non-zero or $i > j$ and $u_{i,j}$ is non-zero.

For a given order of elimination of variables [4] (henceforth called an elimination order), the fill-in(A) is the set of edges of G^* which are not edges of G_A.

---

[4]As commonly assumed in the literature, we assume that no pivoting is necessary, because otherwise the fillin could be increased.

We now solve a system of equations whose non-zero structure corresponds to a δ-near-planar graph. We use the fact shown below that the fill-in for δ-near-planar graphs is not too much more than the fill-in for planar graphs.

Let $Ax = b$ be the system to be solved. Let the graph $G = (V, E)$ correspond to its non-zero structure. Let $G' = (V \cup C, E')$ be its δ-near-planar layout. Recall that G' is a planar graph.

Apply the planar separator theorem to G' to obtain the sets A', B', C'. C' is a separator for G' such that there are no edges in E' from A' to B'. However if we consider the vertices of V in A' and the vertices of V in B', there may be edges in E between them.

Repeat the following for every crossover in C': Let c be a crossover in C' corresponding to edges (v_1, v_2) and (v_3, v_4) of E. If the two vertices of an edge in E are in A' and B', move one of them to C'.

Consider $v_1 \in A'$ and $v_2 \in B'$ where $v_1, v_2 \in V$. If $(v_1, v_2) \in E$, then there is a path in E' between v_1 and v_2 consisting of no other nodes of V. Since there are no edges of E' between vertices in A' and vertices in B', there must be a path of crossover nodes from v_1 to v_2 passing through C', i.e. there is a crossover $c \in C'$ on this path. However, if $c \in C'$ was a crossover, then one of the vertices of each of the edges crossing over at c would be in C', i.e. either v_1 or v_2 would be in C'. Therefore there is no edge in E between a vertex in A' and a vertex in B'. Hence we have the following theorem:

Theorem 4.1 *Let $G = (V, E)$ be presented with a near-planar layout with crossovers $G' = (V \cup C, E')$. Let $n = |V|$ and $c = |C|$. Then, a partition of $V \cup C$ into sets A', B', C' and planar layouts $\mathcal{A}'$ of $A' \cap V$ and $\mathcal{B}'$ of $B' \cap V$ can be found in $O(|V \cup C|)$ time such that*

1. $|A'|, |B'| \leq \frac{2}{3}|V \cup C| = \frac{2}{3}(n + c)$

2. $|C'| \leq 6\sqrt{2}\sqrt{|V \cup C|} = 6\sqrt{2}\sqrt{n + c}$

3. There are no edges in E or in E' between vertices in A' and vertices in B'.

The separator of Theorem 4.1 guarantees that the number of vertices in the individual parts is smaller than the original graph or that the individual parts have fewer crossovers than the original graph or both. It generalizes the planar separator theorem to graphs laid out in the plane with crossovers. For δ-near-planar graphs it gives a $O(\sqrt{n})$ separator. We note again that δ-near-planar graphs are not closed under subgraph and can have arbitrarily large cliques. Hence, this is **not** a recursively applicable $O(\sqrt{n})$ separator theorem. It is, however, recursively applicable to the layout of G, i.e. the planar graph G'. We now give the algorithm to obtain the elimination order.

Algorithm 4.1 (Numbering algorithm) Given a graph $G = (V, E)$ along with its layout with crossovers $G' = (V \cup C, E')$, we number the vertices of V from $a = 1$ to $b = n$ recursively as follows:

If $|V \cup C| \leq n_0$, number the vertices of V arbitrarily.

Otherwise, find sets A', B', C' as described above. Let the number of unnumbered vertices of V in A', B', C' be i, j, k respectively. Number the vertices in $C' \cap V$ arbitrarily from $b - k + 1$ to b. Delete all edges of E with both endpoints in C' (Consequently some edges of E' may have to be deleted). Apply the algorithm recursively to the layout $B' \cup C'$ to number the unnumbered variables of $B' \cap V$ from $b - k - j + 1$ to $b - k$. Apply the algorithm recursively to the layout $A' \cup C'$ to number the unnumbered variables of $A' \cap V$ from a to $a + i - 1$.

Theorem 4.2 *Algorithm 4.1 takes $O(n \log n)$ time to number an n-vertex δ-near-planar graph with n vertices.*

Proof. Let the layout of a near-planar graph $G = (V, E)$ be $G' = (V \cup C, E')$. Let $|V| = n$, $|V \cup C| = n'$. We note that $n' \leq (1 + \delta)n$. The time taken to number the graph can be expressed by the following recurrence relation

$$t(n, n') \leq \begin{cases} c_1(n') + \max\{t(n_1, n_1') + t(n_2, n_2')\} & \text{if } n' > n_0 \\ c_0 & \text{if } n' \leq n_0 \end{cases}$$

where $n_1 + n_2 \leq n$; $n' \leq n_1' + n_2' \leq n' + 6\sqrt{2}\sqrt{n'}$
$\frac{1}{3}n' \leq n_i' \leq \frac{2}{3}n' + 6\sqrt{2}\sqrt{n'}$ $1 \leq i \leq 2$

Here n' is the number of nodes in the layout and n is the number of nodes in the graph being considered. The individual layouts for $A' \cup C'$ and $B' \cup C'$ have sizes n_1' and n_2' respectively. The number of unnumbered vertices of V in $A' \cup C'$ and $B' \cup C'$ are n_1 and n_2 respectively. The solution to this recurrence relation is $t(n, n') = O(n' \log n')$ which is proved by a technique similar to that used in [16]. $\square$

Theorem 4.3 *The fill-in associated with the numbering of Algorithm 4.1 for δ-near-planar graphs is $O(n \log n)$.*

Proof. Suppose that the numbering algorithm is applied to an n-vertex near-planar graph with a layout with c crossovers with l vertices previously numbered. Assume $n' = n + c > n_0$ and let A', B', C' be the vertex partition generated by the algorithm. The maximum number of fill-in edges whose lower numbered endpoint is in C' is

$$|C'|(|C'| - 1)/2 + |C'|l$$

The maximum number of vertices of the graph in C' can be no more than $4\sqrt{2}\sqrt{n'}$. Thus the maximum number of fill-in edges whose lower numbered endpoint is in C' is no more than

$$(4\sqrt{2})^2 n'/2 + 4\sqrt{2}\sqrt{n'}l$$

Two vertices v and w are joined by a fill-in edge iff there is a path from v to w through vertices numbered less than both v and w. Thus no fill-in edge joins a vertex in A' with a vertex in B'. Let $f(l, n, n')$ be the maximum number of fill-in edges whose lower numbered endpoint is numbered by the algorithm. We obtain the following recurrence relation

$$f(l, n, n') \leq \begin{cases} n(n-1)/2 & \text{if } n' \leq n_0 \\ (4\sqrt{2})^2 n'/2 + 4\sqrt{2}\, l\sqrt{n'} + \max\{f(l_1, n_1, n_1') + f(l_2, n_2, n_2')\} & \text{if } n' > n_0 \end{cases}$$

where the maximum is taken over values satisfying

$$l_1 + l_2 \leq l + 8\sqrt{2}\sqrt{n'}$$

$$n \leq n_1 + n_2 \leq n + 4\sqrt{2}\sqrt{n'}$$

$$n_i \leq 2/3n' + 4\sqrt{2}\sqrt{n'} \quad 1 \leq i \leq 2$$

$$1/3n' \leq n_i' \leq 2/3n' + 6\sqrt{2}\sqrt{n'}$$

$$n' \leq n_1' + n_2' \leq n' + 6\sqrt{2}\sqrt{n'}$$

The solution to this recurrence relation is

$$f(l, n, n') = c_1(l + n')\log n' + c_2 l\sqrt{n'}$$

which is proved by a technique similar to that used in [16]. We note that $n' \leq (1+\delta)n$. The required fillin is $f(0, n, n')$. Hence the result. $\square$

Theorem 4.4 *An system of linear equations $\mathcal{L}_{A,b}$ with δ-near-planar graph G_A can be solved in time $O(n^{3/2})$.*

Proof. The number of operations required for the elimination of x_i is the product of the number of non-zero elements in the ith row and the ith column of A_{i-1}. Thus a bound on the number of operations associated with a separator C' generated by one call of the recursive numbering algorithm is

$$\sum_{i=1}^{4\sqrt{2}\sqrt{n'}} (i + l)(i + l) \leq c_1 n'^{3/2} + c_2 l n' + c_3 l^2 \sqrt{n'} + c_4 n' + c_5 l\sqrt{n}$$

Let $g(l, n, n')$ be the number of operations for performing the elimination when l vertices are already numbered, n variables are to be eliminated and the layout for this system has n' vertices. Then the following recurrence is satisfied

$$g(l, n, n') \leq \begin{cases} c_6 n^3 & \text{if } n \leq n' \leq n_0 \\ c_1 n'^{3/2} + c_2 l n' + c_3 l^2 \sqrt{n'} + c_4 n' + c_5 l\sqrt{n'} & \text{otherwise} \\ \quad + \max\{g(l_1, n_1, n_1') + g(l_2, n_2, n_2')\} \end{cases}$$

where the maximum is taken over values satisfying

$$l_1 + l_2 \leq l + 8\sqrt{2}\sqrt{n'}$$

$$n \leq n_1 + n_2 \leq n + 4\sqrt{2}\sqrt{n'}$$

$$n_i \leq 2/3n' + 4\sqrt{2}\sqrt{n'} \quad 1 \leq i \leq 2$$

$$1/3n' \leq n_i' \leq 2/3n' + 6\sqrt{2}\sqrt{n'}$$

$$n' \leq n_1' + n_2' \leq n' + 6\sqrt{2}\sqrt{n'}$$

The solution to this recurrence relation is

$$g(l, n, n') = c_7(n')^{3/2} + c_8 ln' + c_9 l^2 \sqrt{n'}$$

which is proved by a technique similar to [16]. Thus, since $n' \leq (1 + \delta)n$, the total time to eliminate n variables is therefore $g(0, n, n') = O(n^{3/2})$. The substitution takes $O(n \log n)$ time, since there are $O(n \log n)$ elements in L and U and $O(n)$ elements in D. Hence, the overall time to solve a linear system of equations over a field given its layout with c crossovers is $O(n^{3/2})$. $\square$

A^{-1} is found by solving n systems of the form $Ax = b$ where b corresponds to an $n \times 1$ vector of zeros except for the ith element which is a 1 which gives the ith column of A^{-1}. The advantage of solving these systems by LDU decomposition is that the elimination is done once and the solution just involves substitution n times.

Corollary 4.5 *The inverse of an $n \times n$ matrix given the δ-near-planar layout of the graph corresponding to it can be found in $O(n^2 \log n)$.*

Proof. The elimination takes time $O(n^{3/2})$ and each of the n substitutions take $O(n \log n)$ time since the fill-in is $O(n \log n)$. $\square$

4.2   Solution of a system of linear equations over a closed semiring

Following [4], we solve the system of linear equations $\mathcal{L}$ over a closed semiring $\{R, +, \cdot, *, 0, 1\}$ as in [13], of the form

$$x = A \cdot x + b \tag{1}$$

as follows: First, we decompose A into upper and lower triangular matrices U and L over R such that $A^* = U^* \cdot L^*$. Second, we solve the coupled system of equations

$$y = L \cdot y + b \tag{2}$$

$$x = U \cdot x + y \tag{3}$$

The operations used to obtain U and L very closely resemble those used to obtain the LDU decomposition of a matrix A over a field F. This close resemblance enables us to make the two observations below on the relative fill-ins and the operation counts on solving problem **P1** over a field F and over a closed semiring R.

The triangular decomposition is obtained as follows: Writing $M^{(0)} = A$, we compute $M^{(k)}$ successively as

$$m_{i,j}^{(k)} = \begin{cases} m_{i,k}^{(k-1)} (m_{k,k}^{(k-1)})^* & for\ k < i \leq n, j = k \\ m_{i,j}^{(k-1)} + m_{i,k}^{(k-1)} \cdot (m_{k,k}^{(k-1)})^* \cdot m_{k,j}^{(k-1)} & for\ k < i, j \leq n \\ m_{i,j}^{(k-1)} & otherwise \end{cases}$$

L is the strictly lower triangular part of $M^{(n)}$ and U is the upper triangular part of $M^{(n)}$

In comparison, the operations performed in an LDU decomposition are as follows:

$$m_{i,j}^{(k)} = \begin{cases} m_{i,k}^{(k-1)} / m_{k,k}^{(k-1)} & for\ k < i \leq n, j = k \\ m_{k,j}^{(k-1)} / m_{k,k}^{(k-1)} & for\ k < j \leq n, i = k \\ m_{i,j}^{(k-1)} - m_{i,k}^{(k-1)} / m_{k,k}^{(k-1)} \cdot m_{k,j}^{(k-1)} & for\ k < i, j \leq n \\ m_{i,j}^{(k-1)} & otherwise \end{cases}$$

Theorem 4.6 *The fill-in obtained in the triangular decomposition of a matrix over a closed semiring is no more than the fill-in obtained in the LDU decomposition of a matrix A' over a field such that A and A' have the same non-zero structure, provided the same elimination order is used.*

Proof. The fill-in corresponds to the elements of $M^{(n)}$ which are non-zero when the corresponding elements of A are zeros. The operation which contributes to fill-in in the triangular decomposition is the one for the case $k \leq i, j \leq n$, and the fill-in occurs when $m_{i,j}^{(k-1)}$ is zero, $m_{i,k}^{(k-1)}$ is non-zero and $m_{k,j}^{(k-1)}$ is non-zero.

Similarly, the operation which contributes to the fill-in in the LDU decomposition is the one corresponding to the case $k \leq i, j \leq n$, and the fill-in occurs when $m_{i,j}^{(k-1)}$ is zero, $m_{i,k}^{(k-1)}$ is non-zero and $m_{k,j}^{(k-1)}$ is non-zero. Thus the operations in the triangular decomposition have corresponding operations in the LDU decomposition and a fill-in occurs in the triangular decomposition only if a fill-in occurred in the LDU decomposition. Hence the theorem. $\square$

Theorem 4.7 *The number of operations required to obtain the triangular decomposition of a matrix over a closed semiring is no more than the number of operations required for the LDU decomposition of a matrix A' over a field such that A and A' have the same non-zero structure, provided the same elimination order is used.*

Proof. If instead of the fill-in, we are interested in estimating the operation count for obtaining the triangular decomposition, (where the count is only for operations involving non-null elements), we find that a non-trivial operation occurs in the triangular decomposition only if a non-trivial operation occurs in the LDU decomposition. Hence the result. $\square$

The above two theorems enable us to directly translate our results in section 3 for problem **P1** for systems of linear equations over a field to those for problem **P1** for systems of linear equations over a closed semiring as follows.

Corollary 4.8 *A system of linear equations $\mathcal{L}_{A,b}$ over a closed semiring with δ-near-planar graph G_A can be solved in $O(n^{3/2})$ time.*

Shortest path problems have the closed semiring $(\Re, min, +, \infty, 0)$ associated with them.

The single source shortest path problem corresponds to solving a system of equations of the form $x = A \cdot x + b$ where A corresponds to the distance matrix and b is a vector of zeros except for a 1 in the position corresponding to the source. Thus the corollaries obtained earlier can be used to obtain $O(nk^2)$ and $O(n^{3/2})$ algorithms when the graph is a treewidth k graph or a δ-near-planar graph respectively.

The all pairs shortest path problem corresponds to solving n systems of equations of the form $x = A \cdot x + b$ with the same distance matrix A but with different b corresponding to the different sources. Thus apsp can be solved in $O(n^2 \log n)$ when the graph is a δ-near-planar graph.

A^* is found by solving n systems of the form $x = A \cdot x + b$ where b corresponds to an $n \times 1$ vector of zeros except for the ith element which is a 1 which gives the ith column of A^{-1}. Hence the time to compute A^* for a matrix A whose corresponding graph is δ-near-planar is $O(n^2 \log n)$.

The results obtained above can also be applied to the path expression problems in [25] and to the path algebra problems in [11] using the appropriate closed semiring to obtain equivalent results.

5   Structure trees, Weighted Depth, Generalized Satisfiability

This section has three parts. In the first part, we present an efficient algorithm that, given an n-variable formula $F = (V, P)$ along with a δ-near-planar layout of its interaction graph, constructs a structure tree of weighted depth $O(\sqrt{|V|})$. In the second part, we show how to solve an instance of a generalized satisfiability problem, given a structure tree for the instance of weighted depth WD, in number of steps exponential only in WD. When applied to n variable instances with δ-near-planar interaction graphs, these algorithms yield the upper bounds of Table 1 in section

6, for the problems 3SAT,#3SAT, max-3SAT, and more generally for all Nonserial Optimization Problems as well as their counting versions. In the third part, we apply our results to problems for δ-near-planar graphs, defined in terms of predicates corresponding to edges of the graph or to neighborhoods of vertices in the graph.

5.1   Finding a good structure tree

Theorem 5.1 *Let a graph $G = (V, E)$ along with a layout $G' = (V \cup C, E')$ be given. Then a structure tree for G of weighted depth $O(\sqrt{|V \cup C|})$ can be found in $O(|V \cup C| \log |V \cup C|)$ time.*

Proof. We first construct a structure tree S with $T = (N, F)$ for G' as follows: Apply the planar separator theorem to the planar graph G' to obtain the partitions A', B', C'. Construct root node r and set $\alpha(v) = r$ for v in C'. Apply this idea recursively to obtain a tree and an α for both A' and B' and attach these trees to r. Given this α, the function β is given by $\beta(e) = glb(\alpha(v_1), \alpha(v_2))$ in T for $e = \{v_1, v_2\} \in E'$. Each application of the planar separator theorem takes time linear in the size of the graph G' and hence in the size of G. The time taken to obtain the structure tree can be obtained by the recurrence relation:

$$t(n') \leq \begin{cases} c_1 \cdot n' + \max\{t(n'_1) + t(n'_2)\} & \text{if } n' > n_0 \\ c_0 & \text{if } n' \leq n_0 \end{cases}$$

where $n'_1 + n'_2 \leq n'$ and $\frac{1}{3}n' \leq n'_i \leq \frac{2}{3}n'$ $1 \leq i \leq 2$

The solution to this recurrence relation is obtained to be $t(n') = O(n' \log n')$ by a technique similar to [16]. Hence, the structure tree of G' can be found in time $O(|V \cup C| \log |V \cup C|)$. For all nodes n of N, $WD(n) = |\{y \in V \cup C | \alpha(y) = m$, where m occurs on the path from n to the root of $T \}|$

$$\leq 2\sqrt{2}\sqrt{|V \cup C|} + 2\sqrt{2}\sqrt{\tfrac{2}{3}|V \cup C|} + \ldots + 2\sqrt{2}\sqrt{\tfrac{2}{3}^{c \log |V \cup C|}|V \cup C|}$$
$$\leq 2\sqrt{2}\sqrt{|V \cup C|}(1 + \sqrt{\tfrac{2}{3}} + (\sqrt{\tfrac{2}{3}})^2 + \ldots)$$
$$\leq \frac{2\sqrt{2}\sqrt{|V \cup C|}}{1 - \sqrt{\tfrac{2}{3}}}$$

Hence, $WD(S) = \max_{n \in N} WD(n) = O(\sqrt{|V \cup C|})$.

Starting with the structure tree, $S = (T, \alpha, \beta)$ for G', we obtain a structure tree $S' = (T, \alpha', \beta')$ for G as follows: Let (u, v) be an edge of G and $\alpha(u) = x$ and $\alpha(v) = y$. If x and y lie on the same path from the root, then no changes are needed. Otherwise, find a vertex z such that z is an ancestor of both x and y and change the assignment $\alpha(u)$ (or $\alpha(v)$) so that $\alpha(u) = z$. Repeat the last step for all other (u, v) such that $\alpha(u) = x$ and $\alpha(v) = y$ do not lie on the same path from the root. Since moving an assignment up the tree only increases the number of edges in E such that their endpoints lie on the same path from the root, this iteration must stop. Let α' be the revised α restricted to the nodes of G. For $e = \{v_1, v_2\} \in E$,

$\beta'(e) = glb(\alpha'(v_1), \alpha'(v_2))$. We claim the following :

Claim : For each x in T, $|\{v \in V | \alpha'(v) = x\}| \leq 2|\{v \in V \cup C | \alpha(v) = x\}|$.

Proof of claim: The number assigned to a node x is increased only if x has been assigned a node of C. Each node c of C (being of degree 4) can only attract two nodes from V. Since node c will be deleted in restricting α to α', the overall effect is to replace c by at most two nodes.

The lemma now follows by observing that the structure tree $S = (T, \alpha, \beta)$ of G' of weighted depth $O(\sqrt{|V \cup C|})$ was obtained in time $O(|V \cup C| \log |V \cup C|)$ time and that the transformation of S to obtain $S' = (T, \alpha', \beta')$ can be performed in time linear in $|V \cup C|$ increasing the weighted depth by only a constant factor. $\square$

We obtain the following corollary for δ-near-planar graphs.

Corollary 5.2 *A structure tree for an δ-near-planar graph $G = (V, E)$ of weighted depth $O(\sqrt{|V|})$ can be found in $O(|V| \log |V|)$ time.*

In the case of R-formulas whose interaction graphs are δ-near-planar, we can obtain a similar lemma.

Lemma 5.3 *Given a formula $F = (V, P)$, along with a layout $G' = (V \cup C, E')$ of its interaction graph $G = (V, E)$, a structure tree of weighted depth $O(\sqrt{|V \cup C|})$ for F can be found in time $O(|F| + |V \cup C| \log |V \cup C|)$.*

Proof. The structure tree $S = (T, \alpha, \beta)$ of G can be obtained in $O(|V \cup C| \log |V \cup C|)$ time by using the construction of lemma 5.1. We claim the following:

Claim : Let $p(v_1, \ldots, v_k)$ be an R-term of F. There is a branch of T from the root containing all $\alpha(v_i)$ $1 \leq i \leq k$.

Proof of claim: Let there exist two variables v_i and v_j $1 \leq i, j \leq k$, such that they do not occur on the same branch. However since v_i and v_j occur in p, there is an edge between v_i and v_j in the interaction graph G. Therefore $\alpha(v_i)$ and $\alpha(v_j)$ are ancestors of $\beta(\{v_i, v_j\})$. Hence $\alpha(v_i)$ and $\alpha(v_j)$ do lie on the same branch from the root. Thus proving the claim by contradiction.

The structure tree (T, α, β) can now be made into a structure tree (T, α, β') for $F = (V, P)$ by defining for $p_i = \{u_1, \ldots, u_{d_i}\} \in P$, $\beta'(e) = glb(\{\alpha(v_j) | 1 \leq j \leq d_i\})$ Obtaining the structure tree of G takes time $O(|V| \log |V|)$. Obtaining β' can be done in $O(|F|)$ time. $\square$

We now obtain a structure tree with a small number of nodes, since this number plays an important role in algorithms which use structure trees.

Theorem 5.4 *Given the interaction graph $G = (V, E)$ for a formula $F = (V, P)$ along with one of its layouts $G' = (V \cup C, E')$, a structure tree $S = (T, \alpha, \beta)$ for F of weighted depth $O(\sqrt{|V \cup C|})$ can be obtained in $O(|F| + |V \cup C| \log |V \cup C|)$ time, such that $T = (N, F)$ and $|N| \leq 2 \cdot |V|$ and $|N| \leq 2 \cdot |P|$.*

Proof. From Lemma 5.3, a structure tree $S = (T, \alpha, \beta)$ of weighted depth $O(\sqrt{|V \cup C|})$ for F can be found in time $O(|F| + |V \cup C| \log |V \cup C|)$. By the construction of Lemma 5.3, for each $p \in P$, $\beta(p)$ is the lowest node in $\{\alpha(v)|v \in VAR(p)\}$. For each $v \in V$, if $\alpha(v)$ is not the least common ancestor of $\{\beta(p)|v \in VAR(p)\}$, move $\alpha(v)$ lower in the tree so that it is the case. This can be done in $O(|F|)$ time, since finding least common ancestors can be done for a set can be done in time linear in the size of the set after a preprocessing time of linear in the size of the tree. The size of the tree obtained from Lemma 5.1 is $O(|V \cup C|)$. Thus this entire step can be done in $O(|F| + |V \cup C| \log |V \cup C|)$ time. While the root r of T has only one child and $A(r) = B(r) = \phi$, then remove r and make the child of r the root. Traverse the tree T top down and for each node do the following: If node n with parent n_0 has at most one child and $A(n) = B(n) = \phi$, then remove n from the tree and make the child of n(if any) connect directly to n_0. Since the original tree has $O(|V \cup C|)$ nodes, this step can be done in $O(|V \cup C|)$ time. Hence the entire procedure can be executed in $O(|F| + |V \cup C| \log |V \cup C|)$ time.

Observe that the branch variables at any node n in T do not increase. Also, if $n = \alpha(v)$ for some $v \in V$, then $n = \beta(p)$ for at least one $p \in P$. Similarly, if $n = \beta(p)$ for some $p \in P$, then $n = \alpha(v)$ for at least one $v \in V$. Hence every node either has two children or it has the property that $A(n) \neq \phi$ and $B(n) \neq \phi$.

In any tree, at least half the nodes have at most one child. In this case, each such node is associated with at least one v in V and one p in P. Hence the result follows, since the transformation to the smaller structure tree can be done in time linear in $|T| + |G|$. □

The following corollary follows immediately for a formula $F = (V, P)$ presented along with an α-near-planar layout of its interaction graph.

Corollary 5.5 *Given an α-near-planar layout of the interaction graph $G = (V, E)$ for a formula $F = (V, P)$, a structure tree $S = (T, \alpha, \beta)$ for F of weighted depth $O(\sqrt{|V|})$ can be obtained in $O(|F| + |V| \log |V|)$ time, such that $T = (N, F)$ and $|N| \leq 2 \cdot |V|$ and $|N| \leq 2 \cdot |P|$.*

5.2 Solving Generalized Satisfiability Problems

The concepts in the preceding subsection are actually combinatorial concepts having nothing to do with the interpretation or meaning of the predicates. (They depend only on the memberships in the sets $VAR(p)$.) Next, we introduce concepts which connect structure trees with efficient formula evaluation.

Definition 5.6 *Let $S = (T, \alpha, \beta)$ be a structure tree for formula $F = (V, P)$, let n be a node of T, and let γ be any assignment such that (1) $VAR(\gamma) \cap AD(n) = \phi$ and (2) $VAR(\gamma) \cup AD(n) \supset VAR(BD(n))$. Then $E(n, \gamma)$ is the semiring member defined by*
$$E(n, \gamma) = \Sigma_{\gamma' \in \Gamma(AD(n))} \Pi_{p \in BD(n)} p[\gamma + \gamma'].$$

We note that the two conditions of the definition are necessary and sufficient for the expression to be well defined. Condition (1) is needed so that $VAR(\gamma') \cap VAR(\gamma) = \phi$ as required in Definition 2.2. Condition (2) is needed so that $VAR(\gamma+\gamma') \supset VAR(p)$ for all predicates p in the expression as required in Definition 2.3. The value of $\sum_{\gamma \in \Gamma(V)} \Pi_{p \in P} p[\gamma]$ is identical to $E(r, \phi)$ where r is the root of the structure tree.

The formula in the above definition suggests an exhaustive method for computing $E(n, \gamma)$. The next result shows that $E(n, \gamma)$ can be computed from certain values associated with the children of n. This key result is needed to prove that the generalized backtracking algorithm is correct.

Theorem 5.7 *If F, S, γ, and n are as defined in Definition 5.6 and node n has k children $n_1, \ldots n_k$, then $E(n, \gamma) = \sum_{\gamma' \in \Gamma(A(n))}(\Pi_{p \in B(n)} p[\gamma + \gamma']) \cdot \Pi_{i=1}^{k} E(n_i, \gamma + \gamma')$*

Proof. We first prove that the $p[\gamma + \gamma']$ in the given expression are well defined. We know that $VAR(\gamma') = A(n) \subset AD(n)$ and so $VAR(\gamma') \cap VAR(\gamma) = \phi$ follows from (1) of the Definition 5.6 and thus $\gamma + \gamma'$ is well defined. Because $B(n) \subset BD(n)$, (2) of the Definition 5.6 implies $VAR(p) \subset VAR(\gamma + \gamma')$. Thus $p[\gamma + \gamma']$ is well defined.

Now suppose n_i and γ satisfy the constraints of Definition 5.6. Because $AD(n_i) \subset AD(n)$ and $AD(n_i) \cap A(n) = \phi$, Condition (1) for n_i *and γ follows from Condition (1) for n and γ*. From Condition (2) for *n* and γ, and $BD(n_i) \subset BD(n)$, one can derive Condition 2 for n_i and γ.

Now that the right-hand side of the equation in the theorem has been shown to be well defined, we will prove the equality by induction. If n has no children, $A(n) = AD(n)$, $B(n) = BD(n)$, and the equation is identical to Definition 5.6.

Finally, to prove the inductive step, substitute Definition 5.6 into both sides of the equation in the theorem. Applying the distribution and commutative laws to move products inside of sums, and using the two identities given below, the right-hand side can be routinely (but tediously) transformed into the left-hand side. If V_1 and V_2 are disjoint sets of variables and P_1 and P_2 disjoint sets of predicates such that $V_i \supset VAR(P_i)$, the first identity is

$$\sum_{\gamma_1 \in \Gamma(V_1)} \sum_{\gamma_2 \in \Gamma(V_2)} \Pi_{p \in P_1 \bigcup P_2} p[\gamma_1 + \gamma_2] = \sum_{\gamma \in \Gamma(V_1 \bigcup V_2)} \Pi_{p \in P_1 \bigcup P_2} p[\gamma].$$

This identity holds because γ is $\Gamma(V_1 \bigcup V_2)$ if and only if there is γ_1 in $\Gamma(V_1)$ and γ_2 in $\Gamma(V_2)$, such that $\gamma = \gamma_1 + \gamma_2$. For any γ_1 in $\Gamma(V_1)$ and γ_2 in $\Gamma(V_2)$, the second identity is

$$(\Pi_{p \in P_1} p[\gamma_1]) \cdot (\Pi_{p \in P_2} p[\gamma_2]) = \Pi_{p \in P_1 \bigcup P_2} p[\gamma_1 + \gamma_2].$$

This identity holds because $p[\gamma] = p[\gamma_1 + \gamma_2]$ for all p in P_i, the value of p being independent of the values assigned to variables not in VAR(p). $\square$

We now present an algorithm for generalized satisfiability problems. It requires as input a formula F and a structure tree for F. It uses backtracking to evaluates particular assignments by considering all extensions and then backs up to try the next partial assignment. The Generalized Backtracking algorithm is as follows:

Algorithm 5.1 (EVALUATE (n) {n is node of structure tree})
    x, y are local variables.
    {Formula variables are global.
    γ_0 *in* $\Gamma(BV(n) - A(n))$ is already constructed from earlier calls.
    Procedure computes $E(n, \gamma_0)$ from Def. 5.6}

```
    x ← 0;
    FOR γ in Γ(A(n)) DO
    BEGIN
        y ← 1;
        FOR p in B(n) DO
            y ← y · p[γ + γ₀];
        FOR ALL CHILDREN n' of n DO
            y ← y · EVALUATE(n');
        x ← x + y;
    END;
    RETURN(x);
```

Lemma 5.8 *When computation is initiated by calling EVALUATE(r) where EVAL-UATE is defined in Algorithm 5.1 and r is the root of a structure tree, then the following hold for each node n:*

1. *when EVALUATE(n) is called, variables from BV(n)-A(n) have an assignment* γ_0

2. *EVALUATE(n) returns* $E(n, \gamma_0)$

3. *EVALUATE(n) is called* $|\Gamma(BV(n) - A(n))|$ *times.*

4. *Summed over all calls, statements in the outer loop are executed* $|\Gamma(BV(n))|$ *times.*

5. *EVALUATE(r) returns* $\sum_{\gamma \in \Gamma(V)} \Pi_{p \in P} \, p[\gamma]$.

Proof. Statement (1) is proved by induction starting from root r. $BV(r) = A(r)$ by definition so $BV(n) - A(n)$ is empty and trivially has an assignment. The assignment to $BV(n) - A(n)$ is expanded to an assignment to $BV(n)$ when the procedure is applied to child n' of n. But $BV(n)$ is $BV(n') - A(n')$ (immediate from Definition 3.2) and so (1) holds.

Statement (2) is now immediate because Algorithm 5.1 is a straightforward implementation of the formula in Theorem 5.7.

Whenever (3) is true, (4) must also be true since the outer loop is executed $|\Gamma(A(n))|$ time per call and multiplying the number of calls given is (3) by $|\Gamma(A(n))|$ gives (4).

Statement (3) can now be proven by induction. It is true of the root r since EVALUATE(r) is called once and BV(r)-A(r), being empty, has exactly one assignment. Now assume (3) and hence (4) is true for node n and consider any child n' of n. $EVALUATE(n')$ is called only from $EVALUATE(n)$ and is done once each time the procedure goes through its outer loop. By (4) this happens $|\Gamma(BV(n))|$ times and $BV(n) = BV(n') - A(n')$. Thus (3) holds for n'.

Statement (5) is immediate since $E(r, \phi)$ is $\sum_{\gamma \in \Gamma(V)} \Pi_{p \in P} p[\gamma]$ by Definition 5.6. $\square$

From Lemma 5.8, we can bound the number of operations used to solve a generalized satisfiability problem instance as follows:

Theorem 5.9 *Let $F = (V, P)$ be a formula where each variable in V takes on at most D values. Let T be a structure tree for F having m nodes and weighted depth WD. If the procedure EVALUATE is used with F and T, then*

1. *the number of "$\cdot$" operations is at most $(m + |P|) \cdot D^{WD}$;*

2. *the number of "$+$" operations is at most $m \cdot D^{WD}$;*

3. *each p in P is evaluated at most D^{WD} times;*

Proof. We consider part (3) first. A given predicate p is only evaluated when EVALUATE is called at node $\beta(p)$. The evaluation of p is done in the outer loop and so is done $|\Gamma(BV(n))|$ times by Lemma 5.8(4). This quantity is no greater than D^{WD} and part (3) is proven.

The "$\cdot$" operation is performed once for each predicate evaluation and once for every procedure call (except the original call). From part (3), there are at most $|P| \cdot D^{WD}$ predicate evaluations and from Lemma 5.8(3) there are at most $m \cdot D^{WD}$ procedure calls.

The "$+$" operation is done once each time through the outer loop. By Lemma 5.8 (4) this is at most D^{WD} per node and hence (2). $\square$

For δ-near-planar instances, we obtain the following:

Theorem 5.10 *Given the interaction graph $G = (V, E)$ for a formula $F = (V, P)$ along with its δ-near-planar layout $G = (V \cup C, E')$, the generalized satisfiability problem can be solved in time $|F| D^{O(\sqrt{|V|})}$ time, when each variable in V takes on at most D values. The space required is $O(|F| \log D)$.*

Proof. By Theorem 5.4, a structure tree $S = (T, \alpha, \beta)$ for F of weighted depth $O(\sqrt{|V|})$ can be obtained in $O(|V \cup P| \log |V \cup P|)$ time, such that $T = (N, F)$ and $|N| \leq 2 \cdot |V|$ and $|N| \leq 2 \cdot |P|$. Thus the bounds in Theorem 5.9 can be applied with $WD = O(\sqrt{|V|})$, $m \leq 2 \cdot |V|$ and $m \leq 2 \cdot |P|$. Thus the bounds in (1) and (2) can be described as $3 \cdot |P| \cdot D^{O(\sqrt{|V|})}$ and $2 \cdot |V| D^{O(\sqrt{|V|})}$ or $2 \cdot |P| \cdot D^{O(\sqrt{|V|})}$.

Under the unit cost assumptions, the time of the algorithm is $(|F| \cdot D^{O(\sqrt{|V|})})$, the largest contribution being the cost of evaluating predicates at k units per single k-ary function evaluation.

The space requirement is very minimal. In addition to the space needed to store the structure tree, the procedure has a global variable for each variable in V, where each such variable must store a corresponding domain value, and two local variables x and y which must store any semiring elements generated by the procedure. The structure tree and the two variables x and y can be stored using $O(|F|)$ space. Assuming $O(\log D)$ space per domain value, the required space bound is obtained. $\square$

To illustrate the range of applicability of Theorems 5.9 and 5.10, we formulate some algebraic problems as generalized satisfiability problems and use Theorem 5.10 to solve them efficiently.

3SAT: Given a set V of variables, collection C of clauses over V such that each clause $c \in C$ has $|c| \leq 3$, determine if there is a satisfying truth assignment for C.
This problem can also be formulated as follows:
Variables: V.
Domains: $\{F, T\}$.
Semiring: $(\{F, T\}, \cup, \cap, F, T)$(i.e. the Boolean semiring)
Predicates: These are the clauses of C. The predicate p_j corresponding to clause c_j is T iff at least one of the literals in the clause c_j is true.

max-3SAT: Given a set V of variables, collection C of clauses over V such that each clause $c \in C$ has $|c| \leq 3$, determine the maximum number of clauses satisfied by any truth assignment.
This problem can also be formulated as follows:
Variables: V.
Domains: $\{0, 1\}$.
Semiring: $(\{0, 1, 2, \ldots, |C|\}, max, +, 0, 0)$.
Predicates: These are the clauses of C. The predicate p_j corresponding to clause c_j is 1 iff at least one of the literals in the clause c_j is true and 0 otherwise.

#-3SAT: Given a set V of variables, collection C of clauses over V such that each clause $c \in C$ has $|c| \leq 3$, determine the number of satisfying truth assignments for C.
This problem can also be formulated as follows:
Variables: V.
Domains: $\{0, 1\}$.
Semiring: $(N, +, *, 0, 1)$.
Predicates: These are the clauses of C. The predicate p_j corresponding to clause c_j is 1 iff at least one of the literals in the clause c_j is true and 0 otherwise.

Corollary 5.11 *The problems 3SAT, max-3SAT and #-3SAT for a 3CNF formula f with n-variables, presented with a δ-near-planar layout of its interaction graph, can be solved in time $|f| \cdot 2^{O(\sqrt{n})}$ using only $O(|f|)$ space.*

Proof. Follows from Theorem 5.10 by observing that each variable in V takes on at most 2 values. $\square$

Planar-3SAT [14] is the restriction of the problem 3SAT to instances, whose bipartite graphs are planar. (The **bipartite graph** of a 3CNF formula f is defined as follows: The clauses and variables in the formula f are in one to one correspondence with the vertices of the graph. There is an edge between a clause node and a variable node iff the variable appears in the clause.)

Corollary 5.12 *The problems Planar-3SAT, max-Planar-3SAT and #-Planar-3SAT for n-variable 3CNF formulas f, can be solved in time $|f| \cdot 2^{O(\sqrt{n})}$ using only $O(|f|)$ space.*

Another method for solving #-Planar-3SAT appears in [19].

Next, we consider nonserial optimization problems and their associated counting problems defined as follows: Let F be a finite set of finite arity functions f_i with finite domains.

Nonserial Optimization Problems: Given a set $V = \{x_1, x_2, \ldots, x_n\}$ of variables each of which has a finite domain D_i, assign values to $x_1, x_2, \ldots, x_n$ so as to minimize $f(x_1, x_2, \ldots, x_n) = \sum f_i(X_i)$, where $X_i \subseteq \{x_1, x_2, \ldots, x_n\}$ and f_i is a function over the variables of X_i.

This problem can also be formulated as follows:

Variables: $V = \{x_1, x_2, \ldots, x_n\}$.

Domains: D_i.

Semiring: $(N, min, +, \infty, 0)$.

Predicates: These are the terms $f_i(X_i)$. The value of the predicate is the value of $f_i(X_i)$.

Nonserial Optimization Counting Problems: Given a set $V = \{x_1, x_2, \ldots, x_n\}$ of variables each of which has a finite domain D_i, find the number of assignments of values to $x_1, x_2, \ldots, x_n$ which minimize $f(x_1, x_2, \ldots, x_n) = \sum f_i(X_i)$, where $X_i \subseteq \{x_1, x_2, \ldots, x_n\}$ and f_i is a function over the variables of X_i. We define $|f| = \sum_i |X_i|$.

This problem can also be formulated as follows:

Variables: $V = \{x_1, x_2, \ldots, x_n\}$.

Domains: D_i.

Semiring: $((N, N), +, *, (\infty, 0), (0, 1))$ where

$$
\begin{aligned}
(i, n_i) + (j, n_j) \quad &= (i, n_i) \qquad \text{if } i < j \\
&= (j, n_j) \qquad \text{if } i > j \\
&= (i, n_i + n_j) \quad \text{if } i = j \\
(i, n_i) * (j, n_j) \quad &= (i + j, n_i * n_j)
\end{aligned}
$$

Predicates: These are the terms $f_i(X_i)$. The value of the predicate is $(f_i(X_i), 1)$.

The above formulations along with Theorem 5.10 gives the following theorem:

Theorem 5.13 *Both the optimization and the counting versions of any nonserial optimization problem presented with a δ-near-planar layout of its interaction graph, can be solved using number of operations (function calls, semiring additions and multiplications) $|f| \cdot D^{O(\sqrt{n})}$, where D is the size of the largest domain.*

When the functions calls take time and space proportional to the number of parameters, the algorithm runs in $|f| \cdot D^{O(\sqrt{n})}$, using $O(|f| \log D)$ space.

5.3 Graph Problems

We now consider problems for δ-near-planar graphs. First, we consider problems which can be expressed as generalized satisfiability problems with predicates corresponding to edges of the graph, so that the resulting interaction graph has the same structure as the original graph.

Hamiltonian Circuit: Given a graph G = (N, E), find a Hamiltonian circuit (if any).

This problem can be formulated preserving structure as follows:

Variables: v_n for each node n in N.

Domains: v_n has domain $\{1, \ldots, |N|\} \times E_n$ where E_n is the set of edges adjacent to node n.

Semiring: $\{F, T\}$.

Predicate: $p_{(a,b)}(v_a, v_k)$ for $\{a, b\}$ in E where, if v_a has value (k_a, e_a) and v_b has value (k_b, e_a), and the predicate is true if and only if

$$e_a = (a, b) \implies [k_b \equiv k_a + 1(mod|N|)] \wedge [e_b = (a, b) \implies k_a \equiv k_b + 1(mod \ |N|)]$$

Intuitively, v_a is assigned (n_a, e_a) if node a is the n_a-th node on the circuit and e_a is the edge which leads to the next node. The predicate for e_a verifies that the edge for e_a leads to the $n_a + 1$-th node.

#k-colorability: Given a graph $G = (N, E)$ and an integer k, determine the number of ways the nodes of G can be colored with k colors so that no two adjacent nodes have the same colors.

The problem can also be stated as follows:

Variables: v_n for all nodes n in N

Domains: $\{0, \ldots, k - 1\}$.

Semiring: $(N, +, *, 0, 1))$.

Predicates: For all edges (v_i, v_j) in E, the predicates take on the value 1 if $v_i \neq v_j$ and 0 otherwise.

Note the direct correspondence between nodes and variables and between edges and predicates.

Chromatic Polynomial: Given a graph $G = (N, E)$, determine the polynomial $P_G(\lambda)$ such that for any $0 \leq \lambda \leq n$, $P_G(\lambda)$ gives the number of ways of coloring the nodes of G using at most λ colors.

Theorem 5.14 *The problems Hamiltonian Circuit, #k-colorability and Chromatic Polynomial for an n-vertex graph $G = (V, E)$, presented with a δ-near-planar layout can be solved in time $|G| \cdot n^{O(\sqrt{n})}$, $|G| \cdot k^{O(\sqrt{n})}$, and $|G| \cdot n^{O(\sqrt{n})}$ respectively, using space polynomial only in $|G|$.*

Proof. Observe that these graph problems can be expressed as generalized satisfiability problems such that the predicates corresponding to edges of the graph. Hence, the interaction graph of these problems is identical to the original graph. By Theorem 5.10, given the δ-near-planar layout of the interaction graph for a formula $F = (V, P)$, the generalized satisfiability problem can be solved in time $|F|D^{O(\sqrt{|V|})}$ time, when each variable in V takes on at most D values. In the case of Hamiltonian Circuit, each variable can take $|V| \times |E|$ values. Since for a δ-near-planar graph $|E| = O(|V|)$, this problem can be solved in $|G| \cdot n^{O(\sqrt{n})}$ time. In the case of the #k-colorability, each variable can take k values. Hence, this problem can be solved in $|G| \cdot n^{O(\sqrt{n})}$ time. The Chromatic polynomial can be found by finding the number $c(\lambda)$ of λ colorings for $0 \leq \lambda \leq n$ and finding the polynomial passing through $(\lambda, c(\lambda))$ for $0 \leq \lambda \leq n$ by interpolation. Since the time is dominated by the time to find the $c(\lambda)$, this problem can be solved in $|G| \cdot n^{O(\sqrt{n})}$ time. Thus the theorem follows. $\square$

We now consider problems on bounded degree δ-near-planar graphs which can be expressed as generalized satisfiability problems with predicates corresponding to neighborhoods of vertices. By the **neighborhood of a vertex**, we mean vertices and edges within a specified distance of the vertex. We illustrate this technique for Graph Grundy numbering and Dominating set.

Graph Grundy Numbering: Given a directed graph $G = (V, E)$, find a function $f : V \to \{1, \ldots, |V|\}$ such that for each $v \in V$, $f(v)$ is the least non-negative integer not contained in the set $\{f(u)|u \in V, (v, u) \in E\}$.

This problem can also be formulated as follows:

Variables: V.

Domains: $\{1, \ldots, |V|\}$.

Semiring: $(\{F, T\}, \cup, \cap, F, T)$.

Predicates: There is one predicate for each vertex $v \in V$. Its variable set is the set of vertices $\{v\} \cup \{u \in V|(v, u) \in E\}$. The predicate is true if the value assigned to v is the least non-negative integer not assigned to one of the vertices in the set $\{u \in V|(v, u) \in E\}$.

Edge k-coloring: Given a graph $G = (N, E)$ and an integer k, determine if the edges of C can be colored with k colors so that the edges adjacent to any given n in N all have different colors.

Here it is natural to associate variables with edges instead of nodes and to formulate the problem as follows:

Variables: v_e for all edges e in E.

Domains: $\{1, \ldots, k\}$

Semiring: $(\{F, T\}, \cup, \cap, F, T)$.

Predicates: There is one predicate for each vertex $v \in V$. Its variable set is the set of edges incident on v. i.e. $\{e \in E | v \in e\}$. The predicate is true if all the edges incident on v are assigned different values.

Lemma 5.15 *Let Π be a graph problem expressible as a generalized satisfiability problem, such that there is one predicate for each vertex and the variable set of the predicate for a vertex is a subset of the neighborhood of the vertex. Then, the structure tree of weighted depth $O(\sqrt{|V|})$ for the predicate graph for the problem Π on bounded degree δ-near-planar graphs $G = (V, E)$, can be found in $O(n \log n)$ time.*

Proof. We first obtain a structure tree $S = (T = (N, F), \alpha, \beta)$ for G of weighted depth $O(\sqrt{|V|})$ can be obtained in $O(|V| \log |V|)$ time, by Corollary 5.2. Let the predicates corresponding to vertices $v_1, v_2, \ldots, v_n$ be $p_1, p_2, \ldots, p_n$ where each p_i corresponds to a predicate on the neighborhood of v_i. This neighborhood could contain both vertex variables and edge variables. However, since the graph is of bounded degree and the neighborhood is of a fixed distance, the number of variables in the predicate is constantly bounded. Obtain the structure tree for the formula as follows:
Define $\beta'(p_i) = \alpha(v_i)$.
Define $\alpha'(v) = lca(\{p_j | v \text{ occurs in } p_j\})$.
Consider the structure tree $S = (T, \alpha', \beta')$: For any node $n \in N$, the set of branch variables is the set of vertices which were neighbors of some $v_j \in A(n)$. Since these graphs are of bounded degree, the set of variables which are in the neighborhood of any vertex is also bounded. Thus the set of branch variables of any node of the tree can only increase by a constant factor. Thus the lemma follows. $\square$

Theorem 5.16 *The Graph Grundy Numbering and Edge k-coloring problems, for n-vertex bounded degree δ-near-planar graphs $G = (V, E)$, can be solved in time $|G| \cdot 2^{O(\sqrt{n})}$.*

Proof. By Lemma 5.15, given the δ-near-planar layout of the graph, then structure tree of weighted depth $O(\sqrt{|V|})$ for the R-formula for the problem instance can be obtained for Graph Grundy Numbering and Edge k-coloring. Hence by Theorem 5.10, Graph Grundy Numbering can also be solved in $c^{O(\sqrt{n})}$ where c is the maximum degree of any vertex. Similarly, Edge k-coloring can be solved in $k^{O(\sqrt{n})}$. $\square$

Lipton and Tarjan's[15] results hold only for graph problems when the interaction graph itself is planar. However in the case of predicates over neighborhoods of vertices even for bounded degree planar graphs, the interaction graphs are **not** usually planar, but they are δ-near-planar. This is one way, our results extend the range of applicability of the results of [15], even for planar graph problems. Finally, we note that our ideas directly apply to many problems for planar hypergraphs with hyperedges of bounded arity. The interaction graphs of these problems are not usually planar but are again δ-near-planar. Consequently again, the ideas and techniques in [15] do not directly apply to these problems.

6 Summary and Open problems

For each $\delta > 0$, we introduced the class of δ-near-planar graphs. We showed that they are a robust extension of the planar graphs with the following properties:

1. The δ-near-planar graphs have no forbidden subgraphs.

2. Many efficient algorithms for problems for planar graph can be extended to yield efficient algorithms for the corresponding problems for δ-near-planar graphs.

A synopsis of our results appears in Table 1.

Table 1

Problem	δ-*Near-planar graphs*
Systems of equations over a field	$O(n^{3/2})$
Systems of equations over a closed semiring	$O(n^{3/2})$
sssp	$O(n^{3/2})$
apsp	$O(n^2 \log n)$
A^{-1}	$O(n^2 \log n)$
A^*	$O(n^2 \log n)$
3SAT	$\lvert f \rvert \cdot 2^{O(\sqrt{n})}$
#3SAT	$\lvert f \rvert \cdot 2^{O(\sqrt{n})}$
max-3SAT	$\lvert f \rvert \cdot 2^{O(\sqrt{n})}$
Hamiltonian circuit	$n^{O(\sqrt{n})}$
Chromatic Polynomial	$n^{O(\sqrt{n})}$
Nonserial optimization Problems (optimization or counting)	$\lvert f \rvert \cdot D^{O(\sqrt{n})}$

Some interesting questions suggested by our results include the following:

1. What additional efficiently solvable problems, for planar graphs, are also efficiently solvable for δ-Near-planar graphs? In particular, do Frederickson's [8] $O(n\sqrt{\log n})$, $O(n^2)$ and $O(n \log n)$ algorithms for sssp, apsp and maximum flow, respectively, generalize to δ-near-planar graphs.

2. For which NP-hard and #P-hard graph problems Π, do our techniques yield simultaneous $2^{O(\sqrt{n})}$ and $n^{O(\sqrt{n})}$ time and polynomial space bounds on the problem Π restricted to planar and δ-near-planar instances? Candidate problems are those in Bodlaender's [5] classes ECC and LCC.

3. Can Baker's [2] PTAS for planar graph problems be extended to apply to δ-near-planar graph problems?

References

[1] A.V.Aho, J.E.Hopcroft, J.D.Ullman, "The Design and Analysis of Computer Algorithms", Addison-Wesley, 1974.

[2] B.S. Baker, "Approximation algorithms for NP-complete problems on planar graphs", 24th FOCS, IEEE, pp. 265-273, 1983.

[3] U. Bertele and F. Brioschi,"Nonserial Dynamic Programming", Academic Press, New York, 1972.

[4] R.C.Backhouse and B.A.Carre,"Regular algebra applied to pathfinding problems", J. Inst. Math. Appl., 15, pp.161-186, 1974.

[5] H.L. Bodlaender, "Dynamic programming on graphs with bounded treewidth," Technical Report RUU-CS-87-22, Department of Computer Science, University of Utrecht, Utrecht, the Netherlands, 1987.

[6] J.R.Bunch and D.J.Rose,"Partitioning, tearing, and modification of sparse linear systems",J. Math. Anal. Appl. 48,pp.574-593, 1974.

[7] D.E.Comer,"Internetworking with TCP/IP",vol.I, Prentice-Hall,1991.

[8] G.N. Frederickson, "Shortest path problems in planar graphs ", Proc. 24th FOCS, pp. 242-247,1983.

[9] J.R. Gilbert, J.P. Hutchinson, and R.E. Tarjan, "A separator theorem for graphs of bounded genus," J. Algorithms, vol. 5, pp. 391-407, 1984.

[10] M.R.Garey and D.S.Johnson, "Computers and Intractability: A Guide to the Theory of NP-Completeness", W.H.Freeman and Company,1979.

[11] M.Gondran and M.Minoux,"Graphs and Algorithms",John Wiley,1984.

[12] I.N.Herstein and D.J.Winter,"Matrix Theory and Linear Algebra", Macmillan, 1988.

[13] D.J.Lehmann,"Algebraic structures for transitive closure", Theor. Comp. Sc., vol. 4, pp.59-76, 1977.

[14] D.Lichtenstein, "Planar formulae and their uses," SICOMP, vol. 11,2, pp.329-343, May 1982.

[15] R.L.Lipton and R.E.Tarjan, "Applications of a planar separator theorem", SICOMP, vol.9, pp.615-629, 1980.

[16] R.J.Lipton, D.J.Rose and R.E.Tarjan,"Generalized nested dissection", SIAM J. Numer. Analysis 16(2),pp.346-358(1979).

[17] B.M.E. Moret, "Planar NAE3SAT is in P," SIGACT NEWS, No 19, pp 51-54, 1988.

[18] V.Pan and J.Reif,"Parallel nested dissection for path algebra computations", Operation research letters 5(4),pp.177-184,1986.

[19] S.S. Ravi and H.B. Hunt,III, "An application of the planar separator theorem to counting problems", Information Processing Letters 25,pp.317-321,1987.

[20] V.Radhakrishnan, H.B.Hunt,III and R.E.Stearns,"Efficient algorithms for solving systems of linear equations and path problems", Proceedings of the 9th Annual Symposium on Theoretical Aspects of Computer Science, LNCS 577, Springer-Verlag, pp.109-119, 1992.

[21] D.J.Rose,"A graph-theoretic study of the numerical solution of sparse positive definite systems of linear equation",Graph Theory and Computing,R.Read,ed., Academic Press,pp.183-217, 1972.

[22] A. Rosenthal, "Dynamic programming is optimal for nonserial optimization problems," SICOMP, vol. 11, pp. 47-59, 1982.

[23] M.Schwartz,"Computer-communication network design and analysis",Prentice Hall, 1977.

[24] T.J. Schaefer, "The complexity of satisfiability problems," Proc. 10th Annual ACM Symposium on Theory of Computing, pp. 216-226, 1978.

[25] R.E.Tarjan, "A unified approach to path problems", J. ACM 28(3),pp.594-614, 1981.

[26] C.Thomassen, "The graph genus problem is NP-complete", J. of Algorithms 10, pp. 568-576, 1989.

Complexity in Numerical Optimization, pp. 351-386
P.M. Pardalos, Editor
©1993 World Scientific Publishing Co.

Parametric Flows, Weighted Means of Cuts, and Fractional Combinatorial Optimization

Tomasz Radzik
School of Operations Research, Cornell University, Ithaca, NY 14853, USA

Abstract

We consider Newton's method for the linear fractional combinatorial optimization and prove a strongly polynomial bound on the number of iterations. We also consider Newton's method for the maximum mean-weight cut problem, which is an instance of the linear fractional combinatorial optimization. This problem is closely related to the parametric flow problem. We prove that Newton's method solves both these problems in $O(m)$ iterations, and their uniform versions in $\Theta(n)$ iterations, where m and n denote the number of arcs and the number of nodes in the input graph. One iteration is dominated by the standard maximum flow computation. Our bounds improve the previous best bound for the parametric flow problem and the maximum mean-weight cut problem by factor n^2/m, and for the uniform versions of these problems by factor m/n.

Keywords: Fractional programming, parametric network flow, mean-weight cut, minimum-ratio, strongly polynomial.

1 Introduction

Many problems in combinatorial optimization fit the following pattern. Find a combinatorial structure with the maximum (or minimum) cost, where the cost of a structure is the sum of the costs of its elements. We call such problems *linear combinatorial optimization problems*. For example, the minimum spanning tree problem, the minimum cost cycle problem, and the minimum cut problem are linear combinatorial optimization problems. The structures here are the spanning trees, the cycles, and the cuts of a given graph, respectively. The elements of these structures are edges of the graph. It is often the case, however, that besides costs there are also weights associated with the individual elements, and the task is to find a structure with the maximum (or minimum) mean-weight cost. The mean-weight cost of a structure is equal to its cost divided by its weight, where the weight of a structure is the sum of the weights of its elements. We call such a problem a *linear fractional combinatorial optimization problem* (an LFCO problem). Names "0-1 fractional programming problems" and "minimum-ratio problems" are also used in the literature. The minimum-ratio spanning tree problem [2], the minimum-ratio cycle problem [19], and the maximum mean-weight cut problem [20] are examples of LFCO problems. In this paper we consider Newton's method for linear fractional combinatorial optimization in general and for the maximum mean-weight cut problem in particular. We discuss the maximum mean-weight cut problem in connection with the closely related parametric flow problem.

The input instance of the *maximum mean-weight cut problem* is a network with capacity $u(e)$ and weight $b(e)$ on each arc e, and demand $d(v)$ at each node v. Negative $d(v)$ means that node v has supply $-d(v)$. The surplus of a (directed) cut is the difference between the net demand across this cut and the capacity of this cut. The mean-weight surplus of a cut is its surplus divided by its weight, the sum of the weights of its arcs. The goal is to find a cut with the maximum mean-weight surplus. This problem appears, for example, in the context of the minimum cost flow problem [9, 29]. For other motivations we turn to its dual, the parametric flow problem. The *parametric flow problem* is to find the minimum value of parameter δ such that the input network with capacity $u(e) + \delta b(e)$ on each arc e and demand $d(v)$ at each node v is feasible (*i.e.*, all demands can be satisfied without violating the capacities). The maximum mean-weight cut problem and the parametric flow problem can be reduced to each other in time which is negligible in comparison with the best known upper time bounds for these problems.

Consider the following scenario as motivation for the parametric flow problem. We want to ship the commodity from the sources - the nodes with supply, to the sinks - the nodes with demand. The initial capacities of arcs, however, are too tight to satisfy all demand. Increasing the capacity of an arc e by one unit requires $b(e)$ time, but we can simultaneously work on increasing the capacities of all arcs. By how much should we increase the capacity of each arc to make the network feasible in the shortest possible time? This is precisely the parametric flow problem. We face this problem also when it is desirable not to use an arc at its full capacity. Each arc e has some safety level $b(e)$ and the ratio $f(e)/b(e)$ is sought to be minimized, with the general objective of getting the maximal ratio as small as possible (here $u(e)$'s are equal to zero). Another situation captured by the parametric flow problem has dynamic nature. The network is in continuous activity, that is, the sources continuously keep producing commodity and sending it to the sinks. The objective is to maximize the throughput. More formally, suppose $b(e)$ is the amount of the commodity which can be shipped through an arc e in unit time. It means that the shipment of $f(e)$ units of the commodity through an arc e takes $f(e)/b(e)$ time. Further suppose that every source s generates $-d(s)$ units of the commodity every τ units of time, and every sink t consumes $d(t)$ units of the commodity every τ units of time. The problem of computing the minimum possible τ such that no congestion arises at any node can be reduced to the parametric flow problem.

We consider the following algorithm for solving the parametric flow problem and the maximum mean-weight cut problem. We try to satisfy as much demand as possible without violating the initial capacities $u(e)$'s. If not all demand has been satisfied, then a cut with the maximum surplus has been found. We increase the capacities of all arcs in such a way that the surplus of this cut becomes zero. Then we again try to satisfy the demand, but now using the new, bigger, capacities. This process is repeated until the capacities are big enough so all demand can be satisfied. Since we increase the capacities only by as much as we have to, the final capacities give the solution to the parametric flow problem. Moreover, the cut found in the last iteration has the maximum mean-weight surplus in the initial network.

McCormick and Ervolina [20] proposed this algorithm and analyzed it for the uniform version of the maximum mean-weight cut problem, the *maximum mean-cut problem* (all weights are equal to 1). The general scheme lying behind this algorithm can be applied to any linear fractional combinatorial optimization problem. In fact, assuming even wider context, the scheme turns out to be a well known method for the general fractional optimization. Dinkelbach [7] was probably the first who introduced this method, so it is sometimes called the Dinkelbach method. It is best known, however, as the Newton-Raphson method or simply Newton's method for fractional optimization, since it follows the pattern of Newton's root finding technique. To be able to give a comprehensive summary of our results concerning Newton's method and to compare them with the known complexities of other methods, we shall first provide a few more details about linear fractional combinatorial optimization.

An LFCO problem has the following parametric version. The cost of an element is a linear function with parameter δ. The cost of a structure is, as previously, the sum of the costs of its elements, but now this is some linear function of δ. When δ is fixed, the resulting costs of the elements are called *reduced costs*. The maximum cost over all structures is a piecewise linear, convex, decreasing[1] function of δ. We denote this function by $h(\delta)$. We want to compute δ^*, the minimum value of δ for which the maximum reduced cost of a structure is zero, that is, we want to compute the root of $h(\delta)$. We also want to find a structure with this maximum cost. Such a structure turns out to have the maximum mean-weight cost. Computing $h(\delta)$ for a fixed δ amounts to finding a structure with the maximum reduced cost. This is an instance of the *underlying linear (non-parametric) combinatorial optimization problem*. We assume that we have a procedure for the underlying linear problem. We treat this computation as a black box. The parametric version of the maximum mean-weight cut problem is the parametric flow problem. The underlying linear problem is equivalent to the problem of finding a minimum capacity cut.

---

[1] Monotonicity follows from additional assumptions, stated in Section 3.1.

Newton's method for an LFCO problem is Newton's method for finding the root of the corresponding function $h(\delta)$. We prove that Newton's method solves every LFCO problem in a strongly polynomial number of iterations. It means that our bound on the number of iterations is independent of the sizes of the input numbers (the costs and the weights of individual elements) and depends polynomially on the number of the input numbers. This result is somewhat surprising, because function $h(\delta)$ can consist of a superpolynomial number of linear pieces [1].

An immediate corollary of the above result is that Newton's method applied to the maximum mean-weight cut problem gives a strongly polynomial algorithm, since both the number of iterations and the running time of one iteration are strongly polynomial. To obtain a better bound on the number of iterations in this case, we use the combinatorial structure of the underlying network. We show that the number of iterations is $O(m)$ and, since one iteration roughly amounts to one standard maximum flow computation, the overall running time of the algorithm is $O^*(m^2 n)$.[2] Parameters n and m denote the number of nodes and the number of arcs in the input network, respectively. Our bound improves the previous best strongly polynomial bound, due to Megiddo and his parametric search method [22], by factor n^2/m. We also provide a specialized analysis of Newton's method for the maximum mean cut problem. We show that in this case the number of iteration is $\Theta(n)$ and the overall running time is $O^*(mn^2)$. This bound improves the bound shown by McCormick and Ervolina [20] by factor m/n. In the next few paragraphs we compare Newton's method and our bounds with the other two general methods for linear fractional optimization: the binary search and Megiddo's parametric search.

The binary search method for an LFCO problem is simply the bisection method for finding the root of $h(\delta)$. In each iteration we solve an instance of the underlying linear optimization problem to compute the sign of $h(\bar{\delta})$, where $\bar{\delta}$ is the current approximation of the root. If all input numbers are integral, then the binary search method finds the precise value of the root in a polynomial number of iterations. More precisely, the number of iterations is $O(\log(pU))$, where p is the number of the input numbers and U is the biggest absolute value of an input number. Thus, the binary search method can give polynomial algorithms but cannot give strongly polynomial ones. For example, the maximum mean-weight cut problem can be solved with the binary search method in $O^*(mn \log(nU))$ time. Here U is the biggest absolute value of the capacities and demands. We show a bound on the number of iterations in Newton's method which is never greater than $O(\log(pU))$, and is $o(\log(pU))$ if the weights are small in comparison with the costs.

Another method for solving an LFCO problem is Megiddo's parametric search [22], which works as follows. We run an algorithm $\mathcal{A}$ for the underlying linear optimization problem to compute $h(\delta^*)$. Not knowing in advance what is the value of δ^*, we maintain it as an unknown variable and perform arithmetic operations symbolically.

---

[2]Notation $O^*()$ hides factor $\log^c n$ for some constant c. For all bounds in this paper presented in this form, $c \leq 3$.

We assume that every comparison in $\mathcal{A}$ is between two linear functions of the input variables. An algorithm having this property is called a linear algorithm. Thus, whenever there is a comparison "$x : y$" in $\mathcal{A}$, we have to compare $x(\delta^*)$ with $y(\delta^*)$, where $x(\delta)$ and $y(\delta)$ are linear functions of δ. We do not have to know δ^* to resolve such a comparison. It is enough to determine the position of δ^* relatively to the intersection point $\bar{\delta}$ of $x(\delta)$ and $y(\delta)$. This is equivalent to computing the sign of $h(\bar{\delta})$, and can be done by solving an appropriate instance of the underlying linear optimization problem. At the completion of algorithm $\mathcal{A}$, the information gathered about δ^* enables us to compute its precise value by solving one additional instance of the underlying linear optimization problem. If T is the running time of algorithm $\mathcal{A}$, then the total number of instances of the linear optimization problem solved in Megiddo's method is $O(T)$, so it depends on the complexity of the underlying linear problem. Moreover, as we mentioned above, the method requires a linear algorithm for the underlying linear problem, while both Newton's method and the binary search method work with any algorithm. Cohen and Megiddo [4, 5, 6] and Norton, Plotkin, and Tardos [23] extended Megiddo's parametric search method to problems with a fixed number of parameters. Toledo [35] further extended the method by showing how to deal with the situation when $\mathcal{A}$ is a "polynomial" algorithm, that is, the only functions of the input variables $\mathcal{A}$ is allowed to evaluate are bounded degree polynomials.

The bound $O(T)$ on the number of linear subproblems in Megiddo's parametric search method can be reduced by exploring the parallelism in algorithm $\mathcal{A}$. Megiddo [21, 22] applied the parametric search to a number of LFCO problems obtaining efficient algorithms whenever there is an efficient parallel algorithm for the underlying linear problem. Most notably, he obtained an $O(m(\log n)^2 \log \log n)$ algorithm for the minimum-ratio spanning tree problem, and an $O^*(n^3)$ algorithm for the minimum-ratio cycle problem. For the parametric network flow problem, however, he obtained only an $O^*(n^3 m)$ algorithm, since the underlying linear problem, the maximum flow problem, does not have an efficient parallel solution.

We should stress that the subproblems solved in the binary search method and in Megiddo's parametric search method are instances of a "weak" version of the underlying linear optimization problem: find a structure with a positive cost. Newton's method requires a procedure which finds a structure with the maximum cost. This difference can sometimes be quite dramatic. Consider for example the maximum mean-weight cycle problem (the maximum-ratio cycle problem). The binary search method and Megiddo's parametric search method need a procedure for determining if there is a cycle with a positive reduced cost. This computation can be done in $O(mn)$ time. Newton's method needs a procedure for finding a cycle with the maximum cost, and this task is NP-hard. The maximum mean-weight cut problem is an example of the other extreme. Both versions of the underlying linear problem, deciding if a given network is feasible and finding the maximum flow, currently have the same upper time bounds. Newton's method does work with a procedure for the weak version of the underlying linear optimization problem, in a sense that it finds the optimal solution. We do not know, however, how to obtain interesting bounds on the number of iterations in this case.

The rest of the paper is organized as follows. In Section 2 we define the maximum mean-weight cut problem and the related network flow problems, and describe how Newton's method solves them. In Section 3 we discuss Newton's method in the context of linear fractional combinatorial optimization. In Subsection 3.3 we show polynomial bounds on the number of iterations, and in Subsection 3.4 we prove a strongly polynomial bound. In Section 4 we prove an $O(m)$ bound on the number of iterations in Newton's method applied to the maximum mean-weight cut problem. In Section 5 we show a $\Theta(n)$ bound on the number of iterations in Newton's method applied to the maximum mean cut problem. Preliminary versions of some of the results in this paper were earlier reported in [27, 28, 26].

2   The Maximum Mean-weight Cut Problem and Related Network Flow Problems

2.1   The Maximum Mean-weight Cut Problem

A *(transshipment) network* $G = (V, E, u, d)$ is a digraph (V,E) with a capacity function $u : E \longrightarrow \mathbf{R} \cup \{+\infty\}$, and a demand function $d : V \longrightarrow \mathbf{R}$ such that $\sum_{v \in V} d(v) = 0$. We assume for convenience, but without loss of generality, that E is symmetric and u is nonnegative. If $d(v)$ is negative, then v is a *source* − a node with supply. If $d(v)$ is positive, then v is a *sink* − a node with demand. Let n and m denote the cardinality of V and the half of the cardinality of E, respectively. We assume $m \geq n$.

We adopt the convention that any function $\varphi : F \longrightarrow \mathbf{R}$, where F is a finite set, is additively extended to all subsets of F. If $A \subseteq F$, then $\varphi(A) \stackrel{\text{def}}{=} \sum_{x \in A} \varphi(x)$. Thus if $W \subseteq V$, then $d(W) = \sum_{v \in W} d(v)$, the net demand in W. The *total demand* in G is $d(G) \stackrel{\text{def}}{=} \sum_{v:d(v)>0} d(v)$.

An *augmenting path* in G is a directed path from a source to a sink with arcs having positive upper capacities. A *pseudoflow* in G is an antisymmetric function $f : E \longrightarrow \mathbf{R}$, *i.e.*,

$$\forall (v, w) \in E : \; f(v, w) = -f(w, v).$$

Given a pseudoflow f, the *excess* $e^f(v)$ at a node v is equal to the net flow into v:

$$e^f(v) \stackrel{\text{def}}{=} \sum_{w:(w,v)\in E} f(w, v).$$

Observe that flow $f(v, x)$ going out of v along arc (v, x) is indeed subtracted in the above sum, because the antisymmetry condition says that $f(x, v) = -f(v, x)$. If $e^f(v) = d(v)$, then we say that pseudoflow f satisfies the demand at node v. For a pseudoflow f in G, the *residual network* G^f is defined as $(V, E, u^f \overset{\text{def}}{=} u - f, d^f \overset{\text{def}}{=} d - e^f)$. Network $G_{\bar{u}}$ is network G with the capacity function changed to function $\bar{u}$. Thus, for example, $G_{\bar{u}}^f = (V, E, \bar{u}^f, d^f)$.

A *flow* f is a pseudoflow that has the following two properties:

- the capacity constraints hold:

$$\forall (w, v) \in E : \ f(w, v) \le u(w, v),$$

- the excess at each node does not exceed the demand at this node:

$$\forall v \in V : [\, d(v) \ge 0 \ \Rightarrow \ 0 \le e^f(v) \le d(v)\,] \text{ and } [\, d(v) \le 0 \ \Rightarrow \ d(v) \le e^f(v) \le 0\,].$$

The *value of a flow* f is equal to the amount by which f decreases the total demand:

$$value(f) \overset{\text{def}}{=} \sum_{v:\, d(v) > 0} e^f(v).$$

A *maximum flow* in G is a flow in G with the maximum value. The *maximum flow problem* is the problem of finding a maximum flow in a given network. A *satisfying flow* is a flow which satisfies demands at all nodes. G is *feasible* if there exists a satisfying flow in G. The *network feasibility* problem is the problem of verifying if a given network G is feasible.

If S and T partition V, then *cut* (S, T) in G is the set of arcs (v, w) such that $v \in S$ and $w \in T$. The *capacity* and the *surplus* of a cut (S, T) are equal to, respectively,

$$u(S, T) \overset{\text{def}}{=} \sum_{e \in (S,T)} u(e),$$

$$surplus(S, T) \overset{\text{def}}{=} d(T) - u(S, T).$$

It is easy to verify that the surplus of a cut is the same in G as in G^f for any flow f. A positive surplus of a cut (S, T) means that this cut blocks the flow. Any flow must leave at S at least the amount $surplus(S, T)$ of the commodity, which is demanded at T. Therefore we call a cut with a positive surplus a *blocking cut*. A *maximum surplus cut* (*maximum cut, maximum blocking cut*) is a blocking cut with the maximum surplus. The *maximum (surplus) cut* problem is the problem of finding a maximum cut in a given network. A *weight function* is a nonnegative function $b : E \longrightarrow \mathbf{R}$. The *mean surplus* and the *mean-weight surplus* of a cut (S, T) are equal to

$$mean(S, T) \overset{\text{def}}{=} \frac{surplus(S, T)}{|(S, T)|},$$

$$mean\_w(S, T) \overset{\text{def}}{=} \frac{surplus(S, T)}{b(S, T)}.$$

A *maximum mean-weight cut* and a *maximum mean cut* are a blocking cut with the maximum mean-weight surplus and a blocking cut with the maximum mean surplus, respectively. The *maximum mean-weight cut problem* (MMWC) is the problem of finding a maximum mean-weight cut. The *maximum mean cut problem* (MMC) is the problem of finding a maximum mean cut.

Observe that we consider only blocking cuts, so we are interested only in input networks which are not feasible. This is in accordance with the definitions in other papers and with applications (see for example [8, 9, 29]). The problem of finding the maximum mean-weight surplus of a cut in an arbitrary network can be reduced to the problem of finding the maximum mean-weight surplus of a cut in an infeasible network.

The maximum mean cut problem appeared, for example, in the context of the minimum cost flow problem. Goldberg and Tarjan [11] showed a simple strongly polynomial iterative method for solving the minimum cost flow problem. The main task in each iteration was to solve an instance of the minimum mean cycle problem. Ervolina and McCormick [8, 9] showed an analogous method for solving the dual problem to the minimum cost flow problem. Their method solved in each iteration an instance of the maximum mean cut problem. Wallacher [36] used minimum mean-weight cycles in Goldberg and Tarjan's approach. Analogously, maximum mean-weight cuts can be used in Ervolina and McCormick's approach. We should note that $surplus(S, T)$ in our paper is equal to $V(T, S)$, the value of a cut (T, S) in Ervolina and McCormick's papers [8, 9, 20].

2.2 The Parametric Flow Problem and Minimizing the Maximum Cost of an Arc

The MMWC problem is closely related to the *parametric flow problem* (PF) and the *minimum maximum arc cost flow problem* (MMAC). The instance of the PF problem consists of a network G and a weight function $b : E \longrightarrow \mathbf{R}$. The goal is to find the minimum nonnegative δ such that $G_{u+\delta b}$, network G with capacity function $u + \delta b$, is feasible. The instance of the MMAC problem consists of a network G and a nonnegative cost function $c : E \longrightarrow \mathbf{R}$. The goal is to find a satisfying flow which minimizes the maximum arc cost, that is, minimizes $\max_{e \in E}\{f(e)c(e)\}$. Compare it with the classical minimum cost flow problem, where the goal is to minimize $\sum_{e \in E} f(e)c(e)$.

The network feasibility problem, the maximum flow problem and the maximum cut problem are the underlying nonfractional/nonparametric problems. They appear as subproblems in all known methods for solving MMWC, MMAC, and PF. The network feasibility problem and the maximum flow problem can be easily reduced to the standard single-source single-sink maximum flow problem (MF). A maximum flow in a (transshipment) network can be used to find a maximum blocking cut, in a similar way as a maximum flow in a standard single-source single-sink flow network can be used to identify a minimum capacity cut. Let f be a maximum flow in G. If it satisfies all demand, then G is feasible and does not have blocking cuts. Otherwise there are still sources and sinks in G^f but there is no augmenting path. Therefore there is a cut (S, T) such that all sources in G^f belong to S, all sinks belong to T, and all arcs across the cut have zero upper residual capacities. Such a cut is a maximum blocking cut. Its surplus is equal to $d^f(T) - u^f(S, T) = d^f(T)$ and the surplus of any other cut (S', T') is equal to $d^f(T') - u^f(S', T')$ which is not greater than $d^f(T)$, since $d^f(T') \leq d^f(T)$ and $u^f(S', T') \geq 0$. (Observe that we used the fact that the surpluses of cuts are the same in G as in any residual network G^f.) The following fact can be found in the above argument. The surplus of a maximum blocking cut is equal to the total demand left after augmenting the network with a maximum flow, *i.e.*, the total demand in G^f. This observation will be useful in further analysis.

Let $T_P(n, m)$ denote the time complexity of a flow problem P. The best known upper bound on the standard maximum flow problem is

$$T_{\mathrm{MF}}(n, m) = O(\min\{nm \log(n^2/m), nm + n^{2+\epsilon}\}), \tag{1}$$

which is a combination of results from [10] and [18]. The network feasibility problem, the maximum flow problem (in a transshipment network), and the maximum surplus cut problem can be solved in $T_{\mathrm{MF}}(n, m)$ time.

The following theorem shows close relation between the complexities of the MMAC problem and the PF problem.

Theorem 2.1

1. $T_{\mathrm{PF}}(n, m) \leq O(m) + T_{\mathrm{MMAC}}(n, 2m)$.

2. $T_{\mathrm{MMAC}}(n, m) \leq T_{\mathrm{MF}}(n, m)O(\log m) + T_{\mathrm{PF}}(n, m)$.

Proof.

1. For a given instance I of the PF problem construct an instance I' of the MMAC problem in the following way. Replace each arc e in I with two parallel arcs e_1 and e_2. Put $u'(e_1) = u(e)$, $c'(e_1) = 0$, $u'(e_2) = +\infty$, $c'(e_2) = 1/b(e)$. The demand function in I' is the same as in I. A solution (satisfying flow) for I' gives in the natural way a solution for I.

2. If we know δ^* such that

$$\delta^* = \min\{\delta \geq 0 \mid G_{\min\{u, \delta/c\}} \text{ is feasible}\},$$

then a solution to the instance (G, c) of the MMAC problem can be found by computing the maximum flow in $G_{\min\{u, \delta^*/c\}}$. To find δ^*, first identify the numbers in $L = \{u(e)c(e) \mid e \in E\}$ which are bigger than δ^*. This can be done by sorting L and then using binary search. Comparing δ^* with a specific number is equivalent to checking if an appropriate network is feasible. Thus the whole computation takes $T_{\mathrm{MF}}(n,m)O(\log m)$ time. Now construct an instance of PF putting

$$(u'(e), b'(e)) = \begin{cases} (u(e), 0) & \text{if } u(e)c(e) \le \delta^*, \\ (0, 1/c(e)) & \text{if } u(e)c(e) > \delta^*. \end{cases}$$

It can be easily verified that the solution to this instance of PF is equal to δ^*. $\blacksquare$

Let $surplus_\delta(S,T)$ and $mean\_w_\delta(S,T)$ denote the surplus and the mean-weight surplus of (S,T) in $G_{u+\delta b}$. We have

$$surplus_\delta(S,T) \overset{\text{def}}{=} d(T) - (u(S,T) + \delta b(S,T)) = surplus(S,T) - \delta b(S,T), \quad (2)$$

$$mean\_w_\delta(S,T) \overset{\text{def}}{=} \frac{surplus_\delta(S,T)}{b(S,T)} = mean\_w(S,T) - \delta. \quad (3)$$

Since the surplus of a cut is positive if and only if its mean-weight surplus is positive, the above equalities imply the following lemma.

Lemma 2.2 *The maximum mean-weight surplus of a blocking cut in G is equal to the minimum nonnegative δ for which $G_{u+\delta b}$ does not have blocking cuts.*

The following well know theorem gives the dual condition for the feasibility of a network.

Theorem 2.3 [14] *G is feasible if and only if there are no cuts in G with positive surpluses.*

Theorem 2.3 and Lemma 2.2 imply that if G is infeasible, then

$$\min\{\delta \mid G_{u+\delta b} \text{ is feasible}\} = \max\{mean\_w(S,T) \mid (S,T) \text{ cut in } G\}, \quad (4)$$

so if we want to find only the maximum mean-weight surplus of a cut, then the MMWC problem and the PF problem are equivalent. Actually, MMWC and PF are dual to each other in the same sense as the minimum cut and the maximum flow problems are dual to each other. If the maximum mean-weight surplus is known, then a cut with this mean-weight surplus can be found with one maximum flow computation.

Thus the complexities of MMWC, MMAC, and PF are essentially the same. If one of them can be solved in time Q, then the other two can be solved in $O(Q + T_{\mathrm{MF}}(n,m)\log m)$ time. The *uniform* PF problem and the *uniform* MMAC problem are the PF problem with all weights equal to 1 and the MMAC problem with all costs equal to 1, respectively. Problems MMC, uniform PF, and uniform MMAC are related to each other in the same way as problems MMWC, PF, and MMAC.

2.3 Newton's Method for the MMWC/PF Problem

The input instance to the MMWC/PF problem consists of a network G and a (non-negative) weight function b. We assume that G is not feasible. In our presentation we assume that the goal is to solve the PF problem, that is, we want to find the minimum δ such that network $G_{u+\delta b}$ is feasible. We will get a maximum mean-weight cut as a by-product. We use abbreviation G_δ for $G_{u+\delta b}$, and generally subscript δ will indicate that the underlying network is $G_{u+\delta b}$. Thus we want to find

$$\delta^* = \min\{\delta \mid G_\delta \text{ is feasible}\}.$$

If for some δ, $surplus_\delta(S,T)$ is positive, so is $mean\_w_\delta(S,T)$, hence (3) and (4) imply

$$\delta < mean\_w(S,T) \leq \delta^*.$$

This suggests the following iterative scheme for computing δ^*. Let $\bar\delta$ denote the current approximation of δ^*. Initially $\bar\delta = 0$ and always $\bar\delta \leq \delta^*$. If there is no cut with positive surplus in $G_{\bar\delta}$, then Theorem 2.3 says that $G_{\bar\delta}$ is feasible, so $\bar\delta = \delta^*$. Otherwise we find a cut (S,T) with a positive surplus in $G_{\bar\delta}$. We know that $\bar\delta < mean\_w(S,T) \leq \delta^*$, so set $\bar\delta \leftarrow mean\_w(S,T)$ as the next approximation. Increasing $\bar\delta$ should be viewed as increasing the capacities so that the surplus of (S,T) decreases to zero, and (S,T) is no longer a blocking cut. There are finitely many cuts in G, so this process eventually terminates and outputs the precise value of δ^*. The cut found in the last iteration has the maximum mean-weight surplus.

Newton's method is an instance of the above scheme. In each iteration we find a maximum blocking cut in $G_{\bar\delta}$. This is done by computing a maximum flow in $G_{\bar\delta}$ (see the previous section). We will prove that the number of iterations is $O(m)$ for the general PF problem (Section 4) and $\Theta(n)$ for the uniform PF problem (Section 5). These bounds imply the following theorem.

Theorem 2.4 *Newton's method solves MMWC, PF, and MMAC in* $O(T_{\mathrm{MF}}(n,m)m)$ $= O^*(m^2 n)$ *time, and MMC, uniform PF, and uniform MMAC in* $O(T_{\mathrm{MF}}(n,m)n)$ $= O^*(mn^2)$ *time.*

The only previously known strongly polynomial algorithms for the MMWC, PF, and MMAC problems are due to Megiddo and his parametric search method [22]. His method gives an $O^*(n^3 m)$ running time, which can be improved only if a breakthrough in parallel maximum flow computation is achieved. Our $\Theta(n)$ bound on the number of iterations for the uniform cases improves the $O(m)$ bound shown by McCormick and Ervolina [20]. If all input numbers are integral, then the binary search method solves the network problems considered here in $O(T_{\mathrm{MF}}(n,m)\log(nU)) = O(mn\log(nU))$ time, where U is the largest absolute value of the input numbers. The same bound for the Newton's method follows from the bounds we present in Subsection 3.3.

Before analyzing Newton's method for network problems, we put this method in wider context. We show that it runs in strongly polynomial number of iterations for all linear fractional combinatorial optimization problems.

3 Linear Fractional Combinatorial Optimization Problems

3.1 The Class of LFCO Problems

A *linear fractional combinatorial optimization* problem $\mathcal{F}$ is defined as follows. An instance of $\mathcal{F}$ consists of a specification of a set of structures $\mathcal{X} \subseteq \{0,1\}^p$, and two real vectors $\mathbf{a} = (a_1, a_2, \ldots, a_p)$, $\mathbf{b} = (b_1, b_2, \ldots, b_p)$. The task is to

$$\mathcal{F}: \quad \text{maximize} \quad \frac{a_1 x_1 + a_2 x_2 + \cdots + a_p x_p}{b_1 x_1 + b_2 x_2 + \cdots + b_p x_p}, \quad \text{subject to} \quad (x_1, x_2, \ldots, x_p) \in \mathcal{X}.$$

Structures are some special subsets of the underlying set of p elements. A 0-1 vector $\mathbf{x} = (x_1, x_2, \ldots, x_p) \in \mathcal{X}$ is the characteristic vector of the structure it represents. Numbers a_i and b_i are *the cost* and *the weight* of element i. We denote the inner product $c_1 z_1 + c_2 z_2 + \cdots + c_p z_p$ of two vectors $\mathbf{c} = (c_1, c_2, \ldots, c_p)$ and $\mathbf{z} = (z_1, z_2, \ldots, z_p)$ by $\mathbf{cz}$. Numbers $\mathbf{ax}$, $\mathbf{bx}$, and $(\mathbf{ax})/(\mathbf{bx})$ are *the cost, weight,* and *mean-weight cost* of the structure represented by vector $\mathbf{x}$. In this terminology problem $\mathcal{F}$ is to compute the maximum mean-weight cost of a structure in $\mathcal{X}$. We also want to find a structure which has this maximum mean-weight cost. We assume that $\mathbf{ax} > 0$ for some $\mathbf{x} \in \mathcal{X}$, and $\mathbf{bx} > 0$ for all $\mathbf{x} \in \mathcal{X}$, but there may be structures with negative costs and individual elements with negative weights.

Problem $\mathcal{F}$ can be equivalently formulated in the following way.

$$\mathcal{P}: \quad \text{minimize } \delta, \quad \text{subject to} \quad (\mathbf{ax}) - \delta(\mathbf{bx}) \leq 0, \quad \text{for all } \mathbf{x} \in \mathcal{X}.$$

We call $\mathcal{P}$ the parametric version of $\mathcal{F}$. Let δ^* denote the solution to $\mathcal{P}$. Define

$$\begin{aligned} h(\delta) &\stackrel{\text{def}}{=} \max\{(\mathbf{ax}) - \delta(\mathbf{bx}) \mid \mathbf{x} \in \mathcal{X}\} \\ &= \max\{(\mathbf{a} - \delta\mathbf{b})\mathbf{x} \mid \mathbf{x} \in \mathcal{X}\}. \end{aligned}$$

For any fixed δ, vector $\mathbf{a} - \delta\mathbf{b}$ is the vector of the *reduced costs*. Thus $h(\delta)$ is the maximum reduced cost of a structure. Function $h(\delta)$ is convex, piecewise linear, decreasing, and δ^* is its only root. Thus we have another equivalent formulation of an LFCO problem:

$$\mathcal{R}: \quad \text{solve} \quad h(\delta) = 0.$$

The following are the optimization and the decision version of the *underlying linear (non-parametric) problem*. Given are a specification of $\mathcal{X} \subseteq \{0,1\}^p$ and $\mathbf{c} \in \mathbf{R}^p$.

$$\mathcal{L}_{\text{opt}}: \quad \text{maximize} \quad \mathbf{cx}, \quad \text{subject to} \quad \mathbf{x} \in \mathcal{X}.$$

$$\mathcal{L}_{\text{dec}}: \quad \text{is there} \quad \mathbf{x} \in \mathcal{X}, \quad \text{such that} \quad \mathbf{cx} > 0\,?$$

In both cases we also want to find an appropriate $\mathbf{x}$. Problem $\mathcal{P}$ is sometimes called *the parametric extension of problem* $\mathcal{L}_{\text{opt}}$ [3, 6].

Newton's method for an LFCO problem is Newton's method for finding the root of $h(\delta)$. To apply this method we assume that we have a procedure which solves $\mathcal{L}_{\text{opt}}$. We use this computation as a black box to compute for any fixed δ, $h(\delta)$, the maximum reduced cost of a structure, and a structure $\mathbf{x} \in \mathcal{X}$ which has this cost. This structure defines the derivative of h at δ. The binary search method for an LFCO problem is the bisection method for finding the root of $h(\delta)$. To apply the binary search method, it is enough to have a procedure for $\mathcal{L}_{\text{dec}}$.

A *uniform* LFCO problem is

$$\mathcal{U}: \quad \text{maximize} \quad \frac{\mathbf{ax}}{\mathbf{ex}}, \quad \text{subject to} \quad \mathbf{x} \in \mathcal{X}.$$

where $\mathbf{e} \stackrel{\text{def}}{=} (1, 1, \ldots, 1)$. The task here is to find a structure with the maximum mean cost. The *mean cost* of a structure is its cost divided by the number of its elements.

Hansen *et al.* [13] considered the class of unconstrained 0-1 fractional programming problems, which is a subclass of the LFCO problems. In our terminology, an unconstrained 0-1 fractional programming problem is an LFCO problem with $\mathcal{X}$ being the family of all subsets of a given set of p elements which contain the first element. They showed a linear time algorithm for such problems. Karzanov [17] considered the class of uniform LFCO problems and showed that Newton's method solves these problems in at most p iterations. It can be shown that in both of these subclasses functions h consist of at most p linear pieces. In the general case function h can consist of a superpolynomial number of linear pieces [1].

If in the definition of problem $\mathcal{F}$, set $\mathcal{X}$ is some arbitrary[3] subset of $\mathbf{R}^p$, then we get a linear fractional program. Furthermore, if the enumerator and the denominator of the objective function are arbitrary[3] functions on $\mathcal{X}$, then we get a general fractional program. Both fractional programs and linear fractional programs have been extensively studied in the operations research community (see, for example, review papers [32] and [33]). One of the main methods for solving (the parametric versions of) such problems is Newton's method, called also the Newton-Raphson method or the Dinkelbach method [15]. In such general setting, however, most often all we can hope for is an approximate solution, and a natural question to ask is how fast a given iterative method converges. Schaible [31] showed that Newton's method for a fractional program converges superlinearly, assuming continuity of the involved functions and compactness of domain $\mathcal{X}$. The most efficient version of Newton's method for such problems is due to Pardalos and Phillips [25].

The MMWC problem is an LFCO problem. Let a network G and a weight function b constitute an input instance of MMWC. The set of structures $\mathcal{X}_G \subseteq \{0, 1\}^{n+m}$ corresponds to the set of all cuts in G. A vector $\mathbf{x} \in \mathcal{X}_G$ represents cut (S, T) such

---

[3]Additional assumptions are usually made, which reflect applications and/or allow obtaining interesting results.

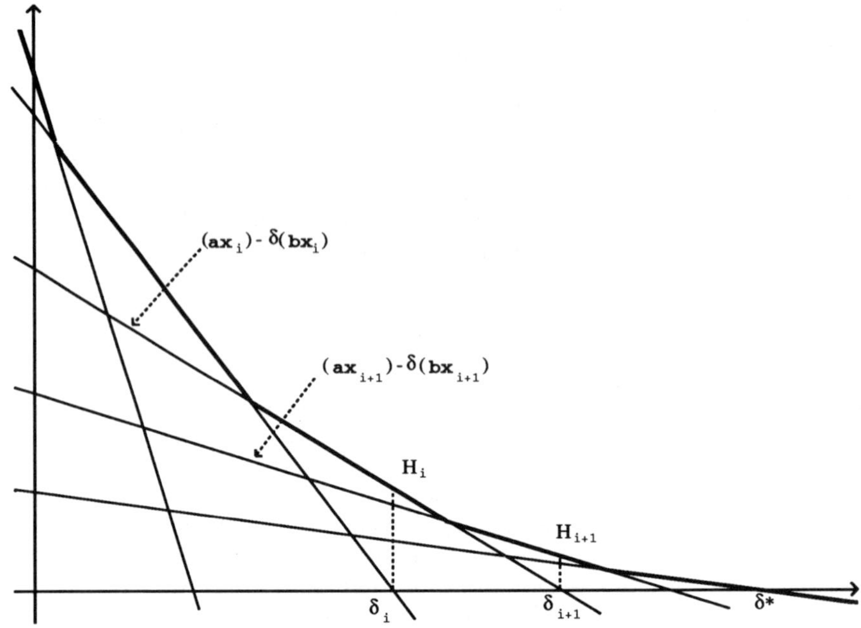

Figure 1: **Newton's method for solving** $h(\delta) = 0$

that for $1 \leq i \leq n$, $x_i = 1$ if and only if $v_i \in T$, and for $1 \leq j \leq m$, $x_{n+j} = 1$ if and only if $e_j \in (S, T)$, where v_i is the ith node and e_j is the jth arc. The cost vector $\mathbf{a}$ is equal to $(\mathbf{d}, -\mathbf{u})$, where $\mathbf{d}$ and $\mathbf{u}$ are the vector representations of the demand function and the capacity function. The weight vector $\mathbf{b}$ represents the weight function b. In this notation, if $\mathbf{x}$ represents cut (S, T), then $\mathbf{ax}$ and $\mathbf{bx}$ are equal to $surplus(S, T)$ and $b(S, T)$. It can be also checked that the PF problem is the parametric version of the MMWC problem in the formal sense defined in this section.

3.2   Newton's Method for an LFCO problem

In this section we describe Newton's method for computing the root δ^* of $h(\delta)$, introduce notation which we will use later in the analysis, and present a basic fact concerning the rate of convergence of this method.

Let $\bar{\delta} \leq \delta^*$ be the current approximation of δ^*. Initially $\bar{\delta} = 0$. Observe that our assumptions guarantee $\delta^* > 0$. During one iteration we compute $h(\bar{\delta})$ and $\overline{\mathbf{x}} \in \mathcal{X}$ such that $h(\bar{\delta}) = (\mathbf{a} - \bar{\delta}\mathbf{b})\overline{\mathbf{x}}$, that is, we maximize linear function $(\mathbf{a} - \bar{\delta}\mathbf{b})\mathbf{x}$ over $\mathcal{X}$ (an instance of problem $\mathcal{L}_{\text{opt}}$). If $h(\bar{\delta}) = 0$, then $\delta^* = \bar{\delta}$ and the algorithm terminates. Otherwise we compute the next approximation $\bar{\delta} \leftarrow \mathbf{a}\overline{\mathbf{x}}/\mathbf{b}\overline{\mathbf{x}}$, the mean-weight cost of

$\overline{\mathbf{x}}$, and go to the next iteration. The process is illustrated in Figure 1.

Let δ_i be the value of $\bar{\delta}$ at the beginning of ith iteration, and $\mathbf{x}_i$, H_i, and B_i, be $\overline{\mathbf{x}}$, $(\mathbf{a} - \delta_i \mathbf{b})\overline{\mathbf{x}}$, and $\mathbf{b}\overline{\mathbf{x}}$ from this iteration. Thus

$$
\begin{aligned}
H_i &= (\mathbf{a} - \delta_i \mathbf{b})\mathbf{x}_i = \max\{(\mathbf{a} - \delta_i \mathbf{b})\mathbf{x} \mid \mathbf{x} \in \mathcal{X}\}, \\
B_i &= \mathbf{b}\mathbf{x}_i, \\
\delta_{i+1} &= \frac{\mathbf{a}\mathbf{x}_i}{\mathbf{b}\mathbf{x}_i},
\end{aligned}
$$

and it can be easily derived that

$$
\delta_{i+1} - \delta_i = \frac{H_i}{B_i}. \tag{5}
$$

The following lemma indicates fast convergence of the above algorithm.

Lemma 3.1
$$
\frac{H_{i+1}}{H_i} + \frac{B_{i+1}}{B_i} \le 1. \tag{6}
$$

Proof. Vector $\mathbf{x}_i$ maximizes $(\mathbf{a} - \delta_i \mathbf{b})\mathbf{x}$, so

$$
(\mathbf{a} - \delta_i \mathbf{b})\mathbf{x}_i \ge (\mathbf{a} - \delta_i \mathbf{b})\mathbf{x}_{i+1}.
$$

Therefore, using the definitions of H_i and B_i and Equality (5),

$$
\begin{aligned}
H_i &= (\mathbf{a} - \delta_i \mathbf{b})\mathbf{x}_i \ge (\mathbf{a} - \delta_i \mathbf{b})\mathbf{x}_{i+1} \\
&= (\mathbf{a} - \delta_{i+1}\mathbf{b})\mathbf{x}_{i+1} + (\delta_{i+1} - \delta_i)\mathbf{b}\mathbf{x}_{i+1} \\
&= H_{i+1} + \frac{H_i}{B_i}B_{i+1}.
\end{aligned}
$$

This implies Inequality (6). $\blacksquare$

Later on we will often need only the following simple consequence of Inequality (6).

$$
\frac{H_{i+1}B_{i+1}}{H_i B_i} \le \frac{1}{4}. \tag{7}
$$

3.3 Polynomial bounds

In this section we assume that costs and weights are integral. Let A and B be the maximum absolute values of the costs and the weights of the individual elements, respectively. Let U denote the maximum of A and B. Lemma 3.1 leads to the following polynomial bound on the number of iterations in Newton's method. This is equal to the bound on the number of iterations in the binary search method.

Theorem 3.2 *Newton's method solves an LFCO problem in $O(\log(pU))$ iterations.*

Proof. For any i, if $H_i > 0$, then

$$(pU)^2 \geq H_i B_i \geq \frac{1}{pU}. \tag{8}$$

The first inequality follows from the fact that $H_i = h(\delta_i) \leq h(0)$ and $h(0)$ is equal to the cost of some structure, so $h(0) \leq pU$. The second inequality holds because

$$
\begin{aligned}
H_i &= (\delta_{i+1} - \delta_i)B_i = \left(\frac{\mathbf{ax}_i}{\mathbf{bx}_i} - \frac{\mathbf{ax}_{i-1}}{\mathbf{bx}_{i-1}} \right) B_i \\
&= \frac{(\mathbf{ax}_i)(\mathbf{bx}_{i-1}) - (\mathbf{ax}_{i-1})(\mathbf{bx}_i)}{(\mathbf{bx}_i)(\mathbf{bx}_{i-1})} B_i \geq \frac{1}{B_i B_{i-1}} B_i \geq \frac{1}{pU}.
\end{aligned} \tag{9}
$$

Inequalities (7) and (8) imply the bound of $O(\log(pU))$ on the number of iteration. ∎

We will show a refined version of this theorem, which implies a better bound if the weights are substantially smaller than the costs. Such a case may occur, for example, in uniform LFCO problems, where all weights are equal to 1. We need the following technical lemma. Sequences (α_i) and (β_i) correspond to the sequences of quotients (H_{i+1}/H_i) and (B_{i+1}/B_i).

Lemma 3.3 *Let μ, α, and β be positive numbers such that $\mu < 2$, $\alpha < 1$, $\beta < 1$. Let (α_i) and (β_i) be two finite positive sequences of length l such that*

$$\alpha_i + \beta_i \leq \mu \ \text{for } i = 1, 2, \ldots, l,$$

$$\prod_{i=1}^{l} \alpha_i \geq \alpha, \quad \text{and} \quad \prod_{i=1}^{l} \beta_i \geq \beta.$$

Then $l \leq L$, where L is the solution to the equation

$$\alpha^{\frac{1}{L}} + \beta^{\frac{1}{L}} = \mu.$$

Proof. Using the fact that the geometrical mean is not greater than the arithmetical mean we have

$$
(\alpha)^{\frac{1}{l}} + (\beta)^{\frac{1}{l}} \leq \left(\prod_{i=1}^{l} \alpha_i \right)^{\frac{1}{l}} + \left(\prod_{i=1}^{l} \beta_i \right)^{\frac{1}{l}} \leq \frac{1}{l} \left(\sum_{i=1}^{l} \alpha_i \right) + \frac{1}{l} \left(\sum_{i=1}^{l} \beta_i \right)
$$

$$
= \frac{1}{l} \left(\sum_{i=1}^{l} (\alpha_i + \beta_i) \right) \leq \mu.
$$

Therefore $l \leq L$ because function $(\alpha)^{\frac{1}{x}} + (\beta)^{\frac{1}{x}}$ increases on $(0, \infty)$. ∎

Theorem 3.4 *Newton's method solves an LFCO problem in*
$O(\frac{\log(pAB)}{1+\log\log(pAB)-\log\log(pB)})$ *iterations.*

Proof. If $B > A$, then this is the same bound as in Theorem 3.2. Therefore we can assume $B \leq A$. Let l denote the number of iterations. Let $\alpha_i = \frac{H_{i+1}}{H_i}$ and $\beta_i = \frac{B_{i+1}}{B_i}$, for $i = 1, 2, \ldots, l-1$. Lemma 3.1 says that $\alpha_i + \beta_i \leq 1$. We have the following bounds.

$$H_1 \leq pA, \quad H_l \geq \frac{1}{pB}, \quad B_1 \leq pB, \quad B_l \geq 1.$$

For the justification of the second inequality see (9). The other inequalities are obvious. These bounds imply

$$\prod_{i=1}^{l-1} \alpha_i = \frac{H_l}{H_1} \geq \frac{1}{p^2 AB} \quad \text{and} \quad \prod_{i=1}^{l-1} \beta_i = \frac{B_l}{B_1} \geq \frac{1}{pB}.$$

According to Lemma 3.3, $l - 1 \leq L$, where L is such that

$$\left(\frac{1}{p^2 AB}\right)^{\frac{1}{L}} + \left(\frac{1}{pB}\right)^{\frac{1}{L}} = 1.$$

For $x > 0$, let

$$h(x) = \left(\frac{1}{p^2 AB}\right)^{\frac{1}{x}} + \left(\frac{1}{pB}\right)^{\frac{1}{x}}.$$

Then

$$h\left(\frac{2\log(p^2 AB)}{\log\log(p^2 AB) - \log\log(pB)}\right) = \left(\frac{\log(pB)}{\log(p^2 AB)}\right)^{\frac{1}{2}} + \left(\frac{\log(pB)}{\log(p^2 AB)}\right)^{\frac{\log(pB)}{2\log(p^2 AB)}} > 1.$$

The inequality holds because $z^{1/2} + z^{z/2} > 1$, for $z > 0$. This fact can be verified using basic calculus. Function $h(x)$ is increasing, so

$$L < \frac{2\log(p^2 AB)}{\log\log(p^2 AB) - \log\log(pB)} = O\left(\frac{\log(pAB)}{1 + \log\log(pAB) - \log\log(pB)}\right).$$

The above bound is valid, because we consider case $B \leq A$. ∎

The above theorem shows that for $A = (pB)^{\omega(1)}$ the number of iterations in Newton's method is asymptotically smaller than the number of iterations in the binary search method, and is smaller by a factor of $\log\log(pA)$, if $A = \Omega((pB)^{\log^\alpha(pB)})$ for any positive constant α.

3.4   Strongly Polynomial Bounds

Before analyzing the general case, we note the complexity of Newton's method for uniform LFCO problems, *i.e.*, costs are arbitrary real numbers, all weights are equal to 1. The following fact was observed by Karzanov [17] and, in the context of the maximum mean cut problem, by McCormick and Ervolina [20].

Theorem 3.5 *Newton's method runs in at most $p+1$ iterations for a uniform LFCO problem.*

Proof. Sequence (B_i), excluding the last iteration, is strictly decreasing. In a uniform LFCO problem B_i is a positive integer not greater than p, because B_i is the cardinality of some structure. ∎

In this subsection we derive a strongly polynomial bound on the number of iterations for the general case of linear fractional combinatorial optimization, when both costs and weights are arbitrary real numbers. We first give some intuition which lies behind the analysis. Lemma 3.1 suggests that there are some sequences, related to the convergence of the method, which tend to zero at least geometrically fast (see Inequality 7). The elements of these sequences are obtained from only $2p$ numbers, the costs $a_1, a_2, \ldots, a_p$, and the weights $b_1, b_2 \ldots, b_p$, using only $O(p)$ additions/subtractions and at most one or two multiplications/divisions. We show that because of limiting use of arithmetical operations, such sequences cannot be long.

To further expand this intuition, let us assume that there is a constant α such that for every i, $B_{i+1}/B_i \leq \alpha < 1$. It means that sequence (B_i) tends to zero at least geometrically fast. Let us also assume that $b_1 \geq b_2 \geq \cdots \geq b_p \geq 0$. Each element of sequence (B_i) is equal to the sum of the elements of a subset of $\{b_1, b_2, \ldots, b_p\}$. Obviously $B_1 \leq p b_1$. Since sequence (B_i) decreases at least geometrically, $B_l < b_1$, for some $l = O(\log p)$. It means that b_1 is not a term in B_l, nor is it in any B_i, for $i \geq l$, so we can exclude b_1 from further considerations. Thus $B_l \leq (p-1)b_2$, and after the next $O(\log p)$ iterations we can exclude b_2, then b_3, and so on. Therefore the length of sequence (B_i), and the number of iterations in the algorithm, is only $O(p \log p)$. (In fact, the length of such a sequence is $O(p)$. See Lemma 4.4.)

There are two reasons why the general case is more complicated. First, we have to deal with positive and negative numbers. Even if both costs and weights are positive, negative numbers appear because subtractions are used in forming the elements of sequence (H_i). Second, if sequence (B_i) does not decrease fast enough, then we have to analyze sequence (H_i), whose elements are not just sums of elements from some small predetermined set of numbers. The following lemma is the tool for dealing with both positive and negative numbers. It says that also in the case when the ground set contains both positive and negative numbers, the length of a geometric sequence of subsums of these numbers is only $O(p \log p)$. In [28], where we first reported our strongly polynomial bound, we showed an $O(p^2 \log p)$ bound on the length of such sequences. The $O(p \log p)$ bound was recently proved by Michel Goemans [personal communication]. We include his original proof. In the statement of the lemma the coordinates of vector $\mathbf{c}$ are numbers which are used to form subsums, and $(\mathbf{y}_k\mathbf{c})_{k\geq 1}$ is a sequence of such subsums.

Lemma 3.6 (Michel Goemans) *Let $\mathbf{c} = (c_1, c_2, \ldots, c_p)$ be a vector with positive real coordinates. Let $\mathbf{y}_1, \mathbf{y}_2, \ldots, \mathbf{y}_q$ be vectors from $\{-1, 0, 1\}^p$. If for all $i = 1, 2, \ldots, q-1$*

$$0 < \mathbf{y}_{i+1}\mathbf{c} \leq \frac{1}{2}\mathbf{y}_i\mathbf{c},$$

then $q = O(p \log p)$.

Proof. The premises of the lemma imply that the following linear program is feasible.

$$(\mathbf{y}_i - 2\mathbf{y}_{i+1})\mathbf{x} \geq 0, \quad \text{for } i = 1, 2, \ldots, q-1,$$

$$\mathbf{y}_q\mathbf{x} = 1,$$

$$\mathbf{x} \geq 0.$$

Let $\mathbf{c}^* \neq \mathbf{0}$ be an extreme point of this LP. There is an invertible matrix A and a vector $\mathbf{b}$, such that $A\mathbf{c}^* = \mathbf{b}$ and each coefficient of A and b is an integer from $[-3, 3]$. Cramer's rule implies that

$$c_i^* = \frac{h_i}{g},$$

where $g, h_1, h_2, \ldots, h_p$ are nonnegative integers not greater than $3^p(p!)$, $g \neq 0$. We know that

$$1 = \mathbf{y}_q\mathbf{c}^* \leq \frac{1}{2^q}\mathbf{y}_1\mathbf{c}^*,$$

and

$$\mathbf{y}_1\mathbf{c}^* \leq p3^p(p!),$$

so $q = O(p \log p)$. ∎

Now we are ready to prove a strongly polynomial bound on the number of iterations. We separately consider the iterations which substantially decrease B_i and those which do not.

Lemma 3.7 *There are at most $O(p \log p)$ iterations k such that $B_{k+1} \leq \frac{2}{3}B_k$.*

Proof. This is immediate consequence of Lemma 3.6. Let $\mathbf{c} = \mathbf{b}$ and sequence $(\mathbf{y}_i)_{1\leq i\leq q}$ be the sequence of $\mathbf{x}_k$'s from every other iteration such that $B_{k+1} \leq \frac{2}{3}B_k$. We have

$$0 \leq \mathbf{y}_{i+1}\mathbf{c} \leq \frac{4}{9}\mathbf{y}_i\mathbf{c},$$

so Lemma 3.6 implies that $q = O(p \log p)$. ∎

Lemma 3.8 *There are at most $O(p \log p)$ consecutive iterations k such that $B_{k+1} \geq \frac{2}{3} B_k$.*

Proof. Consider a sequence of q consecutive iterations such that for each of them

$$B_{k+1} \geq \frac{2}{3} B_k. \tag{10}$$

For convenience we renumber these iteration assigning them indices 1 through q. We show that Lemma 3.6 can be applied to the sequence $(x_k(-a + \delta_{q+1}b))$, $1 \leq k \leq q$. Inequalities (10) and (6) imply that for $k = 1, 2, \ldots, q-1$,

$$H_{k+1} \leq \frac{1}{3} H_k.$$

Therefore

$$\delta_{k+2} - \delta_{k+1} = \frac{H_{k+1}}{B_{k+1}} \leq \frac{1}{2} \frac{H_k}{B_k} = \frac{1}{2}(\delta_{k+1} - \delta_k). \tag{11}$$

This implies

$$\begin{aligned}
\delta_{q+1} - \delta_{k+1} &= (\delta_{q+1} - \delta_q) + (\delta_q - \delta_{q-1}) + \cdots + (\delta_{k+2} - \delta_{k+1}) \\
&\leq \frac{1}{2}(\delta_q - \delta_{q-1}) + \frac{1}{2}(\delta_{q-1} - \delta_{q-2}) + \cdots + \frac{1}{2}(\delta_{k+1} - \delta_k) \\
&= \frac{1}{2}(\delta_q - \delta_k) \\
&\leq \frac{1}{2}(\delta_{q+1} - \delta_k).
\end{aligned}$$

Recall that $\delta_{i+1} = (ax_i)/(bx_i)$, so

$$x_k(-a + \delta_{q+1}b) = (\delta_{q+1} - \delta_{k+1})bx_k.$$

Putting

$$\begin{aligned}
c &= -a + \delta_{q+1}b, \\
y_k &= x_k, \quad \text{for } k = 1, 2, \ldots, q,
\end{aligned}$$

we have for $k = 1, 2, \ldots, q$,

$$y_{k+1}c = (\delta_{q+1} - \delta_{k+2})B_{k+1} < \frac{1}{2}(\delta_{q+1} - \delta_{k+1})B_k = \frac{1}{2}y_k c.$$

Thus vector c and sequence $(y_k)_{1 \leq k \leq q}$ satisfy the conditions of Lemma 3.6, so $q = O(p \log p)$. ∎

Lemmas 3.7 and 3.8 immediately imply the following strongly polynomial bound on the number of iterations.

Theorem 3.9 *Newton's method applied to an LFCO problem finds the optimum in $O(p^2 \log^2 p)$ iterations.*

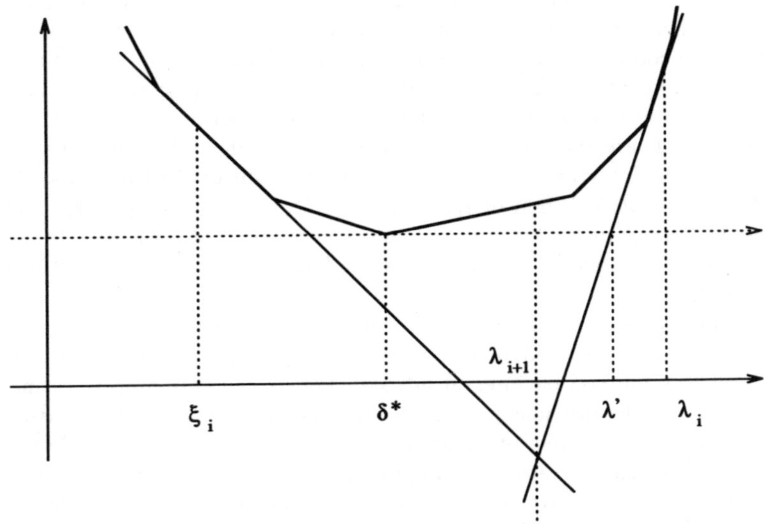

Figure 2: **Newton's method for finding the minimum of** $h(\delta)$

3.5 Minimizing the maximum cost

When we defined the parametric linear combinatorial optimization in Section 3.1 we assumed that the weights of structures were positive. This assumption is natural for fractional optimization, but it is not difficult to imagine a parametric problem for which this condition does not hold. If the parameter corresponds to time, we may have to model a situation when some characteristics improve with the time while some others deteriorate. In this section we discuss the case when the weights of elements are arbitrary, so the weights of some structures may be negative.

We use the same notation as in the previous sections. The cost of element i is equal to $a_i - \delta b_i$. The cost of a structure is a linear function of δ, which now may be increasing, because the weight of a structure may be negative. We want to find the value of the parameter which minimizes the maximum cost over all structures. This is the way Cohen and Megiddo [5, 6] (see also [3]) defined the parametric extension of an optimization problem.[4]

When the value of the parameter is equal to δ, $h(\delta)$ denotes the maximum reduced cost over all structures. Function h is no longer monotonic but is still convex, so it has the global minimum δ^*. (We assume that there is at least one structure with a negative cost.) We find this minimum with the following adaptation of Newton's method. We maintain interval $[\xi, \lambda]$ which contains δ^*, and in each iteration we improve either

---

[4]In [5, 3, 6] parametric extensions with more than one parameter are considered.

ξ or λ. Let $[\xi_i, \lambda_i]$ be this interval at the beginning of iteration i. We compute $h(\xi_i)$ and $h(\lambda_i)$ together with structures $\mathbf{x}_i$ and $\mathbf{y}_i$ which have this maximum reduced costs. Let ψ_i be the δ coordinate of the intersection point of lines $(\mathbf{ax}_i) - \delta(\mathbf{bx}_i)$ and $(\mathbf{ay}_i) - \delta(\mathbf{by}_i)$, the "tangents" to $h(\delta)$ at ξ_i and λ_i. Now we find $h(\psi_i)$ and structure $\mathbf{z}_i$ which has this maximum reduced cost. If the reduced costs, for $\delta = \psi_i$, of all three structures $\mathbf{x}_i$, $\mathbf{y}_i$, and $\mathbf{z}_i$ are the same, then $\delta^* = \psi$ and the algorithms terminates. Otherwise we shrink the interval containing δ^*. Positive weight of $\mathbf{z}_i$ means $\psi_i < \delta^*$, so we set $\xi_{i+1} = \psi_i$. Negative weight of $\mathbf{z}_i$ means $\psi_i > \delta^*$, so we set $\lambda_{i+1} = \psi_i$. The process is illustrated in Figure 2. Actually, only one computation of $h(\delta)$, namely the computation of $h(\psi_i)$, is needed in iteration i. Values $h(\xi_i)$ and $h(\lambda_i)$, and structures $\mathbf{x}_i$ and $\mathbf{y}_i$, are known from previous iterations.

To estimate the convergence of this algorithm, translate the δ-axis such that $h(\delta^*) = 0$. The horizontal dotted line in Figure 2 is the new δ-axis. Now run the standard Newton's method on interval $[\delta^*, \infty)$. Let us call this process $\mathcal{B}$ and the process described in the previous paragraph $\mathcal{A}$. If $\mathcal{A}$ and $\mathcal{B}$ start with the same approximation λ_1, then the number of iterations in $\mathcal{B}$ is not less than the number of those iterations in $\mathcal{A}$ which decrease λ. In Figure 2, from point λ_i process $\mathcal{A}$ moves to λ_{i+1} while process $\mathcal{B}$ would move only to $\lambda' > \lambda_{i+1}$. Analogous argument holds for interval $(-\infty, \delta^*]$. Thus all bounds for our standard Newton's method are also valid for the algorithm described in this section.

4   Analysis of Newton's Method for the PF Problem

Specializing Theorem 3.9 to the MMWC/PF problem we get an $O(m^2 \log n)$ bound on the number of iterations. Using a special structure of the MMWC/PF problem, namely the maximum flow - minimum cut duality, we prove in this section an $O(m)$ bound on the number of iterations.

For any flow f in G, the surpluses of cuts are the same in G as in G^f. It means that in the current iteration we can perform necessary computation on any residual network of $G_{\bar{\delta}}$ instead of network $G_{\bar{\delta}}$. (Recall that $\bar{\delta}$ is the current approximation of the optimum.) In the analysis we assume that the network at the beginning of the current iteration is $G_{\bar{\delta}}^f$, where f is the sum of the maximum flows computed in all previous iterations. In other words, f is a maximum flow in $G_{\delta'}$, where δ' is $\bar{\delta}$ from the previous iteration. Now we view the whole algorithm in the following way. By computing a maximum flow we satisfy as much demand as possible without violating the capacity constraints. If not all demand is satisfied, we identify some specific saturated cut. Then we increase the capacities of the arcs such that the (residual) capacity of this cut becomes equal to the demand left on the head side of this cut. We repeat finding a maximum flow and increasing capacities in the current network until all demand is satisfied.

Let δ_i be equal to $\bar{\delta}$ at the beginning of iteration i, and (S_i, T_i) be the cut found in this iteration. Let f_i be the sum of the flows computed through iteration i. Flow f_i is a maximum flow in G_{δ_i}. The network at the beginning of iteration i is $G_{\delta_i}^{f_{i-1}}$. During this iteration we compute a maximum flow $\hat{f}_i$ in this network and identify a maximum surplus cut (S_i, T_i). After this computation the network is $G_{\delta_i}^{f_{i-1}+\hat{f}_i} = G_{\delta_i}^{f_i}$. Now we increase the capacity function by $\xi_i b$, where $\xi_i = mean\_w_{\delta_i}(S_i, T_i)$, i.e., $\delta_{i+1} \leftarrow \delta_i + \xi_i$. We obtain network $G_{\delta_i+\xi_i}^{f_i} = G_{\delta_{i+1}}^{f_i}$, and go to the next iteration. Observe that according to the description of the algorithm in Section 2.3, we should have $\delta_{i+1} = mean\_w(S_i, T_i)$. This is exactly what we obtain, because $\delta_i + \xi_i = \delta_i + mean\_w_{\delta_i}(S_i, T_i) = mean\_w(S_i, T_i)$.

B_i, and H_i from Section 3.2 have here the following meaning.

$$\begin{aligned} B_i &= b(S_i, T_i), \\ H_i &= h(\delta_i) = surplus(S_i, T_i) - \delta_i b(S_i, T_i) = surplus_{\delta_i}(S_i, T_i). \end{aligned}$$

Recall that H_i, the maximum surplus of a cut in $G_{\delta_i}^{f_{i-1}}$, is equal to the total demand in $G_{\delta_{i+1}}^{f_i}$, that is, the total demand left after iteration i.

Our analysis in this section can be viewed as an instance of the following general paradigm. As computation proceeds, some constraints become unessential and can be dropped. The key part of the analysis lies in bounding the time needed for a new constraint to become unessential. Many strongly polynomial algorithms for the minimum cost flow problem are analyzed in such a way [11, 12, 24, 29, 34]. In our analysis we use Inequality (6) to estimate the progress of the computation. This inequality says that if sequence (B_i) does not decrease fast, then sequence (H_i) has to. After some time, the total demand left H_i should be small enough in comparison with the residual capacities of some arcs, so these capacities are no longer essential and can be dropped. We formalize this idea.

An arc e is *unessential* in G if its capacity is greater than the total demand or its weight is greater than the weight of a maximum surplus cut. When we use expression "e is unessential" in the context of the algorithm, we mean that e is unessential in the current network. By increasing δ we increase the capacities of arcs. By augmenting with a flow we may decrease the capacity of an arc, but by not more than we decrease the total demand. It means that, if an arc is unessential in G_δ^f, for some $\delta \geq 0$ and some flow f, then it is unessential in $G_{\delta+\xi}^{f+f'}$, for any $\xi \geq 0$ and any flow f' in $G_{\delta+\xi}^f$. Therefore, if at some point in the algorithm an arc is unessential, it remains unessential through the end of the computation. An unessential arc cannot belong to a maximum surplus cut. The aim is to show that few iterations are enough to make a new arc unessential. We will need the following lemma.

Lemma 4.1 *From iteration $i + 2$ on, the (current) capacity of cut (S_i, T_i) is greater than H_{i+1}.*

Proof. At the beginning of iteration $i+1$ the capacity of cut (S_i, T_i) is equal to H_i, the remaining total demand. In iteration $i+1$ the capacity of each cut first decreases, but by at most $H_i - H_{i+1}$. Then it increases by $(H_{i+1}/B_{i+1})B$, where B is the weight of this cut. It means that in iteration $i+1$ the capacity of cut (S_i, T_i) first decreases, but does not go below H_{i+1}. Then it increases by more than H_{i+1}, because its weight is greater than B_{i+1}. Therefore at the end of iteration $i+1$ the capacity of (S_i, T_i) is greater than $2H_{i+1}$. From now on the capacity of (S_i, T_i) is always greater than H_{i+1}, because the total decrease of this capacity cannot be greater than the total demand left after iteration $i+1$, which is H_{i+1}.

Putting the above argument in a more formal way we get the following bound on the capacity of (S_i, T_i) at the beginning of iteration $i+l$, where $l \geq 2$. According to our notation, the capacity function at the beginning of iteration $i+l$ is $u_{\delta_{i+l}}^{f_{i+l-1}}$.

$$
\begin{aligned}
u_{\delta_{i+l}}^{f_{i+l-1}}(S_i, T_i) &\geq u_{\delta_i}^{f_i}(S_i, T_i) + (\delta_{i+l} - \delta_i)B_i - (H_i - H_{i+l-1}) \\
&= (\delta_{i+l} - \delta_{i+1})B_i + (\delta_{i+1} - \delta_i)B_i - H_i + H_{i+l-1} \\
&= (\delta_{i+l} - \delta_{i+1})B_i + H_{i+l-1} \\
&\geq (\delta_{i+2} - \delta_{i+1})B_i \\
&> (\delta_{i+2} - \delta_{i+1})B_{i+1} \\
&= H_{i+1}.
\end{aligned}
$$

∎

We first prove an $O(m \log m)$ bound on the number of iterations to show the main idea. To prove an $O(m)$ bound, we will need a finer accounting strategy.

Theorem 4.2 *Newton's method solves PF in $O(m \log m)$ iterations.*

Proof. We use Inequality (7) to show that after $O(\log m)$ iterations a new arc becomes unessential. Let the current iteration be the ith one. Let $l = \lfloor \log m \rfloor + 2$. It follows from Inequality (7) that

$$H_{i+l}B_{i+l} < \frac{1}{m^2}H_{i+1}B_{i+1}. \tag{12}$$

If $B_{i+l} < \frac{1}{m}B_i$, then there exists an arc $e \in (S_i, T_i)$ such that $b(e) > B_{i+l}$. Such an arc is unessential from iteration $i+l$ on.

If $B_{i+l} \geq \frac{1}{m}B_i$, then also $B_{i+l} \geq \frac{1}{m}B_{i+1}$, and Inequality (12) implies that $H_{i+l} < \frac{1}{m}H_{i+1}$. Lemma 4.1 says that the capacity of (S_i, T_i) at the beginning of iteration $i+l+1$ is greater than H_{i+1}, which is greater than mH_{i+l}. It means that the capacity of some arc in (S_i, T_i) in network $G_{\delta_{i+l+1}}^{f_{i+l}}$ (which is the network at the beginning of iteration $i+l+1$) is greater than H_{i+l}, the total demand in $G_{\delta_{i+l+1}}^{f_{i+l}}$. Such an arc is unessential from iteration $i+l+1$ on. Observe that all arc in (S_i, T_i) are not unessential at the beginning of iteration i. ∎

Theorem 4.3 *Newton's method solves PF in $O(m)$ iterations.*

Proof. Lemma 3.1 implies that for each iteration i,

$$\frac{B_i}{B_{i-1}} \leq \frac{1}{2} \tag{13}$$

or

$$\frac{H_i}{H_{i-1}} \leq \frac{1}{2}. \tag{14}$$

We first bound the number of iterations i for which (13) holds. Let q be the number of such iterations. Let $(\mu_i)_{i=1}^q$ be the sequence of B_i's in these iterations, and $(\alpha_j)_{j=1}^p$ be the sequence of all positive b_j's arranged in nonincreasing order. Sequences $(\alpha_j)_{j=1}^p$ and $(\mu_i)_{i=1}^q$ satisfy the conditions of Lemma 4.4, so $q \leq p \leq m$.

Now we bound the number of iterations for which (14) holds. Here the argument is more involved, because H_i's are not just subsums of a set of m elements. They are related to the current demands and capacities, which vary from iteration to iteration.

To avoid towering subscripts, we renumber iterations taking into an account only iterations for which (14) holds. It means that now iteration i is what used to be the ith iteration with (14). From now on we consider only these iterations, and all indices refer to the new numbering.

We assign subsequent iterations to arcs in such a way that at most 5 iterations are assigned to one arc. We stop the process of assigning when all but at most $q + 2 = \lceil \log m \rceil + 5$ iterations have been assigned.

Assume that at least $q + 3$ iterations, i, $i + 1$, $i + 2$, ..., are still unassigned. Consider cut (S_i, T_i). Let p be its cardinality. Let for $1 \leq l \leq q$ and $1 \leq j \leq p$, $\gamma_{l,j}$ be the capacity of the jth arc in (S_i, T_i) at the beginning of iteration $i + l + 2$ (assuming some arbitrary order of the arcs in the cut). It follows from Lemma 4.1 that for each $1 \leq l \leq q$,

$$\sum_{j=1}^p \gamma_{l,j} \geq H_{i+1}. \tag{15}$$

For each arc, the difference between its capacities at the beginning of iterations $k + 1$ and $k + 2$ is not greater than the total demand at the beginning of iteration $k + 1$, which is equal to H_k. It means that

$$|\gamma_{l,j} - \gamma_{l+1,j}| \leq H_{i+l+1} \leq \frac{1}{2^l} H_{i+1}. \tag{16}$$

The definition of q, (15), and (16) imply that matrix $(\alpha_{l,j}) = (\gamma_{l,j}/H_{i+1})$ satisfies the conditions of Lemma 4.5. According to the definition of good elements in the statement of Lemma 4.5, if $\alpha_{l,j}$ is good, then there exists $l' \leq l$ such that $\alpha_{l',j} > 1/2^{l'}$. It means that $\gamma_{l',j} > 1/2^{l'} H_{i+1} \geq H_{i+l'+1}$, so arc j is unessential in iteration $i + l' + 2$ (and in all subsequent iterations). Lemma 4.5 says that there exists $k \geq 3$ such that at least $k - 2$ elements from $\alpha_{k,1}$, $\alpha_{k,2}$, ..., $\alpha_{k,p}$ are good. Therefore the arcs corresponding to these $k - 2$ elements are unessential in iteration $i + k + 2$. We assign iterations i through $i + k + 1$ to these arcs. Notice that none of these arcs was unessential at iteration i, so none of the previous iterations was assigned to any of these arcs. We assign $k + 2$ iterations to $k - 2$ arcs, and $k \geq 3$, so no more than 5 iterations are assigned to one arc.

The above process of assigning iterations to arcs implies that there are at most $5m + O(\log m)$ iterations for which (14) holds. ∎

Lemma 4.4 *Let* $\alpha_1 \geq \alpha_2 \geq \cdots \geq \alpha_p > 0$ *and* $\mu_1 > \mu_2 > \ldots > \mu_q > 0$ *be such that*

1. $\mu_{i+1} \leq \frac{1}{2}\mu_i$, *for* $i = 1, 2, \ldots, q - 1$,

2. $\mu_q \geq \alpha_p$,

3. $\mu_i \leq \sum\{\alpha_j \,|\, \alpha_j \leq \mu_i\}$, *for* $i = 1, 2, \ldots, q$.

Then $q \leq p$.

Proof. Let $\bar{\alpha}_j = \alpha_j + \alpha_{j+1} + \cdots + \alpha_p$. Condition 3 implies that $\mu_1 \leq \bar{\alpha}_1$. Thus, to prove that $q \leq p$, it is sufficient to show that each of the intervals $(0, \bar{\alpha}_p]$, $(\bar{\alpha}_p, \bar{\alpha}_{p-1}]$, $\ldots$, $(\bar{\alpha}_3, \bar{\alpha}_2]$, $(\bar{\alpha}_2, \bar{\alpha}_1]$ contains at most one element from (μ_i). Condition 2 implies that only the last element can be in $(0, \bar{\alpha}_p]$. Let for some $1 \leq j \leq p - 1$ and $1 \leq i \leq q - 1$, $\mu_i \in (\bar{\alpha}_{j+1}, \bar{\alpha}_j]$. We show that $\mu_{i+1} \leq \bar{\alpha}_{j+1}$.

If $\bar{\alpha}_{j+1} \geq \frac{1}{2}\bar{\alpha}_j$, then

$$\mu_{i+1} \leq \frac{1}{2}\mu_i \leq \frac{1}{2}\bar{\alpha}_j \leq \bar{\alpha}_{j+1}.$$

If $\bar{\alpha}_{j+1} < \frac{1}{2}\bar{\alpha}_j$, then

$$\mu_{i+1} \leq \frac{1}{2}\mu_i \leq \frac{1}{2}\bar{\alpha}_j = (\bar{\alpha}_{j+1} - \frac{1}{2}\bar{\alpha}_j) + \alpha_j < \alpha_j.$$

The above inequality and Condition 3 imply $\mu_{i+1} \leq \bar{\alpha}_{j+1}$. ∎

Lemma 4.5 *Let* $(\alpha_{i,j})$ *be a* $q \times p$ *matrix such that*

1. $q \geq \log p + 3$,

2. the sum of each row is not less than 1,

3. $|\alpha_{i+1,j} - \alpha_{i,j}| \leq 1/2^i$, *for* $1 \leq i \leq q$ *and* $1 \leq j \leq p$.

If $\alpha_{i,j} > 1/2^i$, *then we call this and all subsequent elements in column* j *good elements. There exists* k *such that* $3 \leq k \leq q$ *and row* k *contains at least* $k - 2$ *good elements.*

Proof. Condition 3 implies that if $1 \leq i' < i'' \leq q$ and $1 \leq j \leq p$, then

$$\alpha_{i'',j} \;\leq\; \alpha_{i',j} + \frac{1}{2^{i'}} + \frac{1}{2^{i'+1}} + \cdots + \frac{1}{2^{i''-1}}$$

$$< \; \alpha_{i',j} + \frac{1}{2^{i'-1}}. \tag{17}$$

If $\alpha_{i,j}$ is the first good element in column j and $i \geq 2$, then

$$\alpha_{i,j} \leq \alpha_{i-1,j} + \frac{1}{2^{i-1}} \leq \frac{1}{2^{i-1}} + \frac{1}{2^{i-1}} = \frac{1}{2^{i-2}}. \tag{18}$$

Assume that for each $k \geq 3$, row k contains at most $k-3$ good elements. (In particular, there are no good elements in rows 1, 2 and 3.) We will get contradiction by showing that the sum of the last row is less than 1. Let $j_1, j_2, \ldots, j_l$ be the indices of all columns with at least one good element. Let $\alpha_{i_1,j_1}, \alpha_{i_2,j_2}, \ldots, \alpha_{i_l,j_l}$ be the first good elements in these columns. Let $j_1, j_2, \ldots, j_l$ be ordered in such a way that $i_1 \leq i_2 \leq \cdots \leq i_l$. Row i_k, $k = 1, 2, \ldots, l$, contains at least k good elements $(\alpha_{i_k,j_1}, \alpha_{i_k,j_2}, \ldots, \alpha_{i_k,j_k})$ so

$$k \leq i_k - 3. \tag{19}$$

The sum of the last row is at most

$$\frac{p-l}{2^q} + \alpha_{q,j_1} + \alpha_{q,j_2} + \cdots + \alpha_{q,j_l}$$

$$\leq \frac{p}{2^q} + \left(\alpha_{i_1,j_1} + \frac{1}{2^{i_1-1}}\right) + \cdots + \left(\alpha_{i_l,j_l} + \frac{1}{2^{i_l-1}}\right)$$

$$\leq \frac{1}{8} + \left(\frac{1}{2^{i_1-2}} + \frac{1}{2^{i_1-1}}\right) + \cdots + \left(\frac{1}{2^{i_l-2}} + \frac{1}{2^{i_l-1}}\right)$$

$$= \frac{1}{8} + 6\left(\frac{1}{2^{i_1}} + \cdots + \frac{1}{2^{i_l}}\right)$$

$$\leq \frac{1}{8} + 6\left(\frac{1}{2^4} + \frac{1}{2^8} + \cdots\right)$$

$$< 1.$$

The first inequality follows from Inequality (17), the second one from Condition 1 and Inequality (18), and the third one from (19). ∎

5   Analysis of Newton's Method for the Uniform PF Problem

In this section we analyze Newton's method for the maximum mean cut - uniform parametric flow problem. The main content of Section 5.1 is a proof that the number of iterations is only $O(n)$. In Section 5.2 we show networks which require $\Omega(n)$ iterations. We show that the polynomial upper bound implied by Theorem 3.4 is also tight.

5.1   Upper Bounds on the Number of Iterations

We use the notation from Section 4: δ_i is equal to $\bar{\delta}$ at the beginning of iteration i, (S_i, T_i) is the cut found in this iteration, B_i is the weight of (S_i, T_i), which now is equal to the cardinality of (S_i, T_i), and H_i is the surplus of (S_i, T_i) in the current network. We also use

$$\xi_i \stackrel{\text{def}}{=} mean_{\delta_i}(S_i, T_i) = \frac{surplus_{\delta_i}(S_i, T_i)}{|(S_i, T_i)|} = \frac{H_i}{B_i}.$$

Again, it is easier to follow the analysis remembering that H_i is equal to the total demand left after iteration i.

We first state a bound on the number of iterations in the case of integral capacities and demands.

Theorem 5.1 *If all capacities and demands are integers in the range* $[-D, D]$, *then the algorithm terminates after* $O(\frac{\log(nD)}{1+\log\log(nD)-\log\log n})$ *iterations.*

Proof. This is a specialization of Theorem 3.4. Here $p = m + n$, $B = 1$, and $A = D$. ∎

This bound on the number of iterations in Newton's method for the MWC problem with integral data was independently shown by Rote [30]. The bound is asymptotically better than the $O(\log(nD))$ bound on the number of iterations in the binary search method, if $D = n^{\omega(1)}$. If $D = \Omega(n^{\log^{\alpha} n})$ for any positive constant α, the bound for Newton's method is better by a factor of $\log\log(nD)$. The MWC problem with integral data was also studied by Iwano, Misono, Tezuka, and Fujishige [16]. They proposed an algorithm, which runs in $O(nm\log(nD))$ time. Their algorithm essentially follows the pattern of the binary search method, but it computes in each iteration an approximate maximum flow instead of an exact one. The necessary approximation can be computed in $O(mn)$ time, which is slightly better than the best known bound (1) for computing the exact maximum flow. We showed in [27] that this idea of computing only approximate maximum flows can be also used in connection with Newton's method. We obtained an $O(\min\{nm, \frac{\min\{nm+n^2\sqrt{\log n}, nm\log\log n\}}{1+\log\log(nD)-\log\log n}\}\log(nD))$ bound on the overall running time, which is never worse than $O(mn\log(nD))$ and is better for certain values of n, m, and D.

Now we consider the general case, capacities and demands are arbitrary real numbers. The overall scheme of the analysis is the same as in Section 4. We estimate the number of iterations needed to make a new arc unessential. The definition of an unessential arc is slightly different here. An arc is *unessential* if its (current) capacity is greater than or equal to the (current) total demand. An unessential arc will not be in any subsequent blocking cut.

Lemma 5.2 *After at most* $\lfloor \log m \rfloor + 2$ *iterations at least one arc is unessential in every pair of opposite arcs* (v, w) *and* (w, v).

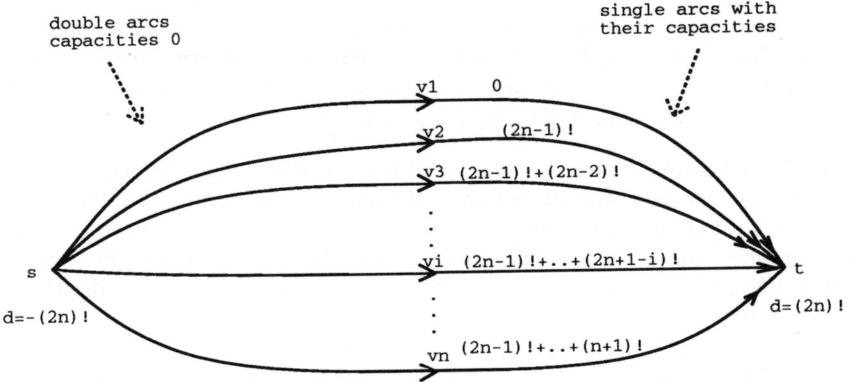

Figure 3: **Network G_n, which requires $\Omega(n)$ iterations**

Proof. Assume there are at least $l = \lfloor \log m \rfloor + 2$ iterations. Using Inequality 7 we have,

$$H_l \leq H_l B_l \leq \frac{1}{4^{l-1}} H_1 B_1 \leq \frac{H_1 B_1}{m^2} \leq \frac{H_1}{m} \leq \xi_1.$$

After the first iteration the sum of the capacities of opposite arcs (v, w) and (w, v) is at least $2\xi_1$ and never decreases. Therefore after the lth iteration one of them has capacity at least ξ_1, which is not less than the total current demand, so this arc is unessential. ∎

Lemma 5.3 *If $B_{i+3} > B_{i+1} - \sqrt{B_{i+1}}$, then at least one arc from (S_i, T_i) is unessential after $(i+3)$rd iteration.*

Proof. Assume that $B_{i+3} > B_{i+1} - \sqrt{B_{i+1}}$. Inequality (6), monotonicity of sequence (B_i), and this assumption imply

$$H_{i+3} \leq H_{i+1} \left(1 - \frac{B_{i+2}}{B_{i+1}}\right)\left(1 - \frac{B_{i+3}}{B_{i+2}}\right) \leq H_{i+1}\left(1 - \frac{B_{i+3}}{B_{i+1}}\right)^2 \leq \frac{H_{i+1}}{B_{i+1}} = \xi_{i+1}.$$

The capacity of cut (S_i, T_i) after the $(i+3)$rd iteration (actually already after the $(i+1)$st iteration) is at least $\xi_{i+1} B_i$, so at least one arc in this cut has capacity at least ξ_{i+1}. Since the total demand after the $(i+3)$rd iteration is at most ξ_{i+1}, such an arc is unessential. Note that arcs in (S_i, T_i) are not unessential before iteration i. ∎

Theorem 5.4 *The algorithm terminates after $O(n)$ iterations.*

Proof. There are at most $3\sqrt{m}$ disjoint sequences of four iterations, $i, i+1, i+2, i+3$, such that $B_{i+3} \leq B_{i+1} - \sqrt{B_{i+1}}$. To see this consider the sequence defined as follows: $a_1 = m$, and $a_{j+1} = a_j - \sqrt{a_j}$ until for some l, $a_l \leq 1$. It is easy to check that $a_{\lceil \sqrt{m/2} \rceil} \leq m/2$. Therefore $l \leq \left\lceil \sqrt{\frac{m}{2}} \right\rceil + \left\lceil \sqrt{\frac{m}{4}} \right\rceil + \left\lceil \sqrt{\frac{m}{8}} \right\rceil + \cdots < 3\sqrt{m}$.

Lemma 5.3 states that every sequence of four iterations such that $B_{i+3} > B_{i+1} - \sqrt{B_{i+1}}$ makes one new arc unessential. If such a sequence occurs after the first $\lfloor \log m \rfloor + 2$ iterations and makes, say, arc (v, w) unessential, then both (v, w) and (w, v) are unessential (arc (w, v) is already unessential by Lemma 5.2). From now on v and w will never be on different sides of a blocking cut. We can consider them as contracted into one super-node. Obviously, there cannot be more than $n - 1$ such contractions.

Putting this all together, there are at most $12\sqrt{m} + \log m + 2 + 4(n-1) = O(n)$ iterations in the algorithm. ∎

5.2   Worst Case Examples

In this section we exhibit examples that show tightness of the bounds from the previous section. To simplify the description, we use multiple arcs in our constructions. Each such arc can be replaced by an appropriate number of paths of length 2, without changing the merit of our results.

Let $G_n = (V, E, u, d)$ be defined as follows.

- $V = \{s, t\} \cup W$, where $W = \{v_1, v_2, \ldots, v_n\}$.

- $E = A \cup B$, where for each $v \in W$, A contains two arcs from s to v and B contains a single arc from v to t.

- $u(e) = 0$, for every $e \in A$, and $u(v_i, t) = \sum_{j=1}^{i-1}(2n - j)!$

- $d(t) = -d(s) = (2n)!$, and $d(v) = 0$, for each $v \in W$.

Network G_n is shown in Figure 3.

Lemma 5.5 *For $1 \leq i \leq n+1$, the blocking cut in G_n found in the ith iteration is $(\{s, v_1, \ldots, v_{i-1}\}, \{v_i, \ldots, v_n, t\})$, and the total demand right after the ith iteration is equal to $(2n + 1 - i)!$.*

Proof. The proof is by induction. It is easy to check that the statement is true for $i = 1$. Let $1 \leq i \leq n$ and assume that the statement is true for all $1 \leq j \leq i$. The demands after iterations $1, 2, \ldots, i$ were $(2n)!, (2n-1)!, \ldots, (2n+1-i)!$, respectively, and the cardinalities of the blocking cuts were $2n, 2n - 1, \ldots, 2n + 1 - i$. Therefore, the capacity of each arc has been increased during iterations 1 through i by $\delta =$

$(2n-1)! + (2n-2)! + \cdots + (2n-i)!$. In $(G_n)_\delta$, the total capacity of two arcs from s to v_l and the capacity of the arc from v_l to t are, respectively,

$$u_\delta(s, v_l) = 2\delta = 2((2n-1)! + (2n-2)! + \cdots + (2n-i)!)$$

and

$$\begin{aligned} u_\delta(v_l, t) = u(v_l, t) + \delta &= [(2n-1)! + (2n-2)! + \cdots + (2n+1-l)!] + \\ &+ [(2n-1)! + (2n-2)! + \cdots + (2n-i)!]. \end{aligned}$$

Therefore, if $i+1 \leq n$, there are two maximum blocking cuts in the $(i+1)$st iteration: $(\{s, v_1, \ldots, v_i\}, \{v_{i+1}, \ldots, v_n, t\})$ and $(\{s, v_1, \ldots, v_{i+1}\}, \{v_{i+2}, \ldots, v_n, t\})$. We assume that $(\{s, v_1, \ldots, v_i\}, \{v_{i+1}, \ldots, v_n, t\})$ is the cut the algorithm detects in this iteration. (We could slightly perturb the capacities and have exactly one maximum blocking cut, the one that is needed for our argument.) The only blocking cut in $(n+1)$st iteration is $(\{s, v_1, \ldots, v_n\}, \{t\})$. The demand after the $(i+1)$st iteration is equal to

$$\begin{aligned} (2n)! - 2\delta(n-i) - \sum_{l=1}^{i}(u(v_l, t) + \delta) &= \\ &= (2n)! - \sum_{l=1}^{i} u(v_l, t) - (2n-i)\delta \\ &= (2n)! - \sum_{l=1}^{i}\sum_{j=1}^{l-1}(2n-j)! - (2n-i)\sum_{j=1}^{i}(2n-j)! \\ &= (2n-i)!. \end{aligned}$$

∎

The above lemma immediately implies the following theorem.

Theorem 5.6 *For every $n \geq 1$, Newton's method runs in $n+1$ iterations on network G_n. The number of nodes and arcs in G_n is $\Theta(n)$.*

This theorem shows that the upper bound proven in Theorem 5.4 is tight. All demands and capacities in G_n are integers not greater than $D = (2n)! = n^{\Theta(n)}$. Let $D = O(n^{O(n)})$, and $k \leq n$ be such that $k = \Theta(\frac{\log D}{\log \log D})$. Adding some insignificant nodes and arcs to G_k we can construct an n node network for which k iterations are required. It means that the combined upper bound from Theorems 5.1 and 5.4 is tight for $D = \Omega(n^{(\log^\alpha n)})$, where α is any positive constant. The next example shows a network with $\Theta(n)$ arcs and $D = O(n)$, that requires $\Omega(\log n)$ iterations. Thus the upper bound is also tight if D is polynomial in n. This construction is similar to the above one, but not being allowed to use big numbers we have to obtain the appropriate mean surpluses of cuts by manipulating with the multiplicities of arcs.

Let $n = k2^k$. $H_n = (V, E, u, d)$ is defined as follows.

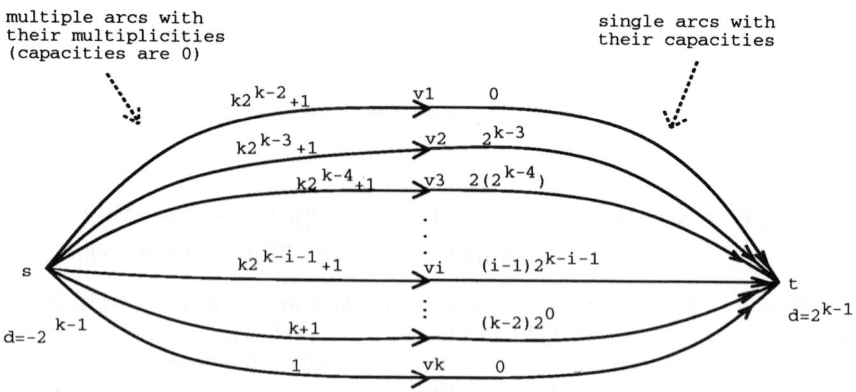

Figure 4: **Network H_n, $n = k2^k$. $\Theta(n)$ arcs, $D = O(n)$, $\Theta(\log n)$ iterations required.**

- $V = \{s, t\} \cup W$, where $W = \{v_1, v_2, \ldots, v_k\}$.

- $E = A_1 \cup A_2 \cup \cdots \cup A_k \cup B$, where for $i = 1, 2, \ldots, k-1$ A_i is the set of $k2^{k-i-1}+1$ arcs from s to v_i, $A_k = \{(s, v_k)\}$, and $B = \{(v, t) : v \in W\}$.

- $u(e) = 0$ for every $e \in A_1 \cup \cdots \cup A_k \cup \{(v_1, t), (v_k, t)\}$, and $u(v_i, t) = (i-1)2^{k-i-1}$, for $i = 2, \ldots, k-1$.

- $d(t) = -d(s) = 2^{k-1}$, and $d(v) = 0$ for each $v \in W$.

Network H_n is shown in Figure 4.

Lemma 5.7 *For $1 \leq i \leq k$, the blocking cut in H_n found in the ith iteration is $(\{s, v_1, \ldots, v_{i-1}\}, \{v_i, \ldots, v_k, t\})$, and the total demand right after this iteration is equal to 2^{k-i}.*

Proof. Similar induction as in the proof of Lemma 5.5. ∎

The construction of H_n and the above lemma imply the following theorem.

Theorem 5.8 *For every $n \geq 1$, Newton's method runs in $\Omega(\log n)$ iterations on network H_n. This network has $\Theta(n)$ arcs and all demands and capacities are integers from $[-n, n]$.*

We believe that by some combination of our two constructions one could show that the upper bound on the number of iterations is also tight for the remaining values of D.

6 Concluding Remarks

We showed a strongly polynomial bound on the number of iterations in Newton's method for any linear fractional combinatorial optimization problem. Is the bound $O^*(p^2)$ the best possible? We have not been able to show an LFCO problem and a family of instances which would require more than $O(p)$ iterations. Another interesting question is how the number of iterations changes if in each iteration we solve only approximately an appropriate instance of the underlying linear problem.

We proved that Newton's method runs in $O(m)$ iterations for the maximum mean-weight cut - parametric flow problem, and in $O(n)$ iterations for the maximum mean cut - uniform parametric flow problem. These bounds show that Newton's method gives the currently fastest algorithms for these problems. We also showed that even in the uniform case Newton's method may run in $\Omega(n)$ iterations. The main question is whether the bound $O(m)$ for the general case is tight. We conjecture that the real bound is $\Theta(n)$.

It is somewhat surprising that the best known algorithms for the maximum mean-weight cut problem and the maximum mean cut problem, with a seemingly rich combinatorial structure, come from the straightforward application of a very general method. Should we expect the existence of methods special for these problems which give more efficient algorithms?

Acknowledgments

We would like to thank Andrew Goldberg and S. Thomas McCormick for helpful comments at various stages of this work. We would also like to thank Michel Goemans for allowing us to include his bound and proof concerning the length of a geometric sequence of subsums (Lemma 3.6).

This research was supported in part by NSF Presidential Young Investigator Grant CCR-8858097 with matching funds from AT&T and DEC, and ONR Young Investigator Award N00014-91-J-1855, while the author was at Computer Science Department, Stanford University, and by NSF grant DMS-8920550 and by the Packard Fellowship of Éva Tardos, while the author was at School of Operations Research, Cornell University.

References

[1] P. J. Carstensen. *The Complexity of Some Problems in Parametric, Linear, and Combinatorial Programming.* PhD thesis, Department of Mathematics, Univ. of Michigan, Ann Arbor. Mich., 1983.

[2] R. Chandrasekaran. Minimum ratio spanning trees. *Networks*, 7:335–342, 1977.

[3] E. Cohen. *Combinatorial Algorithms for Optimization Problems*. PhD thesis, Stanford Univ., June 1991. (Also available as Technical Report STAN-CS-91-1366, Department of Computer Science, Stanford Univ., 1991).

[4] E. Cohen and N. Megiddo. Strongly Polynomial Time and NC Algorithms for Detecting Cycles in Dynamic Graphs. In *Proc. 21st Annual ACM Symposium on Theory of Computing*, pages 523–534, 1989.

[5] E. Cohen and N. Megiddo. Maximizing Concave Functions in Fixed Dimension. Technical Report RJ 7656 (71103), IBM Almaden, 1990.

[6] E. Cohen and N. Megiddo. Algorithms and Complexity Analysis for Some Flow Problems. In *Proc. 2nd ACM-SIAM Symposium on Discrete Algorithms*, pages 120–130, 1991.

[7] W. Dinkelbach. On nonlinear fractional programming. *Management Science*, 13:492–498, 1967.

[8] T. R. Ervolina and S. T. McCormick. A Strongly Polynomial Dual Cancel and Tighten Algorithm for Minimum Cost Network Flow. UBC Faculty of Commerce Working Paper 90-MSC-010, 1990.

[9] T. R. Ervolina and S. T. McCormick. A Strongly Polynomial Maximum Mean Cut Cancelling Algorithm for Minimum Cost Network Flow. UBC Faculty of Commerce Working Paper 90-MSC-009, 1990.

[10] A. V. Goldberg and R. E. Tarjan. A New Approach to the Maximum Flow Problem. *J. Assoc. Comput. Mach.*, 35:921–940, 1988.

[11] A. V. Goldberg and R. E. Tarjan. Finding Minimum-Cost Circulations by Canceling Negative Cycles. *J. Assoc. Comput. Mach.*, 36:388–397, 1989.

[12] A. V. Goldberg and R. E. Tarjan. Finding Minimum-Cost Circulations by Successive Approximation. *Math. of Oper. Res.*, 15:430–466, 1990.

[13] P. Hansen, M. V. Poggi de Aragão, and C. C. Ribeiro. Hyperbolic 0-1 Programming and Query Optimization in Information Retrieval. *Math. Prog. B*, 52:255–263, 1991.

[14] A. J. Hoffman. A generalization of max flow - min cut. *Math. Prog.*, 6:352–359, 1974.

[15] T. Ibaraki. Parametric approaches to fractional programs. *Math. Programming*, 26:345–362, 1983.

[16] K. Iwano, S. Misono, S. Tezuka, and S. Fujishige. A new scaling algorithm for the maximum mean cut problem. Unpublished manuscript (To appear in *Algorithmica*), 1990.

[17] A. V. Karzanov. On minimal mean cuts and circuits in a digraph. In *Methods for Solving Operator Equations*, pages 72–83. Yaroslavl State Univ., Yaroslavl, USSR, 1985. In Russian.

[18] V. King, S. Rao, and R. Tarjan. A Faster Deterministic Maximum Flow Algorithm. In *Proc. 3rd ACM-SIAM Symposium on Discrete Algorithms*, pages 157–164, 1992.

[19] E. L. Lawler. *Combinatorial Optimization: Networks and Matroids*. Holt, Reinhart, and Winston, New York, NY., 1976.

[20] S. T. McCormick and T. R. Ervolina. Computing Maximum Mean Cuts. UBC Faculty of Commerce Working Paper 90-MSC-011, 1990.

[21] N. Megiddo. Combinatorial optimization with rational objective functions. *Math. of Oper. Res.*, 4:414–424, 1979.

[22] N. Megiddo. Applying Parallel Computation Algorithms in the Design of Serial Algorithms. *J. Assoc. Comput. Mach.*, 30:852–865, 1983.

[23] C. Haibt Norton, S. A. Plotkin, and É. Tardos. Using Separation Algorithms in Fixed Dimension. *J. Alg.*, 13:79–98, 1992.

[24] J. B. Orlin. A Faster Strongly Polynomial Minimum Cost Flow Algorithm. In *Proc. 20th Annual ACM Symposium on Theory of Computing*, pages 377–387, 1988.

[25] P. M. Pardalos and A. T. Phillips. Global optimization of fractional programs. *J. Global Opt.*, 1:173–182, 1991.

[26] T. Radzik. *Algorithms for Some Linear and Fractional Combinatorial Optimization Problems*. PhD thesis, Stanford Univ., August 1992. (Also available as Technical Report STAN-CS-92-1451, Department of Computer Science, Stanford Univ., 1992).

[27] T. Radzik. Minimizing capacity violations in a transshipment network. In *Proc. 3rd ACM-SIAM Symposium on Discrete Algorithms*, pages 185–194, 1992.

[28] T. Radzik. Newton's method for fractional combinatorial optimization. In *Proc. 33rd IEEE Annual Symposium on Foundations of Computer Science*, pages 659–669, 1992.

[29] T. Radzik and A. Goldberg. Tight bounds on the number of minimum-mean cycle cancellations and related results. In *Proc. 2nd ACM-SIAM Symposium on Discrete Algorithms*, pages 110–119, 1991. (to appear in *Algorithmica*).

[30] G. Rote. An improved time bound for computing maximum mean cuts. presented at 14th International Symposium on Mathematical Programming, Amsterdam, The Netherlands, August 1991.

[31] S. Schaible. Fractional programming 2. On Dinkelbach's Algorithm. *Management Sci.*, 22:868–873, 1976.

[32] S. Schaible. Fractional programming. *Zeitschrift für Operations Res.*, 27:39–54, 1983.

[33] S. Schaible and T. Ibaraki. Fractional programming. *Europ. J. of Operational Research*, 12, 1983.

[34] É. Tardos. A Strongly Polynomial Minimum Cost Circulation Algorithm. *Combinatorica*, 5(3):247–255, 1985.

[35] S. Toledo. Maximizing non-linear concave functions in fixed dimension. In *Proc. 33rd IEEE Annual Symposium on Foundations of Computer Science*, pages 676–685, 1992.

[36] C. Wallacher. A Generalization of the Minimum-mean Cycle Selection Rule in Cycle Canceling Algorithms. Unpublished manuscript, Institut für Angewandte Mathematik, Technische Universität Carolo-Wilhelmina, Germany, November 1989.

Complexity in Numerical Optimization, pp. 387-405
P.M. Pardalos, Editor
©1993 World Scientific Publishing Co.

Analysis of a Random Cut Test Instance Generator for the TSP

Ronald L. Rardin
School of Industrial Engineering, Purdue University, West Lafayette, IN 47907 USA

Craig A. Tovey
Industrial and Systems Engineering, Georgia Institute of Technology, Atlanta, GA 30332-0205 USA

Martha G. Pilcher
School of Business, University of Washington, Seattle, WA 98105 USA

Abstract

Test Instance Generators (TIG's) are important to evaluate heuristic procedures for NP-hard problems. We analyze a TIG in use for the TSP. This TIG, due to Pilcher and Rardin, is based on a *random cut* method. We show that it generates a class of instances of *intermediate* complexity: not as hard as the entire TSP class unless $NP = co(NP)$; not as easy as P unless $NP = P$. Since the upper bound on complexity must hold for any efficient TIG, our analysis verifies that this random cut TIG is, in a sense, as good as possible a TIG for the TSP. This suggests that the random cut method may be a good basis for constructing TIG's for other problems.

Keywords:
traveling salesman problem, complexity, computational test, test case, problem generation

1   Introduction

A tremendous amount of effort in the past two decades has been directed towards developing good heuristics for NP-hard problems. A good heuristic, classically, has performance ratio v^h/v^* close to 1, where v^h and v^* denote the value of the heuristic and optimal solutions, respectively. Since we have no practical way to find v^* for an arbitrary instance (if we did we wouldn't be resorting to heuristics) we cannot empirically determine v^h/v^* by testing instances generated purely at random. Hence we need a Test Instance Generator (TIG) that supplies instances with known optimal solutions.

Pilcher and Rardin [11, 12] developed a TIG in use for the TSP based on a *random cut* method. This TIG was later extended by Rais and Rardin [13]. We analyze the complexity of the class of instances generated by this TIG. The result is that the TIG essentially does as well as one could hope for. In particular, while it fails to generate a class as difficult as the TSP in general, this is a generic failing of efficient TIG's. We verify that the random cut TIG does generate a class as difficult as possible (for efficient TIG's).

Let us for a moment discuss TIG's in general. These have been considered by Sanchis [14]. Sanchis observes that any TIG designer faces a problem analogous to the cryptographers — how to simultaneously satisfy three conditions:

1. The TIG should generate instances *efficiently*, i.e., it should run in time polynomial in the length of the output.

2. The TIG should generate instances with *known optimal solutions*.

3. The TIG should generate instances that are *hard to solve*.

If one did not enforce condition (2), one could just generate an instance at random. If one did not enforce condition (1), one could just generate an instance at random and solve it by brute force.

Returning to the TSP, let us call the random cut generator the RC TIG. Ideally, to satisfy condition (3), we would like to be able to generate instances as hard as the hardest TSP instances. It is very unlikely that RC accomplishes this; for unless $NP = co(NP)$, its instances are not at the same complexity level as the general TSP. This turns out not to be a specific failing of the RC TIG, but rather an inevitable consequence of satisfying conditions (1) and (2). This is because the language generated by anything satisfying (1) and (2) must be in NP; the "solver" could non-deterministically replicate the generative process and thereby have a succinct proof of optimality. That is, there must be a short computation that solves the instance, namely the computation path of the TIG itself. Therefore, the full complexity of an optimization problem cannot be captured by a TIG satisfying (1) and (2). [See [14] for more details.]

Does this mean that the RC TIG must produce *easy* problems? Surprisingly, the answer is "no". We show that unless $P = NP$, the instances cannot be solved in (deterministic) polynomial time. This is true even if the instance comes with a "promise" that it was created by the generator. At the heart of the analysis is the issue of recognizing valid input.

Taken together, these results imply that the class of instances so generated is of *intermediate* complexity. This depends on distinguishing different questions associated with an instance. The *exact value* question, which asks whether a particular v is the optimal solution value to a given instance, is D^P-complete for general TSP instances (harder than NP-Complete unless $NP = co(NP)$), but only in NP for the instances generated by RC TIG. On the other hand, the search problem, which asks for an optimal solution given an instance, turns out to be NP-hard for the RC TIG instances. Hence, the RC TIG generates instances as hard as can be expected from an efficient TIG.

TIG's with similar properties could be constructed for many other optimization problems for which integer programming formulations are known. We believe this suggests the random cut method is a good basis for effective TIG design, at least according to criteria 1,2,and 3.

2 Versions of the General TSP and Class D^P

The Traveling Salesman Problem is the problem of finding a minimum total weight hamiltonian (vertex-spanning) cycle of a graph. Our interest will always be in the symmetric (undirected) case on a completed graph with $n \triangleq |V|$ vertices and rational edge weights. A formal definition is as follows:

Traveling Salesman Optimization (TSP)

Instance: a complete graph G with rational weights on the edges.

Solution: a minimum total weight hamiltonian cycle of G.

Several different language recognition or decision problems can be derived from this optimization form. The most familiar is the *threshold* decision problem

Traveling Salesman Threshold ($TSP^{\leq}$)

Instance: same as TSP plus a rational threshold v.

Question: Does there exist a hamiltonian cycle of G with total weight less than or equal to v?

A related decision problem important to our development is the *exact value* version

Traveling Salesman Exact Value ($TSP^{=}$)

Instance: same as $TSP^{\leq}$.

Question: Does a minimum total weight hamiltonian cycle of G have weight v?

The rest of this section contains elementary background material related to these three versions of the TSP.

We have already alluded to the well-known facts that $TSP^{\leq} \in NP\text{-}Complete$, $TSP^{\leq} \propto TSP$, and thus $TSP \in NP - Hard$ (here and throughout $\propto$ denotes polynomial reduction).

Also notice that the optimization version of the TSP is qualitatively different from the two decision problems in the type of "solution" it demands. The optimization form requires a full optimal solution. Contrast with decision problems $TSP^{\leq}$ and $TSP^{=}$ that call for only a *yes* or *no* response. An algorithm "solves" the latter problems if it can recognize all *yes* cases, i.e. accepts an input if and only if its is a well formed instance for which the corresponding question is properly answered *yes*.

Still, these decision questions are not equivalent. The former is certainly in NP. The latter is probably not because it is complete for complexity class D^P, introduced by Papadimitriou and Yannakakis [8].

A problem in D^P is formed as the intersection of the set of instances of a member of NP and the set of instances of a member of $co(NP)$ (the collections of complements of problems in NP). One example is $TSP^{=}$. An instance of $TSP^{=}$ with $v = \bar{v}$ has proper answer *yes* if and only if the corresponding instance of $TSP^{\leq}$ can be answered *yes*, and the instance with $v = \bar{v} - \delta$ has answer *no*, where δ is the least common denominator of its edge weight denominators. Informally, $TSP^{=}$ is the intersection of $TSP^{\leq}$ with a "translation" of its own complement.

It is also easy to see that $NP \subseteq D^p$ and $co(NP) \subseteq D^p$. To show any problem in NP (respectively $co(NP)$) belongs to D^p, we need only appeand a vacuous $co(NP)$ question (respectively NP question).

Papadmitriou and Yannakakis [8] also showed there are D^p-complete problems, i.e. members of D^p to which all problems in D^p reduce in polynomial time. Among these "hardest" members of D^p are the exact value versions of many hard discrete optimization problems, including $TSP^{=}$ on which we are focusing.

We may summarize these facts about the three TSP versions we have so far introduced:

Lemma 2.1 $TSP^{\leq} \in NP - Complete$, $TSP^{=} \in D^p - Complete$, $TSP \in NP - Hard$, and $TSP^{\leq} \propto TSP^{=} \propto TSP$.

The following is implicit in [8]:

Lemma 2.2 *If any problem in $D^p - Complete$ also belongs to NP or $co(NP)$, then* $NP = co(NP) = D^p$.

We see from Lemma 2.2 that unless $NP = co(NP)$, $D^p - Complete$ is a higher complexity class that either NP or $co(NP)$. Since $TSP^{=} \in D^p - Complete$, that exact value decision problem is apparently materially different from the threshold version $TSP^{\leq} \in NP$. (Search for TSP is harder still [6].) Figure 1 summarizes in a Venn diagram the containments generally conjectured, and the standing of our three versions of the Traveling Salesman Problem.

3 Polyhedral Relaxations and Random Cut Generators

In this section we present the RC TIG for the TSP. It is based on polyhedral methods, one of the earliest approaches to combinatorial optimization, now experiencing renewed interest (see for example [1]). Given a combinatorial optimization problem OPT, the approach introduces a polynomial-dimension vector of binary decision variables, x, and encodes OPT as the binary linear program:

$$
\begin{aligned}
\text{minimize} \quad & cx \\
\text{subject to} \quad & A_0 x \leq b_0 \\
& 1 \geq x \geq 0 \\
& x \text{ integer}
\end{aligned}
$$

Here $A_0 x \leq b_0$ denotes a system of linear inequalities on solution vectors x satisfied by exactly the binary x feasible for OPT. We make no assumption about the size of the $A_0 x \leq b_0$ system relative to the size of OPT.

The *linear programming* (LP) *relaxation* of this formulation, which we denote $ROPT_0$, is formed by deleting the requirement "x integer". If an optimal solution to this relaxation happens also to be integer, it obviously produces an optimum in OPT. If the relaxation optimum is not integer, we may sharpen the formulation by adding new constraint sets

$$
\begin{aligned}
A_1 x \quad &\leq \quad b_1 \\
&\vdots \\
A_t x \quad &\leq \quad b_t
\end{aligned}
$$

Each system $A_i x \leq b_i$ contains new inequalities or "cuts" valid for all binary solutions, but violated by some x feasible in linear programming relaxations. If any of the sharper relaxations

$$
ROPT_k \qquad
\begin{aligned}
\text{minimize} \quad & cx \\
\text{subject to} \quad & A_i x \leq b_i \quad \text{for all } 0 \leq i \leq k \\
& 1 \geq x \geq 0
\end{aligned}
$$

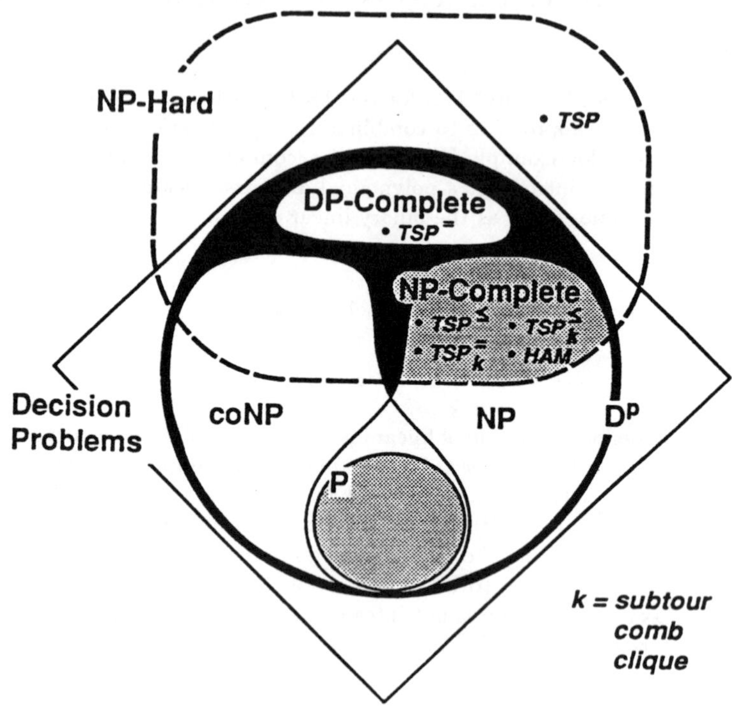

Figure 1: Generally Conjectured Complexity Class Containments

formed over the new systems ($k \leq t$) has an integer optimum, that solution also yields an optimum for OPT.

In the specific case of Traveling Salesman Problems on vertices in V, one integer linear programming formulation is

$$\text{minimize} \quad \sum_{i \in V} \sum_{j > i} c_{ij} x_{ij}$$

$$\text{subject to} \quad \sum_{j < i} x_{ji} + \sum_{j > i} x_{ij} \leq 2 \quad \text{for all } i \in V \tag{1}$$

$$-\sum_{i \in V} \sum_{j > i} x_{ij} \leq -|V| \tag{2}$$

$$\sum_{i < j \in S} x_{ij} \leq |S| - 1 \quad \text{for all } S \subset V \tag{3}$$

$$1 \geq x_{ij} \geq 0 \qquad \text{for all } i < j \in V \tag{4}$$

$$x_{ij} \text{ integer} \qquad \text{for all } i < j \in V$$

The $O(|V|^2)$ variables x_{ij} indicate whether edges (i, j) are part of the tour. Constraints (1) - (2) require the solution to have degree two at each vertex. Inequalities (3) are the famous *subtour elimination* constraints preventing non-spanning cycles. Together (1) - (3) constitute system $A_0 x \leq b_0$ for TSP.

Many other inequalities are known for the TSP that can form the systems $A_i x \leq b_i$, $i \geq 1$. Among these are the *comb inequalities* [3], and the more general *clique tree inequalities* [4].

These polyhedral considerations led us to propose a random cut generation scheme [10, 11] based on creating instances drawn from the large subset that could, in principle, be solved over an appropriate linear programming relaxation. To be more specific, standard linear programming optimality conditions establish that a vector x^* is optimal in $ROPT_k$ above if there exist dual vectors $u_0, u_1 \ldots, u_k, z, w$ satisfying

$$u_i \leq 0 \qquad \text{for all } 0 \leq i \leq k \tag{5}$$

$$u_i(A_i x^* - b_i) = 0 \qquad \text{for all } 0 \leq i \leq k \tag{6}$$

$$z \leq 0, w \geq 0 \tag{7}$$

$$z(1 - x^*) = 0, wx^* = 0 \tag{8}$$

$$c = \sum_{i=0}^{k} u_i A_i + z + w \tag{9}$$

The procedure for generating instances of OPT with known solution x^* exploits (5)–(9) as follows:

Procedure Random Cut [RC]

1. Randomly chose the feasible set for an instance of OPT, and generate any binary vector x^* satisfying the corresponding $A_0 x \leq b_0$ constraints;

2. Randomly select a sample of main and cut inequalities $A_i^* x^* \leq b_i^*, i = 0, 1, \ldots, k$, for the instance such that all are satisfied as equality at x^* (i.e. $A_i^* x^* = b_i^*$);

3. Randomly generate vectors $\{u_i^* < 0 \quad : i = 0, 1, \ldots, k\}$ of dimension consistent with $\{A_i^*\}$;

4. Randomly generate vector $z^* \leq 0$ with $z_j^* = 0$ for all j with $x_j^* \neq 1$, and vector $w^* \geq 0$ with $w_j^* = 0$ for all j with $x_j^* \neq 0$;

5. Compute $c \leftarrow \sum_{i=0}^{k} u_i^* A_i^* + z^* + w^*$;

6. Return an instance of OPT with feasible set as in Step 1 and objective function vector c;

Lemma 3.1 *The solution x^* chosen at Step 1 of procedure [RC] is optimal for the resulting OPT instance.*

Proof: An instance generated by [RC] will have (binary) x^* optimal because the construction clearly assures it satisfies LP optimality conditions (5)–(9) when all dual u_i components not set at Step 3 are taken as zero. $\square$

Pilcher [10] employed subtour and comb inequalities in an implementation of [RC] for generating TSP's. Later Rais and Rardin [13] extended the approach to include clique tree inequalities.

The technical challenge in these implementations of the random cut strategy arises in sampling Step 2. The RC method must only use inequalities that are tight for the prospective optimal solution x^*, because complementary slackness constraints (6) forbid nonzero weighting of nontight constraints at Step 3. Implementation of [RC] on any particular problem thus requires a constructive characterization of the tight set at any optimal solution. Subtour elimination constraints provide an easy illustration.

Lemma 3.2 *Subtour elimination constraint (3) for vertex subset S is satisfied as an equality at the tour indexed by x^* if and only if vertices of S are adjacent on that tour.*

Proof: Consider the subgraph of a tour induced by vertex subset S. This subgraph contains no cycles because S is a proper subset of V. Thus well known results in graph theory show its number of edges $\sum_{i<j} x_{ij}$ will total $|S|$ less the number of its connected components. It follows that the corresponding subtour elimination constraint will be active exactly when the subgraph has only one component, i.e. when vertices of S are adjacent on the tour. $\square$

With the characterization of Lemma 3.2, it is straightforward to implement RC over subtour constraints. Sampling in Step 2 is done simply by choosing intervals off the fixed tour as the S's of tight inequalities (3).

4 Exposed Instances

To analyze RC TIG's, we begin by establishing a simple geometric characterization of the instances that it generates.

Definition. *An instance of a combinatorial optimization problem OPT is exposed for partial polyhedral description k if the corresponding LP relaxation $ROPT_k$ for that instance has an integer optimal solution.*

Informally, an instance is exposed for k it it can be solved by linear programming over $ROPT_k$. Figure 2 illustrates the notion. It shows the complete polytope of a combinatorial optimization problem, along with one LP relaxation $ROPT_0$. The extreme point marked x^* belongs to the underlying OPT feasible set. Every instance of such an OPT consists of a description of its feasible set and a cost vector c. Two different instances are illustrated in Figure 2. The one in part (a) is exposed because c supports the partial polyhedral description at integer solution x^*. Relaxation $ROPT_0$ will yield x^* as an LP optimum. The instance in part (b) is not exposed. Vector x^* is still the integer optimal solution, but a more complete polyhedral description is required before that solution can be found by linear programming.

Although exposed instances can in principle be solved by linear programming, nothing in the definition guarantees it is particularly easy to do so. The number of constraints in $ROPT_k$ need not be polynomial in the size of OPT, even for $k = 0$. Thus, exposed instances are polynomially solvable by linear programming only if suitable separation routines exist for constraints of $ROPT_k$ (see for example [9, chapter 4]). Furthermore, (as we will see), if the $ROPT_k$ optimal solution is not unique, it may still be a difficult task to find an integer optimum among the alternative LP solutions.

Collecting all instances of problem OPT exposed for partial polytope k, we define exposed instance subsets

$$\text{OPT}_k \triangleq \{\text{instances of } OPT : \text{ LP relaxation } ROPT_k \text{ has an integer optimum}\}$$

These exposed instances are exactly the once that can be produced by our RC TIG.

Theorem 4.1 *Each exposed optimization subset OPT_k is exactly the collection of instances of OPT generatable by procedure [RC] over the corresponding $ROPT_k$ polytope.*

Proof: Optimality conditions (5)-(9) are necessary and sufficient for x^* to solve $ROPT_k$. Thus, since procedure [RC] can generate any instance satisfying those conditions, it generates exactly the members of OPT_k. $\square$

To go further we need to be a bit more precise about the nature of constraints $A_i x \leq b_i$. Such constraint systems are termed *nondeterministically recognizable* if the problem of deciding whether a string constitutes one of the constraints belongs to NP. It is easy to check that most well-known constraints for TSP and other discrete

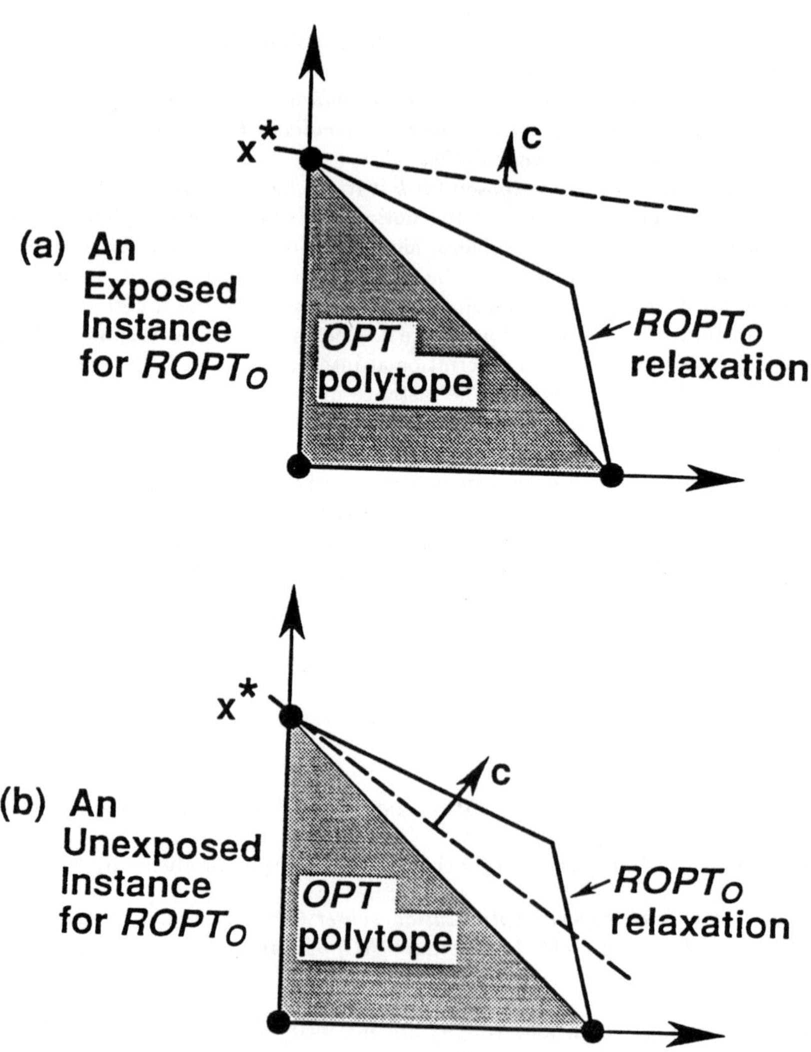

Figure 2: Exposed Instances of Optimization Problems

problems have polynomial-length derivations, which implies they are nondeterministically recognizable. However, some cases are known, notably the hypohamiltonian constraints of Grötschel and Wakabayashi [1987], that are not believed to possess this property.

Parallelling our TSP notation, we can define threshold and exact value decision problems associated with instance subsets OPT_k:

Combinatorial Optimization Threshold ($OPT_k^\leq$)

Instance: an instance in OPT_k plus a rational threshold v.

Question: Does there exist a feasible solution with objective value less than or equal to v?

Combinatorial Optimization Exact Value ($OPT_k^=$)

Instance: same as $OPT_k^\leq$.

Question: Is the optimal objective function value v?

Then we are ready for a main result.

Theorem 4.2 *Given any combinatorial optimization problem OPT for which feasible solutions can be nondeterministically verified, let OPT_k be the exposed subset corresponding to nondeterministically recognizable $ROPT_k$ constraints $A_i x \leq b_i$, $0 \leq i \leq k$. Then the associated threshold and exact value decision problems $OPT_k^\leq$ and $OPT_k^=$ belong to NP.*

Proof: We must exhibit a nondeterministic polynomial algorithm accepting precisely the language of OPT_k instances and v's for which the relevant questions is properly answered *yes*. That is the algorithm should accept an input exactly when both the optimization instance belongs to exposed subset OPT_k, and its optimal value is $\leq v$ (respectively $= v$).

Given any instance of OPT, whether or not a exposed for k, we can emulate generator [RC] to nondeterministically compute and verify its $ROPT_k$ optimal value, say v_k. Standard LP theory establishes that there must exist a corresponding optimal basis for the dual of LP relaxation $ROPT_k$, consisting of polynomially many ($O(|x|)$) active constraints of the primal. To compute value v_k, we need only guess the nondeterministic derivation/recognition of such a polynomial-size collection of binding constraints, compute nonzero parts of the associated dual basic solution, and verify that it is dual feasible with value v_k.

An OPT instance is exposed for k if and only if its integer solution x^* is an extreme-point of the feasible set for $ROPT_k$, i.e. if and only if the complementary primal solution corresponding to a dual nondeterministically solved in this [RC]-like way is primal feasible. Thus our $NDTM$ for $OPT_k^\leq$ (respectively $OPT_k^=$) proceeds by guessing an integer optimum x^* for the given instance of OPT, applying the hypothesized algorithm for nondeterministically verifying its feasibility, computing its objective function value v^*, and nondeterministically solving its $ROPT_k$ dual as just outlined for v_k. We accept exactly when the derived dual solution is complementary with x^* and $v_k \leq v^*$ (respectively $v_k = v^*$). $\square$

5   Intermediate TSP's

Paralleling the above notation, define

$$TSP_{subtour} \triangleq \{TSP \text{ instances exposed for subtour constraints}\}$$
$$TSP_{comb} \triangleq \{TSP \text{ instances exposed for comb inequalities}\}$$
$$TSP_{clique} \triangleq \{TSP \text{ instances exposed for clique tree inequalities}\}$$

Similarly, let $TSP^{\leq}$ and $TSP^{=}$ be the corresponding threshold and exact value decision problems. Then, since all three of the defining constraint forms are nondeterministically recognizable, and feasibility of a TSP solution x can be deterministically verified, we have an immediate corollary to Theorem 4.2.

Corollary 5.1 *Exposed exact value decision problems* $TSP^{=}_{subtour}$, $TSP^{=}_{comb}$, $TSP^{=}_{clique}$ *all belong to NP.*

A subset of instances of an optimization problem can appropriately be termed "intermediate" if it is plausibly neither as general as the full problem nor polynomially solvable. We are now ready to establish the "upper bound" half of the argument that subsets $TSP_{subtour}$, TSP_{comb}, and TSP_{clique} fulfill this definition by distinguishing their exact value forms from that of the full TSP.

Theorem 5.2 $TSP^{=}$ *does not polynomially reduce to* $TSP^{=}_{subtour}$, $TSP^{=}_{comb}$, *or* $TSP^{=}_{clique}$ *unless* $NP = co(NP) = D^p$.

Proof. From Corollary 5.1, exposed decision problems $TSP^{=}_{subtour}$, $TSP^{=}_{comb}$, and $TSP^{=}_{clique}$ all belong to NP. If the D^p-complete problem $TSP^{=}$ polynomially reduced to any of the three, the latter would also be D^p-complete. We know from Lemma 2.2 that a D^p-complete problem can belong to NP only if $NP = co(NP) = D^p$. $\square$

Perhaps the more surprising half of the argument for intermediate status is the "lower bound" fact that all instances of $TSP_{subtour}$, TSP_{comb}, and TSP_{clique} are polynomially solvable only if $P = NP$. Toward that end define the cost vector of graph G, denoted c_G, to be indicator vector of edges in G (1 if the edge belongs to the graph and 0 otherwise).

Lemma 5.3 *For every hamiltonian graph G, the corresponding instance of TSP with weight vector $-c_G$ belongs to $TSP_{subtour}$, TSP_{comb}, and TSP_{clique}.*

Proof: It is sufficient to prove the result only for $TSP_{subtour}$ because subtour constraints are special cases of both comb and clique tree inequalities. Summing degree-two constraints (1) with weights of $-1/2$ and combining with (2) shows that $-\sum_{i<j} x_{ij} = -|V|$ for every x satisfying (1)–(2). Thus for any graph G

$$-|V| \leq \min\{-c_G x : \ x \text{ satisfying (1)-(4)}\} \leq \min\{-c_G x : \ \text{integer } x \text{ satisfying (1)-(4)}\}$$

When G is hamiltonian, the incidence vector of the implied hamiltonian cycle exactly achieves this lower bound. That is, the instance of TSP with weights $-c_G$ has an integer optimal solution over the subtour constraint polytope, exactly what is required for membership in $TSP_{subtour}$. $\square$

Theorem 4.1 tells us any member of $TSP_{subtour}$ should be generatable by random cut procedure [RC]. Thus, under Lemma 5.3 there must exist an [RC] generation sequence based on the subtour LP relaxation that yields the (negative) cost vector of any hamiltonian graph. The actual construction begins with the incidence vector of a hamiltonian cycle/tour and places dual multiplier $-\alpha$ ($\alpha > 0$) on all tight subtour inequalities. Summing as in [RC] Step 5, an interim c will have some integer number of copies of $-\alpha$, say $-q\alpha$, on all tour edges, and another integer number of copies, say $-q'\alpha$, $q' < q$, on nontour edges. Now choosing dual multipliers $q'\alpha/2$ on all degree-two constraints (1), yields the desired cost sum of 0 for nontour edges ($-q'\alpha + q'\alpha/2 + q'\alpha/2$). Corresponding costs on tour edges are $(q' + q')\alpha$. Setting all unmentioned dual variables 0 and fixing $\alpha = 1/(q - q') > 0$ completes the recovery of $-c_G$.

It is interesting to consider what happens for a non-hamiltonian G. The corresponding TSP instance with cost $-c_G$ is still well-defined, but there are two possibilities. If the instance is exposed for subtours, i.e. it belongs to $TSP_{subtour}$, then the LP relaxation optimum over (1)–(4) will be integer, but its value must be strictly worse that $-|V|$. This is the case, for example, if we try non-hamiltonian bipartite graph $K_{2,3}$ as G. The other possibility is that the instance does not belong to $TSP_{subtour}$, i.e. every LP relaxation optimum is fractional. The famous Petersen graph (see for example [7, Figure 11.8]) provides an instance. Its subtour LP relaxation achieves objective value $-|V|$, but the only optimal solution uses $x_{ij} = 2/3$ on all edges.

These ideas lead directly to a reduction from the NP-complete hamiltonian graph problem, HAM (is a given graph hamiltonian?), that proves our exposed subsets are hard.

Theorem 5.4 *Exposed decision problems* $TSP^{\leq}_{subtour}$, $TSP^{\leq}_{comb}$, $TSP^{\leq}_{clique}$, $TSP^{=}_{subtour}$, $TSP^{=}_{comb}$, *and* $TSP^{=}_{clique}$ *belong to* $NP - Complete$.

Proof: We proceed by showing HAM reduces to both $TSP^{\leq}_{subtour}$ and $TSP^{=}_{subtour}$. This is sufficient for all claims because subtour constraints are special cases of both comb and clique tree inequalities, and because Theorem 4.2 already demonstrates all the problems belong to NP.

Consider a graph G and the associated instance of TSP, say I_G, with weight vector $-c_G$. If G is hamiltonian, Lemma 5.3 establishes that I_G is exposed for subtours. It follows that the corresponding threshold and exact value instances with threshold $v = |V|$ are acceptable because the optimal value in I_G will be exactly $-|V|$. If G is not hamiltonian, then (from the above remarks) either I_G is not exposed for subtours, or $I_G \in TSP_{subtour}$, but $\min\{-c_G x : x \text{ satisfies } (1)-(4)\} > -|V|$. Either way input pair $(I_G, |V|)$ would not be accepted in $TSP_{subtour}^{\leq}$ or $TSP_{subtour}^{=}$. $\square$

Theorem 5.5 *If every instance of TSP that belongs to $TSP_{subtour}$ or TSP_{comb} or TSP_{clique} can be solved in polynomial time, then $P = NP$.*

Proof: We proceed as in the proof of Theorem 5.4 to show that an algorithm [A] solving every instance in $TSP_{subtour}$ in time bounded by polynomial $p(n)$ would provide a polynomial algorithm for HAM. Since subtours are special cases of both comb and clique tree inequalities, this will prove all claims.

Given any G, we form the corresponding $-c_G$ and submit to [A]. If G is hamiltonian, [A] would halt with a feasible tour. If G is not hamiltonian, the result is less predictable, but [A] certainly will not halt with a tour; there are none. Thus we will know G is not hamiltonian when [A] either halts with some other outcome or exceeds time limit $p(n)$. $\square$

Readers might be puzzled by the fact that this last result says set $TSP_{subtour}$ is NP-hard, even though well known separation techniques can solve its LP relaxation of in polynomial time (see [7]). By definition instances of $TSP_{subtour}$ have an integer optimal solution over the subtour relaxation. How can they be NP-hard?

The answer hinges on whether the LP optimum over the subtour relaxation is unique. If so, then for any instance of $TSP_{subtour}$ the LP solution will index an optimal tour. But when an instance has alternative relaxation optima, there is no guarantee any LP solver will yield an integer one. Relaxation $(1)-(4)$ does have fractional extreme-points. Thus, even with the optimal value in hand, there remains an NP-hard "rounding" task to find and prove an integer optimum.

6   Well Formed Instances and Promises

It is usual in complexity proofs to take as trivial the issue of recognizing a well formed instance of a problem. For example, define

Traveling Salesman Recognition ($TSP^{\in}$)

Instance: any string

Question: Does the string encode an instance of TSP?

All that is required to answer this question is to decide whether the input string can be viewed as the weight vector of a complete graph.

A corollary of Theorem 5.4 shows the case is quite different for at least the subtour-exposed case:

Subtour-Exposed Traveling Salesman Recognition ($TSP_{subtour}^{\in}$)

Instance: any string

Question: Does the string encode an instance of $TSP_{subtour}$?

Corollary 6.1 *It is NP-hard to determine whether a given instance of TSP is exposed for the subtour polytope, i.e. $TSP_{subtour}^{\in}$ is NP-complete.*

Proof: We will show that $TSP_{subtour}^{\in} \in NP$ and that NP-complete $TSP_{subtour}^{=} \propto NP_{subtour}^{\in}$. The proof of the first is essentially that of Theorem 4.2. When an instance is exposed for the subtour polytope, the implied integer optimum for the LP relaxation must be a tour. By guessing the tour, and then guessing the construction of a corresponding dual-optimal basis, optimality of the tour can be verified in polynomial time.

To show $TSP_{subtour}^{=} \propto TSP_{subtour}^{\in}$ observe that an input is accepted for $TSP_{subtour}^{=}$ if and only if it encodes a subtour-exposed instance of the TSP together with its optimal value. Given any string, our reduction algorithm first invokes any polynomial procedure for $TSP^{\in}$. If the string proves to be a well formed TSP instance, followed by a rational v, we then solve (via separation) the subtour LP relaxation for that instance to obtain its optimal value $\bar{v}$. Strings rejected by $TSP^{\in}$ or having subtour relaxation value $\bar{v}$ equal to the prospective exact value v are submitted directly to an oracle for $TSP_{subtour}^{\in}$. In all other cases, which must have $\bar{v} \neq v$, we submit instead an instance on 10 vertices with weights $-c_G$, where G is the Petersen graph.

We have already remarked that this Petersen graph instance cannot be exposed for subtours. Thus the $TSP_{subtour}^{\in}$ oracle will accept the submitted string exactly when it yields a well formed instance of $TSP_{subtour}$ with optimal value v. $\square$

The proof of Corollary 6.1 depends strongly on the existence of polynomial-time separation procedures for LP optimization over the subtour relaxation (1)-(4). Since separation schemes are not known for comb and clique inequalities, we do not know whether recognition of instances exposed over those polytopes is also NP-hard. Still, we could use Sanchis's reference [14, Proposition 4.4] to conclude that if $TSP_{comb}^{\in}$ or $TSP_{clique}^{\in}$ belongs to P, then $NP = co(NP)$.

Her approach also raises our next issue. We know that threshold versions $TSP_{subtour}^{\leq}$, $TSP_{comb}^{\leq}$ and $TSP_{clique}^{\leq}$ are all NP-complete (Theorem 5.4). What about their complements $co(TSP_{subtour}^{\leq})$, $co(TSP_{comb}^{\leq})$ and $co(TSP_{clique}^{\leq})$? Since the exact optimal value v^* can be nondeterministically computed (proof of Theorem 4.2) for an instance in any of these languages, *no* instances with unattainable threshold $v < v^*$ can also be identified by a $NDTM$. It might seem that say $co(TSP_{subtour}^{\leq}) \in NP$. It would follow that $NP = co(NP)$ because the complement of an NP-complete problem belongs to NP.

One simply may not ignore the "technicality" of recognition of well formed instances. In particular, the complement of $TSP_{subtour}^{\leq}$ is the union of two parts

$$co(TSP_{subtour}^{\leq}) = \{\text{strings acceptable in } TSP_{subtour}^{\in} \text{ with } v \text{ unattainable}\} \quad (10)$$

$$\cup\{\text{strings unacceptable in } TSP^{\in}_{subtour}\}$$

Nondeterministic polynomial recognizability of the first part is not enough to place $co(TSP^{\leq}_{subtour})$ in NP. (Sanchis's proof of Proposition 4.4 simply observes that if any of the instance recognition cases $TSP^{\in}_{-}$ were in P, both halves of (10) could be checked, implying $NP = co(NP)$).

Imagine now receiving some files of test instances (and solutions) from the authors, or generating them yourself. You would know that the instances of say $TSP_{subtour}$ were produced by the RC TIG. They were constructed to be acceptable for $TSP^{\in}_{subtour}$, and the only remaining way a derived threshold instance can be a *no* case is if the threshold is unattainable.

This scenario provides a natural example for that part of complexity theory dealing with *promises* [15, 5]. A promise is an extra bit in the input of a language guaranteeing the instance possesses some mathematical property. The Turing machine is allowed to rely on this promise in deciding whether an input belongs to the language.

What occurs when problems $TSP^{\leq}_{subtour}, TSP^{\leq}_{comb}$ and $TSP^{\leq}_{clique}$ come with a promise that they were constructed by [RC]? In a sense, they become easier. Define

$$
\begin{aligned}
RCTSP_{subtour} &\triangleq \{TSP_{subtour} \text{ instances known generated by [RC]}\} \\
RCTSP_{comb} &\triangleq \{TSP_{comb} \text{ instances known generated by [RC]}\} \\
RCTSP_{clique} &\triangleq \{TSP_{clique} \text{ instances known generated by [RC]}\}
\end{aligned}
$$

Also let threshold ($\leq$), exact value ($=$) decision problems be defined analogously.

Theorem 6.2 *Generated threshold problems* $RCTSP^{\leq}_{subtour}, RCTSP^{\leq}_{comb}$ *and* $RCTSP^{\leq}_{clique}$ *belong to promise* $- (NP \cap co(NP))$.

Proof: The proof is the same for all three forms. It is easy to see that $RCTSP^{\leq}_{subtour} \in$ *promise* $- NP$; we merely ignore the promise bit and use the $NDTM$ for $TSP^{\leq}_{subtour}$. To show the problem is also in *promise-co(NP)*, we must establish that it is the complement of a member of *promise-NP*, i.e. that its complement $co(RCTSP^{\leq}_{subtour})$ is nondeterministically recognizable with the aid of a reliable promise bit. As in (10), the complement $co(RCTSP^{\leq}_{subtour})$ has two parts: those that have the promise bit *off* and those generatable by [RC] that have unattainable thresholds. A mere scan of the bit will settle the first, and we know from discussion above that valid inputs with invalid thresholds can be nondeterministically verified. $\square$

We have already observed several times that subtour form $RCTSP^{\leq}_{subtour}$ is unique among our three classes of exposed instances in that its linear programming relaxation, (1)-(4), is known to be polynomially solvable. This leads to another promise-based result.

Theorem 6.3 *Subtour-generated value problem* $RCTSP^{=}_{subtour}$ *belongs to promise-P*.

Proof: Given an instance of $RCTSP^=_{subtour}$ we first apply the known polynomial-time algorithm to compute $\bar{v} \triangleq$ the optimal value of its LP relaxation. If the promise bit is *off*, or the input threshold v differs from $\bar{v}$, we reject. Otherwise, since the optimal value of subtour-generated TSP's is known to equal $\bar{v}$, we can accept. $\square$

So in one sense, a promise or guarantee that the instances were generated by the RC TIG does reduce their complexity. We conclude by showing that nonetheless they remain formally hard to solve. Return for a moment to the problem of hamiltonian cycles in graphs. One promise-based question in this context is

> **Hamiltonian Cycle Exhibition (HCE)**
>
> Instance: a hamiltonian graph G
>
> Solution: the incidence vector x_G of a hamiltonian cycle in G

As pointed out in [2], the promise of an input's hamiltonicity does not change the fact that the HCE form is hard. For if there were a polynomial time algorithm for HCE, it could be modified to recognize hamiltonian graphs, by incorporating its polynomial time bound as a time limit.

Similarly, full optimization on RC TIG instances remains hard, even with a promise of [RC] generatability.

Theorem 6.4 *Subtour-generated instances $RCTSP_{subtour}$ form an NP-hard set.*

Proof: Reduction from HCE. Given an instance of HCE, we proceed as above to construct one for $RCTSP_{subtour}$ by using graph cost vector $-c_G$. The promise bit can correctly be marked *on* for such an instance because Lemma 5.3 shows that [RC] can generate $-c_G$ so long as the input graph is hamiltonian. The $RCTSP_{subtour}$ output, which will be an integer solution to (4)-(6), then yields the tour required in HCE. $\square$

Even if test instances from the RC TIG come with an absolute seal of authenticity, producing their optimal tours is, in a formal sense, still difficult.

7 Acknowledgement

The authors are grateful to colleagues Eric Allender, Vijay Chandru, Collette Coullard, Andrew Odlyzko, R. Gary Parker and Don Wagner for helpful discussions on the material of this paper, and to the referee of an earlier version for some very useful insights.

References

[1] Crowder, H., E.L. Johnson, and M.S. Padberg [1983], "Solving Large-Scale Zero-One Linear Programming Problems to Optimality," *Operations Research*, 31, 803-834.

[2] Garey, M.R. and D.S. Johnson [1979], *Computers and Intractability: A Guide to the Theory of NP-Completeness*, W.H. Freeman and Company, San Francisco, California.

[3] Grötschel, M. and M.F. Padberg [1979], "On the Symmetric Travelling Salesman Problem I: Inequalities," *Math Programming* 16, 265-280.

[4] Grötschel, M. and W.R. Pulleyblank [1986], "Clique Tree Inequalities and the Symmetric Travelling Salesman Problem," *Math of Operations Research*, 11.

[5] Johnson, D. [1985], "The NP-Completeness Column: An Ongoing Guide," *Journal of Algorithms*, 6, 291-305.

[6] Krentel, M. [1986] "The complexity of optimization problems" Proc. 18th ACM Symp. Th. Comp, 69–76, ACM, New York, 1986.

[7] Lawler, E.L., J.K. Lenstra, A.H.G. Rinnooy Kan, and D.B. Shmoys [1985], editors, *The Traveling Salesman Problem: A Guided Tour of Combinatorial Optimization*, John Wiley and Sons Ltd., London.

[8] Papadimitriou, C.H. and M. Yannakakis [1984], "The Complexity of Facets (and Some Facets of Complexity)", *J. of Computer and System Sciences*, 28, 244-259.

[9] Parker, R.G. and R.L. Rardin [1988], *Discrete Optimization*, Academic Press, Boston, Massachusetts.

[10] Pilcher, M.G. [1985], "Development and Validation of Random Cut Test Problem Generator", Ph.D. dissertation., School of Industrial and Systems Engineering, Georgia Institute of Technology, Atlanta, Georgia.

[11] Pilcher, M.G. and R.L. Rardin [1986], "Partial Polyhedral Description and Generation of Discrete Optimization Problems with Known Optima," report CC-87-4, University Research Initiative in Computational Combinatorics, School of Industrial Engineering, Purdue University, West Lafayette, Indiana, to appear in *Naval Research Logistics*.

[12] Pilcher, M.G. and R.L. Rardin [1986], "Invariant Problem Statistics and Generated Data Validation: Symmetric Traveling Salesman Problems," report CC-87-16, University Research Initiative in Computation Combinatorics, School of Industrial Engineering, Purdue University, West Lafayette, Indiana.

[13] Rais, A., and R. L. Rardin [1988], "Random Generation of Travelling Salesman Problems Using Clique Tree Inequalities, report CC-88-21, University Research Initiative in Computational Combinatorics, School of Industrial Engineering, Purdue University, West Lafayette, Indiana.

[14] Sanchis, L.A., [1990], "On the Complexity of Test Case Generation for NP-Hard Problems," *Information Processing Letters*, 36, 135–140.

[15] Valiant, L. and V. Vazirani, "NP is As Easy As Detecting Unique Solutions," *Proceedings of 17th ACM Symposium on Theory of Computing*, ACM, New York.

Complexity in Numerical Optimization, pp. 406-428
P.M. Pardalos, Editor
©1993 World Scientific Publishing Co.

Some Complexity Issues Involved in the Construction of Test Cases For NP-hard Problems[1]

Laura A. Sanchis
Computer Science Department, Colgate University, Hamilton, NY 13346 USA

Abstract

Approximation algorithms for NP-hard problems must often be evaluated empirically. This creates a need for suitable test cases. Test cases with known answers and varied characteristics can be particularly useful. This paper considers some of the computational complexity issues involved in the generation of instances of NP-hard problems, in such a way that the answers for the instances are known, and such that the instances have certain desirable properties such as hardness and diversity.

Keywords: computational complexity, NP-completeness, generation, approximation algorithms, test cases.

1   Introduction

Many types of problems with important practical applications have been classified as NP-complete or NP-hard. This classification means that the existence of a polynomial-time algorithm for any of these problems would imply that P=NP. It has become very apparent by now that showing this equality or proving its negation, P $\neq$ NP, (which is more generally believed to be the case) is a very difficult problem.

As a consequence, much effort has gone into the design of approximation algorithms for those problems for which finding solutions still seems to be a matter of practical or sometimes theoretical importance. Approximation algorithms, of which there are several types, attempt to find solutions for the problem but have one or

---

[1]This work was supported in part by National Science Foundation grant CCR-9101974. Part of the material in this paper has appeared in [21].

more limitations, such as not providing the optimal answer, or not terminating or not providing an answer for some instances of the problem. In this paper we deal with approximation algorithms that run in polynomial time but which may provide a nonoptimal or incomplete answer.

It is possible to theoretically analyze the performance of some approximation algorithms, and many results have been proved along these lines. Although such results provide much useful and interesting information, it is also often the case that empirical evaluation of an approximation algorithm is necessary or desirable. The reasons for this are varied. Some popular approximation algorithms which have been shown to work well in practice have proved to be not easily amenable to theoretical analysis. Even when such analysis is available, it may be a worst case analysis which does not provide much information about the performance of the algorithm on "typical" instances. In addition, even if an average-case analysis can be obtained (and these are usually harder to derive), the probability distribution assumed for the analysis may not be suitable for consideration of the actual instances of the problem to which the algorithm will be applied. In fact, as applications of a given NP-hard problem vary, each such application may give rise to a different class of instances of the problem, each with their own characteristics.

Empirical testing requires a set of test cases to be available. Test cases for which the answer is known can be particularly useful. For example, if an approximation algorithm for the satisfiability problem is to be evaluated, then it may be desirable to test both formulas known to be satisfiable, and formulas known to be unsatisfiable, of various sizes and characteristics, the latter perhaps depending on the application. For an optimization problem such as the maximum clique problem for graphs, instances of the problem having known maximum cliques and varying edge densities might be required. More will be said later about how the type of approximation algorithm being evaluated affects the kinds of test cases which would be beneficial.

This paper investigates some of the complexity issues involved in the efficient generation of test cases with known answers for NP-complete and NP-hard optimization problems. We consider various desirable properties for test case generators; and present some practical limitations apparently imposed on generators by certain results from complexity theory.

Several authors have considered the problem of generating test instances with known answers for specific NP-hard problems. For example, such generators are described by Krishnamurthy in [11] for network partitioning; by Pilcher and Rardin in [19], [18] for the travelling salesman problem; by Sanchis in [20] for several NP-hard graph problems; and by Khoury, Pardalos, and Du in [9] for the Steiner problem in graphs.

2 Definitions

We use a very general model of test case construction. The only requirements imposed by this model are that a test case constructor must run in polynomial time, must output instances with known answers, and that it should be possible, through the inputs to the program, to control the characteristics of the instances produced. In the following, we let I represent an instance of a problem Π; $|I|$ is the length of some suitable encoding for I.

Definition 2.1 Let Π be a problem and let $l_1, ..., l_k$ be polynomial-time computable functions, $l_j : \Sigma^* \to N$, $1 \le j \le k$. Let q, r be polynomials such that $l_j(I) \le q(|I|)$ for all j and $|I| \le r(l_1(I), ..., l_k(I))$ for all instances I of Π. A test instance construction method (TICM) Π (with respect to $(l_1, ..., l_k)$), denoted as $(l_1, ..., l_k)$-TICM, is a *nondeterministic* polynomial-time program C that given as input $1^{v_1}\#...\#1^{v_k}$, where $v_1, ..., v_k$ are natural numbers, and an answer α, outputs either an instance I of the problem having answer α such that for $1 \le j \le k$, $l_j(I) = v_j$, or the special symbol Λ, denoting that it cannot output any such instance. We denote by $Gen(C, \alpha)$ the set of all instances generated by C with answer α; and by $Gen(C)$ all instances generated by C. When $k = 1$ and $l_1(I) = |I|$, we call the TICM a *length-based* TICM.

Note that the TICM should output Λ only if it cannot output any instance satisfying the input constraints in any of its computations. (Otherwise the time required to actually obtain a test instance could be unbounded). This does not mean however that such instances do not exist, only that this particular TICM does not produce them. Note also that the definition does not require that a certain proportion of existing test instances be generated, or that the test cases be produced uniformly or according to any particular probability distribution. The goal of the definition is to model the construction of instances having certain properties or parameter values similar to those of the real application problem instances (but for which the answer is known).

In the following sections we investigate what types of TICM's can exist for NP-complete or NP-hard problems. There are many features of TICM's which can be considered in this regard. Among these are the number, proportion, and/or variety of instances which can be produced by the TICM, the extent to which uniformity or randomness can be imposed on the generation process, and the NP-hardness of the instances which can be produced.

In this paper we concentrate on some aspects of generability and hardness. Specifically, we investigate whether or not a TICM can generate all instances of a problem and whether or not it can generate at least one instance for each set of parameter values. We also look at the problem of generating hard sets of instances. As will be seen, these issues are closely related to various standard concepts and open questions in complexity theory.

The following observation will be useful in the proofs that follow:

Observation 2.1 *If C is a TICM, then for each answer α, $Gen(C,\alpha)$ is in NP.*

3　Generability

3.1　Definitions

We define three types of TICM's with increasing generating power, and investigate whether such TICM's can exist for various types of problems.

Definition 3.1 Let Π be a problem and let K be an infinite subset of the instances of Π. Let $l_1, ..., l_k$ be as in Definition 2.1.

1. A TICM C is K–<u>extensive</u> if $Gen(C) \cap K$ is infinite.

2. A $(l_1, ...l_k)$–TICM C is K–<u>diverse</u> if the following condition holds: for each tuple $(v_1, ..., v_k)$ and answer α, such that there exists an instance $I \in K$ having answer α and for which $l_j(I) = v_j$ for $1 \le j \le k$, there exists an instance $I' \in K \cap Gen(C)$ having answer α, for which $l_j(I') = v_j$ for $1 \le j \le k$.

3. A TICM C is K–<u>complete</u> if $Gen(C)$ contains all instances in K.

We will sometimes write extensive (diverse, complete) with respect to K instead of K-extensive (diverse, complete).

Clearly if a TICM is K–complete then it is K–diverse, and if it is K–diverse then it is K–extensive. If we are interested in generating instances from K (which may be the set of all instances), then a useful TICM should be at least K–extensive. If it is K–complete, then it generates all instances in K. Diversity guarantees that at least all "types" of instances will be generated, based on the given parameter functions. We wish to investigate which problems have extensive, diverse, and/or complete TICM's with respect to various sets. We consider separately NP-complete decision problems, and NP-hard optimization problems.

3.2　Decision Problems

An NP-complete decision problem has two possible answers for each instance, namely *yes* or *no*, depending on whether or not the instance is in the associated NP-complete language. Two well-known examples of such problems are the Hamiltonian circuit problem for graphs and the satisfiability problem for CNF logic formulas (see [6]).

Since there are only two possible answers for an instance of a decision problem, we can without loss of generality consider TICM's that generate only positive instances, and TICM's that generate only negative instances. In this case it is convenient to think of the TICM as generating elements of a language or set: either the NP-complete language associated with the problem, or its complement. Thus we can reduce the problem of determining what types of TICM's exist for NP-complete languages, to the problem of polynomial-time generability for languages in NP and in co–NP. This latter problem was investigated in [22]. The following definitions are from [22]:

Definition 3.2

1. A <u>polynomial-time constructor (PTC)</u> for a language L is a polynomial-time Turing machine that on input 1^n outputs a string in L of length n, if such a string exists, and outputs Λ otherwise.

2. A <u>polynomial-time generator (PTG)</u> for a language L is a nondeterministic polynomial-time Turing machine that on input 1^n outputs a string in L of length n, if such a string exists, and outputs Λ otherwise. Moreover, for each string x of length n in L, there exists some computation of the generator on input 1^n that outputs x.

Definition 3.3 Let $l_1, ..., l_k$ be as in Definition 2.1.

1. An $(l_1, ..., l_k)$-PTC for a language L is a polynomial-time Turing machine that on input $1^{v_1}\#...\#1^{v_k}$, outputs a string x in L such that for $1 \leq j \leq k$, $l_j(x) = v_j$, if such a string exists, and outputs Λ otherwise.

2. An $(l_1, ..., l_k)$-PTG for a language L is a nondeterministic polynomial-time Turing machine that on input $1^{v_1}\#...\#1^{v_k}$, outputs a string x in L such that for $1 \leq j \leq k$, $l_j(x) = v_j$, if such a string exists, and outputs Λ otherwise. Moreover, for each string $x \in L$ having these parameter values, there exists some computation of the PTG on input $1^{v_1}\#...\#1^{v_k}$ that outputs x.

The relationship between TICM's, PTG's, and PTC's is brought out by the following observation:

Observation 3.1 *Let Π be an NP-complete problem and let K be either the set of positive instances or the set of negative instances of Π.*

1. Π has a K–diverse length-based TICM if and only if K has a PTC.

2. Π has a K–complete length-based TICM if and only if K has a PTG.

3. Π has a K–diverse $(l_1, ..., l_k)$-TICM if and only if K has a $(l_1, ..., l_k)$-PTC.

4. Π has a K–complete $(l_1, ..., l_k)$-TICM if and only if K has a $(l_1, ..., l_k)$-PTG.

Thus the existence of TICM's generating all positive or all negative instances of NP-complete problems, or generating at least one such instance for each possible length or parameter combination, is closely related to the existence of PTG's and PTC's, respectively, for NP-complete languages, and their complements. However, determining which NP-complete languages and/or their complements have PTG's or PTC's is not trivial.

3.2.1 Positive Instances

Turning our attention first to the generation of positive instances, we note that most well-known NP-complete languages appear to have length-based as well as parameter-based PTG's. As an example, consider the following PTG (or equivalently TICM generating all positive instances) for the NP-complete language 3SAT (CNF satisfiability with length 3 clauses) [6]. To construct an instance of 3SAT with n variables and m clauses, randomly assign a truth value T or F to each of $x_1, x_2, ...x_n$. Let $u_i = x_i$ if x_i was assigned T, $u_i = \bar{x}_i$ otherwise. To form each of the m clauses first randomly choose some u_i, thus ensuring that the clause is true, and then randomly choose 2 more variables for the clause from among $x_1, x_2, ...x_n, \bar{x}_1, \bar{x}_2, ...\bar{x}_n$. Clearly each satisfiable formula with n variables and m clauses is generated in this manner. Moreover the number of times a particular satisfiable formula is generated is proportional to the number of different assignments that satisfy it. The reader may easily find similar generators for other NP-complete problems, such as the Hamiltonian circuit problem or set partition.

However, the question of whether such complete TICM's exist for all NP-complete problems is apparently not easy to answer. The following theorem brings together several results from [22] which indicate this.

Theorem 3.1 *Let $l_1, ..., l_k$ be as in Definition 2.1.*

1. *A language in NP has a $(l_1, ..., l_k)$–PTG if and only if it has a $(l_1, ..., l_k)$–PTC.*

2. *If a language has a $(l_1, ..., l_k)$–PTG it is in NP.*

3. *All languages in P have PTG's if and only if all languages in NP have PTG's if and only if all NP-complete languages have PTG's.*

4. *If all NP languages have PTG's then there are no sparse languages in $NP - P$.*

5. *If there are no sparse languages in $D^p - P$, then all languages in NP have PTG's.*

6. *If all NP languages have PTG's, then all NP languages have $(l_1, ..., l_k)$–PTG's.*

(The class D^p is defined as $D^p = \{L_1 - L_2 | L_1, L_2 \in NP\}$ [16]. A language L is sparse if there exists a polynomial p such that there are at most $p(n)$ strings of length n in L, for each $n \geq 1$.)

The existence of sparse languages in NP-P or in D^p-P, even under the assumption that $P \neq NP$, are currently open questions in complexity theory (see [8], [7], [12], [2],[3]). Hence parts (4) and (5) of the above theorem show that determining whether or not all NP languages (or all NP-complete languages, by part (3)) have PTG's or PTC's is a difficult problem.

From the above theorem and Observation 3.1 we have the following:

Corollary 3.2 *Let $l_1, ..., l_k$ be as in Definition 2.1.*

1. *An NP-complete problem has a $(l_1, ..., l_k)$–TICM which is diverse with respect to the set of positive instances, if and only if it has a $(l_1, ..., l_k)$–TICM which is complete with respect to this set.*

2. *If all NP-complete problems have length–based TICMs which are diverse with respect to the set of positive instances, then there are no sparse languages in $NP - P$.*

3. *If there are no sparse languages in $D^p - P$, then all NP-complete problems have $(l_1, ..., l_k)$–TICM's which are diverse (complete) with respect to the set of positive intances.*

Thus determining whether or not all NP-complete problems have TICM's capable of generating all positive instances of the problem, is an open question, related to the existence of PTG's for languages in NP. In practice, however, such TICM's appear to be easy to find, for most commonly known NP-complete problems.

Part (1) of the above corollary says that the existence of complete and of diverse TICM's with respect to the set of positive instances, is equivalent. We should note, however, that the complete TICM which can be obtained from the diverse one is somewhat unsatisfying, as it provides no mechanism for producing, on demand, instances not generated by the original (diverse) TICM. Briefly, given a diverse TICM C, a complete TICM C' can be constructed which operates as follows. Assume without loss of generality that C is length-based. On input 1^n, C' nondeterministically constructs an encoding I of length n, and then runs a polynomial-time nondeterministic program to determine whether I is in the NP-complete language. If this program accepts, then C' outputs I; otherwise, C' simulates C and produces its output. (Note that if C outputs Λ, this means that there exist no instances of length n). Clearly in most cases, instances in $Gen(C)$ will be generated much more often than other instances, by C'.

Finally, we remark that even showing that all NP problems have extensive TICM's with respect to the set of positive instances, would resolve an open problem in complexity theory. An infinite set is said to be P–immune if it does not contain an infinite subset in P. It is an open question whether NP has any P-immune sets (assuming that $P \neq NP$) (see for example [1]). Note that any TICM can be made into a deterministic machine which on a given input always outputs the same instance, for example, that obtained by always taking the first branch in any nondeterministic choice of the original TICM. The set of instances produced by this deterministic machine is in P. We therefore have the following.

Proposition 3.3 *If all NP problems have extensive TICMs with respect to the set of positive instances, then NP has no P-immune sets.*

Notice that the above proposition refers to all NP problems rather than merely all NP-complete problems.

3.2.2　Negative Instances

Although it is not uncommon to find TICM's for NP-complete problems that can generate all of the positive instances of the problem, the situation with regards to negative instances is quite different. In fact, it is easily seen that an NP-complete problem cannot have a TICM generating all negative instances of the problem unless NP=co-NP (which is considered unlikely). This is because, as seen from part (2) of Theorem 3.1, the existence of such a TICM would imply that the set of negative instances is in NP. However, if the complement of an NP-complete language is in NP then NP = co–NP (Theorem 7.2 in [6]). Thus we have the following.

Proposition 3.4 *Let* Π *be an NP-complete decision problem. Then* Π *cannot have a TICM which is complete with respect to the set of negative instances, unless NP = co-NP.*

Since completeness with respect to the set of negative instances probably cannot be achieved even for a single NP-complete problem, we turn our attention to diversity. As was the case for completeness with respect to positive instances, most NP-complete problems appear to be diversely generable with respect to the set of negative instances, at least for natural sets of parameter functions. But the question as to whether all NP-complete problems have this property is again related to open questions in complexity theory, as shown by the following results about PTC's which follow from results in [22] or using similar techniques.

Theorem 3.5

1. *If all co-NP languages have PTC's, then there are no sparse languages in* $(co--NP) - P$.

2. *Let* $l_1, ..., l_k$ *be as in Definition 2.1. If all co–NP languages have PTC's, then all co–NP languages have* $(l_1, ..., l_k)$*-PTC's.*

3. *All co–NP languages have PTC's if and only if all (co–NP)–complete languages have PTC's.*

Corollary 3.6

1. *If all NP-complete problems have length-based TICM's which are diverse with respect to the set of negative instances, then there are no sparse languges in* $(co–NP) - P$.

2. *Let* $l_1, ..., l_k$ *be as in Definition 2.1. If all NP-complete problems have length-based TICM's which are diverse with respect to the set of negative instances, then all NP-complete problems have* $(l_1, ..., l_k)$*–TICM's which are diverse with respect to the set of negative instances.*

Although TICM's which are diverse with respect to the set of negative instances seem easy to construct for most well-known NP-complete problems, the instances produced by these TICM's may tend to be trivial (hardness will be addressed in the next section).

We know that for any such diverse TICM C, $Gen(C)$ is in NP. Therefore $Gen(C)$ most likely does not contain all of the negative instances, as seen by Proposition 3.4. At first glance it appears that most such TICM's can be improved upon by incorporating into the TICM some other method of generating negative instances not included in the original TICM. This intuition turns out to be correct in most cases. We need to make precise the notion of "improving" a TICM to make it generate more instances. A *maximal* TICM is defined to be a TICM which cannot be improved in this manner:

Definition 3.4 Let Π be a problem and let K be an infinite subset of the instances of Π. A TICM C for Π is <u>maximal</u> with respect to K if for all other TICM's C' for Π, $(Gen(C') - Gen(C)) \cap K$ is finite.

In other words, a TICM C for Π is maximal with respect to K if no other TICM can generate infinitely more elements of K. The following definitions identify properties that most well-known NP-complete problems appear to possess, and which will ensure that a TICM generating instances of the problem cannot be maximal with respect to the set of negative instances.

Definition 3.5 Let Π be a problem and let K be a subset of instances of Π. We say that Π is $(l_1, ..., l_k)$–<u>diversely generable</u> with respect to K if it possesses a $(l_1, ..., l_k)$–TICM which is diverse with respect to K.

Definition 3.6 A set L is <u>paddable</u> if there exists a polynomial-time computable function $P : \Sigma^* \times \Sigma^* \to \Sigma^*$ that is one-to-one in its second argument, such that for all $x, y \in \Sigma^*$, $x \in L$ if and only if $P(x, y) \in L$. If there exists a polynomial p such that $p(|P(x, y)|) \geq |x| + |y|$, then L is <u>honestly paddable</u>.

Definition 3.7 A set L is <u>augmentable</u> if there is a polynomial-time computable function S such that $x \in L \Leftrightarrow S(x) \in L$, and $|S(x)| > |x|$ for all x.

A proof of the following lemma may be found in [15].

Lemma 3.7 *If a set is honestly paddable then it is augmentable.*

Observation 3.2 *A set is honestly paddable if and only if its complement is honestly paddable.*

(The proof of the following proposition uses a technique used in [15] to show that sets that are not in P and that are honestly paddable have no maximal P-subsets).

Proposition 3.8 *Let* Π *be an NP-complete problem. Let* K *be the set of negative instances of* Π. *Let* $l_1, ..., l_k$, q, *and* r *be as in Definition 2.1. Suppose that* K *is honestly paddable and that* Π *is* $(l_1, ..., l_k)$*–diversely generable with respect to* K. *Then* Π *cannot have a maximal* $(l_1, ..., l_k)$*–TICM with respect to* K *unless* $NP = co\text{–}NP$.

Proof: Note that if $NP \neq co\text{–}NP$, then K is not in NP. Let C be any $(l_1, ..., l_k)$–TICM for Π; we will show that C is not maximal with respect to K. Let C_0 be a $(l_1, ..., l_k)$–TICM for Π which is diverse with respect to K. Since K is honestly paddable, by Lemma 3.7 there exists a function S such that $|S(x)| > |x|$ for all instances x, and $x \in K$ if and only if $S(x) \in K$.

Since $Gen(C)$ is in NP, and we are assuming that K is not in NP, the set $K - Gen(C)$ must be infinite. Let x_0 be any element in this set. Consider the sequence $L_{x_0} = \{x_0, x_1, ...\}$ where $x_i = S(x_{i-1})$ for $i \geq 1$. Since $x_0 \in K$, L_{x_0} is an infinite subset of K. Using C, we can construct another $(l_1, ..., l_k)$–TICM C_{x_0} for Π as follows. Given input $z = 1^{v_1}\#...\#1^{v_k}$, and answer *no*, C_{x_0} computes $x_0, x_1, ...$ until it finds an element x_t such that $|x_t| \geq r(v_1, ..., v_k)$. It then tests each of $x_1, ..., x_t$ to see for which i (if any), $l_j(x_i) = v_i$ for $1 \leq j \leq k$. If there are any such x_i, then C_{x_0} nondeterministically chooses one of them and outputs this element. If there are no such x_i, then C_{x_0} simulates C on z and answer *no*, and produces C's output. Note that $Gen(C_{x_0})$ contains L_{x_0}. Therefore if $L_{x_0} - Gen(C)$ is infinite, then C cannot be maximal.

We have in fact shown that if there exists $x \in K - Gen(C)$ such that $L_x - Gen(C)$ is infinite, then C cannot be maximal. Suppose on the other hand that for all $x \in (K - Gen(C))$, $L_x - Gen(C)$ is finite. Define

$$H = \{x \,|\, x \in K, x \notin Gen(C), \text{ and } S(x) \in Gen(C)\}$$

Given any string in $H \subset K - Gen(C)$, there exists another larger string in $K - Gen(C)$, and hence another string in H, derived if necessary by using S. Therefore H is infinite.

We define a TICM C_1 for Π which generates all elements of H. Given input $z = 1^{v_1}\#...\#1^{v_k}$ and answer *no*, C_1 nondeterministically constructs a string y of length at most $r(v_1, ..., v_k)$, and checks whether $l_j(y) = v_j$ for $1 \leq j \leq k$. If this is not the case, then C_1 simulates C_0 on z and produces C_0's output. If this is the case, then C_1 computes $x = S(y)$ and runs C on input $1^{w_1}\#...\#1^{w_k}$, where $w_j = l_j(x)$ for $1 \leq j \leq k$. If C outputs x, then C_1 outputs y, otherwise it again simulates C_0 on z. Note that $Gen(C_1)$ includes H, and hence again C cannot be maximal. $\square$

(We remark that the question of maximality does not arise with respect to the diverse generation of positive instances, since by Corollary 3.2, the existence of a TICM which is diverse with respect to positive instances immediately implies the existence of a TICM generating all positive instances; this TICM is obviously maximal).

Finally, even the existence of TICM's that are merely extensive with respect to the set of negative instances is again related to an open question in complexity theory. An <u>NP–simple</u> set is a set in NP whose complement is infinite but does not contain an infinite subset in NP. Recall that the set of negative instances generated by a TICM for an NP-complete problem Π is an NP subset of the complement of the NP-complete language. It follows that if Π has a TICM generating an infinite number of negative instances, then the associated NP-complete language cannot be NP–simple. The existence of NP–simple sets is an open question (see for example [1]). We have the following proposition.

Proposition 3.9 *If all NP problems have extensive TICMs with respect to the set of all negative instances, then there are no NP–simple sets.*

Corollary 3.10 *If all NP-complete problems have length-based diverse TICMs with respect to the set of all negative instances, then there are no NP–simple sets.*

Proof: By Theorem 3.5 and Observation 3.1, if all NP-complete problems have length-based diverse TICM's with respect to the set of negative instances, then all NP problems have such TICM's, implying that they have extensive TICM's. $\Box$

3.3   Optimization Problems

An NP-hard optimization problem Π consists of instances having solutions with various values; for each instance a solution with optimum value is sought. If Π is a maximization problem, an algorithm for solving Π must find for each instance I a solution with the largest possible value; this value is denoted by $OPT(I)$. Likewise, for a minimization problem a solution with the smallest value is sought. We assume that all possible solution values are nonnegative integers. We also assume that there exists a polynomial s such that $|OPT(I)| \leq s(|I|)$ for all instances I. Hence $OPT(I) \leq 2^{cs(|I|)}$ for some constant c. In terms of our terminology for TICM's, the *answer* for instance I is $OPT(I)$.

See [6] for a more detailed definition of NP-hard optimization problems. Two examples of such problems are the maximum clique problem for graphs and the travelling salesman problem.

As Leggett and Moore show in [13], NP-hard optimization problems as defined in this section have the property that the set of tuples $(I, OPT(I))$ is not in NP unless NP=co-NP. Hence it follows from the next proposition that these problems cannot have TICM's generating all instances of the problem unless NP=co-NP.

Proposition 3.11 *Let Π be an NP-hard optimization problem. Let L be the language consisting of all tuples $(I, OPT(I))$ where I is an instance of Π. If there exists a TICM generating all instances of Π, then L is in NP.*

Proof: If there exists a TICM generating all instances of Π, then this TICM can be used as a nondeterministic polynomial-time recognizer for the language L. $\Box$

Corollary 3.12 *Let Π be an NP-hard optimization problem. Then Π cannot have a TICM which is complete with respect to the set of all instances of the problem, unless $NP = co{-}NP$.*

Whether all NP-hard optimization problems have TICM's which are diverse with respect to the set of all instances, is again an open problem. Although some optimization problems do possess diverse TICM's, finding such TICM's in general appears to be harder than finding TICM's for decision problems which are diverse with respect to the set of negative instances, as discussed in the last section. For example, in [20] may be found descriptions of TICM's for some NP-hard graph optimization problems; these generators are diverse with respect to the set of all instances, and are based on some natural graph parameters. But the construction of these TICM's is apparently not trivial, since it relies on extremal graph theory results which determine for which parameter values and answers it is possible to produce the desired instance.

Since any decision problem may be viewed as a maximization problem (with $OPT(I) = 1$ if I is a positive instance, and $OPT(I) = 0$ if I is a negative instance), the consequences presented in part (1) of Corollary 3.6 and in Proposition 3.9 or Corollary 3.10, would follow as well from the existence of diverse or extensive TICM's for all NP or NP-hard optimization problems.

In addition, we can again show that under proper conditions, an NP-hard optimization problem which is diversely generable cannot have a maximal TICM with respect to the set of all instances. The proof is similar to that of Proposition 3.8 and is omitted.

Proposition 3.13 *Let Π be an NP-hard optimization problem. Let K be the set of all instances of Π and let L be the set of all tuples $(I, OPT(I))$ where I is an instance of Π. Let $l_1, ..., l_k$, q, and r be as in Definition 2.1. Suppose that L is honestly paddable, and that Π is $(l_1, ..., l_k)$–diversely generable with respect to K. Then Π cannot have a maximal $(l_1, ..., l_k)$–TICM with respect to K, unless $NP{=}co{-}NP$.*

4　Generation of Hard Instances

Since it is unlikely that a TICM can generate all instances of an NP-hard problem, we wish to avoid TICM's which produce subsets of instances which are particularly easy to solve. There are several ways in which the "hardness" of a set of instances can be defined, the most basic of which is that the set of instances should not be solvable by a polynomial-time algorithm, unless P=NP.

We first discuss in some more detail what is meant by a polynomial-time approximation algorithm.

4.1   Approximation Algorithms

Consider first an NP-hard optimization problem Π. Recall that an instance I of such a problem has an optimal solution with value $OPT(I)$. An approximation algorithm A provides an approximation for this value, which we denote by $A(I)$. The value $A(I)$ may or may not be optimal. (This notation follows that in [6]).

The definition of approximation algorithm for a decision problem is somewhat different, but, as will be seen shortly, closely related to the definition from the previous paragraph. If an approximation algorithm for a decision problem runs in polynomial time, then when it halts it will either have determined that its input is positive or negative, or it will have failed to make a decision. Accordingly, we assume that an approximation algorithm for a decision problem answers either *yes*, *no*, or *?* (don't know), always answering correctly in the first two cases. This definition of approximability for NP-complete problems can be found in [10] and [15].

If Π is a decision problem, we say that an approximation algorithm A gives an <u>exact answer</u> for instance I if $A(I)$ equals *yes* or *no*. If Π is an optimization problem, then A gives an <u>exact answer</u> for I if $A(I) = OPT(I)$. In either case, if A does not give an exact answer for I (that is, if $A(I) =?$ or if $A(I) \neq OPT(I)$), then we say that A gives an <u>inexact answer</u>.

Most approximation algorithms A used in practice for optimization problems, provide not only the value $A(I)$ but also a (possibly suboptimal) solution for I having this value. For example, the Nearest Neighbor heuristic (NN) for the Travelling salesman problem [6] produces a route whose length $NN(I)$ is greater than or equal to the length $OPT(I)$ of the shortest route. Thus these types of algorithms provide a lower bound for the optimal answer, for maximization problems, and an upper bound in the case of minimization problems.

The type of algorithm discussed in the last paragraph is the most widely used, since it does generally provide a (suboptimal) solution for the problem instance. One can however envision as well a process which provides an upper bound on the optimal answer for maximization problems, or a lower bound for minimization problems. One situation where these types of bounds arise is in branch and bound algorithms (see [17]). A branch and bound algorithm for a minimization problem like the travelling salesman problem will, if allowed to run to completion, find an optimal solution; however, the worst case running time is of course exponential. If the branch and bound process is stopped before completion (say within a polynomial amount of time), then the process may yield a suboptimal solution, which entails an upper bound on the optimal tour; it will also yield a lower bound, derived from the bounding mechanism used in the branch and bound procedure. The bounding function may also be used on an instance of the problem to obtain a lower bound on the optimal solution, without necessarily going through the branch and bound process. Depending on the type of function used, useful information may be provided about an instance in this way.

Note that the first type of algorithm provides a "positive" type of answer, by asserting that there does exist a route of length less than or equal to $A(I)$; while the second type of algorithm provides a "negative" answer, by asserting that there is no route shorter than $A(I)$. We will distinguish between these two types of algorithm behavior, since it turns out that generation of hard instances appears to be much easier for the first than for the second type. Although the positive type of approximation is the one most commonly used, it is interesting to note the difficulties inherent in producing hard instances for testing the second type of approximation.

The two types of approximation behavior are actually more clearly distinguished in regards to decision problems. An approximation algorithm for an NP-complete problem may either attempt to prove that a given instance is positive (i.e. in the NP-complete language), or to show that a given instance is negative (i.e. not in this language). An example of the first type of behavior can be seen in the greedy search algorithm for the satisfiability problem described in [23]. This algorithm attempts to find a satisfying assignment for the input formula by doing a "greedy" search, which continues for a specified number of actions and number of trials; these numbers are entered as parameters to the procedure. As mentioned in [23], only satisfiable formulas are useful for the testing of this algorithm, since the algorithm never does conclude that a formula is unsatisfiable.

Another well-known algorithm for the satisfiability problem is the Davis–Putnam procedure [4]. This algorithm does a backtracking search in the space of all truth assignments. It may be turned into an approximation algorithm by stopping it within a given time bound. The result will thus be either *yes*, if a satisfying assignment is found before termination, *no* if it the search completed before termination and no satisfying assignment was found, or *?* if the process did not terminate. It can be seen that both satisfiable and unsatisfiable formulas would be useful for testing either how long the procedure takes on different inputs, or how often it terminates within a reasonable amount of time.

It is useful to consider the "positive" and "negative" behavior of approximations for decision problems separately, in part because they are closely related to the "positive" and "negative" types of approximation algorithms for optimization problems, as discussed above.

Definition 4.1 Let Π be an NP-complete decision problem. A polynomial-time approximation algorithm for Π is <u>positive</u> if it always answers either *yes* (indicating that the input is in the NP-complete language) or *?* (don't know). The algorithm is <u>negative</u> if it always answers *no* (indicating that the input is not in the language) or *?*.

Definition 4.2 Let Π be an NP-hard maximization (minimization) problem. A polynomial-time approximation algorithm A for Π is <u>positive</u> if $A(I) \leq OPT(I)$ $(A(I) \geq OPT(I))$ for all instances I of Π. The algorithm A is <u>negative</u> if $A(I) \geq OPT(I)$ $(A(I) \leq OPT(I))$ for all instances I.

Again, note that both types of approximation behavior could be present in a given algorithm.

We now consider the relationship between approximation algorithms for decision and for optimization problems. The optimization problem Π can be associated with the NP-complete decision problem Π', which has instances of the form (I, k) where I is an instance of Π and k is a nonnegative integer. The element (I, k) is a positive instance of Π' if and only if $k \geq OPT(I)$ for a minimization problem, or $k \leq OPT(I)$ for a maximization problem.

Definition 4.3 Let Π be an NP-hard optimization problem and B a set of instances of Π. Let s be a polynomial and c a constant such that $OPT(I) \leq 2^{cs(|I|)}$ for all instances I. Define

$$Dec(B) = \{(I, k) | I \in B, 0 \leq k \leq 2^{cs(|I|)}\}$$

Definition 4.4 Let Π be an NP-hard maximization (minimization) problem. Let Π' be the associated NP-complete decision problem. An approximation algorithm A for Π' is <u>reasonable</u> if whenever A answers *yes* for input (I, k), it answers *yes* as well for all inputs (I, l) where $l < k$ $(l > k)$; and whenever A answers *no* for input (I, k) it answers *no* as well for all inputs (I, l) where $l > k$ $(l < k)$.

There is an obvious correspondence between approximation algorithms for an optimization problem, and for its associated NP-complete decision problem. The following discussion is in terms of an NP-hard minimization problem Π, although similar arguments may be made for maximization problems. Let A be a positive approximation algorithm for Π. A positive algorithm A' for Π' may be defined as follows. On input (I, k), A' runs A on input I to obtain the value $A(I)$. Since A is positive, $A(I) \geq OPT(I)$, and hence A' should output *yes* if $k \geq A(I)$, and *?* otherwise. If on the other hand A were a negative algorithm for Π, then A' would output *no* if $k < A(I)$, and *?* if $k \geq A(I)$. In this case A' would of course be negative.

We can also make the opposite transformation starting from a positive *reasonable* algorithm A' for Π'. A positive algorithm A for Π may be defined as follows. Since A' is reasonable, for a given I, there exists a value T such that $A'((I, t)) = yes$ for $t \geq T$, and $A'((I, t)) = ?$ for $t < T$. Note that $T \geq OPT(I)$. On input I, A does a binary search on the pairs (I, t), $0 \leq t \leq 2^{cs(|I|)}$, using A', to determine T. A then outputs T. A similar argument shows how to obtain a negative approximation algorithm for Π from a negative reasonable approximation algorithm for Π'.

We therefore have the following:

Lemma 4.1 *Let Π be an NP-hard optimization problem and let Π' be its associated NP-complete decision problem. Let L be the set of positive instances of Π'.*

1. *If A is a positive (negative) approximation algorithm for Π, then there exists a positive (negative) approximation algorithm A' for Π' such that for any set B of instances of Π, A gives the exact answer for all instances in B if and only if A' gives the exact answer for all instances in $Dec(B) \cap L$ $(Dec(B) \cap \overline{L})$.*

2. *If A' is a positive (negative) approximation algorithm for Π', then there exists a positive (negative) approximation algorithm A for Π such that for any set B of instances of Π, A gives the exact answer for all instances in B if and only if A' gives the exact answer for all instances in $Dec(B) \cap L$ $(Dec(B) \cap \overline{L})$.*

4.2 Hardness

Definition 4.5 Let Π be an NP-complete problem.

1. A set B of positive instances of Π is <u>hard</u> with respect to Π if no polynomial-time positive approximation algorithm for Π gives the exact answer for all of the instances in B, unless P=NP.

2. A set B of negative instances of Π is <u>hard</u> with respect to Π if no polynomial-time negative approximation algorithm for Π gives the exact answer for all of the instances in B, unless P=NP.

Definition 4.6 Let Π be an NP-hard optimization problem.

1. A set B of instances of Π is <u>positively hard</u> with respect to Π if no polynomial-time positive approximation algorithm for Π gives the exact answer for all of the instances in B, unless P=NP.

2. A set B of instances of Π is <u>negatively hard</u> with respect to Π if no polynomial-time negative approximation algorithm for Π gives the exact answer for all of the instances in B, unless P=NP.

Note that these definitions do not assume that a polynomial approximation algorithm for Π would be able to determine whether its input is from B or not; in general this will not be the case (see Proposition 4.9 below).

By the above definitions, the hardness of a set of positive instances of an NP-complete problem depends only on positive approximation algorithms; and likewise the hardness of a set of negative instances depends only on negative approximation algorithms. The hardness of a set of instances of an optimization problem, however, may be measured either in terms of positive or negative approximation algorithms, yielding the two types of hardness defined above.

There is an easy characterization of hard sets of positive and negative instances. The following definition is from [5].

Definition 4.7 Two sets A and B are <u>p–separable</u> if there exists a set C in P such that $A \subseteq C$ and $B \cap C = \emptyset$. The sets A and B and <u>p–inseparable</u> if they are not p–separable.

Proposition 4.2 *Let Π be a decision problem. Let L be the set of positive instances of Π. Assume that $P \neq NP$. A set B of positive instances is hard with respect to Π if and only if B and $\overline{L}$ are p–inseparable. A set B' of negative instances is hard with respect to Π if and only if B' and L are p–inseparable.*

Proof: If A is a positive approximation algorithm for Π, let $Yes(A)$ be the set of instances for which A answers *yes*. Note that $Yes(A)$ is a set in P. If B is not hard, then there exists an algorithm A such that $B \subseteq Yes(A)$, and hence B and $\overline{L}$ are p–separable. Suppose conversely that B and $\overline{L}$ are p–separable by set C. That is, $B \subseteq C$ and $C \cap \overline{L} = \emptyset$. Define $A(I) = yes$ if $I \in C$, and $A(I) =?$ if $I \notin C$. Then A is a positive approximation algorithm for Π and since $B \subseteq C = Yes(A)$, B is not hard. A similar argument works for B' and L. $\Box$

The following proposition follows by Lemma 4.1.

Proposition 4.3 *Let Π be an NP-hard optimization problem and let Π' be the corresponding decision problem. Let L be the set of positive instances of Π'. Let B be a set of instances of Π. Then B is positively hard with respect to Π if and only if $Dec(B) \cap L$ is hard with respect to Π'; and B is negatively hard with respect to Π if and only if $Dec(B) \cap \overline{L}$ is hard with respect to Π'.*

In other words, the positive hardness of the instances in B depends on the hardness of the positive instances in $Dec(B)$, while the negative hardness of the instances in B depends on the hardness of the negative instances in $Dec(B)$.

As we shall see, generating hard positive instances appears to be much easier than generating hard negative instances. The following proposition presents a type of reduction which can be used to show that a set of positive instances is hard.

Proposition 4.4 *Let Π be an NP-complete decision problem and let B be a set of positive instances of Π. Let L be the set of all positive instances of Π. Suppose that M is an NP-complete language and that f is a polynomial-time many-one reduction from M to L, such that if $x \in M$, then $f(x) \in B$. Then B is hard with respect to Π.*

Proof: Suppose that A is a positive approximation algorithm for Π which answers *yes* for all instances from B. Then the following polynomial-time procedure could be used to decide whether $x \in M$: compute $A(f(x))$; if $A(f(x)) = yes$, then $f(x) \in L$, which implies that $x \in M$; if $A(f(x)) =?$, then $f(x) \notin B$, which implies that $x \notin M$. $\Box$

Unfortunately such a reduction apparently cannot be used to show that a set of generated negative instances of an NP-complete problem is hard, without at the same time showing that NP=co–NP. Suppose that $B = Gen(C, no)$ where C is a TICM for Π. Suppose we have a many-one reduction from M to L as described in the above proposition, such that if $x \in \overline{M}$, then $f(x) \in B$. But since $B = Gen(C, no)$ is an NP set, this would imply that $\overline{M}$ is in NP, and hence that NP=co–NP, since we are assuming that M is NP-complete. We cannot have the reduction go from $\overline{M}$ into L either, as this would again imply that $\overline{M}$ is in NP, and hence that NP=co–NP.

Notice that not even a Turing reduction could be employed in this case. Suppose that we had a polynomial-time procedure for solving M, which worked by querying an oracle for L. Assume that whenever the oracle must be queried about a string y, $y \in \overline{L}$ implies that $y \in B$. Again, because both L and B are sets in NP, this would imply that $\overline{M}$ is in NP, with the same consequences as above.

We turn now to optimization problems. A reduction similar to that used in the last proposition can be used to show that a set of instances of an NP-hard optimization problem is positively hard.

Proposition 4.5 *Let Π be an NP-hard maximization (minimization) problem and let B be a set of instances of Π. Let Π' be the associated NP-complete decision problem, and let L be the set of all positive instances of Π'. If M is an NP-complete language and f is a polynomial-time many-one reduction from M to L, such that $x \in M$ implies that $f(x) \in Dec(B)$, then B is hard with respect to Π.*

Proof: By Proposition 4.4, the reduction f shows that the set $Dec(B) \cap L$ is hard with respect to Π'. Hence by Proposition 4.3, B is positively hard with respect to Π.
□

Again, proving that a set of instances of an NP-hard optimization problem is hard with respect to negative type approximation algorithms apparently cannot be done using reductions of the type just described.

At this point we should say a few words about the existence of concrete examples of hard generable sets. Obviously, by definition, the set of all positive instances of an NP-complete decision problem is hard; likewise the set of all negative instances is hard. As discussed in a previous section, most well-known NP-complete problems seem to have TICM's generating the set of all positive instances, so obtaining hard positive instances for these problems appears to be quite feasible. On the other hand the set of all negative instances cannot be produced by a TICM unless NP = co–NP, and we do not know of a method to prove that a subset of negative instances produced by a TICM is hard. It is an open question whether such sets exist for any NP-complete problem.

Turning to optimization problems, examples of TICM's that generate hard sets of instances for certain NP-hard graph problems (in relation to positive approximation algorithms) may be found in [20]. The generated sets are shown to be hard using Proposition 4.5. As mentioned before, these TICM's can also be shown to generate diverse sets of instances. Again, we have not found a way of efficiently generating instances of NP-hard optimization problems (with answers) that can be shown to be hard with respect to negative approximation algorithms.

4.3 Structure of Generated Hard Sets

In this section we investigate the structure of sets that are both generable (i.e. produced by a TICM) and hard, with respect to an NP-hard problem.

Any set produced by a TICM is in NP. An interesting question is whether such a set could actually be in P; or, in other words, whether there is a polynomial-time procedure for deciding whether a given instance belongs to the set generated by a TICM. The answer is that there probably is not, if the generated set can be shown to be hard as well.

Let B be a set of hard positive instances for an NP-complete decision problem, and let L be the set of all positive instances of the problem. If B were in P, then B and $\overline{L}$ would be p–separable, implying that B is not hard. A similar argument holds if B is a hard set of negative instances. Note that for this proof we do not use the fact that B is generated by a TICM.

Proposition 4.6 *Let Π be an NP-complete problem. Any hard set of positive instances of Π cannot be in P unless P=NP. Any hard set of negative instances of Π cannot be in P unless P=NP.*

If B is a set containing both positive and negative instances, then the above argument cannot be used. But we can show the following.

Proposition 4.7 *Let Π be an NP-complete problem and C a TICM for Π. Let $B = Gen(C)$. If the positive instances of B have been shown to be hard using Proposition 4.4, then B cannot be in P unless NP=co–NP.*

Proof: Let B^{pos} and B^{neg} denote the set of positive and negative elements of B, respectively. The reduction of Proposition 4.4 actually shows that B^{pos} is an NP-complete language. Note that its complement is $\overline{B} \cup B^{neg}$ and that $B^{neg} = Gen(C, no)$ is in NP. Thus if B were in P, then $\overline{B} \cup B^{neg}$ would be in NP, implying that NP=co–NP.
□

There is a similar result pertaining to optimization problems. First we need the following lemma:

Lemma 4.8 *Let Π be an NP-hard optimization problem, and Π' its associated NP-complete decision problem. Let L be the set of positive instances of Π'. Let C be a TICM for Π and $B = Gen(C)$. Then $Dec(B) \cap L$ and $Dec(B) \cap \overline{L}$ are sets in NP.*

Proof: Assume without loss of generality that Π is a minimization problem. The TICM C can be used as a nondeterministic polynomial-time recognizer for both sets. Specifically, to determine whether $(I, l) \in Dec(B) \cap L$, run C on input $1^{v_1} \# ... \# 1^{v_k}$, with answer α, where $v_j = l_j(I)$ for $1 \leq j \leq k$, and α is chosen nondeterministically from among those integers between 0 and l. If C outputs I, then (I, l) is in $Dec(B) \cap L$. A similar argument works for $Dec(B) \cap \overline{L}$. □

Proposition 4.9 *Let $B = Gen(C)$ where C is a TICM for an NP-hard problem Π. If B has been shown to be hard with respect to Π using Proposition 4.5, then B cannot be in P unless NP=co-NP.*

Proof: Let Π' be the NP-complete decision problem associated with Π, and L the set of positive instances of Π'. Clearly B is in P if and only if $Dec(B)$ is in P. As shown in the proof of Proposition 4.5, the reduction used in that proposition actually shows that $Dec(B) \cap L$ is NP-complete. Since $B = Gen(C)$, by Lemma 4.8 $Dec(B) \cap \overline{L}$ is in NP. Hence if $Dec(B)$ were in P, then the complement of $Dec(B) \cap L$, namely $\overline{Dec(B)} \cup (Dec(B) \cap \overline{L})$ would be in NP, implying that NP $=$ co–NP. $\square$

Recall that a TICM cannot generate all instances of an NP-complete or NP-hard problem unless NP=co–NP. Another interesting and related fact which is not hard to derive, is that a TICM cannot generate all of the "negatively hard" instances of an NP-hard problem, unless NP=co–NP. In other words the set of instances not generated by a TICM must contain hard negative instances, unless NP=co–NP.

Proposition 4.10 *Let Π be an NP-complete problem, C a TICM for Π, and K the set of all negative instances of Π. Then $K - Gen(C)$ is hard with respect to Π, unless NP=co–NP.*

Proof: Let A be a negative approximation algorithm for Π and let F be the set of all instances for which A answers *no*. Note that F is in P. If A answers *no* for all instances in $K - Gen(C)$, then $K = Gen(C, no) \cup F$ and hence K is in NP, implying that NP $=$ co–NP. $\square$

Proposition 4.11 *Let Π be an NP-hard optimization problem, C a TICM for Π, and K the set of all instances of Π. Then $K - Gen(C)$ is negatively hard with respect to Π, unless NP=co–NP.*

Proof: Let Π' be the NP-complete decision problem associated with Π, and H the set of negative instances of Π'. Let A be a negative approximation algorithm for Π. As previously discussed, from A we can obtain a negative approximation algorithm A' for Π'. Let F be the set of all negative instances of Π' for which A' answers *no*. Note that if A gives the exact answer for all instances in $K - Gen(C)$, then all negative instances in $Dec(K - Gen(C))$ are in F, and hence $H = F \cup (Dec(Gen(C)) \cap H)$. By Lemma 4.8, $Dec(Gen(C)) \cap H$ is in NP, and hence H is in NP, implying that NP $=$ co–NP. $\square$

If a set of instances B for a problem Π is hard according to Definition 4.5 or 4.6, then no polynomial-time algorithm can work correctly on all elements of B (unless P $=$ NP). In fact, it is clear that any polynomial-time algorithm must give an inexact answer on an infinite subset of B. In fact, a stronger statement can be made.

A polynomial complexity core [14] for a language L is an infinite recursive set X such that for all polynomials p and all algorithms T that recognize L, T takes more than $p(|x|)$ time on all but a finite number of strings x in X. In other words, a complexity core is an infinite set on which all polynomial approximations for L do poorly. It is shown in [14] that any language L that is not in P has a complexity core. Thus any NP-complete language has a complexity core if P $\neq$ NP. Not surprisingly, such polynomial complexity cores also arise in sets that are hard according to Definitions 4.5 or 4.6.

Proposition 4.12 *Let Π be an NP-complete problem, and let B be a set of positive (negative) instances of Π which is hard with respect to Π. Then unless $P = NP$, there exists an infinite set $X \subseteq B$ such that any positive (negative) approximation algorithm for Π gives an inexact answer for all but a finite number of the instances in X.*

Proof: We give the proof assuming that B consists of positive instances. It is not difficult to see that the hardness of B implies that if T is any algorithm for solving Π, and p is any polynomial, then it is not the case that T terminates within $p(|I|)$ steps for all $I \in B$. A modification of the proof in [14] of the existence of complexity cores for languages not in P shows that, unless P=NP, there is a set $X \subseteq B$ such that for any polynomial p and algorithm T that solves Π, T takes more than $p(|I|)$ steps on all but a finite number of elements of X. This in turn implies that if A is any positive approximation algorithm for Π, then A must give an inexact answer for all but a finite number of the elements of X. $\square$

Proposition 4.13 *Let Π be an NP-hard optimization problem, and let B be a set of instances of Π which is positively (negatively) hard with respect to Π. Then unless $P = NP$, there exists an infinite set $X \subseteq B$ such that any positive (negative) approximation algorithm for Π gives an inexact answer for all but a finite number of the instances in X.*

Proof: Assume without loss of generality that B is positively hard. Let Π' be the NP-complete decision problem associated with Π and let L be the set of positive instances of Π'. Define $B' = Dec(B) \cap L$. By Proposition 4.3, B' is hard with respect to Π'. It follows by Proposition 4.12 that there exists a set $X' \subseteq B'$ such that any positive approximation algorithm for Π' must give an inexact answer ($?$) for all but a finite number of the instances in X'. Define $X = \{I | (I, k) \in X'\}$. Note that $X \subseteq B$. By Lemma 4.1, it follows that any positive approximation algorithm for Π must give an inexact answer for all but a finite number of the instances in X. $\square$

References

[1] Jose L. Balcazar. Simplicity, relativizations, and nondeterminism. *SIAM Journal on Computing*, 14(1), February 1985.

[2] Jin-Yi Cai, Thomas Gundermann, Juris Hartmanis, Lane Hemachandra, Vivian Sewelson, Klaus Wagner, and Gerd Wechsung. The boolean hierarchy I: Structural properties. *SIAM Journal on Computing*, 17:1232–1252, 1988.

[3] Jin-Yi Cai, Thomas Gundermann, Juris Hartmanis, Lane Hemachandra, Vivian Sewelson, Klaus Wagner, and Gerd Wechsung. The boolean hierarchy II: Applications. *SIAM Journal on Computing*, 18:95–111, 1989.

[4] M. Davis and H. Putnam. A computing procedure for quantification theory. *Journal of the Association for Computing Machinery*, 7:201–215, 1960.

[5] Devdatt Dubhashi. On p-separability. Technical Report TR 89-973, Cornell University, Department of Computer Science, February 1989.

[6] Michael R. Garey and David S. Johnson. *Computers and Intractability — A Guide to the Theory of NP-Completeness*. W.H. Freeman and Company, 1979.

[7] Juris Harmanis, Vivian Sewelson, and Neil Immerman. Sparse sets in NP-P: EXPTIME versus NEXPTIME. In *Proceedings of the 15th Annual ACM Symposium on Theory of Computing*, pages 382–391, April 1983.

[8] Juris Hartmanis. On sparse sets in NP-P. *Information Processing Letters*, 16:55–60, February 1983.

[9] B.N. Khoury, P.M. Pardalos, and D.-Z Du. A test problem generator for the Steiner problem in graphs. *ACM Transactions on Mathematical Software*, to appear 1993.

[10] Ker-I Ko and Daniel Moore. Completeness, approximation, and density. *SIAM Journal on Computing*, 10(4):787–796, November 1981.

[11] Balakrishnan Krishnamurthy. Constructing test cases for partitioning heuristics. *IEEE Transactions on Computers*, C–36(9):1112–1114, September 1987.

[12] Stuart A. Kurtz. Sparse sets in NP-P: Relativizations. *SIAM Journal on Computing*, 14(1):113–119, February 1985.

[13] E.W. Leggett-Jr. and Daniel J. Moore. Optimization problems and the polynomial hierarchy. *Theoretical Computer Science*, 15:279–289, 1981.

[14] Nancy Lynch. On reducibility to complex or sparse sets. *Journal of the Association for Computing Machinery*, 22(3):341–345, July 1975.

[15] Pekka Orponen, David A. Russo, and Uwe Schoning. Optimal approximations and polynomially levelable sets. *SIAM Journal on Computing*, 15(2):399–408, May 1986.

[16] C.H. Papadimitriou and M. Yannakakis. The complexity of facets (and some facets of complexity). *Journal of Computer and System Sciences*, 28(2):244–259, 1984.

[17] Christos H. Papadimitriou and Kenneth Steiglitz. *Combinatorial Optimization – Algorithms and Complexity.* Prentice Hall, 1982.

[18] Martha G. Pilcher and Ronald L. Rardin. Invariant problem statistics and generated data validation: Symmetric traveling salesman problems. Technical Report CC-87-16, Purdue University, Institute for Interdisciplinary Engineering Studies, April 1987.

[19] Martha G. Pilcher and Ronald L. Rardin. A random cut generator for symmetric traveling salesman problems with known optimal solutions. Technical Report CC-87-4, Purdue University, Institute for Interdisciplinary Engineering Studies, February 1987.

[20] Laura A. Sanchis. Test case construction for NP-hard problems. In *Proceedings of the 26th Annual Allerton Conference on Communication, Control, and Computing*, September 1988.

[21] Laura A. Sanchis. On the complexity of test case generation for NP-hard problems. *Information Processing Letters*, 36:135–140, November 1990.

[22] Laura A. Sanchis and Mark A. Fulk. On the efficient generation of language instances. *SIAM Journal on Computing*, 19(2):281–296, April 1990.

[23] Bart Selman, David Mitchell, and Hector Levesque. A new method for solving hard instances of satisfiability. In *Proceedings of AAAI-92*, San Jose, August 1992, 440–446.

Complexity in Numerical Optimization, pp. 429-447
P.M. Pardalos, Editor
©1993 World Scientific Publishing Co.

Maximizing Non-Linear Concave Functions in Fixed Dimension[1]

Sivan Toledo
Laboratory for Computer Science
Massachusetts Institute of Technology
Cambridge, Massachusetts 02139

Abstract

Consider a convex set $\mathcal{P}$ in $\mathbb{R}^d$ and a piecewise polynomial concave function $F\colon\mathcal{P} \to \mathbb{R}$. Let $\mathcal{A}$ be an algorithm that given a point $x \in \mathbb{R}^d$ computes $F(x)$ if $x \in \mathcal{P}$, or returns a concave polynomial p such that $p(x) < 0$ but for any $y \in \mathcal{P}$, $p(y) \geq 0$. We assume that d is fixed and that all comparisons in $\mathcal{A}$ depend on the sign of polynomial functions of the input point. We show that under these conditions, one can find $\max_{\mathcal{P}} F$ in time which is polynomial in the number of arithmetic operations of $\mathcal{A}$. Using our method we give the first strongly polynomial algorithms for many *non-linear* parametric problems in fixed dimension, such as the parametric max flow problem, the parametric minimum s-t distance, the parametric spanning tree problem and other problems. We also present an efficient algorithm for a very general convex programming problem in fixed dimension.

Keywords: Convex programming, parametric searching, parametric optimization, network flow.

1   Introduction

Consider a convex set $\mathcal{P}$ in $\mathbb{R}^d$ and a piecewise polynomial concave function $F\colon\mathcal{P} \to$ $\mathbb{R}$. Let $\mathcal{A}$ be an algorithm that given a point $x \in \mathbb{R}^d$ computes $F(x)$ if $x \in \mathcal{P}$, or returns a *separation polynomial*, a concave polynomial p such that $p(x) < 0$ but for

---

[1]An extended abstract of this paper appeared in [15].

any $y \in \mathcal{P}$, $p(y) \geq 0$. We assume that d is fixed and that all comparisons in $\mathcal{A}$ depend on the sign of polynomial functions of the input point which are called *comparison polynomials*. Our main result is that under these conditions, one can find $\max_{\mathcal{P}} F$ and a maximizer in time which is polynomial in the number of arithmetic operations of $\mathcal{A}$.

The algorithm is based on the ingenious parametric searching technique invented by Megiddo [12]. This technique can be directly applied to one dimensional concave maximization problems, that is, when the domain of F is an interval on the real line. The technique, which is described in detail in Section 2, can be summarized as follows. We simulate the execution of $\mathcal{A}$ on a maximizer x^* even tough we do not know the location of x^*. Therefore we handle it in the algorithm as a symbolic constant. When we need to determine the sign of some polynomial p at x^*, we compute the roots of p. Since F is concave, it is fairly easy to determine the location of the maximum of F with respect to each root. By determining between which two roots x^* lies, we can determine the sign of $p(x^*)$. In other words, we compute a decomposition of $\mathbb{R}$ such that in every cell the sign of p does not change, and determine which cell of the decomposition contains a maximizer of F.

Cohen and Megiddo [4, 5] and independently Norton, Plotkin and Tardos [14] generalized the technique to handle the multidimensional case, but only when F is piecewise linear and the separation and comparison polynomials are all affine. When a comparison polynomial p is affine, a space decomposition which is invariant for the sign of p is a hyperplane H and two open half spaces. Such a hyperplane is called a *critical* hyperplane. The restriction of F to H is a concave function in one dimension lower. By induction, the maximum of F on H can be found. The assumption that the comparison polynomials are affine also makes it relatively easy to explore the neighborhood of H, and to determine on which side of it F is increasing, thereby resolving the comparison. Since it is assumed that F is piecewise linear, there is a maximizer which is a vertex of its graph. Such a maximizer can be found using linear programming.

In the non-linear case the problem of resolving comparisons becomes much harder. The comparison polynomials are not necessarily concave, and it is hard to compute a sign invariant space decomposition and to locate the cell in the decomposition which contain the maximizer. Our algorithm computes the decomposition and locates the cell, but we prefer to describe it in a slightly different form which is more amenable to recursive breakdown. We use a searching technique which is based on the weighted Euclidean 1-center algorithm of Megiddo [13]. In this technique, the d dimensional algorithm works by simulating the $d-1$ dimensional one on a hyperplane that contains a maximizer of F. Three new tools are required in order to apply this searching technique to general concave maximization problems. The first is a sign invariant space decomposition which will enable us to simulate the algorithm in one dimension lower. The decomposition technique we use is Cylindrical Algebraic Decomposition [3]. We use *critical hyperplanes* to locate a maximizer, so we need a way to decide on which side of a hyperplane there is a maximizer. In Section 2.1 we describe a very general algorithm for doing so, which is based on Fibonacci search. Finally, we use Lagrange multipliers to find a maximizer.

Using our new technique we obtain the first strongly polynomial algorithms to a wide variety of non-linear parametric problems in fixed dimension. For example, given a graph in which the edges have concave polynomial weights, we can maximize the max-flow in the graph, the minimum spanning tree, the minimum s-t distance and so on. As is the case with other applications of Megiddo's parametric search technique, when there is a fast parallel algorithm for evaluating F, a fast parallel algorithm for maximizing F can be obtained, and the efficiency of the sequential maximization algorithm can be greatly improved. Using this improvement, we obtain an efficient algorithm for optimizing a linear function under convex polynomial constraints in fixed dimension. Our running time analysis is in the Random Access Machine model [2].

Megiddo's parametric searching technique is a *lifting* transformation. An algorithm for evaluating a function F is simulated, and the point in which the function is evaluated is handled as a symbolic constant λ. Megiddo's lifting technique applies to a very specific class of algorithms. Variables that are functions of λ are assumed to be polynomial functions of λ, and conditional statements involving λ are assumed to depend on the sign of such variables. Unfortunately, parametric searching algorithms do not belong to this class, since they also find roots of polynomials. This poses a difficulty in trying to lift such an algorithm, in order to generate a two dimensional searching algorithm. Our algorithm can be viewed as an extension of the Megiddo's technique to algorithms that are also allowed to find roots of polynomials. Interestingly, Megiddo's technique has been applied to many algorithms that perform polynomial root finding, especially in geometric optimization (see for example [1]). The implicit assumption in such applications is that the algorithm, which includes root finding, is implemented using symbolic algebra tools that allow computation in algebraic extensions and eliminate the need to handle algebraic numbers explicitly. If so implemented, the algorithm does not perform root finding and can be lifted using Megiddo's technique. In this paper we show explicitly how to lift an algorithm that performs root finding.

The next section describes in detail the application of Megiddo's technique to concave maximization problems. The general case is presented in Section 4. The two dimensional case is presented separately in Section 3 in order to provide a full description of an easy to visulize case. We conclude the paper in Section 5 with several applications of our results.

2   Maximizing One Dimensional Concave Functions

We begin with a brief review of Megiddo's parametric search technique [12] and how to use it to solve parametric maximization problems. We first define the class of algorithms that can be used as evaluators of F.

Definition Let $\mathcal{A}$ be an algorithm that gets as a part of its input a point $x \in \mathbb{R}^d$, and returns a real number depending on x, denoted $F(x)$. We say that $\mathcal{A}$ is *polynomial in x with degree δ* if the only dependencies on x are:

1. $\mathcal{A}$ is allowed to evaluate the polynomials $p_1(x), \ldots, p_k(x)$ of degree at most δ, where δ does not depend on the input.

2. The only operations on variables in $\mathcal{A}$ that depend on x are addition of such variables, addition of constants, and multiplication by constants.

3. The conditional branches in $\mathcal{A}$ that depend on x depend only on signs of variables that depend on x.

Definition A point $x_0 \in \mathcal{P}$ is called a *non-singular point of F* if there is an $\epsilon > 0$ such that the restriction of F to $\mathcal{P} \cap \{x : |x - x_0| < \epsilon\}$ is a polynomial function. If x_0 is a non-singular point of F, this restriction of F is called the *piece of F* at x_0.

Corollary 2.1 *If $\mathcal{A}$ is polynomial in x with degree δ, then all the variables in $\mathcal{A}$ that depend on x contain polynomials of degree of at most δ, and $F(x)$ is piecewise polynomial whose pieces are polynomials of degree at most δ.*

Assume that we have an efficient algorithm $\mathcal{A}$ for evaluating $F(x)$, which is polynomial in x with some fixed degree δ, and that F is a concave function. Megiddo's main idea is to simulate $\mathcal{A}$ at a maximizer of F, denoted $x^\star$. As long as no comparisons are made, that is, no conditional branches that depend on the input point are to be executed, it is easy to simulate the algorithm, by treating the variables as polynomials and performing polynomial arithmetic. How do we resolve a conditional branch that depend on the sign of a variable? We find the roots of the polynomial stored in that variable, and locate $x^\star$ among them as follows. We evaluate F at each of the roots, and determine the location of a maximizer with respect to each of the roots. (For now we assume that if we can evaluate F at a point we can also decide the direction to $x^\star$, and in section 2.1 we justify this assumption.) In other words, for every root, we test whether it is $x^\star$, or else whether $x^\star$ is to its left or to its right. Given this information, we can easily decide which way should the branch take, since the sign of a polynomial is constant between its roots. We thus obtain a smaller and smaller interval that is known to contain $x^\star$, and finally the algorithm terminates. In section 2.1 we show how to obtain at this stage a maximizer of F and the two pieces of F to its left and right. These pieces allow us to generalize the algorithm to higher dimensions, and they provide a certificate of optimality for the maximizer.

In more abstract terms, given a comparison polynomial p, we decompose the space (here $\mathbb{R}$) into cells which are invariant for the sign of p. In the one dimensional case, the cells are points, which are the roots of p, and open intervals. Given this decomposition, we decide in which cell there is a maximizer of F, and thus resolve the comparison.

Running time analysis. Assuming that the algorithm $\mathcal{A}$ runs in T_0 time, the one dimensional maximization algorithm runs in time $T_1 = O(T_0^2)$, since whenever the algorithm makes a comparison, we evaluate the function at each of the roots. Megiddo [12] noticed that if we also have a parallel algorithm that evaluates the function, we can exploit the parallelism to obtain faster maximization algorithm. Assume that the parallel algorithm uses P processors and runs in T_p parallel time. We simulate the algorithm sequentially. In each parallel step there are at most P independent comparisons. Instead of evaluating the function at each of the roots of all the associated polynomials, we perform a binary search over the set of $O(P)$ roots to locate x^* among them. This results in $O(\log P)$ evaluations of F, and $O(P)$ overhead for performing the binary search by repeatedly finding the median of the set of unresolved roots. Having done this, we can determine the sign of each of the $O(P)$ roots at x^* and proceed to the next parallel step. The total cost of this procedure is $T_1 = O(PT_p + T_0T_p \log P)$. Since we only require that comparisons will be made in parallel, we can use Valiant's weak model of parallel computation [16].

2.1 Where is F Maximized?

Given a point x_1, we need to determine the location of x^* relative to x_1. The techniques for doing so in one dimension and the techniques that were used by [4, 5, 14] do not seem to generalize to non-linear comparison polynomials and higher dimensions. This section describes a new technique which is easy to generalize. We evaluate $F(x_1)$. If we have previously encountered a point x_0 such that $F(x_0) \geq F(x_1)$, we can safely assume that there is a maximizer in the direction of x_0. Otherwise, we do not resolve the comparison. We duplicate the state of the simulated algorithm, and in one copy resolve the comparison as if there is a maximizer to the left of x_1, and in the other copy as if there is a maximizer to the right of x_1. We run those two copies in parallel (by interleaving their execution on a sequential machine). For each root of a comparison polynomial we obtain (from either copies; we do not know which one of them is correct), we evaluate F at that point. As long as we do not encounter a value of F larger than $F(x_1)$, we can determine which side of a given root contains a maximizer. If we run into a point x_2 where the value of F is larger than $F(x_1)$, we again will not be able to resolve the comparison. But in this case, the maximizer is on the same side of x_1 as x_2, so now we can resolve the comparison that involved x_1. In particular, we can decide which copy of the algorithm was given the correct answer and discard the other. Of course, we now must run two copies of the algorithm in which we resolve the comparison involving x_2 in different ways. There are always two copies of the algorithm executing. Eventually, both of our copies will terminate. Each one of them returns $F(x)$ as a polynomial. One of them corresponds to the piece of F to the right of the point x_k with the highest F value encountered, and the other corresponds to the piece of F to the left of this point. We maximize these polynomials over the corresponding intervals. If one of them attains a maximum higher than the other inside its interval, this is the optimum, and this polynomial is the piece of F on both sides of the maximum. Otherwise, they both attain the same maximum, and in that case the point x_k is a maximizer, and these two polynomials are the pieces of F on its two sides. Since in most cases the cost of evaluating F dominates the cost of duplicating the state of the algorithm we ignore this cost in the running time analysis.

The same idea works in any dimension. Let $F: \mathbb{R}^d \to \mathbb{R}$ be a concave function. Suppose that we already know the value of F at some points in $\mathbb{R}^d$, and that the highest value we computed is $F(x_0)$. Given a hyperplane H, let the maximum of F on H be $F(x_1)$. If $F(x_0) \geq F(x_1)$, we can safely assume that there is a maximizer in the direction of x_0. Otherwise there is another point x on the other side of H with $F(x) > F(x_1)$. Hence at the intersection of the line segment $\overline{x\,x_1}$ with H the value of F must be higher than $F(x_1)$ due to the concavity of F, a contradiction.

2.2   Finding a Feasible Point

In many cases the domain $\mathcal{P}$ is either all of $\mathbb{R}^d$, or easy to compute as the intersection of a small number of constraints $p_i(x) \geq 0$, where the p_i's are concave polynomials, such as in the parametric max flow problem. But there are cases in which the domain of F is defined by an exponential number of constraints, such as the parametric minimum s-t distance in a connected graph. Using ideas from [4, 5, 14], we describe how to deal with this problem in the non-linear case.

We assume that there is an algorithm $\mathcal{A}_f$ for testing whether a point x belongs to $\mathcal{P}$, which either declares that $x \in \mathcal{P}$, or declares that $x \notin \mathcal{P}$ and provides in addition a violated constraint $p(x) < 0$, where p is a concave polynomial. We use this algorithm to either find a point x_f in $\mathcal{P}$ or decide that $\mathcal{P}$ is empty. If $\mathcal{P}$ is empty we report this and halt. Otherwise, given a critical point x_0, we test whether $x_0 \in \mathcal{P}$, and if not, we know that there is a maximizer of F in the direction of x_f.

We simulate the feasibility testing algorithm $\mathcal{A}_f$ on x_f. During the simulation, we maintain an interval $[a, b]$, which is known to contain $\mathcal{P}$. In addition, for each endpoint $z \in \{a, b\}$ of the interval, if it is finite, we also maintain a constraint p_z that is violated if we pass this endpoint. (We begin with the interval $(-\infty, \infty)$.) When we must resolve a comparison, we find the roots of the comparison, and determine whether one of them is a feasible point, in which case we return this point and halt. If a given root being tested is not in $\mathcal{P}$, a violated constraint p is returned. Since for each $x \in \mathcal{P}$, $p(x) \geq 0$, we know that $\mathcal{P}$ must lie in $[a, b]$ and also in the interval $\{x : p(x) \geq 0\}$. We therefore update $[a', b'] = [a, b] \cap \{x : p(x) \geq 0\}$. If this new interval is empty, we conclude that $\mathcal{P}$ is empty. Note that this event actually carries more information, since if we assume without loss of generality that $\{x : p(x) \geq 0\}$ is to the left of $[a, b]$, then the two constraints p and p_a provide a certificate that $\mathcal{P}$ is empty. If the new interval in not empty and not equal to $[a, b]$, we replace the polynomials associated with the updated endpoints with p. It is easy to see that if we have a parallel feasibility testing algorithm, we can exploit the parallelism and obtain a faster algorithm using Megiddo's scheme.

If at no point of the simulation the feasible interval becomes empty, then our simulated algorithm terminates, and returns an answer. In addition, we have an interval $[\hat{a}, \hat{b}]$ where $\mathcal{P}$ must lie. If the algorithm returns "yes", it means that every point in $[\hat{a}, \hat{b}]$ is a feasible point. Otherwise, it returns "no" and a violated constraint p. It follows that this constraint is violated for all points of $[\hat{a}, \hat{b}]$, so this constraint together with either $p_{\hat{a}}$ or $p_{\hat{b}}$ provide proof of emptiness for $\mathcal{P}$.

3   A Two Dimensional Algorithm

We now describe the parametric maximization algorithm in two dimensions, that is, when $F: \mathcal{P} \to \mathbb{R}$ where $\mathcal{P} \subseteq \mathbb{R}^2$. Let $(x^\star, y^\star)$ be a maximizer of F. The main idea of the algorithm is to simulate the one dimensional algorithm on F restricted to a line $x = x^\star$. Since a concave function restricted to a line (or a hyperplane in higher dimensions) is still concave, the problem of maximizing F restricted to a line is a one dimensional problem. If we can simulate the one-dimensional algorithm on such a line, we can find a maximizer $y^\star$ on the line, which is also a global maximizer, and we are done. The problem of course is how to make decisions during the simulation. Let p be a comparison polynomial in the simulated non-parametric algorithm. We compute a cylindrical decomposition of $\mathbb{R}^2$ which is invariant for the sign of p. This decomposition is constructed by computing the self intersections of the curve $p = 0$ and the points of vertical tangency of the curve. Those points are projected on the x-axis, and the plane is decomposed into vertical slabs between those points. A vertical slab (which is a generalized cylinder) may intersect the curve $p = 0$, but the roots of p do not intersect each other inside the slab (see Figure 1).

We execute the one dimensional algorithm on the vertical lines that decompose the plane into slabs, and we decide in which slab there is a maximizer (recall that in Section 2.1 we have shown that if we can maximize F on a line, we can also decide on which of its sides there is a maximizer).

The crucial point is that in each slab there is a constant number of roots to the polynomial $p(x, y)$ as a one dimensional polynomial in y. The location of those roots depends on x, but the dependency is continuous. Hence we can simulate the one

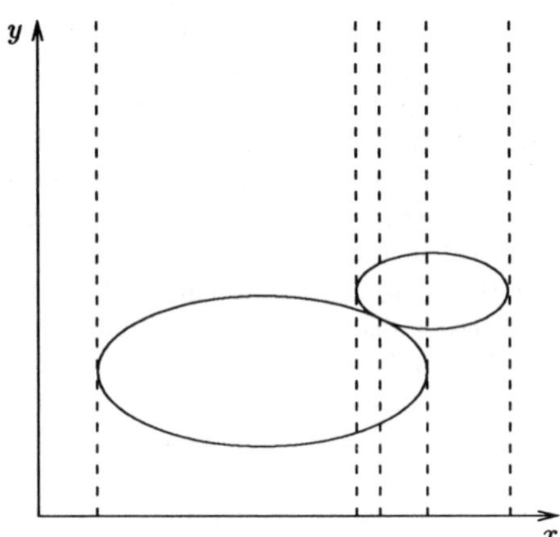

Figure 1: A cylindrical decomposition of $\mathbb{R}^2$ invariant for the sign of a polynomial. The solid curve is the root of the polynomial.

dimensional algorithm in a consistent manner. However, the one dimensional algorithm executes the non-parametric algorithm on the roots of p. We cannot perform this directly, since the location of the roots depend on x. However, we can simulate the non-parametric algorithm. When the non-parametric algorithm performs a comparison involving a polynomial q, we compute a cylindrical decomposition which is invariant for both p and q. Again, we determine the slab in this decomposition that contains a maximizer. We examine the root of p on which we simulated the non-parametric algorithm in this slab, and determine the sign of q in that cell of the decomposition, which is possible since the decomposition is invariant for the sign of q.

Once all the executions of the non-parametric algorithm terminate, we must compare the returned values, which are polynomials, to each other in order to decide which one is higher. To compare a value $r_1(x, y)$ on a root $y_1(x)$ of p with a value $r_2(x_y)$ on a root $y_2(x)$, we compute (again using cylindrical decomposition) the x-coordinates of the intersections of $\{r_1(x, y_1) = r_2(x, y_2), p(x, y_1) = p(x, y_2) = 0\}$, and decide in which slab there is a maximizer. Using this information, we can compare r_1 to r_2. Using a similar approach we can also find the maximum of a polynomial in a certain interval, which is required in the technique of Section 2.1. Finally the simulated one dimensional algorithm terminates. In fact, two copies of it terminate, one which is a simulation to the left of the vertical line l on which the highest value of F was found, and the other to its right. Each returns a curve $p_i(x, y) = 0$, $i \in \{L, R\}$ on which the maximum is obtained, and the two pieces of F above and below this curve, $p_{i,A}$ and $p_{i,B}$. We find the maximum of F to the left of the line L by solving

$$\max\ p_{L,A}(x,y)$$
$$p_{L,A}(x,y) = p_{L,B}(x,y)$$

using Lagrange multipliers. All the functions involved are polynomial, so this can be solved using cylindrical decomposition (since the problem reduces to finding all the solutions to a system of polynomial equations). This method establishes the global maximum, and in addition generates four pieces of F that prove that the point found is indeed a maximizer. Using Helly's theorem, it is easy to show that we can reduce the number of these pieces to three.

When there exist a parallel non parametric algorithm, we can obtain a parallel one dimensional algorithm. To use it, we also need to sort the list of roots that are obtained in each step of the one dimensional algorithm. We could do it by constructing the cylindrical decomposition invariant for the signs of *all* the comparison polynomials, but this would be too expensive. Instead, we simulate a sorting algorithm, and whenever it compares two roots, we construct the decompositions invariant for the signs of both polynomials, locate the slab that contains a maximizer, and test which of our two roots is higher in that slab. If we use Megiddo's technique and sort the roots using a simulation of a parallel sorting algorithm such as Cole's parallel merge sort [7], the number of calls to the one dimensional algorithm during the sorting algorithm will be only $O(\log^2 P)$, where P is the number of roots we have to sort.

Running time analysis. We first note that the cost of constructing the cylindrical algebraic decomposition of a constant number of polynomials of bounded degree in fixed dimension is only a constant (in fact, the cost is polynomial in the number of polynomials, the degree, and the binary encoding of the coefficients, but not in the dimension). Denoting the running time of the non-parametric algorithm $\mathcal{A}$ by T_0 and of the d-dimensional algorithm by T_d, the running time is $T_2 = O(T_0(T_1 + T_0T_1)) = O(T_0^4)$. If there exists a parallel non-parametric algorithm that runs in T_p parallel time and uses P processors, we can again improve the running time. The total running time is $O(PT_p + T_p(\log PT_1 + \log^2 PT_1 + \log PT_p \log PT_1))$ and since $T_1 = O(PT_p + T_p \log PT_0)$, the running time is $T_2 = O(T_0(T_p \log P)^3)$. The breakdown of the running time into terms is as follows. In each parallel step of the non-parametric algorithm P comparison polynomials are generated and a decomposition invariant for each one of them is constructed. Then the one dimensional algorithm is called $\log P$ times to locate the slab in which there is a maximizer in the combined list of critical values. As explained before, sorting all the y critical values requires $\log^2$ more calls to the one dimensional algorithm. Finally a binary search is performed using a simulated non-parametric algorithm. It is simulated on $\log P$ y roots. On each of them the algorithm proceeds in T_p steps, and in each one P comparison polynomials are generated, each requiring a constant number of calls to the one dimension algorithm.

3.1   Finding a Feasible Point

We now extend the technique of section 2.2 to two dimensions; the same technique works in any dimension. We use the technique we have just described for simulating the one dimensional feasibility testing algorithm on a vertical line that intersects $\mathcal{P}$. All we need to show is how to decide on what side of a given line $\mathcal{P}$ lies, since if $\mathcal{P}$ intersects the line, the one dimensional algorithm will detect this.

Suppose we are given a line in the plane, and must test on which of its sides $\mathcal{P}$ lies. Since the one dimensional algorithm determines that the line does not intersect $\mathcal{P}$, it reports two constraints that are contradictory on that line, p and q. Since we simulate the one dimensional algorithm and perform its arithmetic on polynomials, the violated constraints are polynomials in x and y. Since for any $(x, y) \in \mathcal{P}$ both $p(x, y) \geq 0$ and $q(x, y) \geq 0$, we find a point in the intersection of $p(x, y) \geq 0$ and $q(x, y) \geq 0$. If there is such a point x_0 and $\mathcal{P}$ is not empty, then $\mathcal{P}$ lies on the same side of the line as x_0, and we maintain p and q as a certificate for this fact. If $p(x, y) \geq 0 \cap q(x, y) \geq 0 = \emptyset$, then $\mathcal{P}$ is empty.

If the simulated algorithm terminates without either finding a point in $\mathcal{P}$, or deciding that $\mathcal{P}$ is empty, we have two cases. The answer it gives is valid for any line $x = x_0$ we might run it on, as long as x_0 is in the interval $[x_a, x_b]$ which is known to contain $\mathcal{P}$. When the simulated algorithm terminates, if it returns a point in $\mathcal{P}$, we are done. Otherwise it declares that the intersection of $\mathcal{P}$ with any vertical line $x = x_0$ in the vertical slab which is known to contain $\mathcal{P}$ is empty, and supplies a pair of constraints p and q as a certificate. We compute a point in $p(x, y) \geq 0 \cap q(x, y) \geq$. If there is no such point, then p and q are a proof that $\mathcal{P}$ is empty. Otherwise, if they intersect above to the left of the slab for example (i.e. to the left of x_a), then p and q together with the two constraints that assert that $\mathcal{P}$ is to the right of x_a, provide a proof that $\mathcal{P}$ is empty. Those four contradictory constraints allow us to generalize the algorithm to higher dimension, in the same way we used the one dimensional certificates of emptiness for constructing the two dimensional algorithm in this section. The number of constraints is a certificate of emptiness can be brought down to at most $d + 1$ in dimension d, by Helly's Theorem.

4   The General Algorithm

Before we describe the algorithm and prove its correctness, we need some definitions and lemmas.

Definition A *semi-algebraic cell* of $\mathbb{R}^d$ is a set of points satisfying a finite set of polynomial equalities and inequalities.

Definition A *semi-algebraic variety* is either a semi-algebraic cell, or one of the sets $A \cap B$, $A \cup B$ and $A \setminus B$, where A and B are two semi-algebraic varieties.

Definition A *decomposition* of $\mathbb{R}^d$ is the representation of $\mathbb{R}^d$ as the union of a finite number of disjoint and connected semi-algebraic varieties.

Definition A decomposition of $\mathbb{R}^d$ is *invariant for the signs* of a family of polynomials if, over each cell of the decomposition, each polynomial is always positive, always negative, or always zero.

Definition A decomposition D_d of $\mathbb{R}^d$, that is $\mathbb{R}^d = E_1 \cup \cdots \cup E_N$ is *cylindrical* if $n = 0$ (the trivial case) or if $n > 0$ and:

1. $\mathbb{R}^{d-1}$ has a cylindrical decomposition D_{d-1} which can be written $\mathbb{R}^{d-1} = F_1 \cup \cdots \cup F_M$, and

2. For each cell E_i of D_d there is a cell F_j of D_{d-1} such that E_i can be written in one of the following forms

$$\{(\boldsymbol{x}, y) : \boldsymbol{x} \in F_j \;\wedge\; y < f_k(\boldsymbol{x})\} \quad \text{(a segment)}$$
$$\{(\boldsymbol{x}, y) : \boldsymbol{x} \in F_j \;\wedge\; y = f_k(\boldsymbol{x})\} \quad \text{(a section)}$$
$$\{(\boldsymbol{x}, y) : \boldsymbol{x} \in F_j \;\wedge\; f_k(\boldsymbol{x}) < \\ y < f_l(\boldsymbol{x})\} \quad \text{(a segment)}$$
$$\{(\boldsymbol{x}, y) : \boldsymbol{x} \in F_j \;\wedge\; y > f_k(\boldsymbol{x})\} \quad \text{(a segment)}$$

where the f_k's are the solutions of polynomial equations ($\boldsymbol{x}$ denotes $x_1, \ldots, x_{d-1}$ and y denotes x_d).

Theorem 4.1 (Collins) *There exists an algorithm that computes a cylindrical decomposition of $\mathbb{R}^d$ invariant for the sign of a family of n polynomials. If the polynomials are all of degree δ or less, and the length of the binary encoding of their coefficients is bounded by H, the running time of this algorithm is bounded by*

$$(2\delta)^{2^{2d+8}} n^{2^{d+6}} H^3.$$

Lemma 4.2 *Let D_d be a cylindrical decomposition of $\mathbb{R}^d$ invariant under a family P of polynomials, and let H be a hyperplane in $\mathbb{R}^d$ specified by $x_1 = \alpha$ for some real α. Then the intersection of D_d with H is a cylindrical decomposition of $\mathbb{R}^{d-1}$ (with the natural mapping of $\mathbb{R}^{d-1}$ onto H) invariant under the restriction of the polynomials in P to H.*

Proof: The intersection of D_d with H is obviously a decomposition of $\mathbb{R}^{d-1}$ and invariant under the signs of the family of polynomials. We prove that it is also cylindrical. The proof uses induction on the recursive structure of the cylindrical decomposition. We assume that the intersection of D_{d-1} with H is cylindrical, and we prove that the intersection of D_d with H is cylindrical. The claim is obviously for $d = 1$, because D_1 is a decomposition of the x_1 axis which is invariant for the sign of some family of polynomials P_1. The intersection of H with the x_1-axis is only a point, and the decomposition of a point is always cylindrical and invariant for the signs of P_1.

We now assume that the claim is true for D_{d-1}. Let C' be a cell of the intersection of D_d with H, which is the intersection of the cell C with H. Let us assume that C is of the form

$$\{(x_1, x_2, \ldots, x_d) : \ (x_1, x_2, \ldots, x_{d-1}) \in F \ \wedge$$
$$x_d > f_k(x_1, x_2, \ldots, x_{d-1})\}$$

where $F \in D_{d-1}$. Let F' be the intersection of F with H. Then C' can be written as

$$\{(\alpha, x_2, \ldots, x_d) : \ (\alpha, x_2, \ldots, x_{d-1}) \in F' \ \wedge$$
$$x_d > f_k(\alpha, x_2, \ldots, x_{d-1})\}.$$

The other cases are similar. We note that the crucial point in the proof is that H is parallel to all the projection axes. ∎

Lemma 4.3 *Using the same notation, let*

$$D_1 = \{-\infty = \alpha_0, \alpha_1, \ldots, \alpha_k, \alpha_{k+1} = \infty\}$$

(that is the α's are the points in the one dimensional decomposition). Then the intersection of D_d with H depends continuously on α as long as $\alpha_i < \alpha < \alpha_{i+1}$.

Proof: It is obvious that the intersection of H with D_1 changes continuously. Let us assume that the intersection of H with D_d changes continuously but that the intersection with D_{d+1} does not. This can only happen if for some $\alpha_i < \alpha < \alpha_{i+1}$ two sections of D_{d+1} intersect, which contradicts the previous lemma. ∎

The algorithm[2]. We construct the algorithm inductively. The induction hypothesis describes the structure and correctness of the $d-1$ dimensional algorithm. We assume that the d dimensional algorithm work by constructing a sequence of cylindrical decompositions (CADs for short) in $\mathbb{R}$ through $\mathbb{R}^{2d}$ of up to $2d+1$ polynomials, and tests the sign of one of the polynomials in various cells of the decomposition. The algorithm returns a maximizer of F. The location of the maximizer is returned as a specific zero dimensional cell in a d dimensional CAD of up to d polynomials. The value of the maximizer is returned as a polynomial.

Let us prove that the induction hypothesis holds for the one dimensional parametric searching algorithm. The one dimensional algorithm finds the roots of polynomials. Finding the roots of a polynomial is equivalent to computing a CAD invariant for the sign of it. Then the parametric searching algorithm evaluates the sign of other polynomials on the roots. Finding the sign of a polynomial q at a root of a

---

[2]The notation p' in this section means some arbitrary polynomial and not the derivative of a function p.

polynomial p can certainly be done by constructing a CAD invariant for the sign of p and q and evaluating the sign of q at the root of p, which is a cell of the CAD. The one dimensional algorithm also compares values of F at various roots. The values of F are all polynomials. Suppose we need to compare the value of q' at a root α' of p' with the value of q'' at a root α'' of p''. In other words, we need to test the sign of the polynomial $q'(\alpha') - q''(\alpha'')$ at a point (α', α'') in which $p'(\alpha') = 0$ and $p''(\alpha'') = 0$. We could certainly do this by constructing the CAD invariant for the sign of $q'(\alpha') - q''(\alpha'')$, $p'(\alpha')$ and $p''(\alpha'')$, and testing the sign of $q'(\alpha') - q''(\alpha'')$ in some particular cell. This is a CAD of 3 polynomials in $\mathbb{R}^2$. Finally, we need to maximize F over two open intervals in which F does not have a breakpoint. This is done by maximizing two concave polynomials over the intervals, which can be done be finding the roots of their derivatives, which again amounts to computing CADs. We now compare the two maxima using the method just described. The higher is the global maximum. The maximum is returned as a polynomial p at a root α of another polynomial q.

Let us describe the d dimensional algorithm. The algorithm works by simulating the $d - 1$ dimensional algorithm on a hyperplane $x_1 = x_1^\star$, where $x_1^\star$ is a projection of a maximizer $x^\star = (x_1^\star, \ldots, x_d^\star)$. Suppose the simulated algorithm constructs a CAD D of n polynomials in $\mathbb{R}^m$ and tests the sign of one of the polynomial in some cell of the CAD. The polynomials are of dimension $m + 1$. We therefore constructs the CAD D' in $\mathbb{R}^{m+1}$ (we consider the additional variable to be x_1). Let the critical values of the decomposition be $\alpha_1, \ldots, \alpha_k$, the roots of some one dimensional polynomial $p(x_1)$. We locate a slab containing a maximizer $\alpha_j < x_1^\star < \alpha_{j+1}$ by performing a binary search (or a Fibonacci search, which would result in a slightly better constant in the running time). To determine whether the slab is to the left or to the right of some critical value α_i, we call the $d - 1$ dimensional algorithm on the hyperplane $x_1 = \alpha_i$. In the called $d - 1$ dimensional algorithm we add to the CADs constructed the polynomial $p(x_1)$. The number of polynomials and the dimension of each CAD are increased by one. We compare the values returned from different calls to the $d - 1$ dimensional algorithm, in order to find a slab containing a maximizer. Suppose we need to compare a value $q'(\alpha', x_2, \ldots, x_d)$ on a zero dimensional cell in the CAD of

$$p'_1(\alpha', x_2, \ldots, x_d), \ldots, p'_{d-1}(\alpha', x_2, \ldots, x_d)$$

where α' is some root of $p'(x_1)$, with a value $q''(\alpha'', x_2, \ldots, x_d)$ on a zero dimensional cell in the CAD of

$$p''_1(\alpha'', x_2, \ldots, x_d), \ldots, p''_{d-1}(\alpha'', x_2, \ldots, x_d)$$

where α'' is some root of $p''(x_1)$. We construct a CAD invariant for the signs of

$$q'(\alpha', x_2', \ldots, x_d') - q''(\alpha'', x_2'', \ldots, x_d'')$$
$$p_1'(\alpha', x_2', \ldots, x_d')$$
$$\vdots$$
$$p_{d-1}'(\alpha', x_2', \ldots, x_d')$$
$$p'(x_1')$$
$$p_1''(\alpha'', x_2'', \ldots, x_d'')$$
$$\vdots$$
$$p_{d-1}''(\alpha'', x_2'', \ldots, x_d'')$$
$$p''(x_1'').$$

This is a CAD in $\mathbb{R}^{2d}$ of $2(d-1) + 3 = 2d + 1$ polynomials. The space $\mathbb{R}^{2d}$ here is basically the cartesian product of two d dimensional spaces, so we resolve the comparison by testing the sign of

$$q'(\alpha', x_1', \ldots, x_d') - q''(\alpha'', x_1'', \ldots, x_d'')$$

on the zero dimensional cell which is the cartesian product of the two cells in $\mathbb{R}^d$ returned by the $d - 1$ dimensional algorithm. Now we have a slab in the original CAD which is known to contain a maximizer of F. In this slab the intersection of any hyperplane $x_1 = \alpha$ with the CAD D' changes continuously with α, so we can determine the sign of any of the polynomials in any of the cells. Since the sign is constant for any x_1, the sign equals the sign for $x_1 = x_1^\star$ which is what we need to determine in order to continue the simulation of the $d - 1$ dimensional algorithm. When the simulation of the $d-1$ dimension algorithm terminates we end up with two values, one valid in a slab $\alpha_a < x_1 < \alpha_b$ and the other valid in a slab $\alpha_b < x_1 < \alpha_c$. Let us describe how we find the maximum of F in the slab $\alpha_a < x_1 < \alpha_b$. The maximum of $q(x_1, \ldots, x_d)$ on some one dimensional cell c of the CAD of

$$p_1(x_1, x_2, \ldots, x_d), \ldots, p_{d-1}(x_1, x_2, \ldots, x_d)$$

is found using Lagrange multipliers. We need to solve the equations

$$\nabla q - \sum_{i=1}^{d-1} \lambda_i \nabla p_i = 0$$

and

$$p_1(x_1, x_2, \ldots, x_d) = 0, \ldots, p_{d-1}(x_1, x_2, \ldots, x_d) = 0$$

for both $x_1, \ldots, x_d$ and $\lambda_1, \ldots, \lambda_{d-1}$. We can do so by constructing a CAD in $\mathbb{R}^{2d-1}$ with $2d-1$ polynomials. The variables are ordered $x_1, \ldots, x_d, \lambda_1, \ldots, \lambda_{d-1}$. The space $\mathbb{R}^{2d-1}$ is a cartesian product of two spaces, the $x_1, \ldots, x_d$ space and the $\lambda_1, \ldots, \lambda_{d-1}$

space. Our one dimensional cell c is mapped into a d dimensional manifold c_d. If there is a point on this manifold in which the sign of the polynomials

$$\nabla q - \sum_{i=1}^{d-1} \lambda_i \nabla p_i$$

is zero, then this is an extremum of q in c. If q is constant on c we are done. Otherwise there is an extremum point, and points which are not extrema. Suppose there is a point x_1' which is an extremum, that is, for small enough $\epsilon \neq 0$, $q(x_1' + \epsilon) < q(x_1')$ (the notation $q(x)$ here denotes the value of q on a point $x_1 = x$ in c). We claim that in this case, x_1' is one of the critical values in the one dimensional decomposition which is the base of the CAD. Assume for contradition that it is not. Consider the intersection of a hyperplane $x_1 = x_1' + \epsilon$ with the CAD for small ϵ. The intersection is a CAD which changes continuously. But we know that at $\epsilon = 0$ there is a set of Lagrange multipliers, or a zero for the polynomials in the CAD, where as for any $\epsilon \neq 0$, there isn't, a contradiction. Assuming that $\alpha_a < x_1' < \alpha_b$ is the root of the polynomial $p(x_1)$, the location of a maximizer of F can be defined by some particular zero dimensional cell in the CAD of

$$p(x_1), p_1(x_1, x_2, \ldots, x_d), \ldots, p_{d-1}(x_1, x_2, \ldots, x_d).$$

This concludes our inductive proof.

The running time. Let us count the number of CADs constructed by the algorithm. We denote the number of CADs by C_d. If the evaluation algorithm $\mathcal{A}$ runs in time T_0, it performs no more then T_0 comparisons, so $C_0 \leq T_0$. For each CAD performed by the $d-1$ dimensional algorithm, the d dimensional algorithm constructs the same CAD in dimension one higher. If this CAD has N critical values, the d dimensional algorithm calls the $d-1$ dimensional one $\log N$ times during the binary search. From Theorem 4.1 and from the fact that all CADs constructed contain at most $2d+1$ polynomials, we conclude that N is a function of d, $N = N(d)$. Each call to the $d-1$ dimensional algorithm constructs C_{d-1} CADs. To find a slab containing a maximizer, $\log N$ comparisons between returned values need to be performed. Every comparison is resolved by constructing a CAD. Finally, two sets of Lagrange multipliers need to be found, which results in the construction of two more CADs. Therefore we have

$$\begin{aligned}
C_d &\leq C_{d-1}(C_{d-1} \log N(d) + \log N(d)) + 2 \\
&\leq 2C_{d-1}^2 \log N(d) \, .
\end{aligned}$$

Solving the recurrence we obtain

$$C_d \leq (2 \log N(d))^{2^d - 1} C_0^{2^d} = K(d) C_0^{2^d}$$

where $K(d)$ is some constant depending on d. Therefore the running time of the d dimensional algorithm is $O(C_0^{2^d}) = O(T_0^{2^d})$.

Let us examine the use of a parallel evaluation algorithm in the construction of a more efficient optimization algorithm. Let as assume that the d_1 dimensional algorithm computes C_{d-1} batches of at most P_{d-1} CADs each. If there is an evaluation algorithm that runs in T_p parallel time and uses P processors, we set $C_0 = T_p$ and $P_0 = P$. The d dimensional algorithm simulates the construction of a batch of CADs in the following way. P_{d-1} CADs are constructed, and the combined list of critical values of all the CADs is sorted. Sorting is done using a parallel sorting algorithm, which works in $O(\log NP_{d-1})$ parallel steps. In each step $O(NP_{d-1})$ pairs of roots are compared. A comparison between a root of $p'(x_1)$ and a root of $p''(x_1)$ is resolved by constructing the CAD invariant for the signs of p', p'' which totally orders all their roots. Then a binary search is performed over the sorted list. If each CAD generates N critical values, the d dimensional algorithm calls the $d - 1$ dimensional algorithm $\log(NP_{d-1})$ times. Otherwise the algorithm is similar to the case of one CAD per batch (a sequential algorithm). The number of CADs constructed per batch is therefore unchanged, $P_d = P_{d-1}$. The number of batches is

$$C_d \leq C_{d-1}\left(C_{d-1}\log(NP_{d-1}) + \log(NP_{d-1}) + O(\log(NP_{d-1}))\right) + 2.$$

Again, $N = N(d)$, so solving the recurrence we obtain

$$\begin{aligned} C_d &\leq K(d)\left(C_0 \log P_0\right)^{2^d-1} \\ &= K(d)\left(T_p \log P\right)^{2^d-1} \\ &= O\left(\left(T_p \log P\right)^{2^d-1}\right). \end{aligned}$$

This concludes the proof of our main result.

Theorem 4.4 *Let $\mathcal{A}$ be a polynomial algorithm in x with degree δ, where $x \in \mathbb{R}^d$. Let $F: \mathcal{P} \to \mathbb{R}$ be a concave function, and let $\mathcal{P} \subseteq \mathbb{R}^d$ be a convex set. Assume that for any $x \in \mathcal{P}$, $\mathcal{A}(x) = F(x)$, and for any $x \notin \mathcal{P}$, $\mathcal{A}(x) = p_x$ where p_x is a concave polynomial such that $p_x(x) < 0$, but for any $y \in \mathcal{P}$, $p_x(y) \geq 0$. Assume that $\mathcal{A}$ runs in time T_0, and that there is an equivalent parallel (in Valiant's model of parallel computation [16]) algorithm $\mathcal{A}_p$ that runs in time T_p and uses P processors. Then there is an algorithm $\mathcal{A}_d$ that runs in time $O(T_0(T_p \log P)^{2^d-1})$ and either decides that $\mathcal{P} = \emptyset$, or decides that F is unbounded on $\mathcal{P}$, or finds the maximum of F on $\mathcal{P}$.*

5 Applications

Convex Programming.   Consider the convex programming problem

$$
\begin{aligned}
&\text{Minimize } y \\
&\text{subject to} \\
&y \geq p_1(x_1, \ldots, x_d) \\
&\qquad \vdots \\
&y \geq p_n(x_1, \ldots, x_d) \\
&p_1, \ldots, p_n \text{ are convex.}
\end{aligned}
$$

The function

$$
F(x_1, \ldots, x_d) = \max_i \{ p_i(x_1, \ldots, x_d) \}
$$

can be evaluated at a point in parallel time $O(\log \log n)$ using n processors [16]. Using Theorem 4.4 we conclude that the above convex programming problem can be solved in time $O(n(\log n \log \log n)^{2^d - 1})$. Dyer [10, 11] showed how to solve some special cases in $O(n)$ time, but no efficient algorithm was known for the general case.

Functions defined in terms of graphs.   Let $G = (V, E)$ be a graph, let s and t be two vertices of G, and let W be a function mapping edges of G to real numbers. We use the notation W_e to denote $W(e)$. Let $S \subset 2^E$ (for example, the set of all spanning trees, all minimum s-t cuts etc.). A *minimization problem* on G is the problem of finding $\min_{s \in S} \sum_{e \in s} W_e$, and usually finding a minimizer is also desirable.

Now assume that W maps edges to concave polynomial functions over some convex set $\mathcal{P} \subseteq \mathbb{R}^d$, instead of to numbers. For every point $x \in \mathcal{P}$, we get an induced minimization problem obtained by mapping every element $e \in E$ to a real number $W_e(x)$. We define a function $F \colon \mathcal{P} \to \mathbb{R}$ by

$$
F(x) = \min_{s \in S} \sum_{e \in s} W_e(x).
$$

Lemma 5.1 *The function F is a concave function.*

Proof: Since for all $e \in E$, the function W_e is concave, so is the function $\sum_{e \in s} W_e$ for any subset s of E, and therefore the minimum of such functions is also concave. ∎

Lemma 5.2 *If the edge weights W are all polynomial of degree at most δ, then F is a piecewise polynomial function, and its pieces are of degree at most δ.*

Proof: Obvious. ∎

When S is the collection of all the edge-cuts separating s from t in G and the weights W_e are interpreted as capacities, then the associated minimization problem is the max flow problem in G, by the Max-Flow Min-Cut Theorem. The parametric max flow function is hence a concave function on $\mathcal{P} = \cap_{e \in E}\{x : W_e(x) \geq 0\}$. The definition of $\mathcal{P}$ ensures that all the edge capacities are non-negative. Since each W_e is concave, the regions $\{x : W_e(x) \geq 0\}$ are convex, and therefore their intersection $\mathcal{P}$ is convex.

When S is the collection of all paths between s and t, the minimization problem is the problem of finding the minimum s-t distance. In this case the domain of the parametric function is the convex region $\mathcal{P} = \cap_C\{x : \sum_{e \in C} W_e(x) \geq 0\}$ where the intersection is over all the simple cycles in G. The combinatorial complexity of $\mathcal{P}$ may be super-polynomial, but fortunately there is a *separation algorithm* for $\mathcal{P}$. The Bellman-Ford algorithm can be modified so that it either decides that the graph does not contain a negative cycle and finds the shortest path, or finds a negative cycle C (see [8]). Summing the weights of the edges of the cycle as polynomials, we find a concave violated constraint $p(x) = \sum_{e \in C} W_e(x) < 0$ which is *not* violated for any $y \in \mathcal{P}$. Therefore the conditions of Theorem 4.4 are satisfied, and we can find the maxi-min s-t distance in strongly polynomial time.

Cohen and Megiddo [4, 6] showed how to solve such problems when the edge weights are affine functions. Again, an algorithm for the general concave polynomial case was not known until now.

Acknowledgments

Thanks to Pankaj K. Agarwal for reading and commenting on a preliminary version of this paper. Thanks to Esther Jesurum, Mauricio Karchmer, Nimrod Megiddo, Boaz Patt-Shamir and Serge Plotkin for helpful discussions. My research was supported in part by the Defense Advanced Research Projects Agency under Grant N00014–91–J–1698.

References

[1] P.K. Agarwal, M. Sharir and S. Toledo, Applications of parametric searching in geometric optimization, *Proc. 3nd ACM-SIAM Symp. on Discrete Algorithms*, 1992, 72–82.

[2] A.V. Aho, J.E. Hopcroft and J.D. Ullman, *The Design and Analysis of Computer Algorithms*, Addison-Wesley, 1974.

[3] D.S Arnon, G.E. Collins, S. McCallum, Cylindrical algebraic decomposition I: the basic algorithm, *SIAM J. Comput.* **13** (1984), 865–877.

[4] E. Cohen and N. Megiddo, Strongly polynomial time and NC algorithms for detecting cycles in dynamic graphs, *Proc. 21st ACM Symp. on Theory of Computing*, 1989, 523–534.

[5] E. Cohen and N. Megiddo, Maximizing concave functions in fixed dimension, Research Report RJ 7656 (71103), IBM Almaden Research Center, San Jose, 1990. (Also in this volume.)

[6] E. Cohen and N. Megiddo, Algorithms and complexity analysis for some flow problems, *Proc. 2nd ACM-SIAM Symp. on Discrete Algorithms*, 1991, 120–130.

[7] R. Cole, Parallel merge sort, *SIAM J. Comput.* **17** (1988), 770–785.

[8] T.H. Cormen, C.E. Leiserson and R.L. Rivest, *Introduction to Algorithms*, MIT Press, Cambridge, MA, 1990.

[9] J.H. Davenport, Y. Siret and E. Tournier, *Computer Algebra* Academic Press, 1988.

[10] M. Dyer, On a multidimensional search procedure and its application to the Euclidean one-centre problem, *SIAM J. Comput.* **13** (1984), 31–45.

[11] M. Dyer, A class of convex programs with applications to computational geometry, *Proc. 8th ACM Symp. on Computational Geometry*, 1992, 9–15.)

[12] N. Megiddo, Applying parallel computation in the design of serial algorithms, *J. ACM* **30** (1983), 852–865.

[13] N. Megiddo, The weighted Euclidean 1-center problem, *Math. of Operations Research*, **8** (1983), 498–504.

[14] C.H. Norton, S.A. Plotkin and É. Tardos, Using separation algorithms in fixed dimension, *J. of Algorithms*, **13** (1992), 79–98.

[15] S. Toledo, Maximizing non-linear concave functions in fixed dimension, *Proc. 33rd Annual Symp. on Foundations of Computer Science*, 1992, 676–685.

[16] L. Valiant, Parallelism in comparison problems, *SIAM J. Comput.* **4** (1975), 345-348.

Complexity in Numerical Optimization, pp. 448-461
P.M. Pardalos, Editor

A Note On the Complexity of Fixed-point Computation for Noncontractive Maps

C. W. Tsay[1] and K. Sikorski[2]
Department of Computer Science, The University of Utah, Utah 84112, USA

Abstract

We study the information complexity (minimal number of function evaluations) of computing an ε-approximation to a fixed point of a noncontractive function $f : [0, 1]^n \rightarrow [0, 1]^n$ with the Lipschitz constant $L \geq 1$. This is done for the *absolute* error criterion. For $n = 1$, we show that the complexity is $\lceil \log(0.5\varepsilon^{-1}) \rceil$, which is the cost of the bisection algorithm. For $n > 1$, we show that the complexity is infinitive whenever $0 < \varepsilon < 0.5$. That is, the multi-dimensional problem is unsolvable in the worst case. We remark that the same problem with the residual error criterion has finite but exponential complexity [4].

1 Introduction

Fixed-point computation has become an important research topic since 1967, when Scarf [10] first demonstrated a simplicial algorithm to approximate fixed points. Several new algorithms have been proposed since then, such as restart methods [6, 7] and homotopy methods [2, 3]. Overviews of fixed point algorithms are given in [1, 5, 16]. The fixed-point computation establishes an important tool in applications such as the computation of equilibria [5, 11] and the solution of partial differential equations [15].

The (information) complexity of a fixed-point problem is defined as the minimal number $m(\varepsilon)$ of function evaluations needed to approximate a fixed point of any function in a given class to within a prescribed tolerance ε. In this paper we consider the class of functions: $f : [0, 1]^n \rightarrow [0, 1]^n$ that satisfy the Lipschitz condition with the constant $L \geq 1$ with respect to the Euclidean norm. We use the *absolute error* criterion, that is, we are interested in computing an ε-*approximation* $\hat{\mathbf{x}}$ such that

$$\|\hat{\mathbf{x}} - \mathbf{f}^*\| \leq \varepsilon < 0.5.$$

Here, the notation $\mathbf{f}^*$ (or f^* in univariate case) represents a fixed point of f, and $\|\cdot\|$ is the Euclidean norm.

We first summarize complexity results for contractive functions (i.e., $L < 1$).

For univariate functions, Sikorski [13] constructed a Fixed Point Envelope (FPE) algorithm to compute an ε-approximation to the fixed point. The algorithm is optimal, i.e., it minimizes the number of function evaluations. As a result, the complexity in this case is equal to the cost of the FPE algorithm which is given by

$$m(\varepsilon) = \left\lceil \frac{\log(0.5\varepsilon^{-1})}{\log(L^{-1}(1 + L))} \right\rceil \leq b(\varepsilon),$$

where $b(\varepsilon) = \lceil \log(0.5\varepsilon^{-1}) \rceil$ (independent of L) is the cost of the bisection algorithm.

For multivariate contractive functions that transform unit ball into itself, the simple iteration algorithm (SIA) given by

$$\mathbf{x}_k = f(\mathbf{x}_{k-1}), \quad k \geq 1, \mathbf{x}_0 = 0,$$

computes an ε-approximation with $s(L, \varepsilon) = \lceil \log(1/\varepsilon)/\log(1/L) \rceil$ function evaluations. If $n \geq s(L, \varepsilon)$, Nemirovsky's result [9] yields that the SIA is optimal and hence $m(\varepsilon) = s(L, \varepsilon)$. For moderate/small n, $m(\varepsilon)$ is still unknown. It is conjectured in [14] that $m(\varepsilon)$ is of the form $c(n)(\log(1/\varepsilon) + g(L))$, where $c(n)$ depends only on n and $g(L) \to \infty$ slowly as $L \to 1^-$. An upper bound presented in [14] shows that

$$m(\varepsilon) < n^3(\log \frac{1}{\varepsilon} + \log \frac{1}{1 - L^2} + \log n).$$

In this paper we consider the complexity problem for noncontractive functions with $L \geq 1$. Observe that f and g, $g(\mathbf{x}) = \lambda \mathbf{x} + (1 - \lambda)f(\mathbf{x}), \lambda \in (0, 1)$, have the same set of fixed points. Moreover, the Lipschitz constant of g is $\lambda - (\lambda - 1)L$, which is arbitrarily close to 1 as $\lambda \to 1$. It is then evident that the complexity is independent of L when $L > 1$. This paper further proves that the complexity is the same for both $L = 1$ and $L > 1$. We summarize our results as follows.

For $n = 1$, we show that the bisection is an optimal algorithm. Therefore, the complexity $m(\varepsilon)$ is equal to $b(\varepsilon)$.

For $n > 1$, we show that no finite algorithms exist whenever $\varepsilon < 0.5$. We illustrate this with 2-dimensional functions with $L = 1$. We demonstrate that no matter how many evaluation points are adaptively chosen, there exist two functions that coincide at all evaluation points but their fixed points are apart with distance equal to one. Consequently, there exist no algorithms able to approximate fixed points to within $\varepsilon < 0.5$. We stress that although the Euclidean norm is chosen in this paper for technical reasons, our results could be generalized for other norms.

Remark 1. Our result for $n > 1$ indicates that to guarantee finite algorithms, one must further restrict the class of functions. This approach is indeed taken by Natarajan [8]. He defined the *condition number*

$$K_\epsilon = K_\epsilon(f) = \max_{x \in X_\epsilon} \min_{z \in Z} \frac{\|\mathbf{x} - \mathbf{z}\|}{\|f(\mathbf{x}) - \mathbf{x}\|},$$

where $X_\epsilon = \{\mathbf{x} \mid \|f(\mathbf{x}) - \mathbf{x}\| \geq \epsilon\}$ and $Z = \{\mathbf{z} \mid f(\mathbf{z}) = \mathbf{z}\}$. For the class of functions with K_ϵ bounded by ε^{-1}, his algorithm can compute an ε-approximation with at most $O(\log(\varepsilon^{-1})(2K_\epsilon A)^n)$ function evaluations, where A $(L - 1 \leq A \leq L + 1)$ is the Lipschitz constant of the function $f(\mathbf{x}) - \mathbf{x}$. This means that the complexity is finite and has an upper bound of the form $O(\log(\varepsilon^{-1})(2K_\epsilon A)^n)$.

Remark 2. Hirsch et al. [4] considered residual error criterion to approximate fixed points, i.e., to compute an approximation $\hat{\mathbf{x}}$ satisfying $\|f(\hat{\mathbf{x}}) - \hat{\mathbf{x}}\| \leq \epsilon$. They established that the lower bound on the number of function evaluations is an exponential function of n. We stress that their construction of functions can be used to show that the complexity with the absolute error criterion is infinitive whenever $\varepsilon < 0.25$ and $L > 1$. This means our result (L = 1) is stronger.

This paper is organized as follows. Section 2 formulates the problem and main theorems. Section 3 proves that bisection is optimal. Section 4 conducts the proof of infinitive complexity for 2-dimensional nonexpanding functions.

2   Main Results

A function $f : [0,1]^n \to [0,1]^n$ is noncontractive if it satisfies

$$\|f(\mathbf{x}) - f(\mathbf{y})\| \leq L\|\mathbf{x} - \mathbf{y}\| \quad \forall \, \mathbf{x}, \mathbf{y} \in [0,1]^n, \tag{1}$$

where the Lipschitz constant $L \geq 1$. (f is called a *nonexpanding* function if $L = 1$.) Brouwer Fixed Point Theorem ensures that f has at least one *fixed point* $\mathbf{f}^* \in [0,1]^n$ such that $\mathbf{f}^* = f(\mathbf{f}^*)$. Let $F_{n,L}$ denote the class of functions that satisfy (1) and have *exactly one* fixed point.

Our problem is to compute an ε-approximation $\hat{\mathbf{x}}$ to $\mathbf{f}^*$ for any $f \in F_{n,L}$ such that

$$\|\hat{\mathbf{x}} - \mathbf{f}^*\| \leq \varepsilon < 0.5.$$

To solve this problem we use the *adaptive information operators* defined as:

$$N_M(f) = [f(\mathbf{x}_1), f(\mathbf{x}_2), \ldots, f(\mathbf{x}_M)] \tag{2}$$

where

$$\mathbf{x}_i = \alpha_i(f(\mathbf{x}_1), f(\mathbf{x}_2), \ldots, f(\mathbf{x}_{i-1})), \quad 1 \leq i \leq M,$$

with each $\alpha_i : ([0,1]^n)^{i-1} \to [0,1]^n$ being an arbitrary function (see [17]). Note that the index M indicates the number of function evaluations in N_M.

An *algorithm ϕ using information N_M* is an arbitrary transformation

$$\phi : N_M(F_{n,L}) \to [0,1]^n \tag{3}$$

which generates an approximation to a fixed point. The *error* of ϕ is defined by

$$e(\phi) = \sup_{f \in F_{n,L}} \|f^* - \phi(N_M(f))\|.$$

An algorithm ϕ is called *optimal* if $e(\phi) \leq \varepsilon$ and it needs minimal number of function evaluations (i.e., minimal M). We remark that the combination of (2) and (3) formulates the concept of *algorithms based on function evaluations*, which was used in [4, 8].

We now survey several useful notions. The *radius of information* is defined by

$$r(N_M) = \sup_{f \in F_{n,L}} \mathrm{rad}(U(f)), \tag{4}$$

where $\mathrm{rad}(U(f))$ is the radius of the smallest ball containing the set

$$U(f) = \{z \in [0,1]^n \mid z = \tilde{f}^*, N_M(f) = N_M(\tilde{f})\}, \tag{5}$$

i.e., $U(f)$ is the set of fixed points of functions in $F_{n,L}$ that share the same information with f. Then, it is known [17] that

$$\inf_{\phi \in \Phi(N_M)} e(\phi) = r(N_M), \tag{6}$$

where $\Phi(N_M)$ is the class of algorithms using N_M. In other words, no algorithms in $\Phi(N_M)$ are able to approximate fixed points with error less than $r(N_M)$ for the class $F_{n,L}$.

The following two theorems will be proved in Section 3 and in Section 4, respectively.

Theorem 1 *Given the class $F_{1,L}$ and $0 < \varepsilon < 0.5$, for every information N_M,*

$$r(N_M) > \varepsilon \quad \text{if} \quad M < b(\varepsilon).$$

Theorem 2 *Given the class $F_{2,1}$, for every information N_M, $r(N_M) \geq 0.5$.*

We shall prove both theorems with the same technique. More precisely, for the class $F_{1,L}$ and any information N_M with $M < b(\varepsilon)$, we explicitly construct two functions $g, h \in F_{1,L}$ such that $N_M(g) = N_M(h)$ and $|g^* - h^*| > 2\varepsilon$, which implies $r(N_M) > \varepsilon$. Similarly, for the class $F_{2,1}$ and any information N_M, we construct $g, h \in F_{2,1}$ such that $N_M(g) = N_M(h)$ and $\|g^* - h^*\| = 1$, which implies $r(N_M) \geq 0.5$ in this case.

Theorem 1 yields that bisection is optimal. Hence, the complexity of computing ε-approximations for the class $F_{1,L}$ is equal to $b(\varepsilon)$.

By (6) and Theorem 2, we obtain that there *exist no* algorithms able to approximate fixed points with error less than 0.5 for the class $F_{2,1}$. This negative result also holds for $F_{2,L}$ with $L > 1$, since $F_{2,1} \subset F_{2,L}$ by definition. Moreover, observe that any $f \in F_{2,L}$ can be easily extended to n-dimensions by setting $\hat{f} \in F_{n,L}$ as:

$$\hat{f}(x_1, x_2, \ldots, x_n) = (f_1(x_1, x_2), f_2(x_1, x_2), \underbrace{0, \ldots, 0}_{(n-2) \text{ 0's}})$$

Following this, the constructions of g and h for $F_{2,1}$ can be easily adapted for higher dimensional cases. As a result, we conclude that no algorithms exist to compute ε-approximations of fixed points for the class $F_{n,L}$ whenever $n \geq 2$ and $L \geq 1$.

3 Complexity for the Class $F_{1,L}$

We prove Theorem 1 as follows. Suppose N_M is any information operator with $M < b(\varepsilon)$; this implies $2^{-M} > 2\varepsilon$. Apparently, there exists δ, $2^{-M} \geq \delta > 0$, such that

$$2^{-M} - 2\delta/(1 + L) > 2\varepsilon.$$

We are ready to show that there are two functions $g, h \in F_{1,L}$ such that $N_M(g) = N_M(h)$ and $|g^* - h^*| > 2\varepsilon$. First, suppose that N_M adaptively selects x_1, x_2, ..., and x_M as evaluation points. We use the following flowchart to assign $y_i = g(x_i) = h(x_i)$ for $1 \leq i \leq M$ (g and h defined below).

```
        ℓ = 0; r = 1;
        for i = 1, 2, ..., M do begin
            m = (ℓ + r)/2;
            if xᵢ ≤ m then
                    yᵢ = xᵢ + δ;
                    ℓ = xᵢ;
            else
                    yᵢ = xᵢ − δ;
                    r = xᵢ;
            endif
        end.
```

After this assignment, $r - \ell \geq 2^{-M}$, each x_i is outside of the interval (ℓ, r), and

$$y_i = \begin{cases} x_i + \delta & \text{if } x_i \leq \ell \\ x_i - \delta & \text{if } x_i \geq r \end{cases}$$

Define

$$g(x) = \begin{cases} x + \delta & 0 \leq x \leq \ell \\ \alpha x + (\ell + \delta - \alpha\ell) & \ell \leq x \leq g^* \\ -Lx + (1 + L)r - \delta & g^* \leq x \leq r \\ x - \delta & r \leq x \leq 1 \end{cases}$$

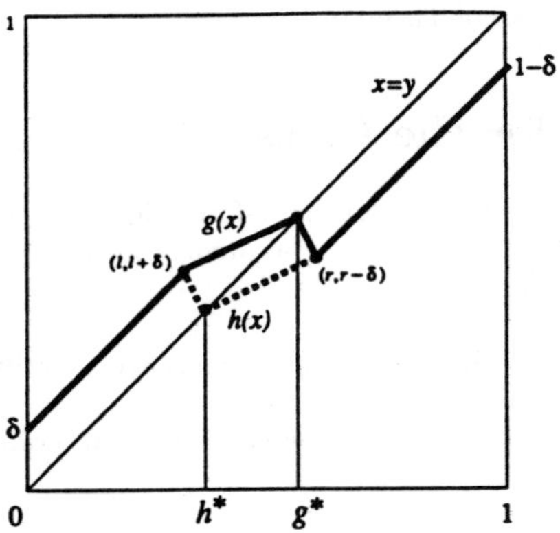

Figure 1

and

$$
h(x) = \begin{cases}
x + \delta & 0 \le x \le \ell \\
-Lx + (1+L)\ell + \delta & \ell \le x \le h^* \\
\alpha x + (r - \delta - \alpha r) & h^* \le x \le r \\
x - \delta & r \le x \le 1
\end{cases}
$$

where

$$
\alpha = \frac{g^* - (\ell + \delta)}{g^* - \ell} < 1,
$$

and

$$
\begin{aligned}
g^* &= r - \delta/(1+L), \\
h^* &= \ell + \delta/(1+L)
\end{aligned}
$$

are the unique fixed points of g and h, respectively (see Fig. 1). We obtain

$$
|g^* - h^*| = (r - \ell) - 2\delta/(1+L) \ge 2^{-M} - 2\delta/(1+L) > 2\varepsilon
$$

It remains to show that $g, h \in F_{1,L}$, i.e., (i) they are Lipschitz with constant L and (ii) they map $[0, 1]$ into itself. The condition (i) is satisfied since they are piecewise linear functions and are Lipschitz with constant L in each subinterval. We now show that $0 \le g(x) \le 1$ for all $0 \le x \le 1$. We check this for $x \in [0, h^*]$ and $x \in [h^*, 1]$. First, observe that $0 \le \ell < r \le 1$

and $r - \ell \geq 2^M \geq \delta$. For $0 \leq x \leq h^*$, we have $\delta \leq g(x) \leq \ell + \delta \leq r$, and for $h^* \leq x \leq 1$, we have $\ell \leq r - \delta \leq g(x) \leq x$. Hence, $0 \leq g(x) \leq 1$. The proof for h is similar and omitted here. This completes the proof of Theorem 1.

4   Complexity for the Class: $F_{2,1}$

We prove Theorem 2 by showing that for any information N_M, there exist $g, h \in F_{2,1}$ such that $N_M(g) = N_M(h)$ and $\|g^* - h^*\| = 1$. The proof, which is similar to that for Theorem 1, consists of the following three steps:

(I) assign function values at evaluation points which are adaptively generated by N_M;

(II) explicitly define g and h such that they coincide at all evaluation points with preassigned values and $\|g^* - h^*\| = 1$;

(III) prove that both g and h are members of $F_{2,1}$.

   (I). Choose $\delta = 2^{-(M+1)}$ for the information N_M. We assign $(\alpha_i, \beta_i) = g(x_i, y_i) = h(x_i, y_i)$ for $1 \leq i \leq M$ by the following flowchart:

```
ℓ = 0; r = 1;
for i = 1, 2, ..., M do begin
        m = (ℓ + r)/2;
        if xᵢ ≤ m then
                (αᵢ, βᵢ) = (xᵢ + δ, yᵢ);
                ℓ = xᵢ;
        else
                (αᵢ, βᵢ) = (xᵢ − δ, yᵢ);
                r = xᵢ;
        endif
    end
```

   After this assignment, we have $0 \leq \ell < m = (\ell + r)/2 < r \leq 1$ and $r - \ell \geq 2^{-M}$. Let $d = (r - \ell)/2$. Then, $d > 2^{-(M+1)} = \delta$ and

$$0 < \ell + \delta \leq m \leq r - \delta < 1. \tag{7}$$

Note that all (x_i, y_i) are outside of the set $(\ell, r) \times [0, 1]$, and

$$(\alpha_i, \beta_i) = \begin{cases} (x_i + \delta, y_i) & \text{if } x_i \leq \ell \\ (x_i - \delta, y_i) & \text{if } x_i \geq r \end{cases} \tag{8}$$

(II). We now explicitly define g and h as follows. First, partition the domain $[0,1]^2$ into six subdomains:

$$
\begin{aligned}
D_1 &= \{(x,y) \in [0,1]^2 \mid 0 \le x \le \ell\} \\
D_2 &= \{(x,y) \in [0,1]^2 \mid \ell \le x \le \ell + yd\} \\
D_3 &= \{(x,y) \in [0,1]^2 \mid \ell + yd \le x \le m\} \\
D_4 &= \{(x,y) \in [0,1]^2 \mid m \le x \le r - yd\} \\
D_5 &= \{(x,y) \in [0,1]^2 \mid r - yd \le x \le r\} \\
D_6 &= \{(x,y) \in [0,1]^2 \mid r \le x \le 1\}
\end{aligned}
$$

as illustrated in Fig. 2 (a). Then, the function $g(x,y) = (g_1(x,y), g_2(x,y))$ is defined by:

$$
g(x,y) = \begin{cases}
(x + \delta, y), & (x,y) \in D_1 \\
(x - \frac{x-m}{d}\delta, y), & (x,y) \in D_2 \text{ or } (x,y) \in D_5 \\
(x - \frac{x-m}{d}\delta, y + \frac{x-(l+yd)}{d}\xi), & (x,y) \in D_3 \\
(x - \frac{x-m}{d}\delta, y + \frac{(r-yd)-x}{d}\xi), & (x,y) \in D_4 \\
(x - \delta, y), & (x,y) \in D_6
\end{cases}
\tag{9}
$$

where

$$
\xi = -1 + \sqrt{1 + \delta(2d - \delta)}
\tag{10}
$$

is chosen to be the positive solution of the quadratic equation (15). (The purpose of such a particular choice of ξ is to assure that g and h is nonexpanding in both D_3 and D_4.) The function $h(x,y) = (h_1(x,y), h_2(x,y))$ is defined by:

$$
h_1(x,y) = g_1(x,y) \quad \text{and} \quad h_2(x,y) = 1 - g_2(x, 1-y)
\tag{11}
$$

Fig. 2 (b)–(d) illustrates $g_1(x,y)$, $g_2(x,y)$ and $h_2(x,y)$.

According to (8), (9), and (11), $g(x_i, y_i) = h(x_i, y_i) = (\alpha_i, \beta_i)$ for all $1 \le i \le M$, i.e., $N_M(g) = N_M(h)$. By solving the equation $g(x,y) = (x,y)$, we easily determine the unique fixed point $g^* = (m, 1)$. Similarly, $h^* = (m, 0)$ is the unique fixed point of h. Apparently, the distance $\|g^* - h^*\| = 1$.

We remark that if $(x,y) \in D_4$, then there exists $(x', y) = (2m - x, y) \in D_3$ such that

$$
g(x,y) = (2m - g_1(x',y), g_2(x',y))
\tag{12}
$$

(III). To complete the proof, we must show that both g and h are members of $F_{2,1}$. Observe that it is enough to show this for g, since $h_1 = g_1$ and h_2 is a *reflection* of g_2. The following two lemmas establish that $g \in F_{2,1}$.

Lemma 1 *Let g be defined as (9). Then g maps $[0,1]^2$ to $[0,1]^2$.*

Proof: It is enough to show that (a) $0 \le g_1(x,y) \le 1$ and (b) $0 \le g_2(x,y) \le 1$ for every $(x,y) \in [0,1]^2$. The condition (a) is satisfied by the following observations.

(i) For $0 \leq x \leq \ell$, $0 < x < g_1(x,y) = x + \delta \leq \ell + \delta \leq m < 1$.

(ii) For $\ell \leq x \leq r$, $g_1(x,y)$ monotonically increases from $\ell + \delta$ to $r - \delta$ for any y. Therefore,

$$\ell + \delta \leq g_1(x,y) \leq r - \delta \tag{13}$$

which along with (7) guarantees $0 < g_1(x,y) < 1$.

(iii) For $r \leq x \leq 1$, we get $0 < m \leq r - \delta \leq g_1(x,y) = x - \delta < x \leq 1$.

Now we show that (b) is satisfied. Since $0 \leq g_2(x,y) = y \leq 1$ whenever (x,y) belongs to D_1, D_2, D_5, or D_6, it remains to prove (b) for $(x,y) \in D_3$ or $(x,y) \in D_4$. If $(x,y) \in D_3$, then $g_2(x,y)$ can be written as:

$$g_2(x,y) = (1 - \xi)y + \frac{x - \ell}{d}\xi,$$

and for a fixed x, it monotonically increases from $(x - \ell)\xi/d$ to $(x - \ell)/d$ when y goes from 0 to $(x - \ell)/d$. Since $\ell + yd \leq x \leq m$, we have

$$0 \leq y\xi \leq (x - \ell)\xi/d \leq \xi$$

and

$$0 \leq y \leq (x - \ell)/d \leq 1$$

Hence, $0 \leq y\xi \leq g_2(x,y) \leq 1$.

If $(x,y) \in D_4$, then by (12), $g(x,y) = (2m - g_1(x',y), g_2(x',y))$, where $(x',y) = (2m - x, y)$ belongs to D_3. We already proved $0 \leq g_2(x',y) \leq 1$, thus $0 \leq g_2(x,y) \leq 1$. Since

$$\ell + \delta \leq g_1(x',y) \leq m$$

we have

$$m \leq 2m - g_1(x',y) \leq 2m - (\ell + \delta)$$

which becomes

$$0 < m \leq 2m - g_1(x',y) = g_1(x,y) \leq r - \delta < 1$$

by substituting $m = (\ell + r)/2$ and by (7). We thus completed the proof.   **Q.E.D.**

The proof that g is nonexpanding in $[0,1]^2$ can be simplified to showing that g is nonexpanding in any D_i, $i = 1,\ldots,6$. More precisely, suppose $P \in D_1$ and $Q \in D_2$. Let R be the intersection point of the segment $\overline{PQ}$ and the line $x = \ell$. Since $R \in D_1 \cap D_2$,

$$\|g(P) - g(R)\| \leq \|P - R\|$$

and

$$\|g(R) - g(Q)\| \leq \|R - Q\|$$

Then, by the triangle inequality,

$$
\begin{aligned}
\|g(P) - g(Q)\| &= \|g(P) - g(R) + g(R) - g(Q)\| \\
&\leq \|g(P) - g(R)\| + \|g(R) - g(Q)\| \\
&\leq \|P - R\| + \|R - Q\| \\
&= \|P - Q\|
\end{aligned}
$$

This means that g is nonexpanding in $D_1 \cup D_2$. Following the same argument, g is nonexpanding in $D_1 \cup D_2 \cup D_3$, ..., and finally in the domain $[0,1]^2 = \cup_{i=1}^6 D_i$. Consequently, we only need to prove the following lemma to show g is nonexpanding in $[0,1]^2$.

Lemma 2 *Suppose that both $P = (p_1, p_2)$ and $Q = (q_1, q_2)$ belong to one of the subdomains: $D_1, D_2, \ldots,$ or D_6. Then*

$$
\|g(P) - g(Q)\| \leq \|P - Q\|
$$

Proof: Assume without losing generality that $p_2 \geq q_2$. Let

$$
\Delta x = p_1 - q_1 \quad \text{and} \quad \Delta y = p_2 - q_2 \geq 0
$$

Then,

$$
g(P) - g(Q) = \begin{cases}
(\Delta x, \Delta y) & P, Q \in D_1 \text{ or } P, Q \in D_6 \\
((1 - \delta/d)\Delta x, \Delta y) & P, Q \in D_2 \text{ or } P, Q \in D_5 \\
((1 - \delta/d)\Delta x, (1 - \xi)\Delta y + \xi\Delta x/d) & P, Q \in D_3 \\
((1 - \delta/d)\Delta x, (1 - \xi)\Delta y - \xi\Delta x/d) & P, Q \in D_4
\end{cases}
$$

(Note that $0 < 1 - \delta/d < 1$.) Clearly,

$$
g(P) - g(Q) \begin{cases}
= \|(\Delta x, \Delta y)\| & P, Q \in D_1 \text{ or } P, Q \in D_6 \\
\leq \|(\Delta x, \Delta y)\| & P, Q \in D_2 \text{ or } P, Q \in D_5
\end{cases}
$$

Hence, the remaining cases to prove are: (i) $P, Q \in D_3$ or (ii) $P, Q \in D_4$.

(*Case* i). In this case we have

$$
\|(\Delta x, \Delta y)\|^2 - \|g(P) - g(Q)\|^2 =
$$
$$
\frac{2d\delta - \delta^2 - \xi^2}{d^2}\Delta x^2 + \xi(2 - \xi)\Delta y^2 - \frac{2\xi(1 - \xi)}{d}\Delta x \Delta y \tag{14}
$$

We further divide this case into two sub-cases: $\Delta x \leq 0$ or $\Delta x > 0$.

If $\Delta x \leq 0$, then the term

$$
\frac{2\xi(1 - \xi)}{d}\Delta x \Delta y \leq 0
$$

due to $0 < \xi < 1$ and $\Delta y \geq 0$. Since the other two terms in the right hand side of (14) are already non-negative due to $0 < \xi < \delta \leq d < 1$, we obtain

$$
\|(\Delta x, \Delta y)\|^2 - \|g(P) - g(Q)\|^2 \geq 0
$$

i.e., $\|g(P) - g(Q)\| \le \|(\Delta x, \Delta y)\|$.

If $\Delta x > 0$, then the second component of $g(P) - g(Q)$: $(1 - \xi)\Delta y + \xi \Delta x/d$ is positive. Furthermore, if this component is not greater than Δy, then $\|g(P) - g(Q)\| \le \|(\Delta x, \Delta y)\|$, for that the first component of $g(P) - g(Q)$ is positive and less than Δx. Thus, we only need to consider $(1 - \xi)\Delta y + \xi \Delta x/d > \Delta y$ which is equivalent to $\Delta y < \Delta x/d$. We obtain:

$$
\begin{aligned}
& \|(\Delta x, \Delta y)\|^2 - \|g(P) - g(Q)\|^2 \\
= {}& \Delta x^2 + \Delta y^2 - (1 - \delta/d)^2 \Delta x^2 - ((1 - \xi)\Delta y + \xi \Delta x/d)^2 \\
> {}& \Delta x^2 + \Delta y^2 - (1 - \delta/d)^2 \Delta x^2 - (\Delta y + \xi \Delta x/d)^2 \\
= {}& \left(\frac{2\delta}{d} - \frac{\delta^2}{d^2} - \frac{\xi^2}{d^2} \right) \Delta x^2 - 2\frac{\xi}{d}\Delta x \Delta y \\
> {}& \left(\frac{2\delta}{d} - \frac{\delta^2}{d^2} - \frac{\xi^2}{d^2} \right) \Delta x^2 - 2\frac{\xi}{d}\Delta x \frac{1}{d}\Delta x \\
= {}& \frac{2d\delta - \delta^2 - 2\xi - \xi^2}{d^2} \Delta x^2
\end{aligned}
$$

Since ξ is already chosen to be the positive root of the quadratic equation:

$$
2d\delta - \delta^2 - 2x - x^2 = 0 \tag{15}
$$

we obtain

$$
\frac{2d\delta - \delta^2 - 2d\xi - \xi^2}{d^2} \Delta x^2 = 0,
$$

and hence $\|g(P) - g(Q)\| \le \|(\Delta x, \Delta y)\|$ when $P, Q \in D_3$.

(*Case* ii). Suppose both P and Q belong to D_4. By (12), there exist $P' = (2m - p_1, p_2)$ and $Q' = (2m - q_1, q_2)$ in D_3 such that

$$
g(P) = (2m - g_1(P'), g_2(P'))
$$

and

$$
g(Q) = (2m - g_1(Q'), g_2(Q')).
$$

Hence

$$
\begin{aligned}
\|g(P) - g(Q)\| &= \|(g_1(Q') - g_1(P'), g_2(P') - g_2(Q'))\| \\
&= \|(g_1(P') - g_1(Q'), g_2(P') - g_2(Q'))\| \\
&= \|g(P') - g(Q')\|.
\end{aligned}
$$

Since g is already proved to be nonexpanding in D_3,

$$
\|g(P') - g(Q')\| \le \|P' - Q'\| = \|P - Q\|.
$$

This shows $\|g(P) - g(Q)\| \le \|P - Q\|$ for every $P, Q \in D_4$.   **Q.E.D.**

We thus finish the proof of Theorem 2.

5   Conclusion

We remark that the FPE algorithm constructed by Sikorski [13] can be easily adapted for the class $F_{1,L}$. The cost of the FPE algorithm is also $b(\varepsilon)$ in the worst case. However, due to the additional knowledge of L, the FPE algorithm always generates intervals of shorter length compared with the intervals generated by bisection. Hence, the FPE algorithm could be more efficient than the bisection for many functions. In fact, it could be shown that the FPE algorithm has the smallest *local error* for every function in $F_{1,L}$ since it is a *central* algorithm (as defined in [17]).

For $n > 1$, we established the infinity complexity even for $L = 1$. We believe that the negative result also holds for more general information operators:

$$N_M(f) = [L_1(f), L_2(f;,\mathbf{y}_1), \ldots, L_M(f;\mathbf{y}_1,\ldots,\mathbf{y}_{M-1})]$$

where

$$\mathbf{y}_i = L_i(f;\mathbf{y}_1,\ldots,\mathbf{y}_{i-1})$$

and

$$L_{i,f}(\cdot) \equiv L_i(\cdot;\mathbf{y}_1,\ldots,\mathbf{y}_{i-1}) : F_{n,L} \to \Re$$

is any linear functional.

References

[1] ALLGOWER, E. L., AND GEORG, K. (1990), *Numerical Continuation Methods.* Springer-Verlag, New York.

[2] EAVES, B. C. (1972), Homotopies for computation of fixed points. *Math. Programming*, 1–22.

[3] EAVES, B. C., AND SAIGAL, R. (1972), Homotopies for computation of fixed points on unbounded regions. *Mathematical Programming*, 225–237.

[4] HIRSCH, M. D., PAPADIMITRIOU, C., AND VAVASIS, S. (1989), Exponential lower bounds for finding brouwer fixed points. *J. Complexity* 5, 379–416.

[5] GARCIA, C. B., AND ZANGWILL W. I. (1981), *Pathways to Solutions, Fixed Points, and Equilibria.* Prentice-Hall.

[6] MERRILL, O. H. (1972), *Applications and Extensions of an Algorithm that Computes Fixed Points of a Certain Upper Semi-continuous Point to Set Mapping.* PhD thesis, University of Michigan, Ann Arbor, MI.

[7] MERRILL, O. H. (1972), A summary of techniques for computing fixed points of continuous mapping. In *Mathematical Topics in Economic Theory and Computation*, R. H. Day and S. M. Robinson, Eds. SIAM, Philadelphia, PA, pp. 130–149.

[8] NATARAJAN, B. K. (1992) A Note on Condition-Sensitive Computation of Approximate Fixed Points. (submitted for publication).

[9] NEMIROVSKY, A. S. (1991) On Optimality of Krylov's Information When Solving Linear Operator Equations. *J. Complexity* 7, 121–130.

[10] SCARF, H. (1967), The approximation of fixed point of a continuous mapping. *SIAM J. Appl. Math.*, 1328–1343.

[11] SCARF, H. E., AND T. HANSEN. (1973), *Computation of Economic Equilibria.* Yale University Press.

[12] SIKORSKI, K., AND WOŹNIAKOWSKI, H. (1987), Complexity of fixed points I. *J. Complexity* 3, 388–405.

[13] SIKORSKI, K. (1989), Fast algorithms for the computation of fixed points. In *Robustness in Identification and Control*, M. Milanese, R. Tempo, and A. Vicino, Eds. Plenum Press, pp. 49–59.

[14] SIKORSKI, K., TSAY, C., AND WOŹNIAKOWSKI, H. (1993), An Ellipsoid Algorithm for the Computation of Fixed Points. (to appear in *J. Complexity*).

[15] SWAMINATHAN, S., Ed. (1976), *Fixed Point Theory and Its Applications.* Academic Press, New York.

[16] TODD, M. J. (1976), *The Computation of Fixed Points and Applications.* Spring-Verlag, New York.

[17] TRAUB, J. F., WASILKOWSKI, G. W., AND WOŹNIAKOWSKI, H. (1988), *Information-Based Complexity.* Academic Press, New York.

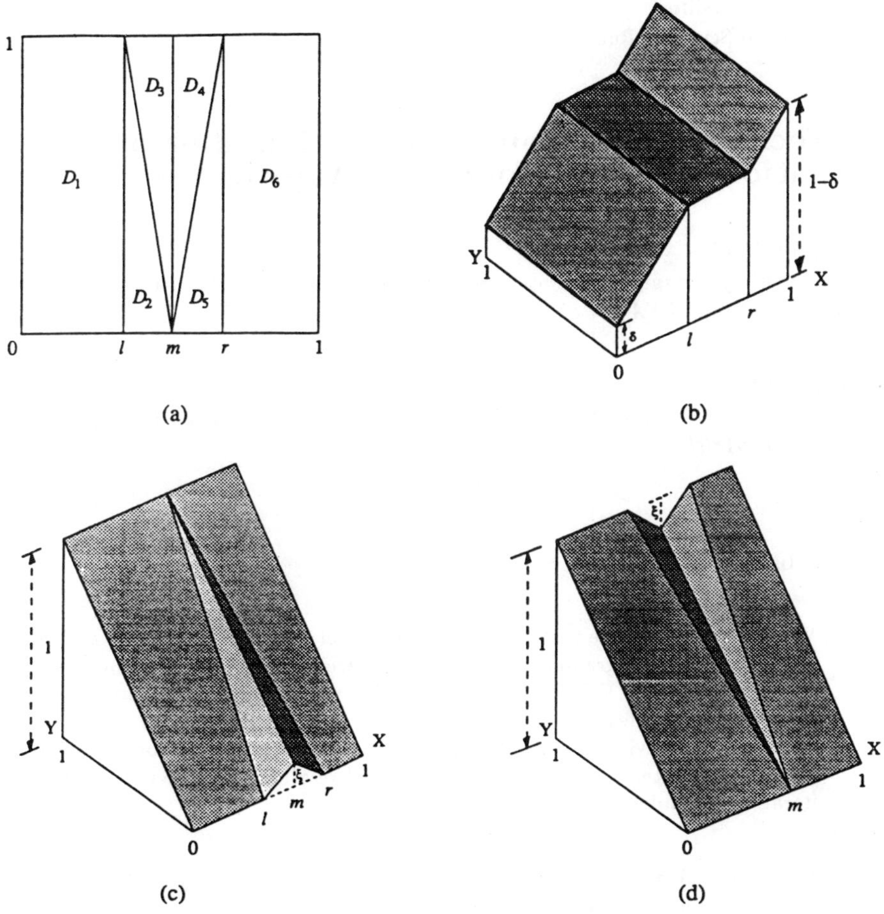

Figure 2

a) Domain Partition

b) $g_1(x, y) = h_1(x, y)$

c) $g_2(x, y)$

d) $h_2(x, y)$

Complexity in Numerical Optimization, pp. 462-489
P.M. Pardalos, Editor
©1993 World Scientific Publishing Co.

A Technique for Bounding the Number of Iterations in Path Following Algorithms

Pravin M. Vaidya
Department of Computer Science, University of Illinois at Urbana-Champaign, Urbana, IL 61801 USA

David S. Atkinson
Department of Computer Science, University of Illinois at Urbana-Champaign, Urbana, IL 61801 USA

Abstract

We present a technique that bounds the number of iterations required in a path following linear programming algorithm via a combination of two measures based on the size of $\nabla f^T (\nabla^2 f)^{-1} \nabla f$ where f is a convex barrier function with positive definite Hessian. We also present a new barrier function that is a hybrid of two previously studied barrier functions. Our technique for bounding the number of iterations shows that the hybrid results in a smaller number of iterations than does either of its components.

Keywords: Linear programming, complexity, interior point algorithm, convex barrier function, hybrid barrier function.

1   Introduction

Until the mid-1980's, essentially all linear programming (LP) algorithms followed the basic scheme of improving the linear objective while moving from point to point on the surface of the polytope formed by the linear constraints. The simplex algorithm and the primal-dual algorithm are the obvious examples. Unfortunately, none of the surface (exterior) algorithms has a polynomial bound on the number of iterations that might be required. Interior point methods grew out of the research into polynomially bounded techniques for the LP problem. Interior point methods have since been extended into more general mathematical programming problems, but LP remains the canonical example, and it is the subject of much continuing research.

One formulation of LP—sometimes called the dual formulation—is

$$\max \ c^T x$$
$$s.t. \ Ax \geq b \tag{1}$$

where $A \in R^{m \times n}$, $c, x \in R^n$, and $b \in R^m$. We assume that the polytope P defined by $P = \{x \in R^n \mid Ax \geq b\}$ is bounded and has non-empty interior. All the known interior point algorithms for solving LP are iterative algorithms [4, 6, 10, 12, 15]; each iteration typically consists of computations such as solving linear systems of equations, inverting matrices, and matrix-vector multiplications.

The computation within an iteration is obviously an important factor in the complexity of an algorithm. The types of computations within an iteration vary little from one interior point algorithm to another, however, so that it is usually the number of iterations that distinguishes algorithms. The preceding statement is especially true if we consider a parallel implementation. In the EREW PRAM model (or any other reasonable model) of parallel computation, the types of computations mentioned in the preceding paragraph are easily implemented in polylogarithmic time given a polynomial number of processors—the computations in an iteration can be performed in NC. Thus, the bottleneck in the parallel complexity of an interior point LP algorithm is the number of iterations. The primary goal of this paper is to demonstrate a new path following interior point algorithm with a smaller number of iterations (whenever $n = o(m)$) than those currently available.

With the constant L defined as

$$L = \log_2(\det_{\max}) + \log_2(c^T c + b^T b) + \log_2(m + n)$$

where $\det_{\max}$ is equal to the largest absolute value of the determinant of any square submatrix of A, the best known bound on the number of iterations for an interior point LP algorithm is $O(\sqrt{m}\, L)$ [12]. Implementations suggest, however, that the number of iterations grows *very* slowly with m and n—on the order of $O((\log m)\, L)$ [1, 7, 8, 9, 13]. There is thus reason to hope that the theoretical bounds can be improved. In this paper, we take a step in the direction of improvement. We give an interior point algorithm that solves LP in $O((mn)^{1/4}\, L)$ iterations. Our algorithm uses a *hybrid* barrier function, which combines properties of the logarithmic barrier function [15, 12] and the log determinant barrier function [16]. The computations within an iteration take polylogarithmic time in the parallel case, and so our algorithm leads to a faster parallel algorithm for $n = o(m)$. The hybrid barrier was introduced by Vaidya [17] in a conference paper. We give here a full treatment of the hybrid barrier, we develop the properties of the quantities $\delta(\cdot)$ and $\Lambda(\cdot)$ introduced in [17], and we show how those quantities lead to the conclusion that the hybrid barrier is an improvement over older barrier functions. We also investigate a property we call the *uniform ellipsoid property,* and we relate this property to the quantity $\delta(\cdot)$ and thus to the bound on the number of iterations. The uniform ellipsoid property is not a new concept. It is, in fact, equivalent to a particular case of Nesterov and Nemirovsky's self-concordance condition [11]. We prefer the term uniform ellipsoid property because it is more descriptive. Our definition also avoids the unnecessary introduction of the third differential into the picture.

We will be careful not to claim too much of our algorithm in the sequential case since it does involve a more complicated barrier function than the algorithms commonly in use. A 'dumb' sequential implementation of our algorithm requires $O(mn^2)$ computations per iteration. This leads to an overall sequential time bound of $O(m^{5/4}n^{9/4}L)$. The bound compares unfavorably to most current sequential interior point methods for LP for modest sizes of m, c.f. [12, 15]. Even the sequential implementation, however, is superior for large enough values of m; the lower power on m in the number of iterations is eventually felt. There is also the hope that our technique can be refined into not-so-dumb sequential implementations. The difficulty is that accurate gradient and Hessian calculations are expensive. It is an open question whether some amortization or other simplification might improve the sequential time bound.

In Section 2 we review the well-known basic idea of a path following algorithm for LP. We give a generic algorithm for path following and introduce the two quantities $\delta(\cdot)$ and $\Lambda(\cdot)$ for a strictly convex barrier function. These two quantities determine the number of iterations of a path following algorithm that uses the specified barrier function. In Section 3, we build the mathematical framework around path following. We develop properties of $\delta(\cdot)$ and $\Lambda(\cdot)$, and we discuss the uniform ellipsoid property of a barrier function as a measure of the region of linear convergence of Newton's method to the minimizer of the function. After this general development, in Section 4 we take two previously existing barrier functions and consider the quantities $\delta(\cdot)$ and $\Lambda(\cdot)$ for them. Then, in Section 5, we form the hybrid barrier function from the two existing functions and show that $\delta(\cdot)$ and $\Lambda(\cdot)$ for the hybrid are such that the claimed improvement in the number of iterations is obtained.

## 2	Generic path following

Consider the linear programming problem (1). Let g be a strictly convex, continuously twice differentiable function over the interior of P such that $g(x)$ approaches ∞ as x approaches the boundary of P. The implicit function theorem implies that the equation

$$\nabla g(x) = t\,c, \quad t \in R$$

implicitly defines x as a function of t. (Here $\nabla(\cdot)$ denotes the gradient.) As t varies continuously from $-\infty$ to ∞, $x = x(t)$ sweeps a continuous trajectory in int$P$. The two limit points of the trajectory are the points that minimize and maximize $c^T x$ over P.

Path following algorithms fix a value t^k and approximate ω^k (where $\nabla g(\omega^k) = t^k c$) by some x^k. The measure of closeness of x^k to ω^k can be important in the complexity analysis—compare Vaidya's [15] path following algorithm with Renegar's [12]. After x^k is found, t^k is advanced to some value t^{k+1}, and a new approximation x^{k+1} must be found.

In general, define

$$g^k(x) = g(x) - t^k c^T x.$$

During the kth iteration, we advance t^{k-1} to t^k and then (approximately) minimize $g^k(x)$. The algorithm halts when t^k becomes large enough to isolate the limit point of the trajectory. For a linear programming problem, we need $t^k = 2^{O(L)}$ where

$$L = \log_2(\det_{\max}) + \log_2(c^T c + b^T b) + \log_2(m + n)$$

with $\det_{\max}$ equal to the largest absolute value of the determinant of any square submatrix of A.

The convergence rate of a path following algorithm depends upon the rate at which the t^k values can be advanced. The time complexity of the algorithm also takes into account the work needed to move from x^{k-1} to x^k. Usually, some variant of Newton's method is used to find x^k. We could shift work from one ledger column to another by making large advances in t^k and doing more work to find x^k. Newton's method, however, is difficult to analyze when x^{k-1} is not known to be relatively close to ω^k. To avoid difficulties in analyzing Newton's method, we choose to advance t^k as much as possible, but with the restriction that we can prove that only $O(1)$ Newton steps will be required to locate the new approximation x^k.

Let $\beta(n, m)$ be a (small) constant depending upon n and m. $\beta(n, m)$ represents the rate at which we advance t^k. At the beginning of the kth iteration of a path following algorithm, we have a parameter t^{k-1}, and a feasible point x^{k-1} (a close approximation to ω^{k-1}). During the kth iteration of a generic path following algorithm, we execute the following steps.

1. $t^k \leftarrow (1 + \beta(n, m))t^{k-1}$

2. Compute x^k from x^{k-1} by executing $O(1)$ Newton steps of the form

$$\begin{aligned}
&z \leftarrow x^{k-1}; \\
&\text{For } j = 1 \text{ to } N \text{ (where } N = O(1)) \text{ do} \\
&\qquad z \leftarrow z - (\nabla^2 g^k(z))^{-1} \nabla g^k(z); \\
&x^k \leftarrow z.
\end{aligned}$$

In practice, we might choose to take steps that are some constant multiple less than 1 of a full Newton step. Such a strategy would not increase the number of Newton steps beyond $O(1)$ and might enhance stability. The main question in the above algorithm is the size of $\beta(n,m)$. That is to say, how fast can we allow the t^k's to grow, and still ensure only $O(1)$ Newton steps in step 2 of the algorithm.

How fast can t^k be increased?

For any strictly convex function f with positive definite Hessian, define the function $\Psi(f,x) = \nabla f(x)^T (\nabla^2 f(x))^{-1} \nabla f(x)$. We will show in Section 3 that $g^k(x^{k-1}) - g^k(\omega^k)$ is small if and only if $\Psi(g^k, x^{k-1})$ is small. We will also show that a small value of $g^k(x^{k-1}) - g^k(\omega^k)$ guarantees that $O(1)$ Newton steps suffice. Suppose for the moment that we can calculate a quantity $\delta(g)$ with the property that for the strictly convex function g, if

$$\Psi(g^k, x^{k-1}) \leq \delta(g), \tag{2}$$

then x^k is computable in $O(1)$ Newton steps from x^{k-1}.

Define the quantity

$$\Lambda(f) = \sup_{x \in \mathrm{int} P} \Psi(f,x).$$

We can establish a condition for the allowable increase in t^k in terms of $\delta(g)$ and $\Lambda(g)$. For simplicity, assume that $x^{k-1} = \omega^{k-1}$.

Theorem 2.1 *If*

$$t^k \leq \left(1 + \left(\frac{\delta(g)}{\Lambda(g)}\right)^{1/2}\right) t^{k-1},$$

then the approximation x^k can be obtained in $O(1)$ Newton steps.

Proof:

$$
\begin{aligned}
\Psi(g^k, \omega^{k-1}) &= \nabla g^k(\omega^{k-1})^T (\nabla^2 g^k(\omega^{k-1}))^{-1} \nabla g^k(\omega^{k-1}) \\
&= (\nabla g(\omega^{k-1}) - t^k c)^T (\nabla^2 g(\omega^{k-1}))^{-1} (\nabla g(\omega^{k-1}) - t^k c) \\
&= \Psi(g, \omega^{k-1}) - 2 t^k \, c^T (\nabla^2 g(\omega^{k-1}))^{-1} \nabla g(\omega^{k-1}) + (t^k)^2 c^T (\nabla^2 g(\omega^{k-1}))^{-1} c.
\end{aligned}
$$

But, by definition of the trajectory, $c = (1/t^{k-1}) \nabla g(\omega^{k-1})$. Therefore,

$$
\begin{aligned}
\Psi(g^k, \omega^{k-1}) &= \Psi(g, \omega^{k-1}) - 2 \frac{t^k}{t^{k-1}} \Psi(g, \omega^{k-1}) + \left(\frac{t^k}{t^{k-1}}\right)^2 \Psi(g, \omega^{k-1}) \\
&= \left(1 - 2 \frac{t^k}{t^{k-1}} + \left(\frac{t^k}{t^{k-1}}\right)^2\right) \Psi(g, \omega^{k-1}) \\
&= \left(1 - \frac{t^k}{t^{k-1}}\right)^2 \Psi(g, \omega^{k-1}). \tag{3}
\end{aligned}
$$

By hypothesis,

$$t^k \leq \left(1 + \left(\frac{\delta(g)}{\Lambda(g)}\right)^{1/2}\right) t^{k-1}.$$

Thus,

$$\left(1 - \frac{t^k}{t^{k-1}}\right)^2 = \left(\frac{t^k}{t^{k-1}} - 1\right)^2$$
$$\leq \frac{\delta(g)}{\Lambda(g)}.$$

It thus follows by (3) that

$$\Psi(g^k, \omega^{k-1}) \leq \frac{\delta(g)}{\Lambda(g)} \Psi(g, \omega^{k-1})$$
$$\leq \delta(g),$$

which, by definition of $\delta(g)$, implies that x^k is computable in $O(1)$ Newton steps. ∎

The justification of using ω^{k-1} in place of x^{k-1} in the above theorem is a simple, though laborious, task of showing error bounds on $\Psi(g^{k-1}, x)$ for all x in a 'small enough' neighborhood of ω^{k-1}. We refer the reader to [3] for a similar development.

3   A mathematical framework

The two quantities $\delta(g)$ and $\Lambda(g)$ introduced in the preceding section both involve conditions on the function $\Psi(\cdot, \cdot)$. In this section, we have four primary goals.

1. Develop some properties of the function Ψ in a general setting.

2. Consider the concept of a Hessian ellipsoid about a feasible point in the polytope.

3. Discuss the uniform ellipsoid property for a barrier function and develop its relation to linear convergence of Newton's method.

4. Relate the uniform ellipsoid property to the quantity $\delta(\cdot)$.

The ideas in these four goals are all closely related, and we hope to make clear the relationships among them. We work in as general a setting as possible in this section, although we will assume strict convexity and positive definite Hessian for our functions. We will not need to assume our functions look like the g^k functions of Section 2 except when considering the relationship of the uniform ellipsoid property to $\delta(\cdot)$.

Throughout this section, we assume that f is a strictly convex function with positive definite Hessian from some open subset $\mathcal{O} \subset R^n$ to R. Let z in $\mathcal{O}$ be fixed. Define the function $K_z : \mathcal{O} \times R \to R$ as follows:

$$K_z(x, c) = \frac{c^T \nabla^2 f(x) c}{c^T \nabla^2 f(z) c}.$$

Since f is a smooth convex function and $\nabla^2 f(x)$ is positive definite, it is clear that K_z is a continuous function on $\mathcal{O} \times R$. It is further clear that $K_z(z, c) = 1$ for all c in R^n. It follows that given $\epsilon > 0$ there exists an open neighborhood $N_{(z,c)} \subset \mathcal{O} \times R^n$ about (z, c) such that

$$(x, d) \in N_{(z,c)} \text{ implies } |K_z(x, d) - 1| < \epsilon.$$

Since the function K_z is homogeneous of degree 0 in c, the neighborhood $N_{(z,c)}$ may be chosen to be of the form $N_z \times R^n$ for some open neighborhood $N_z \subset \mathcal{O}$ about the point z. The following lemma summarizes the above discussion.

Lemma 3.1 *Given $\epsilon > 0$, for every z in $\mathcal{O}$ there exists a neighborhood N_z of z such that for x in N_z and for every c in R^n,*

$$(1 - \epsilon) c^T \nabla^2 f(z) c < c^T \nabla^2 f(x) c < (1 + \epsilon) c^T \nabla^2 f(z) c.$$

The following lemma is also useful. Details can be found in [2], [5], or [16].

Lemma 3.2 *If A and B are positive definite $n \times n$ matrices such that $\xi^T A \xi \leq \theta \, \xi^T B \xi$ for some constant θ and for all ξ in R^n, then $\xi^T A^{-1} \xi \geq (1/\theta) \xi^T B^{-1} \xi$ for all ξ in R^n.*

Corollary 3.3 *Given $\epsilon > 0$, for every z in $\mathcal{O}$ there exists a neighborhood N_z of z such that for x in N_z and for every c in R^n,*

$$\frac{1}{(1 + \epsilon)} \, c^T (\nabla^2 f(z))^{-1} c < c^T (\nabla^2 f(x))^{-1} c < \frac{1}{(1 - \epsilon)} \, c^T (\nabla^2 f(z))^{-1} c.$$

Define a *Hessian ellipsoid* centered at z in $\mathcal{O}$ by

$$E(\nabla^2 f(z), z, r) := \{ x \in R^n \mid (x - z)^T \nabla^2 f(z)(x - z) \leq r^2 \}.$$

The non-negative quantity r is called the *radius* of the ellipsoid. It is clear that about each point z in $\mathcal{O}$ we can select a radius $r_z > 0$ such that

$$E(\nabla^2 f(z), z, r_z) \subset N_z. \tag{4}$$

The following corollary (presented without proof) is the result of simple arguments for maximizing linear functionals over ellipsoids. Details may be found in [2] or [16], among others.

Corollary 3.4 *Let $E(\nabla^2 f(z), z, r)$ be the Hessian ellipsoid defined by the positive definite matrix $\nabla^2 f(z)$, and let w be any vector in R^n. Then*

$$\max_{x \in E(\nabla^2 f(z), z, r)} w^T(x - z) = r\sqrt{w^T(\nabla^2 f(z))^{-1}w},$$

and

$$\min_{x \in E(\nabla^2 f(z), z, r)} w^T(x - z) = -r\sqrt{w^T(\nabla^2 f(z))^{-1}w}.$$

If ∇f plays the role of w, we see the maximum value of $\nabla f(z)^T(x - z)$ over the given ellipsoid is

$$r\sqrt{\nabla f(z)^T(\nabla^2 f(z))^{-1}\nabla f(z)} = r\sqrt{\Psi(f, z)}.$$

Let ω be the unique minimizer of the strictly convex function f. We have the following theorem:

Theorem 3.5 *If z is a point at which $\Psi(f, z) \leq ((1 - \epsilon)/2)^2 r_z^2$, then*

$$\frac{1}{2(1 + \epsilon)} \, \Psi(f, z) \leq f(z) - f(\omega) \leq \frac{1}{2(1 - \epsilon)} \, \Psi(f, z).$$

Proof: Consider the trajectory $x(t)$ implicitly defined by the equation

$$\nabla f(x) = t \, \nabla f(z). \tag{5}$$

Define the function

$$
\begin{aligned}
\alpha(t) &= f(x(t)) - f(z) \\
&= -\int_t^1 \nabla f(x(s)) \cdot \dot{x}(s)\, ds \\
&= -\int_t^1 \nabla f(x(s)) \cdot (\nabla^2 f(x(s)))^{-1}\nabla f(z)\, ds \quad \text{(implicit diff. of (5))} \\
&= -\int_t^1 s\nabla f(z)^T(\nabla^2 f(x(s)))^{-1}\nabla f(z)\, ds
\end{aligned}
$$

Since $(\nabla^2 f(x))^{-1}$ is positive definite over the entire domain of f, it follows that $\alpha(t)$ is a monotone increasing function on $[0, 1]$ with $\alpha(1) = 0$.

We claim that $x(t) \in E(\nabla^2 f(z), z, r_z)$ for every t in $[0, 1]$. Suppose not. Then there exists t' in $[0, 1]$ such that $x' := x(t')$ lies on the boundary of $E(\nabla^2 f(z), z, r_z)$. Then

$$
\begin{aligned}
0 &> \alpha(t') \\
&= f(x') - f(z)
\end{aligned}
$$

$$\begin{aligned}
&= \nabla f(z)^T(x' - z) + \frac{1}{2}(x' - z)^T \nabla^2 f(y)(x' - z) \quad \text{(some } y \text{ between } z \text{ and } x') \\
&\geq \nabla f(z)^T(x' - z) + \frac{1 - \epsilon}{2} r_z^2 \quad \text{((4) and Lemma 3.1)} \tag{6} \\
&\geq -r_z\sqrt{\Psi(f, z)} + \frac{1 - \epsilon}{2} r_z^2 \tag{7} \\
&\geq 0 \tag{8}
\end{aligned}$$

Contradiction. It follows that $x(t) \in E(\nabla^2 f(z), z, r_z)$ for every t in $[0, 1]$. Therefore,

$$\begin{aligned}
-\alpha(0) &= f(z) - f(\omega) \\
&= \int_0^1 t\, \nabla f(z)^T (\nabla^2 f(x(t)))^{-1} \nabla f(z)\, dt \\
&\leq \int_0^1 t\, \Psi(f, z)/(1 - \epsilon)\, dt \quad \text{(by Cor. 3.3 and (4))} \\
&= \frac{\Psi(f, z)}{2(1 - \epsilon)}
\end{aligned}$$

The other inequality in the conclusion of the theorem follows from an identical argument except that the left inequality in Corollary 3.3 is used. ∎

The role of the ellipsoid $E(\nabla^2 f(z), z, r_z)$ inside N_z (see (4)) can sometimes be played by other types of regions. For example, if f is a logarithmic barrier function over a polytope $\{x \mid Ax \geq b\}$, then the region

$$\Sigma(z, r) := \{x \mid |a_i^T x - b_i|/|a_i^T z - b_i| \leq r \text{ for all } i\},$$

where a_i is the ith row of the matrix A, defines a polytope neighborhood about z that allows similar results to those above to be derived [2, 3, 16]. The crucial fact is that we need some neighborhood which is in N_z and on which we can bound from below the change in f itself, as we did in (6) above.

Almost linear convergence

By 'almost linear' convergence, we mean that we will develop a result that has the feel of a linear convergence result. We will have to make another assumption about the function f in order to prove truly linear convergence. Theorem 3.5 implies that for any strictly convex function with positive definite Hessian, a small enough value of $\Psi(f, z)$ implies that $f(z) - f(\omega)$ is sandwiched between constant multiples of $\Psi(f, z)$. Our ultimate goal is to show that this means taking Newton steps beginning at z will lead to linear convergence (measured in terms of function values) to ω. The ultimate goal is unattainable until we can remove the obvious difficulties presented by the dependence on r_z. Such difficulties limit the usefulness of these results in practice to functions f where r can be chosen independently of z—the uniform ellipsoid property. We proceed with some preliminary results that result in almost linear convergence.

Theorem 3.6 *Suppose $z \in \mathcal{O}$, and define $\eta = (\nabla^2 f(z))^{-1} \nabla f(z)$ and a parametrized line $y(t) = z - t\eta$. Suppose that t is restricted to a small enough range $[0, a]$ to ensure that $z - t\eta \in N_z$ for all t in $[0, a]$. Then*

$$f(z) - f(y(t)) \geq \Psi(f, z) \left(t - \frac{1+\epsilon}{2} t^2 \right).$$

Proof: Define $\alpha(t) = f(y(t)) = f(z - t\eta)$. Then

$$\alpha'(t) = \nabla f(y(t))^T y'(t) = -\nabla f(z - t\eta)^T \eta$$

$$\alpha''(t) = y'(t)^T \nabla^2 f(y(t)) y'(t) = \eta^T \nabla^2 f(z - t\eta)\eta,$$

so that

$$\begin{aligned}
\alpha''(t) &= \eta^T \nabla^2 f(z - t\eta)\eta & (9)\\
&\leq (1+\epsilon) \eta^T \nabla^2 f(z)\eta & (10)\\
&= (1+\epsilon) \Psi(f, z), & (11)
\end{aligned}$$

since $z - t\eta$ is assumed to lie in N_z.

Therefore,

$$\begin{aligned}
\alpha'(t) &= \alpha'(0) + \int_0^t \alpha''(s)\, ds\\
&= -\Psi(f, z) + \int_0^t \alpha''(s)\, ds\\
&\leq -\Psi(f, z) + (1+\epsilon)\Psi(f, z)\, t\\
&= -(1 - (1+\epsilon)t)\Psi(f, z),
\end{aligned}$$

and

$$\begin{aligned}
\alpha(0) - \alpha(t) &= -\int_0^t \alpha'(s)\, ds\\
&\geq \Psi(f, z) \int_0^t (1 - (1+\epsilon)s)\, ds\\
&= \Psi(f, z) \left(t - \frac{1+\epsilon}{2} t^2 \right) \quad \blacksquare
\end{aligned}$$

We need two more short lemmas, which we will combine with Theorem 3.6 to prove almost linear convergence of Newton's method in Theorem 3.9. True linear convergence will require an additional assumption that we will make later.

Lemma 3.7 *If $f(z) - f(\omega) \leq ((1 - \epsilon)/2)r_\omega^2$, then $z \in E(\nabla^2 f(\omega), \omega, r_\omega)$.*

Proof: Suppose not. Connect z to ω by a line segment and let x' be the intersection of the line segment with $E(\nabla^2 f(\omega), \omega, r_\omega)$. By strict convexity of f,

$$
\begin{aligned}
f(z) - f(\omega) \;&>\; f(x') - f(\omega) \\
&=\; \nabla f(\omega)^T (x' - \omega) + \frac{1}{2}(x' - \omega)^T \nabla^2 f(y)(x' - \omega) \quad \text{(some y between x' and ω)} \\
&=\; 0 + \frac{1}{2}(x' - \omega)^T \nabla^2 f(y)(x' - \omega) \\
&\geq\; \frac{1 - \epsilon}{2} r_\omega^2. \quad \text{((4) and Lemma 3.1)}
\end{aligned}
$$

Contradiction. Thus $z \in E(\nabla^2 f(\omega), \omega, r_\omega)$. ∎

The following lemma has (half of) the same conclusion as in Theorem 3.5. The hypothesis, however, is now an upper bound on $f(z) - f(\omega)$ rather than an upper bound on $\Psi(f, z)$.

Lemma 3.8 *If* $f(z) - f(\omega) \leq ((1 - \epsilon)/2)r_\omega^2$, *then* $\Psi(f, z) \geq 2(1 - \epsilon)(f(z) - f(\omega))$.

Proof: Define the trajectory $x(t)$ implicitly by the equation $\nabla f(x) = t \nabla f(z)$. Then $\dot{x}(t) = (\nabla^2 f(x))^{-1} \nabla f(z)$, and

$$
\begin{aligned}
f(z) - f(\omega) \;&=\; \int_0^1 t \nabla f(z)^T (\nabla^2 f(x(t)))^{-1} \nabla f(z) \, dt \\
&\leq\; \frac{1}{2(1 - \epsilon)} \Psi(f, z)
\end{aligned}
$$

where the inequality follows by Corollary 3.3 and Lemma 3.7. ∎

The following theorem is the 'almost linear' convergence result promised. We speak now not of t being small enough to ensure $z - t\eta \in N_z$, as we did in Theorem 3.6, but rather of t being small enough to ensure

$$
z - t\eta \in E(\nabla^2 f(z), z, r_z) \subset N_z.
$$

By definition of $E(\nabla^2 f(z), z, r_z)$ and η, this condition is the same as $t \leq r_z / \sqrt{\Psi(f, z)}$.

Theorem 3.9 *Suppose* $f(z) - f(\omega) \leq ((1 - \epsilon)/2)r_\omega^2$. *If* $t \leq r_z / \sqrt{\Psi(f, z)}$, *then*

$$
f(y(t)) - f(\omega) \leq \left(1 - 2(1 - \epsilon)t + (1 - \epsilon)(1 + \epsilon)t^2\right)(f(z) - f(\omega)).
$$

Proof: By Theorem 3.6, the hypotheses allow us to conclude that

$$
f(z) - f(y(t)) \geq \Psi(f, z)\left(t - \frac{1 + \epsilon}{2}t^2\right).
$$

We now invoke Lemma 3.8 to get

$$f(z) - f(y(t)) \geq 2(1 - \epsilon)(f(z) - f(\omega))(t - \frac{1 + \epsilon}{2} t^2).$$

Simple re-arrangement gives the result of the theorem. ∎

Theorem 3.9 *almost* states that if $f(z) - f(\omega)$ is small enough, then Newton's Method leads to linear convergence to ω (linear relative to function value decrease). We need to remove the 'almost.'

The uniform ellipsoid property and linear convergence

A point x_ω is considered a suitable approximation to ω if for some suitable $\delta > 0$ we have

$$f(x_\omega) - f(\omega) \leq \delta^2.$$

In applications, f is usually a strictly convex barrier function over a convex set in R^n defined as the set of points simultaneously satisfying m convex constraints. Most properly, then, we will say that r_z (or r_ω) is $r_{(n,m,z)}$. As the hypotheses of the preceding lemmas and theorems suggest, usually we will have $\delta = \rho r_{(n,m,\omega)}$ for some (small) constant ρ. That is, δ has no dependence on n or m that is not already present in $r_{(n,m,\omega)}$. Most barrier functions in current use, as well as the new hybrid barrier to be presented later, make this assumption about δ.

Theorem 3.5 tells us that if $\Psi(f, z) \leq ((1 - \epsilon)/2)^2 r_{(n,m,z)}^2$ then

$$\begin{aligned}
f(z) - f(\omega) &\leq \frac{1}{2(1 - \epsilon)} \Psi(f, z) \\
&\leq \frac{1}{2(1 - \epsilon)} \left(\frac{(1 - \epsilon)}{2}\right)^2 r_{(n,m,z)}^2 \\
&= \frac{1 - \epsilon}{8} r_{(n,m,z)}^2.
\end{aligned}$$

Since the hypotheses of Theorem 3.9 require $f(z) - f(\omega)$ to be bounded above by a constant multiple of $r_{(n,m,\omega)}^2$, we are led to consider the idea of a *uniform* ellipsoid radius $\bar{r}$. When it is true that we can select a radius $\bar{r}$ such that for every z in the domain of f, including $z = \omega$,

$$r_{(n,m,z)} = \bar{r}, \tag{12}$$

f will be said to have the *uniform ellipsoid property*. As mentioned in the introduction, the uniform ellipsoid property is equivalent to a particular case of Nesterov and Nemirovsky's self-concordance condition in which their parameter a is equal to 1 [11]. The uniform ellipsoid definition—without reference to third differentials—is sufficient to capture the important property, and we believe that uniform ellipsoid property is a more descriptive term.

For a function with the uniform ellipsoid property, two very important simplifications occur. First, the hypothesis of Theorem 3.5 becomes "If z is a point at which $\Psi(f,z) \leq ((1-\epsilon)/2)^2 \bar{r}^2$." That is, there is a constant *independent of z* such that $\Psi(f,z)$ less than or equal to that constant implies $f(z) - f(\omega)$ is small. This suggests that the hypothesis in Theorem 3.9—that $f(z) - f(\omega)$ be less than or equal to a constant multiple of r_ω^2—can be satisfied by bounding $\Psi(f,z)$ above by a constant! The second simplification resulting from a uniform radius is in the conclusion of Theorem 3.9. With

$$t \leq \frac{r_z}{\sqrt{\Psi(f,z)}} = \frac{\bar{r}}{\sqrt{\Psi(f,z)}}, \tag{13}$$

and with $\Psi(f,z) \leq ((1-\epsilon)/2)^2 \bar{r}^2$, the multiplicative decrease in the conclusion of Theorem 3.9 becomes a constant. That is, we have linear convergence.

Our discussion is summarized in the following theorem.

Theorem 3.10 *If there is a uniform radius $\bar{r}$ (which may depend upon n and m) such that*

$$E(\nabla^2 f(z), z, \bar{r}) \subset N_z$$

for every z *in the domain, then the condition*

$$\Psi(f,z) \leq \left(\frac{1-\epsilon}{2}\right)^2 \bar{r}^2$$

implies $O(1)$ Newton steps give us a satisfactory approximation of ω.

Proof: Follows from above discussion. ∎

An interesting property of $\delta(g)$

Comparison of the hypothesis of Theorem 3.10 and of (2) in Section 2 makes the following lemma fairly obvious.

Lemma 3.11 *Let g be a strictly convex function with positive definite Hessian, and suppose g has the uniform ellipsoid property. Then*

$$\delta(g) = \Omega(\bar{r}^2).$$

Proof: We have exhibited that $((1-\epsilon)/2)^2 \bar{r}^2$ is a sufficiently small upper bound on $\Psi(g^k, x^{k-1})$ to guarantee linear convergence to x^k. Since ϵ is a constant of our choosing, the result follows. ∎

Note that by the definition of the function $g^k(x)$, we have $\nabla^2 g(x) = \nabla^2 g^k(x)$. Finding a uniform radius $\bar{r}$ resulting from the function g is therefore equivalent to finding one resulting from g^k. This explains why the condition (2) is written in terms of $\delta(g)$ and not $\delta(g^k)$.

We have shown that the obvious mathematical properties of a smooth convex function $f(z)$ with positive definite Hessian and of the function $\Psi(f, z)$ lead to sufficient conditions for effective minimization via Newton steps (Thorem 3.9). These sufficient conditions are of no practical value, however, without the additional condition of the uniform ellipsoid property. With the uniform ellipsoid property, maintaining guaranteed easy Newton minimization of f from a point z near its minimum becomes a simple matter of checking that z is a point at which $\Psi(f, z)$ is bounded above by a constant. Finally, our definition of $\delta(\cdot)$ meshes very nicely into the discussion of the uniform ellipsoid radius, as we saw in Lemma 3.11.

We will use two different functions with the uniform ellipsoid property in the development of our hybrid algorithm: the logarithmic barrier function and the log determinant barrier function.

4 Two old barrier functions

Our goal in this section is to re-introduce the logarithmic barrier function and the log determinant barrier function and to consider the quantities $\delta(\cdot)$ and $\Lambda(\cdot)$ for them. These two barrier functions will be combined in the next section to give a new hybrid barrier. We will be able to use the work in this section in finding $\delta(\cdot)$ and $\Lambda(\cdot)$ for the hybrid barrier. In the calculations of $\delta(\cdot)$, we will make much use of Lemma 3.11.

The path of analytic centers

The majority of path following algorithms in the literature [10, 12, 14, 15] let the function g be the logarithmic barrier function over P. The logarithmic barrier function $\phi(x)$ over P is

$$\phi(x) = -\sum_{i=1}^{m} \ln(a_i^T x - b_i)$$

where a_i denotes the ith row of A. These algorithms follow the path of *analytic centers* [10, 12, 14] defined as the points x that satisfy

$$\nabla \phi(x) = t\,c, \quad t \in R.$$

The following simple theorem exhibits a bound for $\Lambda(\phi)$.

Theorem 4.1 *For the logarithmic barrier function $\phi(x)$ over a polytope P, we have*

$$\Psi(\phi, x) \leq m$$

for all x in intP. Thus $\Lambda(\phi) \leq m$ for the logarithmic barrier function.

Proof: Define the $m \times m$ matrix D to be the diagonal matrix whose ith diagonal entry is $1/(a_i^T x - b_i)$. Let $\mathbf{e}$ be the m-vector of all 1's. Then we have

$$
\begin{aligned}
\nabla \phi(x) &= -\sum_{i=1}^{m} \frac{a_i}{a_i^T x - b_i} \\
&= -A^T D \mathbf{e} \\
\nabla^2 \phi(x) &= \sum_{i=1}^{m} \frac{a_i a_i^T}{(a_i^T x - b_i)^2} \\
&= A^T D^2 A.
\end{aligned}
$$

(14)

(15)

It follows that

$$
\begin{aligned}
\nabla \phi(x)^T (\nabla^2 \phi(x))^{-1} \nabla \phi(x) &= \mathbf{e}^T (DA)(\nabla^2 \phi(x))^{-1}(DA)^T \mathbf{e} \\
&= \mathbf{e}^T (DA)[A^T D^2 A]^{-1}(DA)^T \mathbf{e} \\
&= \mathbf{e}^T (DA)[(DA)^T (DA)]^{-1}(DA)^T \mathbf{e}
\end{aligned}
$$

(16)

However, $(DA)[(DA)^T(DA)]^{-1}(DA)^T$ is the projection matrix onto the column space of the matrix DA. The inner product of a vector with its projection onto a subspace is certainly less than or equal to the inner product of that vector with itself. Thus

$$
\begin{aligned}
\nabla \phi(x)^T (\nabla^2 \phi(x))^{-1} \nabla \phi(x) &= \mathbf{e}^T (DA)[(DA)^T (DA)]^{-1}(DA)^T \mathbf{e} \\
&\leq \mathbf{e}^T \mathbf{e} \\
&= m.
\end{aligned}
$$

(17)

It follows by definition of Λ that we have $\Lambda(\phi) \leq m$. $\blacksquare$

Theorem 4.2 *For the logarithmic barrier function* ϕ,

$$
\delta(\phi) = \Omega(1).
$$

Proof: It follows from Lemma 3.11 that we need only show that ϕ has the uniform ellipsoid property and the uniform radius $\bar{r}$ for $\phi(x)$ satisfies $\bar{r} = \Omega(1)$. From (15) it follows that for any $\xi \in R^n$

$$
\begin{aligned}
\xi^T \nabla^2 \phi(x) \xi &= \sum_{i=1}^{m} \frac{(a_i^T \xi)^2}{(a_i^T x - b_i)^2} \\
&= \sum_{i=1}^{m} \frac{(a_i^T \xi)^2}{(a_i^T z - b_i)^2} \cdot \frac{(a_i^T z - b_i)^2}{(a_i^T x - b_i)^2}
\end{aligned}
$$

(18)

Suppose now that for some constant α, $0 < \alpha < 1/4$, we have $x \in E(\nabla^2\phi(z), z, \alpha)$. By definition of the ellipsoid,

$$
\begin{aligned}
\alpha^2 &\geq \sum_{i=1}^{m} \frac{(a_i^T(x-z))^2}{(a_i^T z - b_i)^2} \\
&= \sum_{i=1}^{m} \frac{((a_i^T x - b_i) - (a_i^T z - b_i))^2}{(a_i^T z - b_i)^2} \\
&= \sum_{i=1}^{m} \left(\frac{a_i^T x - b_i}{a_i^T z - b_i} - 1 \right)^2,
\end{aligned}
$$

which implies

$$
\left| \frac{a_i^T x - b_i}{a_i^T z - b_i} - 1 \right| \leq \alpha \tag{19}
$$

for every i. Or, in other words,

$$
1 - \alpha \leq \frac{a_i^T x - b_i}{a_i^T z - b_i} \leq 1 + \alpha. \tag{20}
$$

In (20), select $\epsilon = 1/(1-\alpha)^2 - 1$ (note that $0 < \epsilon < 1$ since $0 < \alpha < 1/4$) take reciprocals, and square. We have

$$
1 - \epsilon \leq \left(\frac{a_i^T z - b_i}{a_i^T x - b_i} \right)^2 \leq 1 + \epsilon. \tag{21}
$$

It now follows from (18) that for every $\xi \in R^n$

$$
(1 - \epsilon)\xi^T \nabla^2 \phi(z)\xi \leq \xi^T \nabla^2 \phi(x)\xi \leq (1 + \epsilon)\xi^T \nabla^2 \phi(z)\xi. \tag{22}
$$

Therefore, $\phi(x)$ has the uniform ellipsoid property, and, since α was chosen as a constant, $\bar{r} = \Omega(1)$. $\blacksquare$

Now Theorem 2.1 implies that we can set

$$
t^k \approx \left(1 + \frac{1}{\sqrt{m}} \right) t^{k-1}
$$

and be assured that x^k can be found in $O(1)$ Newton steps. We stated that t typically needs to become as large as $2^{O(L)}$. It follows that the number of iterations for the path following algorithm using analytic centers is $O(\sqrt{m}\, L)$.

The path of volumetric centers

Vaidya [16] has introduced another barrier function for path following algorithms on polytopes: the *log-determinant barrier*

$$
V(x) = \frac{1}{2} \ln(\det(\nabla^2 \phi(x))).
$$

$\phi(x)$ is the logarithmic barrier as before. The log-determinant barrier is strictly convex over $\mathrm{int}\,P$. The minimizer of V is called the *volumetric center* of the polytope P. A path following algorithm based on $V(x)$ follows the path of volumetric centers defined by

$$\nabla V(x) = \nabla \left(\frac{1}{2}\ln(\det(\nabla^2 \phi(x))) \right) = t\,c, \quad t \in R.$$

We will show that for the function $V(x)$, the parameter $\delta(V)$ can be chosen to be approximately $1/\sqrt{m}$. That is, the r_ω in Theorem 3.9 is not a constant for V (as it was for ϕ), but rather depends upon the number of constraints defining P. Define

$$\sigma_i(x) = \frac{a_i^T(\nabla^2\phi(x))^{-1}a_i}{(a_i^T x - b_i)^2}, \quad 1 \le i \le m.$$

Vaidya showed [16, Claim 3] that

$$\sigma_i(x) \le 1 \tag{23}$$

and

$$\sum_{i=1}^{m} \sigma_i(x) = n \tag{24}$$

for every x in $\mathrm{int}\,P$. Define also

$$Q(x) = \sum_{i=1}^{m} \sigma_i(x)\frac{a_i a_i^T}{(a_i^T x - b_i)^2}. \tag{25}$$

Then it can be shown ([16, Lemma 1] and [16, Lemma 3]) that

$$\nabla V(x) = -\sum_{i=1}^{m} \sigma_i(x)\frac{a_i}{a_i^T x - b_i} \tag{26}$$

and

$$\xi^T Q(x)\xi \le \xi^T \nabla^2 V(x)\xi \le 5\,\xi^T Q(x)\xi \quad \text{for all } \xi \in R^n. \tag{27}$$

It follows from (27) and Lemma 3.2 that

$$\frac{1}{5}\xi^T(Q(x))^{-1}\xi \le \xi^T(\nabla^2 V(x))^{-1}\xi \le \xi^T(Q(x))^{-1}\xi \quad \text{for all } \xi \in R^n. \tag{28}$$

The following lemma is therefore significant.

Lemma 4.3 *For all x in $\mathrm{int}P$,*

$$\nabla V(x)(Q(x))^{-1}\nabla V(x) \le n.$$

Proof: The proof is extremely similar to the proof of Lemma 4.1. Suppose x is fixed in $\text{int}\,P$. Define the matrix D to be the diagonal matrix whose ith diagonal entry is $\sqrt{\sigma_i(x)}/(a_i^T x - b_i)$. Define the vector $\mathbf{s} \in R^m$ by $\mathbf{s}_i = \sqrt{\sigma_i(x)}$. Observe that

$$\nabla V(x) = -\mathbf{s}^T D A$$

and

$$Q(x) = A^T D^2 A.$$

Then

$$
\begin{aligned}
\nabla V(x)(Q(x))^{-1}\nabla V(x) &= \mathbf{s}^T(DA)[A^T D^2 A]^{-1}(DA)^T \mathbf{s} \\
&= \mathbf{s}^T(DA)[(DA)^T(DA)]^{-1}(DA)\mathbf{s} \\
&\leq \mathbf{s}^T \mathbf{s} \quad \text{(the projection matrix argument)} \\
&= \sum_{i=1}^{m} \sigma_i(x) \\
&= n. \quad \blacksquare
\end{aligned}
$$

The following theorem is now immediate.

Theorem 4.4 *For the log-determinant barrier function V, we have*

$$\Lambda(V) \leq n.$$

Proof: An application of (28) with $\xi = \nabla V(x)$ followed by an application of Lemma 4.3 gives

$$
\begin{aligned}
\nabla V(x)^T(\nabla^2 V(x))^{-1}\nabla V(x) &\leq \nabla V(x)^T(Q(x))^{-1}\nabla V(x) \\
&\leq n.
\end{aligned}
$$

The result follows by definition of $\Lambda(\mathrm{V})$. $\blacksquare$

We will now sketch a proof that for the log-determinant barrier $\bar{r} = \Omega(1/m^{1/4})$, so that $\delta(V) = \Omega(1/m^{1/2})$.

Lemma 4.5 ([16] Lemma 4) *For every $\xi \in R^n$,*

$$\xi^T \nabla^2 \phi(x)\xi \geq \xi^T Q(x)\xi \geq \frac{1}{4m}\xi^T \nabla^2 \phi(x)\xi.$$

Proof: The left inequality follows trivially by the definition of $Q(x)$ (25) and by the fact that $\sigma_i(x) \leq 1$ (23).

Let $I = \{1, \ldots, m\}$ and $S(x) = \{i \in I \mid \sigma_i(x) \geq 1/(2m)\}$. Then

$$
\begin{aligned}
\xi^T Q(x)\xi &\geq \sum_{i \in S(x)} \sigma_i(x) \frac{(a_i^T \xi)^2}{(a_i^T x - b_i)^2} \\
&\geq \frac{1}{2m} \sum_{i \in S(x)} \frac{(a_i^T \xi)^2}{(a_i^T x - b_i)^2}.
\end{aligned}
\tag{29}
$$

Moreover,

$$
\begin{aligned}
\sum_{i \in S(x)} \frac{(a_i^T \xi)^2}{(a_i^T x - b_i)^2} &= \xi^T \nabla^2 \phi(x)\xi - \xi^T \left(\sum_{i \in I \setminus S(x)} \frac{a_i a_i^T}{(a_i^T x - b_i)^2} \right) \xi \\
&= \xi^T \nabla^2 \phi(x)\xi - \xi^T (\nabla^2 \phi(x))^{1/2} \mathbf{M} (\nabla^2 \phi(x))^{1/2} \xi
\end{aligned}
\tag{30}
$$

where $\mathbf{M}$ is the matrix defined by

$$
\mathbf{M} = \sum_{i \in I \setminus S(x)} \frac{(\nabla^2 \phi(x))^{-1/2} a_i a_i^T (\nabla^2 \phi(x))^{-1/2}}{(a_i^T x - b_i)^2}.
$$

The matrix $\mathbf{M}$ is symmetric positive semidefinite; its largest eigenvalue is bounded by the trace $\mathrm{Tr}(\mathbf{M})$ of $\mathbf{M}$. But, since $\mathbf{M}$ is a sum of vector outer products,

$$
\begin{aligned}
\mathrm{Tr}(\mathbf{M}) &= \sum_{i \in I \setminus S(x)} \frac{a_i^T (\nabla^2 \phi(x))^{-1} a_i}{(a_i^T x - b_i)^2} \\
&= \sum_{i \in I \setminus S(x)} \sigma_i(x).
\end{aligned}
$$

It follows from (30) that

$$
\begin{aligned}
\sum_{i \in S(x)} \frac{(a_i^T \xi)^2}{(a_i^T x - b_i)^2} &\geq \xi^T \nabla^2 \phi(x)\xi \left(1 - \sum_{i \in I \setminus S(x)} \sigma_i(x) \right) \\
&\geq \frac{1}{2} \xi^T \nabla^2 \phi(x)\xi.
\end{aligned}
$$

Now, from (29) we conclude

$$
\xi^T Q(x)\xi \geq \frac{1}{4m} \xi^T \nabla^2 \phi(x)\xi. \quad \blacksquare
$$

It immediately follows from Lemma 3.2 that

$$
\xi^T (Q(x))^{-1} \xi \leq 4m \, \xi^T (\nabla^2 \phi(x))^{-1} \xi
\tag{31}
$$

for every ξ in R^n. Dividing both sides of (31) by $(a_i^T x - b_i)^2$ gives

$$\frac{a_i^T(Q(x))^{-1}a_i}{(a_i^T x - b_i)^2} \le 4m \frac{a_i^T(\nabla^2\phi(x))^{-1}a_i}{(a_i^T x - b_i)^2} = 4m\,\sigma_i(x). \tag{32}$$

We now derive a companion inequality to (32) that will allow a sharper upper bound for $(a_i^T(Q(x))^{-1}a_i)/(a_i^T x - b_i)^2$—a bound we will need presently.

If B is a positive definite matrix which can be written as $B = \hat{B} + ww^T$ for some positive semidefinite matrix $\hat{B}$ and some vector w, then

$$\begin{aligned}
w^T B^{-1}w &= w^T B^{-1}BB^{-1}w \\
&= w^T B^{-1}(\hat{B} + ww^T)B^{-1}w \\
&= w^T B^{-1}\hat{B}B^{-1}w + (w^T B^{-1}w)^2.
\end{aligned} \tag{33}$$

Therefore, since $\hat{B}$ is positive semidefinite,

$$\begin{aligned}
w^T B^{-1}w(1 - w^T B^{-1}w) &= w^T B^{-1}\hat{B}B^{-1}w \\
&\ge 0.
\end{aligned} \tag{34}$$

Since positive definiteness of B implies positive definiteness of B^{-1}, it now follows that

$$w^T B^{-1}w \le 1. \tag{35}$$

The point of developing (35) is that the matrix $Q(x)$ may be written in the required format simply by splitting off the ith term:

$$Q(x) = \hat{Q}(x) + \sigma_i(x)\frac{a_i a_i^T}{(a_i^T x - b_i)^2}.$$

With $w = \sqrt{\sigma_i(x)}a_i/(a_i^T x - b_i)$, it follows that

$$\sigma_i(x)\frac{a_i^T(Q(x))^{-1}a_i}{(a_i^T x - b_i)^2} \le 1. \tag{36}$$

The results in (32) and (36) will combine to give the following lemma.

Lemma 4.6 *For each i, $1 \le i \le m$,*

$$\frac{a_i^T(Q(x))^{-1}a_i}{(a_i^T x - b_i)^2} \le 2\sqrt{m}.$$

Proof: The inequalities in (32) and (36) together imply

$$\frac{a_i^T(Q(x))^{-1}a_i}{(a_i^T x - b_i)^2} \le \min\{4m\,\sigma_i(x), 1/\sigma_i(x)\}.$$

482   *P.M. Vaidya and D.S. Atkinson*

Since the first argument is strictly increasing as a function of $\sigma_i(x)$ and the second is strictly decreasing as a function of $\sigma_i(x)$, we simply need to notice that the two arguments are equal at $\sigma_i(x) = 1/(2\sqrt{m})$. $\blacksquare$

The following lemma gives a bound on $|a_i^T(x - z)/(a_i^T z - b_i)|$ that looks much like the bound derived in the proof of Theorem 4.2. The reason is that we are again working toward the goal of showing a uniform radius for a barrier function, and both $\nabla^2 \phi(x)$ and $Q(x)$ have squares of distances $(a_i^T x - b_i)^2$ in their denominators.

Lemma 4.7 *Suppose for some constant $\alpha > 0$ we have $x \in E(Q(z), z, \alpha/m^{1/4})$. Then for every i,*

$$\left| \frac{a_i^T(x - z)}{a_i^T z - b_i} \right| \leq \alpha\sqrt{2}.$$

Proof:

$$
\begin{aligned}
|a_i^T(x - z)| &\leq \frac{\alpha}{m^{1/4}}\sqrt{a_i^T(Q(z))^{-1}a_i} \quad &\text{(Corollary 3.4)} \\
&\leq \frac{\alpha}{m^{1/4}}\sqrt{2}\, m^{1/4}(a_i^T z - b_i) \quad &\text{(Lemma 4.6)} \\
&= \alpha\sqrt{2}(a_i^T z - b_i), & (37)
\end{aligned}
$$

which implies, upon dividing both sides by $a_i^T z - b_i$ (and recognizing that $a_i^T z - b_i > 0$ by definition),

$$\left| \frac{a_i^T(x - z)}{a_i^T z - b_i} \right| \leq \alpha\sqrt{2}. \quad \blacksquare \tag{38}$$

Corollary 4.8 *If $x \in E(Q(z), z, \alpha/m^{1/4})$, then for every $\xi \in R^n$,*

$$\frac{1}{(1 + \alpha\sqrt{2})^2} \xi^T \nabla^2 \phi(z)\xi \leq \xi^T \nabla^2 \phi(x)\xi \leq \frac{1}{(1 - \alpha\sqrt{2})^2} \xi^T \nabla^2 \phi(z)\xi,$$

and also

$$(1 - \alpha\sqrt{2})^2 \xi^T(\nabla^2 \phi(z))^{-1}\xi \leq \xi^T(\nabla^2 \phi(x))^{-1}\xi \leq (1 + \alpha\sqrt{2})^2 \xi^T(\nabla^2 \phi(z))^{-1}\xi.$$

Proof: The first statement follows exactly as in the development in (19)–(22) in the proof of Theorem 4.2. The second statement follows from the first with an application of Lemma 3.2. $\blacksquare$

Theorem 4.9 *For the log-determinant barrier function V,*

$$\delta(V) = \Omega(1/\sqrt{m}).$$

Proof: It follows from Lemma 3.11 that we need only show that $V(x)$ has the uniform ellipsoid property and the uniform radius $\bar{r}$ for $V(x)$ satisfies $\bar{r} = \Omega(1/m^{1/4})$. The inequalities in (27) imply that for every $r > 0$,

$$E(Q(z), z, r/\sqrt{5}) \subset E(\nabla^2 V(z), z, r) \subset E(Q(z), z, r).$$

It follows that if we can show the ellipsoid about z defined by $Q(z)$ has uniform radius $\bar{r}$ satisfying $\bar{r} = \Omega(1/m^{1/4})$, that will suffice.

Suppose now that $x \in E(Q(z), z, \alpha/m^{1/4})$ for some constant $\alpha > 0$. Following the development in the proof of Theorem 4.2, we have for every $\xi \in R^n$,

$$
\begin{aligned}
\xi^T Q(x)\xi &= \sum_{i=1}^m \sigma_i(x)\frac{(a_i^T \xi)^2}{(a_i^T x - b_i)^2} \\
&= \sum_{i=1}^m \sigma_i(z)\frac{(a_i^T \xi)^2}{(a_i^T z - b_i)^2} \frac{\sigma_i(x)}{\sigma_i(z)} \frac{(a_i^T z - b_i)^2}{(a_i^T x - b_i)^2} \\
&= \sum_{i=1}^m \sigma_i(z)\frac{(a_i^T \xi)^2}{(a_i^T z - b_i)^2} \frac{a_i^T(\nabla^2\phi(x))^{-1}a_i}{a_i^T(\nabla^2\phi(z))^{-1}a_i} \frac{(a_i^T z - b_i)^4}{(a_i^T x - b_i)^4}
\end{aligned}
\tag{39}
$$

Since $x \in E(Q(z), z, \alpha/m^{1/4})$ and since

$$\left|\frac{a_i^T(x-z)}{a_i^T z - b_i}\right| = \left|\frac{a_i^T x - b_i}{a_i^T z - b_i} - 1\right|,$$

Lemma 4.7 implies that

$$\frac{1}{(1 + \alpha\sqrt{2})^4} \leq \frac{(a_i^T z - b_i)^4}{(a_i^T x - b_i)^4} \leq \frac{1}{(1 - \alpha\sqrt{2})^4}.
\tag{40}$$

Moreover, the second conclusion of Corollary 4.8 implies

$$(1 - \alpha\sqrt{2})^2 \leq \frac{a_i^T(\nabla^2\phi(x))^{-1}a_i}{a_i^T(\nabla^2\phi(z))^{-1}a_i} \leq (1 + \alpha\sqrt{2})^2.
\tag{41}$$

Thus, (39) implies that

$$\frac{(1 - \alpha\sqrt{2})^2}{(1 + \alpha\sqrt{2})^4} \xi^T Q(z)\xi \leq \xi^T Q(x)\xi \leq \frac{(1 + \alpha\sqrt{2})^2}{(1 - \alpha\sqrt{2})^4} \xi^T Q(z)\xi.
\tag{42}$$

Of course, given $\epsilon > 0$, we can choose $\alpha > 0$ so that

$$1 - \epsilon < \frac{(1 - \alpha\sqrt{2})^2}{(1 + \alpha\sqrt{2})^4} < \frac{(1 + \alpha\sqrt{2})^2}{(1 - \alpha\sqrt{2})^4} < 1 + \epsilon.$$

In other words, the quadratic form on Q varies by a multiple bounded between $1 - \epsilon$ and $1 + \epsilon$ for x in an ellipsoid about z of radius $\Omega(1/m^{1/4})$. But this is precisely the definition of $\bar{r} = \Omega(1/m^{1/4})$. ∎

With $\Lambda(V) \le n$ (Theorem 4.4) and $\delta(V) = \Omega(1/m^{1/2})$ (Theorem 4.9), it now follows by Theorem 2.1 that we can set

$$t^k \approx \left(1 + \frac{1}{m^{1/4}n^{1/2}}\right) t^{k-1}$$

and be assured that x^k can be found in $O(1)$ Newton steps. We can isolate a minimizer when t becomes as large as $2^{O(L)}$. It follows that the number of iterations for the path following algorithm using volumetric centers is $O(m^{1/4}n^{1/2}L)$. This bound is better than the $O(\sqrt{m}\,L)$ bound for the path of analytic centers when $m > n^2$, but worse when $m \le n^2$.

5   The hybrid barrier function

The bound $\Lambda(V) \le n$ for the log determinant barrier function is much better than the bound $\Lambda(\phi) \le m$ for the logarithmic barrier function. Unfortunately, the lower bound $\delta(V) = \Omega(1/m^{1/2})$ is worse than the lower bound $\delta(\phi) = \Omega(1)$—the log determinant barrier has a small uniform ellipsoid radius. The log determinant barrier is therefore 'more non-linear' than the logarithmic barrier. We present now a barrier function that is a hybrid of $\phi(x)$ and $V(x)$. The hybrid will have nearly as good a value of $\Lambda(\cdot)$ as V has, but will damp out some of the non-linearity in V with a small multiple of ϕ to achieve a much better (larger) value of $\delta(\cdot)$.

Let $H(x)$ be the hybrid barrier function defined as

$$H(x) = V(x) + \frac{n}{m}\,\phi(x), \tag{43}$$

and let the path of hybrid centers be defined by

$$\nabla H(x) = t\,c, \quad t \in R. \tag{44}$$

Lemma 5.1 *For every ξ in R^n and every x in the polytope P,*

$$\xi^T(\nabla^2 H(x))^{-1}\xi \le \xi^T(\nabla^2 V(x))^{-1}\xi \quad \text{and} \quad \xi^T(\nabla^2 H(x))^{-1}\xi \le \frac{m}{n}\,\xi^T(\nabla^2\phi(x))^{-1}\xi.$$

Proof: The result follows simply from the definition of $H(x)$, the fact that $(\nabla^2 V(x))^{-1}$ and $(\nabla^2\phi(x))^{-1}$ are positive definite, and Lemma 3.2.   ∎

Theorem 5.2 *For the hybrid barrier function H, we have $\Lambda(H) \le 4n$.*

Proof: For every x in P we have

$$
\begin{aligned}
&\nabla H(x)(\nabla^2 H(x))^{-1}\nabla H(x) \\
&= (\nabla V(x) + \frac{n}{m}\nabla\phi(x))^T(\nabla^2 H(x))^{-1}(\nabla V(x) + \frac{n}{m}\nabla\phi(x)) \\
&= \nabla V(x)^T(\nabla^2 H(x))^{-1}\nabla V(x) + 2\frac{n}{m}\nabla\phi(x)^T(\nabla^2 H(x))^{-1}\nabla V(x) \\
&\quad + \frac{n^2}{m^2}\nabla\phi(x)^T(\nabla^2 H(x))^{-1}\nabla\phi(x) \\
&\leq 2\left(\nabla V(x)^T(\nabla^2 H(x))^{-1}\nabla V(x) + \frac{n^2}{m^2}\nabla\phi(x)^T(\nabla^2 H(x))^{-1}\nabla\phi(x)\right) \qquad (45)\\
&\leq 2\left(\nabla V(x)^T(\nabla^2 V(x))^{-1}\nabla V(x) + \frac{n^2}{m^2}\frac{m}{n}\nabla\phi(x)^T(\nabla^2\phi(x))^{-1}\nabla\phi(x)\right) \quad \text{(by Lemma 5.1)}\\
&= 2\left(\nabla V(x)^T(\nabla^2 V(x))^{-1}\nabla V(x) + \frac{n}{m}\nabla\phi(x)^T(\nabla^2\phi(x))^{-1}\nabla\phi(x)\right) \\
&\leq 2(n + n) \\
&= 4n,
\end{aligned}
$$

where the inequality in (45) follows simply because $2\mathbf{a}\cdot\mathbf{b} \leq \mathbf{a}\cdot\mathbf{a} + \mathbf{b}\cdot\mathbf{b}$ with the substitutions $\mathbf{a} = (n/m)\nabla\phi(x)^T(\nabla^2 H(x))^{-1/2}$ and $\mathbf{b} = \nabla V(x)^T(\nabla^2 H(x))^{-1/2}$. $\blacksquare$

As in the preceding section, one of our goals is to find a uniform ellipsoid radius for our barrier function. We again want to develop a bound on $|a_i^T(x - z)/(a_i^T z - b_i)|$. The next three lemmas give us that bound.

Lemma 5.3 *For every ξ in R^n and every x in the polytope P,*

$$
\xi^T\left(Q(x) + \frac{n}{m}\nabla^2\phi(x)\right)\xi \leq \xi^T\nabla^2 H(x)\xi \leq \xi^T\left(5Q(x) + \frac{n}{m}\nabla^2\phi(x)\right)\xi.
$$

Proof: Follows immediately from the definition of $H(x)$ and (27). $\blacksquare$

Lemma 5.4 *For each i, $1 \leq i \leq m$, and for every $x \in P$,*

$$
\frac{a_i^T(\nabla^2 H(x))^{-1}a_i}{(a_i^T x - b_i)^2} \leq \sqrt{\frac{m}{n}}.
$$

Proof: We have

$$
\begin{aligned}
\frac{a_i^T(\nabla^2 H(x))^{-1}a_i}{(a_i^T x - b_i)^2} &\leq \frac{a_i^T(\nabla^2 V(x))^{-1}a_i}{(a_i^T x - b_i)^2} \quad \text{(Lemma 5.1)} \qquad (46)\\
&\leq \frac{a_i^T(\nabla^2 Q(x))^{-1}a_i}{(a_i^T x - b_i)^2} \quad \text{(see (28))} \qquad (47)
\end{aligned}
$$

and

$$\frac{a_i^T(\nabla^2 H(x))^{-1}a_i}{(a_i^T x - b_i)^2} \leq \frac{m}{n}\frac{a_i^T(\nabla^2 \phi(x))^{-1}a_i}{(a_i^T x - b_i)^2} = \frac{m}{n}\sigma_i(x). \quad \text{(Lemma 5.1)} \tag{48}$$

Thus, using (32) and (36), we can say

$$\frac{a_i^T(\nabla^2 H(x))^{-1}a_i}{(a_i^T x - b_i)^2} \leq \min\{4m\,\sigma_i(x), 1/\sigma_i(x), (m/n)\sigma_i(x)\}. \tag{49}$$

Since $n \geq 1$, the third argument will always be smaller than the first. The second argument is a decreasing function of $\sigma_i(x)$ and the third argument is an increasing function of $\sigma_i(x)$. The two arguments are equal if $\sigma_i(x) = \sqrt{n/m}$. It follows that in all cases

$$\frac{a_i^T(\nabla^2 H(x))^{-1}a_i}{(a_i^T x - b_i)^2} \leq \sqrt{\frac{m}{n}}. \quad \blacksquare$$

Lemma 5.5 *Suppose for some constant $\alpha > 0$ we have $x \in E(\nabla^2 H(z), z, \alpha n^{1/4}/m^{1/4})$. Then for every i,*

$$\left|\frac{a_i^T(x - z)}{a_i^T z - b_i}\right| \leq \alpha.$$

Proof:

$$\begin{aligned}
|a_i^T(x - z)| &\leq \alpha\frac{n^{1/4}}{m^{1/4}}\sqrt{a_i^T(\nabla^2 H(z))^{-1}a_i} \quad \text{(Corollary 3.4)} \\
&\leq \alpha\frac{n^{1/4}}{m^{1/4}}\frac{m^{1/4}}{n^{1/4}}(a_i^T z - b_i) \quad \text{(Lemma 5.4)} \\
&= \alpha(a_i^T z - b_i),
\end{aligned} \tag{50}$$

which implies, upon dividing both sides by $a_i^T z - b_i$ (and recognizing that $a_i^T z - b_i > 0$ by definition),

$$\left|\frac{a_i^T(x - z)}{a_i^T z - b_i}\right| \leq \alpha. \quad \blacksquare \tag{51}$$

With the result of Lemma 5.5 in hand, the calculation of $\delta(H)$ proceeds in a manner very similar to the calculations of $\delta(\phi)$ and $\delta(V)$ in Section 4.

Theorem 5.6 *For the hybrid barrier function H,*

$$\delta(H) = \Omega(n^{1/2}/m^{1/2}).$$

Proof: Again we use Lemma 3.11 and prove the theorem by showing that $H(x)$ has a uniform radius $\bar{r}$ satisfying $\bar{r} = \Omega(n^{1/4}/m^{1/4})$.

The inequalities of Lemma 5.3 have the obvious extension to

$$\xi^T \left(Q(x) + \frac{n}{m} \nabla^2 \phi(x) \right) \xi \leq \xi^T \nabla^2 H(x) \xi \leq 5 \, \xi^T \left(Q(x) + \frac{n}{m} \nabla^2 \phi(x) \right) \xi.$$

(The 5 is pulled outside the parentheses.) Therefore, we have for every $r > 0$,

$$E(Q(z) + \frac{n}{m} \nabla^2 \phi(z), z, r/\sqrt{5}) \subset E(\nabla^2 H(z), z, r) \subset E(Q(z) + \frac{n}{m} \nabla^2 \phi(z), z, r).$$

Just as in the proof of Theorem 4.9, it now suffices to show that the matrix $Q(z) + \frac{n}{m} \nabla^2 \phi(z)$ has uniform radius $\bar{r}$ satisfying $\bar{r} = \Omega(n^{1/4}/m^{1/4})$.

Suppose $x \in E(Q(z) + \frac{n}{m} \nabla^2 \phi(z), z, \alpha n^{1/4}/m^{1/4})$. We have seen in Lemma 5.5 that

$$\left| \frac{a_i^T(x - z)}{a_i^T z - b_i} \right| \leq \alpha. \tag{52}$$

We have dealt with $\nabla^2 \phi(x)$ in Theorem 4.2 and with $Q(x)$ in Theorem 4.9. We can use the constructions in those two proofs along with (52) to conclude

$$\begin{aligned}
\frac{(1-\alpha)^2}{(1+\alpha)^4} &\xi^T Q(z)\xi + \frac{1}{(1+\alpha)^2} \frac{n}{m} \xi^T \nabla^2 \phi(z)\xi \\
&\leq \xi^T \left(Q(x) + \frac{n}{m} \nabla^2 \phi(x) \right) \xi \\
&\leq \frac{(1+\alpha)^2}{(1-\alpha)^4} \xi^T Q(z)\xi + \frac{1}{(1-\alpha)^2} \frac{n}{m} \xi^T \nabla^2 \phi(z)\xi,
\end{aligned} \tag{53}$$

which, of course, implies

$$\begin{aligned}
\frac{(1-\alpha)^2}{(1+\alpha)^4} &\xi^T \left(Q(z) + \frac{n}{m} \nabla^2 \phi(z) \right) \xi \\
&\leq \xi^T \left(Q(x) + \frac{n}{m} \nabla^2 \phi(x) \right) \xi \\
&\leq \frac{(1+\alpha)^2}{(1-\alpha)^4} \xi^T \left(Q(z) + \frac{n}{m} \nabla^2 \phi(z) \right) \xi.
\end{aligned} \tag{54}$$

The remainder of the proof follows exactly the proof of Theorem 4.9 from (42) onward. We conclude that for $\nabla^2 H$ we have a uniform radius $\bar{r}$ satisfying $\bar{r} = \Omega(n^{1/4}/m^{1/4})$.
∎

With $\Lambda(H) \leq 4n$ and $\delta(H) = \Omega(n^{1/2}/m^{1/2})$, it now follows by Theorem 2.1 that we can set

$$t^k \approx \left(1 + \frac{1}{2n^{1/4}m^{1/4}} \right) t^{k-1}$$

and be assured that x^k can be found in $O(1)$ Newton steps. The path following algorithm using hybrid centers thus has number of iterations bounded from above by $O((mn)^{1/4} L)$.

6 Conclusions

The hybrid barrier function allows us to further close the gap between the observed number of iterations needed for interior point linear programming algorithms and the theoretical upper bounds. Unless the experimental work has fortuitously avoided pathological cases, the opportunity for further improvement is substantial.

We believe the quantities $\delta(\cdot)$ and $\Lambda(\cdot)$ provide a useful construct for judging the convergence rates of interior point algorithms based on barrier functions. The relationship between the uniform ellipsoid radius $\bar{r}$ and $\delta(\cdot)$ adds an intuitive feel for the meaning of $\delta(\cdot)$ and, via Lemma 3.11, a useful analytical tool for determining $\delta(\cdot)$.

The most obvious direction for improvement in our results is to find a barrier function with still more beneficial values of $\delta(\cdot)$ and $\Lambda(\cdot)$.

References

[1] I. Adler, N. K. Karmarkar, M. G. C. Resende, and G. Veiga, "An implementation of Karmarkar's algorithm for linear programming," *Mathematical Programming* 44(1989)297–335.

[2] D. S. Atkinson and P. M. Vaidya, "A scaling technique for finding the weighted analytic center of a polytope," *Mathematical Programming* 57(1992)163–192.

[3] D. S. Atkinson, "A cutting plane algorithm that uses analytic centers," chapter in Ph. D. thesis, tech. report UILU-ENG-92-2222 ACT 121, University of Illinois, Coordinated Science Laboratory, 1992.

[4] D. A. Bayer and J. C. Lagarias, "The non-linear geometry of linear programming I: Affine and projective scaling trajectories," *Trans. Amer. Math. Soc.* 314(1989)499–526.

[5] M. Grötschel, L. Lovász, and A. Schrijver, *Geometric algorithms and combinatorial optimization* (Springer-Verlag, Berlin, 1988).

[6] N. K. Karmarkar, "A new polynomial time algorithm for linear programming", *Combinatorica* 4(1984)373–395.

[7] N. K. Karmarkar and K. G. Ramakrishnan, "Further developments in the new polynomial time algorithms for linear programming," Talk at the 12th International Symposium on Mathematical Programming, Boston, Aug. 1985.

[8] N. K. Karmarkar and K. G. Ramakrishnan, "Implementation and computational results of Karmarkar's algorithm for linear programming using an iterative method for computing projections," Technical Memorandum, Mathematical Sciences Center, AT&T Bell Laboratories, Murray Hill, NJ, Nov. 1989.

[9] N. K. Karmarkar and K. G. Ramakrishnan, "Computational results of an interior point algorithm for large scale linear programming," *Mathematical Programming* 52(1991)555–586.

[10] N. Megiddo, ed., *Progress in mathematical programming: interior point and related methods* (Springer-Verlag, Berlin, 1989).

[11] J. E. Nesterov and A. S. Nemirovsky, *Self-concordant functions and polynomial time methods in convex programming*, manuscript, USSR Academy of Sciences (Moscow, 1989).

[12] J. Renegar, "A polynomial-time algorithm, based on Newton's method, for linear programming", *Mathematical Programming* 40(1988)59–93.

[13] D. Shanno and C. Monma, "Computational experience with the primal-dual method," Talk at ORSA/TIMS conference, Washington, D. C., April 1988.

[14] G. Sonnevend, "An analytic center for polyhedrons and new classes of global algorithms for linear (smooth, convex) programming", preprint, Department of Numerical Analysis, Institute of Mathematics, Eötvös University (Budapest, 1989).

[15] P. M. Vaidya, "An algorithm for linear programming which requires $O(((m + n)n^2 + (m + n)^{1.5}n)L)$ arithmetic operations", *Mathematical Programming* 47(1990)175–201.

[16] P. M. Vaidya, "A new algorithm for minimizing convex functions over convex sets," in: Proceedings of 30th annual IEEE symposium on foundations of computer science (IEEE Computer Society Press, Los Alamitos, CA, 1989) pp. 338–343, to appear in *Mathematical Programming*.

[17] P. M. Vaidya, "Reducing the parallel complexity of certain linear programming problems," in: Proceedings of 31st annual IEEE symposium on foundations of computer science (IEEE Computer Society Press, Los Alamitos, CA, 1990), pp. 583–589.

Complexity in Numerical Optimization, pp. 490-500
P.M. Pardalos, Editor
©1993 World Scientific Publishing Co.

Polynomial Time Weak Approximation Algorithms for Quadratic Programming

Stephen A. Vavasis[1]
Department of Computer Science, Cornell University, Ithaca, NY 14853 USA

Abstract

We consider the problem of computing an approximate solution to quadratic programming problems. We focus on algorithms whose running time is polynomial. We show that a solution satisfying weak bounds with respect to optimality can be computed in polynomial time. We also show that for a special case, choosing a random feasible point satisfies the same bound. Recent complexity theory suggests that computing solutions with much stronger bounds cannot be done in polynomial time.

Keywords: Complexity, quadratic programming, approximation algorithms.

1 Nonconvex quadratic programming

Quadratic programming (QP) is a nonlinear optimization problem of the following form:

$$\text{minimize} \quad \tfrac{1}{2}\mathbf{x}^T H\mathbf{x} + \mathbf{h}^T\mathbf{x}$$
$$\text{subject to} \quad W\mathbf{x} \geq \mathbf{b}. \tag{1}$$

In this formulation, $\mathbf{x}$ is the n-vector of unknowns. The remaining variables stand for data in the problem instance: H is an $n \times n$ symmetric matrix, $\mathbf{h}$ is an n-vector, W is an $m \times n$ matrix, and $\mathbf{b}$ is an m-vector. The relation '$\geq$' in the constraint $W\mathbf{x} \geq \mathbf{b}$ is the usual componentwise inequality.

---

[1]Supported by an NSF Presidential Young Investigator Award. Part of this work was also supported in part by the National Science Foundation, the Air Force Office of Scientific Research, and the Office of Naval Research, through NSF grant DMS 8920550.

Quadratic programming, a generalization of linear programming, has applications in economics, planning, and many kinds of engineering design. In addition, more complicated kinds of nonlinear programming problems are often simplified into QP problems (see Gill, Murray and Wright [5]).

No efficient algorithm is known to solve the general case of (1). The lack of an efficient algorithm is not surprising, since QP is known to be NP-hard, a result due to Sahni [14]. More recently, Vavasis [15] showed that the decision version of the problem lies in NP, and hence is NP-complete.

Many avenues for addressing (1) have been pursued in the literature. For example, efficient algorithms are known for the special case in which H is positive semidefinite, known as the *convex* case. See Kozlov, Tarasov and Hačijan [8] for the first polynomial-time algorithm for the convex case. See Kapoor and Vaidya [7] or Ye and Tse [21] for efficient interior point algorithms for this problem.

The traditional approach for nonconvex QP in the optimization literature is to find a "local" solution. In general, however, a local solution does not satisfy any nontrivial bounds in terms of how well it minimizes the objective function; it may be arbitrarily close to the worst feasible point.

Accordingly, in this work we pursue what we call "approximate solutions" that are guaranteed to satisfy some bound on how far off from optimal they are. First it is necessary to give a definition of ϵ-approximation:

Definition 1 *Consider an instance of quadratic programming written in the form (1). Let $f(\mathbf{x})$ denote the objective function $\frac{1}{2}\mathbf{x}^T H\mathbf{x} + \mathbf{h}^T\mathbf{x}$. Let $\mathbf{x}^*$ be an optimum point of the problem. We say that $\mathbf{x}^\circ$ is an ϵ-approximate solution if there exists another feasible point $\mathbf{x}^\#$ such that*

$$f(\mathbf{x}^\circ) - f(\mathbf{x}^*) \leq \epsilon[f(\mathbf{x}^\#) - f(\mathbf{x}^*)].$$

Notice that we may as well take $\mathbf{x}^\#$ in Definition 1 to be the point where the objective function is maximized. Thus, another way to interpret this definition is as follows. Let P denote the feasible region, and let interval $[a, b]$ be $f(P)$. Then $f(\mathbf{x}^\circ)$ should lie in the interval $[a, a + \epsilon(b - a)]$.

Observe that any feasible point is a 1-approximation by this definition, and only an optimum is a 0-approximation. Thus, the definition makes sense only for ϵ in the interval $(0, 1)$. Our definition of approximation also appears in other places such as Nemirovsky and Yudin [11].

This definition has some useful properties. First, it is insensitive to translations or dilations of the objective function. In other words, if the objective function $f(\mathbf{x})$ is replaced by a new function $g(\mathbf{x}) = af(\mathbf{x}) + b$ where $a > 0$, a vector $\mathbf{x}^\circ$ that was previously an ϵ-approximation will continue to have that property. A second useful property is that ϵ-approximation is preserved under affine linear transformations of the feasible region.

In our earlier work [17, 18] we showed the following result: Assume that the feasible region $\{\mathbf{x} : W\mathbf{x} \geq \mathbf{b}\}$ is compact. Let t be the number of negative eigenvalues of H. There is an algorithm to find an ϵ-approximate solution to (1) in

$$O\left(\left\lceil \frac{n(n+1)}{\sqrt{\epsilon}} \right\rceil^t \ell\right)$$

steps. In this formula, ℓ denotes the time to solve a convex quadratic programming problem of the same size as (1).

Unfortunately, if we think of t as being as large as n, this running bound is not polynomial in the input size. Is it possible to construct a polynomial-time approximation algorithm for indefinite QP? Recent results suggest that such an algorithm, if it exists, could only satisfy weak approximation bounds. Specifically, Bellare and Rogaway [1] show the following theorem is true for some constant $\delta > 0$.

Theorem 2 *Suppose there were an algorithm to approximate quadratic programming with* $\epsilon = \left(2^{(\ln n)^\delta} - 1\right) / \left(2^{(\ln n)^\delta} + 1\right)$. *Then any problem in* NP *can be solved in quasi-polynomial time, that is, time* $O(n^{(\ln n)^k})$.

This theorem is based partly on complexity results by Feige et al [3]. Since the concluding statement of the theorem is thought to be unlikely, the supposition is probably false. In other words, we cannot hope to approximate QP in polynomial time unless we are willing to accept an approximation factor that tends to 1 asymptotically as the problem gets larger.

In Section 2 we propose a polynomial-time approximation algorithm for QP that satisfies a weak approximation bound of this sort. In Section 3 we argue that, at least for the special case of simple-bound constraints, a randomly selected feasible point is also an approximation satisfying the same asymptotic bounds

2　A weak approximation in polynomial time

The goal of this section is to prove the following theorem.

Theorem 3 *Assume the feasible set for* (1) *is compact. Then an approximate solution for* (1) *can be computed in polynomial time, where the approximation factor is* $1 - \Theta(n^{-2})$.

The first part of the proof is a sequence of basis changes. First, we test whether the constraint set $\{\mathbf{x} : W\mathbf{x} \geq \mathbf{b}\}$ is full dimensional. This can be done by solving a single linear programming problem as shown by Freund, Roundy and Todd [4]. If not, a linear change of basis lowers the dimension of the problem and ensures without loss of generality that the feasible set is full dimensional.

Let P denote the constraint set $P = \{\mathbf{x} : W\mathbf{x} \geq \mathbf{b}\}$. For the rest of this section, we assume that set P is compact.

The next step is to compute a *weak Löwner-John* pair for set P. Recall that a *Löwner-John* pair for a convex body $P \subset \mathbf{R}^n$ is a pair of concentric ellipsoids E_1, E_2 such that $E_1 \subset P \subset E_2$ and E_1 is obtained from E_2 by shrinking each dimension by $1/n$. Such a pair always exists. A weak Löwner-John pair is defined analogously, except that the shrinking factor is $1/(n+1)$. Lovász [10] shows how to compute a weak Löwner-John pair for a convex body that is a system of linear inequalities in polynomial time.

Let us assume that the interior and exterior ellipsoids are defined by:

$$E_1 = \{\mathbf{x} \in \mathbf{R}^n : (\mathbf{x} - \mathbf{c})^T M (\mathbf{x} - \mathbf{c}) \leq 1\},$$
$$E_2 = \{\mathbf{x} \in \mathbf{R}^n : (\mathbf{x} - \mathbf{c})^T M (\mathbf{x} - \mathbf{c}) \leq (n+1)^2\},$$

where M is a symmetric positive definite matrix, and $\mathbf{c}$ is some n-vector.

The next change of basis is to translate $\mathbf{x}$ by $\mathbf{c}$, and factorize M, thereby making the Löwner-John pair be the two spheres:

$$S_1 = \{\mathbf{x} \in \mathbf{R}^n : \mathbf{x}^T \mathbf{x} \leq 1\},$$
$$S_2 = \{\mathbf{x} \in \mathbf{R}^n : \mathbf{x}^T \mathbf{x} \leq (n+1)^2\}.$$

We will continue to assume that the problem takes the form of (1), but because of all the changes of basis we can assert that $S_1 \subset P \subset S_2$.

Next we minimize the quadratic objective function over S_1. Minimizing a quadratic function over a sphere can be done in polynomial time. In the optimization literature this is known as a "trust region" method and goes back to Levenberg [9] and Marquardt [12]. More recently Ye [20] and Karmarkar [6] both have argued that approximating a solution to this problem can be done in polynomial time, and furthermore, that solving this problem is a useful subproblem for indefinite QP algorithms. Finally, Vavasis and Zippel [19] showed that, when posed as a language-recognition problem, minimization of a quadratic function over a sphere lies in P. (Part of the difficulty here is that the exact solution is usually irrational.)

Let $\mathbf{x}^\circ$ be the solution computed by this algorithm. Since S_1 lies in the feasible set, clearly $\mathbf{x}^\circ$ is itself feasible. Let $\mathbf{x}^\#$ be the worst point (maximizer of the objective function of (1)) on S_1. We claim that

$$f(\mathbf{x}^\#) - f(\mathbf{x}^\circ) \geq \max(\|H\|_2/2, 2\|\mathbf{h}\|_2) \tag{2}$$

where, as above, f is the objective function in (1). Recall that $\|H\|_2$ is equal to the maximum absolute value among the eigenvalues of H. Let λ be this eigenvalue. We assume λ is negative; if not we interchange the role of $\mathbf{v}$ and $\mathbf{0}$ in the chain of inequalities below. Let $\mathbf{v}$ be the corresponding unit-length eigenvector, with its sign selected so that $\mathbf{h}^T\mathbf{v} \leq 0$ (or the opposite if $\lambda > 0$). Then we see that $\mathbf{v} \in S_1$ so

$$
\begin{aligned}
f(\mathbf{x}^\#) - f(\mathbf{x}^\circ) &\geq f(\mathbf{0}) - f(\mathbf{v}) \\
&\geq 0 - \frac{1}{2}\mathbf{v}^T H \mathbf{v} - \mathbf{h}^T \mathbf{v} \\
&\geq -\lambda/2 - \mathbf{h}^T \mathbf{v} \\
&\geq -\lambda/2 \\
&= \|H\|_2/2.
\end{aligned}
$$

This proves one part of (2).

Next, let $\mathbf{u}$ be the unit vector pointing in the same direction as $\mathbf{h}$. Then we see that:

$$
\begin{aligned}
f(\mathbf{x}^\#) - f(\mathbf{x}^\circ) &\geq f(\mathbf{u}) - f(-\mathbf{u}) \\
&\geq \mathbf{h}^T\mathbf{u} - (-\mathbf{h}^T\mathbf{u}) \\
&= 2\|\mathbf{h}\|_2.
\end{aligned}
$$

This proves (2).

Next, let $\mathbf{x}^*$ be the optimum for (1). We claim

$$
f(\mathbf{x}^\#) - f(\mathbf{x}^*) \leq 4(n+1)^2 \max(\|H\|_2/2, 2\|\mathbf{h}\|_2) \tag{3}
$$

In fact we claim more generally that for any $\mathbf{v} \in S_2$,

$$
|f(\mathbf{v})| \leq 2(n+1)^2 \max(\|H\|_2/2, 2\|\mathbf{h}\|_2).
$$

If we could show this, then (3) would obviously follow since $\mathbf{x}^\#, \mathbf{x}^*$ are both feasible and hence both in S_2. To show the latter bound, observe that

$$
\begin{aligned}
|f(\mathbf{v})| &\leq \frac{1}{2}|\mathbf{v}^T H \mathbf{v}| + |\mathbf{h}^T \mathbf{v}| \\
&\leq \frac{1}{2}\|\mathbf{v}\|_2^2\|H\|_2 + \|\mathbf{h}\|_2\|\mathbf{v}\|_2 \\
&\leq \frac{1}{2}(n+1)^2\|H\|_2 + (n+1)\|\mathbf{h}\| \\
&\leq 2(n+1)^2 \max(\|H\|_2/2, 2\|\mathbf{h}\|_2).
\end{aligned}
$$

Now finally, we combine (2) and (3) to obtain:

$$
4(n+1)^2(f(\mathbf{x}^\#) - f(\mathbf{x}^\circ)) \geq f(\mathbf{x}^\#) - f(\mathbf{x}^*).
$$

Rearranging, we obtain:

$$
f(\mathbf{x}^\circ) - f(\mathbf{x}^*) \leq \frac{4(n+1)^2 - 1}{4(n+1)^2}(f(\mathbf{x}^\#) - f(\mathbf{x}^*)).
$$

This proves the theorem.

3   Picking a random point

In this section we show, for a special case, that selecting a random feasible point also yields a $1 - \Theta(n^{-2})$ approximate solution. The special case is minimization of a quadratic function over the unit-volume cube

$$I^n = \{\mathbf{x} : -1/2 \le x_i \le 1/2, i = 1, \ldots, n\}.$$

We remark that the cube-constrained case of QP is still NP-hard—see Vavasis [16].

Let $\mathbf{x}^\circ$ be this randomly selected point. The expected value of $f(\mathbf{x}^\circ)$, where $f(\mathbf{x})$ denotes $\mathbf{x}^T H \mathbf{x}/2 + \mathbf{h}^T \mathbf{x}$, is easily found by integration to be

$$E\left[f(\mathbf{x}^\circ)\right] = (H_{11} + \cdots + H_{nn})/24.$$

Here, H_{ii} denotes the ith diagonal entry of H.

The analysis of the expected quality of $\mathbf{x}^\circ$ is divided into cases. Let QDQ^T be a diagonalization of H, where Q is an orthogonal matrix. Observe that

$$E\left[f(\mathbf{x}^\circ)\right] = (D_{11} + \cdots + D_{nn})/24$$

since similarity transforms do not affect the trace. Recall also that $\|H\|_2$ is the maximum absolute value among $D_{11}, \ldots, D_{nn}$.

Define three quantities:

$$\alpha = \|H\|_2 = \max\{|D_{ii}| : i = 1, \ldots, n\},$$
$$\beta = n(|H_{11}| + \cdots + |H_{nn}|),$$
$$\gamma = (|h_1| + \cdots + |h_n|)/n.$$

We will take three cases depending on which of these three quantities is largest. First, though, we analyze the range of f on the feasible set. Arguments similar to those of the previous section show that for $\mathbf{x} \in I^n$,

$$|f(\mathbf{x})| \le \alpha n/8 + \gamma n/2.$$

Thus, if $\mathbf{x}^*$ is the optimal point and $\mathbf{x}^\#$ is some arbitrary point,

$$f(\mathbf{x}^\#) - f(\mathbf{x}^*) \le \alpha n/4 + \gamma n. \tag{4}$$

Now we take three cases.

Case 1, α is the largest. In this case, let i be the index where $|D_{ii}|$ is maximized. There are two subcases: $D_{ii} \ge 0$ and $D_{ii} < 0$. In Subcase A, if $D_{ii} \ge 0$ then we see by the assumption that $\alpha \ge \beta$:

$$D_{ii} \ge n(|H_{11}| + \cdots + |H_{nn}|). \tag{5}$$

In Subcase B, if $D_{ii} < 0$, define $p = \max\{D_{jj} : j = 1,\ldots,n\}$. Then we have the calculations:

$$H_{11} + \cdots + H_{nn} = D_{11} + \cdots + D_{nn}$$

so

$$H_{11} + \cdots + H_{nn} \le D_{ii} + (n-1)p$$

i.e.,

$$-H_{11} - \cdots - H_{nn} \ge -D_{ii} - (n-1)p.$$

On the other hand, because $\alpha \ge \beta$ for this case and $D_{ii} < 0$, we have:

$$-D_{ii}/n \ge |H_{11}| + \cdots + |H_{nn}|.$$

Combining the two previous inequalities gives:

$$-D_{ii}/n \ge -D_{ii} - (n-1)p.$$

Simplifying this last inequality yields:

$$p \ge -D_{ii}/n.$$

Let j be the index such that $p = D_{jj}$. Then we have $D_{jj} \ge -D_{ii}/n$. In particular,

$$D_{jj} \ge |H_{11}| + \cdots + |H_{nn}|. \tag{6}$$

Thus, we see that in either subcase we can assume there is an index j such that (6) holds (in Subcase A take $j = i$; in Subcase B take j as in the last paragraph).

Let $\mathbf{q}$ be the jth column of Q, so that $\mathbf{q}$ is the eigenvector of H corresponding to D_{jj}. Now we let $\mathbf{x}^\# = \pm\mathbf{q}/2$ where the sign is chosen so that $\mathbf{h}^T\mathbf{x}^\# \ge 0$. (Note that this vector is clearly in I^n.) Then we see that

$$\begin{aligned}
f(\mathbf{x}^\#) &= \frac{1}{2}\mathbf{x}^{\#T}H\mathbf{x}^\# + \mathbf{h}^T\mathbf{x}^\# \\
&\ge \frac{1}{8}\mathbf{q}^T Q D Q^T \mathbf{q} \\
&= D_{jj}/8.
\end{aligned}$$

Then

$$\begin{aligned}
f(\mathbf{x}^\#) - E\left[f(\mathbf{x}^\circ)\right] &\ge D_{jj}/8 - (H_{11} + \cdots + H_{nn})/24 \\
&= D_{jj}/12 + D_{jj}/24 - (H_{11} + \cdots + H_{nn})/24 \\
&\ge D_{jj}/12.
\end{aligned} \tag{7}$$

Note that (6) was used to derive the last line.

On the other hand, we have that $nD_{jj} \geq \alpha$. (In Subcase A, $D_{jj} = D_{ii} = \alpha$ and in Subcase B, $D_{jj} \geq -D_{ii}/n = \alpha/n$.) Furthermore, by the assumption for this case, $\alpha \geq \gamma$. Plugging these facts into (4) yields:

$$f(\mathbf{x}^{\#}) - f(\mathbf{x}^{*}) \leq D_{jj} \cdot (5n^2/4).$$

Combining this with (7) and using calculations like those at the end of the previous section, we conclude that $\mathbf{x}^{\circ}$ is expected to be a $1 - \Theta(n^{-2})$ approximation.

Case 2, β is largest. In this case, we choose $\mathbf{x}^{\#}$ "randomly" as follows. If $H_{ii} \leq 0$, then the ith coordinate of $\mathbf{x}^{\#}$ is chosen to be zero. If $H_{ii} > 0$, then the ith coordinate of $\mathbf{x}^{\#}$ is chosen to be $\pm 1/2$, with each choice of sign having probability $1/2$. Then we can easily check that the expected value of $f(\mathbf{x}^{\#})$ is

$$\sum_{\substack{i=1 \\ H_{ii}>0}}^{n} H_{ii}/8,$$

i.e.,

$$\sum_{i=1}^{n} \max(0, H_{ii}/8).$$

Since this is an expected value, there must be one particular choice of $\mathbf{x}^{\#}$ whose objective function value is at least as large. Then we have:

$$\begin{aligned}
f(\mathbf{x}^{\#}) - E\left[f(\mathbf{x}^{\circ})\right] &\geq \sum_{i=1}^{n} \max(0, H_{ii}/8) - \sum_{i=1}^{n} H_{ii}/24 \\
&\geq \sum_{i=1}^{n} |H_{ii}|/24. \\
&= \beta/(24n).
\end{aligned}$$

On the other hand, since β is the largest of α, β, γ, we have from (4):

$$f(\mathbf{x}^{\#}) - f(\mathbf{x}^{*}) \leq \beta \cdot (5n/4).$$

As above we can conclude that $f(\mathbf{x}^{\circ})$ is expected to be a $1 - \Theta(n^2)$ approximation.

Case 3, γ is largest. In this case choose $\mathbf{x}^{\#}$ so that its ith coordinate entry is $\pm 1/2$, where the sign is chosen to agree with the sign of h_i. Then we have:

$$\begin{aligned}
f(\mathbf{x}^{\#}) &= \frac{1}{2}\mathbf{x}^{\#T} H \mathbf{x}^{\#} + \mathbf{h}^{T}\mathbf{x}^{\#} \\
&\geq -\alpha n/8 + |h_1|/2 + \cdots + |h_n|/2 \\
&= -\alpha n/8 + \gamma n/2 \\
&\geq \gamma \cdot (3n/8)
\end{aligned}$$

Furthermore,

$$E\left[f(\mathbf{x}^{\circ})\right] \leq \beta/(24n) \leq \gamma/(24n).$$

Combining,

$$f(\mathbf{x}^{\#}) - E\left[f(\mathbf{x}^{\circ})\right] \geq \gamma \cdot (3n/8) - \gamma/(24n) \geq \gamma n/4.$$

Since $\gamma \geq \alpha$, from (4) we have:

$$f(\mathbf{x}^{\#}) - f(\mathbf{x}^{*}) \leq \gamma \cdot (5n/4).$$

In this case, we can conclude that $\mathbf{x}^{\circ}$ is expected to be a 4/5 approximate solution.

In all three cases, we have shown that the expected value of the randomly chosen point is a $1 - \Theta(n^{-2})$ solution (the last case being much better). If we want to claim that we can build a good randomized approximation algorithm by choosing random points, we further need to argue that if a sequence of points is chosen at random, then with high probability it satisfies these approximate bounds. Standard arguments show that if a random variable chosen from $[0, 1]$ has expected value $1 - \delta$, then with probability at least $\delta/2$ the variable will be less than $1 - \delta/2$. Therefore, we expect to find a $1 - \Theta(n^{-2})$ approximation after $c \cdot n^2$ samples with very high probability.

4   Conclusions

We have argued that there is an approximation algorithm for quadratic programming running in polynomial time and satisfying weak bounds. As mentioned earlier, it does not seem that we can do much better because of complexity results. There is still, however, a gap between our approximation result and the bound of Bellare and Rogaway—it would be useful to close that gap.

We also have argued that for a special case, an extremely simple randomized algorithm returns an approximate solution. It would be interesting to generalize this result from cubes to arbitrary polytopes. For more general polytopes, selecting a random feasible point has been addressed by Dyer, Frieze and Kannan [2]. But we do not know how to compute the expected value of the objective function.

Another open question is as follows. Our approximation algorithms proposed above have an approximation factor that depends on n. If H is positive definite, however, then we can find the optimal solution (a 0-approximation) in polynomial time. Therefore, we might conjecture that the best possible approximation factor actually depends on t, the number of negative eigenvalues of H, rather than n, which is always at least as large as t. Is it possible to find a $1 - \Theta((t+1)^{-2})$ approximate solution in polynomial time?

Finally, we remark that our definition of approximation, although designed for nonlinear programming, can be applied to combinatorial optimization with some interesting effects. For example, the two NP-hard problems, Minimum Vertex Cover and Maximum Independent Set, are generally thought of as "the same problem" as far as NP-hardness goes, since a vertex cover of an undirected graph is the complement of an independent set.

In terms of traditionally-defined approximation algorithms, however, they have very different properties (see, e.g., Papadimitriou and Steiglitz [13]). If we adopt the definition of approximation algorithm proposed here, then the anomaly is resolved and the ϵ-approximate solutions of the problems are in correspondence with each other.

References

[1] M. Bellare and P. Rogaway (1993), "The complexity of approximating a nonlinear program," this volume.

[2] M. Dyer, A. Frieze, and R. Kannan (1989), "A random polynomial time algorithm for approximating the volume of convex bodies," *Proceedings of the 21st Annual ACM Symposium on the Theory of Computing,* pp. 375–381.

[3] U. Feige, S. Goldwasser, L. Lovász, S. Safra, M. Szegedy (1991), "Approximating clique is almost NP-complete," *Proceedings of the 32nd Symposium on Foundations of Computer Science,* pp. 2–12.

[4] R. M. Freund, R. Roundy, and M. J. Todd (1985), "Identifying the set of always active constraints in a system of linear inequalities by a single linear program," Working Paper 1674-85, Sloan School of Management, MIT, Cambridge, MA.

[5] P. E. Gill, W. Murray, and M. H. Wright (1981), *Practical Optimization,* Academic Press, London.

[6] N. Karmarkar (1989), "An interior-point approach to NP-complete problems—part I," in: J. C. Lagarias and M. J. Todd, eds., *Mathematical Developments Arising from Linear Programming,* Contemporary Mathematics v. 114, American Mathematical Society, Providence, RI, pp. 297–308.

[7] S. Kapoor and P. M. Vaidya (1986), "Fast algorithms for convex quadratic programming and multicommodity flows," *Proceedings of the 18th Annual ACM Symposium on Theory of Computing,* pp. 147–159.

[8] M. K. Kozlov, S. P. Tarasov and L. G. Hačijan (1979) "Polynomial solvability of convex quadratic programming," *Doklad Akademii Nauk SSSR* 248:1049–1051. Translated in *Soviet Mathematics Doklady* 20 (1979) 1108–1111.

[9] K. Levenberg (1944), "A method for the solution of certain non-linear problems in least squares," *Quarterly Appl. Math.* 2:164–168.

[10] L. Lovász (1986), *An Algorithmic Theory of Numbers, Graphs and Convexity,* SIAM, Philadelphia.

[11] A. S. Nemirovsky and D. B. Yudin (1983), *Problem Complexity and Method Efficiency in Optimization,* John Wiley and Sons, Chichester, translated by E. R. Dawson from *Slozhnost' Zadach i Effektivnost' Metodov Optimizatsii* (1979).

[12] D. W. Marquardt (1963), "An algorithm for least-squares estimation of nonlinear parameters," *J. SIAM* 11:431–441.

[13] C. H. Papadimitriou and K. Steiglitz (1982), *Combinatorial Optimization: Algorithms and Complexity,* Prentice-Hall, New Jersey.

[14] S. Sahni (1974), "Computationally related problems," *SIAM Journal on Computing* 3:262–279.

[15] S. A. Vavasis (1990), "Quadratic programming is in NP," *Information Processing Letters* 36:73–77.

[16] S. A. Vavasis (1991), *Nonlinear Optimization: Complexity Issues,* Oxford Univ. Press, Oxford.

[17] S. A. Vavasis (1992), "On approximation algorithms for concave quadratic programming," in: C. A. Floudas and P. M. Pardalos, eds., *Recent Advances in Global Optimization,* Princeton Univ. Press, Princeton, NJ, pp. 3–18.

[18] S. A. Vavasis (1992), "Approximation algorithms for indefinite quadratic programming," *Math. Progr.* 57:279–311.

[19] S. A. Vavasis and R. Zippel (1990), "Proving polynomial-time for sphere-constrained quadratic programming," Technical Report 90-1182, Department of Computer Science, Cornell University, Ithaca, New York.

[20] Y. Ye (1992), "On affine scaling algorithms for nonconvex quadratic programming," *Math. Progr.* 56:285–300.

[21] Y. Ye and E. Tse (1989), "An extension of Karmarkar's projective algorithm for convex quadratic programming," *Math. Progr.* 44:157–179.

Complexity in Numerical Optimization, pp. 501-511
P.M. Pardalos, Editor
©1993 World Scientific Publishing Co.

Complexity Results
for a Class of Min-Max Problems
with Robust Optimization Applications

Gang Yu
Department of Management Science and Information Systems, Graduate School of Business, University of Texas, Austin, TX 78712 USA

Panagiotis Kouvelis
The Fuqua School of Business, Duke University, Durham, NC 27706 USA

Abstract

In this paper we discuss the complexity of a class of min-max versions of classical optimization problems such as assignment, minimum cost network flow, transportation, shortest path, knapsack, resource allocation, and flow shop scheduling problem. Such formulations arise naturally in decision environments with significant data uncertainty for which a robustness approach is used to hedge against the worst contingency.

Keywords: Complexity, min-max optimization, robust decisions.

1   Introduction

In this paper we discuss the complexity of a class of min-max problems. A common characteristic of all these problems is that they are motivated by decision making in the presence of significant data uncertainty. Lack of complete knowledge about the random state of nature has been a pervasive characteristic of most applied economic and decision models. For environments with such lack of knowledge and in the presence of considerable input data uncertainty, a *robustness* approach, which assumes inadequate knowledge of the decision maker about the random state of nature and develops a decision that hedges against the worst contingency that may arise, is appropriate (see White [13]). Such an approach leads to min-max formulations, as our discussion below indicates.

Let us look at the following decision problem. The set of our decision variables is denoted by X and D is the set of our input data. The input data uncertainty is described through a set of scenarios S. Each scenario $s \in S$ corresponds to an instance of the input data, which can be realized with some positive (but perhaps unknown) probability. We use the notation D^s to denote the instance of the input data that corresponds to scenario s. Let F_s denote the set of all feasible decisions when scenario s is realized, and suppose that the quality of the decision $X \in F_s$ is evaluated using the performance measure $f(X, D^s)$ (note the dependence of the performance measure on both the decision X and the input data instance). Then, an optimal decision X_s^* given the input data instance D^s must satisfy

$$f(X_s^*, D^s) = \min_{X \in F_s} f(X, D^s) \tag{1}$$

In the literature there exist multiple definitions of decision robustness (see Gupta and Rosenhead [5], Rosenhead et al. [10], Kouvelis et al. [9], Sengupta [11] and Kouvelis and Yu [14]). The definition of decision robustness we will use in this paper, and to which we refer to as *absolute robustness*, is the following:

The absolute robust decision X_R is defined as the one that exhibits the best worst case performance (with respect to the used performance measure $f(\cdot, \cdot)$) among all feasible decisions over all realizable input data scenarios, i.e.

$$\max_{s \in S} f(X_R, D^s) = \min_{X \in \cap_{s \in S} F_s} \max_{s \in S} f(X, D^s) \tag{2}$$

This can be rewritten as the following program:

$$\min\{y | y \geq f(X, D^s), s \in S; X \in \cap_{s \in S} F_s\} \tag{3}$$

We restrict our attention to the nontrivial cases $\cap_{s \in S} F_s \neq \emptyset$. The above concept has widespread applications to managerial decision making, with significant uncertainty, where the manager is judged on a performance measure regardless of the actual realization of operational parameters in his/her environment.

The above discussions motivated the min-max formulations of classical optimization problems. The complexity of such problems has not been studied in the literature and this is the main research task of this paper. We report the computational complexity of the min-max versions of the following problems: (i) assignment, (ii) transportation, (iii) minimum cost network flow, (iv) shortest path, (v) resource allocation, (vi) knapsack, and (vii) flow shop scheduling problem.

Since the 2-partition problem is frequently refered in later sections of this paer, we now give a formal definition.

The 2-partition problem:

Instance: Finite set I and a size $s_i \in Z_+$ for $i \in I$.

Question: Is there a subset $I' \subseteq I$ such that $\sum_{i \in I'} s_i = \sum_{i \in I \setminus I'} s_i$?

It is well known that the 2-partition problem is NP-hard (Karp [8]).

2 The Min-Max Assignment Problem

The Assignment Problem (ASP) is defined as:

$$(ASP) \quad z_{ASP} = \min \sum_{i,j} c_{ij} x_{ij}$$

subject to

$$\sum_i x_{ij} = 1 \qquad j = 1, ..., n$$
$$\sum_j x_{ij} = 1 \qquad i = 1, ..., n$$
$$x_{ij} \in \{0, 1\} \qquad i = 1, ..., n; j = 1, ..., n.$$

(ASP) can be solved in polynomial time (an $O(n^3)$ algorithm was provided by Balinski [2]). The robust version of the assignment problem is as follows.

$$(\overline{ASP}) \quad z_{\overline{ASP}} = \min y$$

subject to

$$y \geq \sum_{i,j} c_{ij}^s x_{ij} \quad s \in S$$
$$\sum_i x_{ij} = 1 \qquad j = 1, ..., n.$$
$$\sum_j x_{ij} = 1 \qquad i = 1, ..., n.$$
$$x_{ij} \in \{0, 1\} \qquad i = 1, ..., n; j = 1, ..., n.$$

where c^s is the cost matrix corresponding to scenario $s \in S$.

Theorem 1 $(\overline{ASP})$ *is NP-hard.*

Proof: We reduce the $(\overline{ASP})$ to the 2-partition problem.

Construct the following 2-scenario ($|S| = 2$) robust assignment problem. Let $n = 2|I|$. Let the cost matrices be:

$$c_{ij}^1 = \begin{cases} s_i & j = i \text{ and } i \leq |I| \\ 0 & j = |I| + i \text{ and } i \leq |I| \\ 0 & |I| + 1 \leq i \leq n \\ \infty & \text{otherwise} \end{cases}$$

$$c_{ij}^2 = \begin{cases} 0 & j = i \text{ and } i \leq |I| \\ s_i & j = |I| + i \text{ and } i \leq |I| \\ 0 & |I| + 1 \leq i \leq n \\ \infty & \text{otherwise} \end{cases}$$

The finite elements in each of the first $|I|$ rows of c^1 (or c^2) take on one of 2 different values. A feasible assignment solution selects one of the two from each row. If $x_{ii} = 1, i \leq |I|$, c^1 contributes a value of s_i to the sum while c^2 contributes 0. On the other hand in the case of $x_{i,i+|I|} = 1$, c^1 contributes a value of 0 and c^2 contributes s_i. Let I' be the set of nonzero finite elements selected from c^1 by an assignment solution. Then $I \setminus I'$ must be the set of nonzero finite elements contributed by c^2. By definition of $(\overline{ASP})$, we have $y \geq \sum_{i \in I'} s_i$ and $y \geq \sum_{i \in I \setminus I'} s_i$. Clearly, there exists a 2-partition if and only if the $(\overline{ASP})$ finds a solution with $z_{\overline{ASP}} = \frac{1}{2} \sum_{i \in I} s_i$.

$\blacksquare$

Corollary 1 *The min-max version of the transportation problem and the min-max version of the minimum cost network flow problem are NP-hard.*

Proof: This is because $(\overline{ASP})$ is a special case of the min-max version of the transportation problem and the well known fact that the transportation problem is a special case of the minimum cost network flow problem.

$\blacksquare$

3   The Min-Max Shortest Path Problem

Given a graph $G = (V, E)$ with a nonnegative length c_e associated with each edge $e \in E$, an origin node $s \in V$ and a destination node $t \in V$, the Shortest Path Problem (SPP) is defined as finding a path of minimum total length from s to t. The (SPP) can be solved in polynomial time (an $O(n^2)$ labeling algorithm has been given by Dijkstra [3]). The robust version of the shortest path problem is referred to as $(\overline{SPP})$, and is defined as finding the minimum of the maximum length from s to t over all possible scenarios where each scenario corresponds to a predetermined set of edge lengths.

Theorem 2 $(\overline{SPP})$ *is NP-hard.*

Proof: We reduce $(\overline{SPP})$ to the 2-partition problem.

Construct the graph shown in figure 1. Dashed lines are zero length edges for crossing from one side to the other without traversing additional distances. Define the 2-scenario $(\overline{SPP})$ lengths for the solid edges as follows: $c_i^1 = s_i, d_i^1 = 0, i \in I$; $c_i^2 = 0, d_i^2 = s_i, i \in I$.

With the above mapping, if a path from s to t passes through a set I' of solid edges from the top, it must pass through $I \setminus I'$ solid edges from the bottom. The total length of the path for scenario 1 is $\sum_{i \in I'} c_i^1 + \sum_{i \in I \setminus I'} d_i^1 = \sum_{i \in I'} s_i$, while for scenario 2 the total length is $\sum_{i \in I'} c_i^2 + \sum_{i \in I \setminus I'} d_i^2 = \sum_{i \in I \setminus I'} s_i$. Thus, a 2-partition

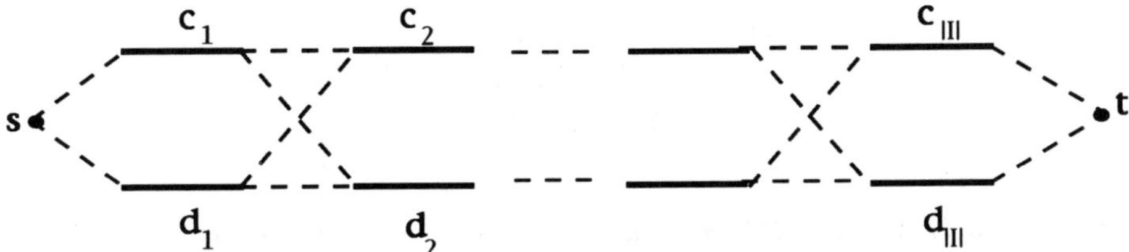

Figure 1: Graph construction for the shortest path problem.

exists if and only if the robust shortest path problem has a solution with total length $z = \frac{1}{2} \sum_{i \in I} s_i$.

∎

4 The Min-Max Resource Allocation Problem

The Resource Allocation Problem (RAP) is defined as follows. N units of a given resource are to be allocated to n activities. The operation of each activity incurs a cost. Let x_i be the amount of the resource allocated to activity i. Let $f_i(x_i)$ be the cost incurred from activity i by allocating x_i units of the resource to activity i. It is desirable to find an optimal allocation of the resource to minimize the total cost. The (RAP) can then be defined by the following nonlinear integer program:

$$(RAP) \quad z_{RAP} = \min \sum_{i=1}^{n} f_i(x_i)$$

subject to

$$\sum_{i=1}^{n} x_i \leq N$$
$$x_i \in Z_+ \qquad i = 1, ..., n.$$

In many applications, the functions $f_i(\cdot), i = 1, ..., n$, are nonincreasing and convex to reflect the fact that the more resources we allocate to an activity, the less cost will be generated and the marginal decrease in cost is diminishing. One such an application is to allocate workers to production lines to minimize total production time.

For nonincreasing convex cost functions $f_i(\cdot), i = 1, ..., n$, the (RAP) can be solved in polynomial time by a simple greedy algorithm (for a detailed discussion of the minisum resource allocation problem see Ibaraki and Katoh [6]).

The robust version of the resource allocation problem is defined as follows.

$$(\overline{RAP}) \qquad z_{\overline{RAP}} = \min y$$

subject to

$$y \geq \sum_{i=1}^{n} f_i^s(x_i) \quad s \in S$$
$$\sum_{i=1}^{n} x_i \leq N$$
$$x_i \in Z_+ \qquad i = 1, ..., n.$$

Theorem 3 $(\overline{RAP})$ *is NP-hard in the strong sense even in the case that all the functions $f(\cdot)$ are linear and decreasing.*

Proof: We reduce the strongly NP-complete Set Covering Problem (SCP) to the $(\overline{RAP})$. Define the set-element incidence matrix for the (SCP) as: $a_{is} = 1$ if element s is covered by (included in) set i; 0 otherwise. The (SCP) tries to answer the question that if there exists a solution x such that

$$\sum_{i=1}^{n} x_i \leq N$$
$$\sum_{i=1}^{n} a_{is} x_i \geq 1 \quad s \in S$$
$$x_i \in Z_+ \qquad i = 1, ..., n.$$

where $x_i > 0$ if set i is selected and 0 otherwise. The above program tries to select no more than N sets to cover all elements of S. Note that the extension of the range of values of x_i from $\{0, 1\}$ to general nonnegative integers will not change the yes/no answer to the problem. This is due to the fact that elements of the set-element incidence matrix can only take values 0 or 1. For a given instance of (SCP), we define the following reduction:

$$f_i^s(x_i) = 1/n - a_{is} x_i$$

Thus $f(\cdot)$ is linear and decreasing. The corresponding $(\overline{RAP})$ is:

$$z_{\overline{RAP}} = \min y$$

subject to

$$1 - \sum_{i=1}^{n} a_{is} x_i \leq y \quad s \in S$$
$$\sum_{i=1}^{n} x_i \leq N$$
$$x_i \in Z_+ \qquad i = 1, ..., n.$$

If there exists a solution with $z_{\overline{RAP}} \leq 0$ for $(\overline{RAP})$ then there exists a solution for (SCP) and vice versa.

∎

5   The Min-Max Knapsack Problem

Let the Knapsack Problem (KP) be

$$(KP) \quad z_{KP} = \min \sum_{i=1}^{n} v_i x_i$$

subject to

$$\sum_{i=1}^{n} a_i x_i \geq b$$
$$x_i \in \{0,1\} \qquad i = 1, ..., n$$

The knapsack problem is well known to be NP-complete, however it can be solved in pseudo-polynomial time $O(nb)$ with the use of a dynamic programming algorithm (see Toth [12]). The robust version of the knapsack problem is defined as

$$(\overline{KP}) \quad z_{\overline{KP}} = \min y$$

subject to
$$y \geq \sum_{i=1}^{n} v_i^s x_i \quad s \in S$$
$$\sum_{i=1}^{n} a_i x_i \geq b$$
$$x_i \in \{0,1\} \quad i = 1, ..., n.$$

Theorem 4 $(\overline{KP})$ *is NP-hard in the strong sense.*

Proof: We reduce the $(\overline{KP})$ to the Weighted Set Packing (WSP) problem. The (WSP) problem is known to be NP-hard in the strong sense (see Garey and Johnson [4]).

The feasibility version of the (WSP) problem can be described as follows.

Instance: Element set J and a collection I of finite subsets of J; weight $w_i \geq 0$ associated with each set $I_i \in I$; number $W \geq 0$.

Question: Does I contain a subcollection of mutually disjoint sets such that the total weight of this subcollection is at least W?

We may also define the (WSP) as follows. Let the set-element incidence matrix be: $\delta_{ij} = 1$ if element j is included in set $I_i \in I$; 0 otherwise. Let $x_i = 1$ if set I_i is selected; 0 otherwise. The (WSP) problem searches for a solution x such that

$$\sum_{i=1}^{|I|} \delta_{ij} x_i \leq 1 \quad j \in J$$
$$\sum_{i=1}^{|I|} w_i x_i \geq W$$
$$x_i \in \{0,1\} \qquad i = 1, ..., |I|$$

Define the following mapping:

$$n = |I|$$
$$S = J$$

$$v_i^s = \delta_{is} \quad i = 1, ..., n; s \in S$$

$$a_i = w_i \quad i = 1, ..., n$$

$$b = W$$

The (WSP) problem has a feasible solution if and only if the $(\overline{KP})$ has a solution with $z_{\overline{KP}} \leq 1$.

$\blacksquare$

6   The Min-Max Flow Shop Scheduling Problem

Consider the permutation Flow Shop Scheduling (FSS) problem $PFm||C_{max}$. A set $\{1, ..., n\}$ of independent jobs require processing on m machines. All jobs have to pass through all the machines according to a prespecified machine sequence and jobs pass through machines in the same order. The goal is to minimize the makespan C_{max}. The $PF2||C_{max}$ problem can be solved efficiently and a $O(n \log n)$ algorithm was provided by Johnson [7] which is considered as the first major result in scheduling theory (for a textbook exposure of scheduling theory see Baker [1]). However, the $PFm||C_{max}$ problem for $m \geq 3$ is well known NP-hard [4].

The robust version of the (FSS) problem is refered to as $(\overline{FSS})$, and is defined as finding the minimum makespan over all possible scenarios, i.e. $z_{\overline{FSS}} = \min_\pi \max_{s \in S} C_{max}^s(\pi)$, where $C_{max}^s(\pi)$ is the makespan for scenario s and for a given job permutation π. The minimization is over all possible job permutations π. Each scenario is fully described by the processing time vector defined below.

Denote $P_i^s = (p_{i1}^s, p_{i2}^s, ..., p_{im}^s)$ as the processing time vector, where p_{ij}^s is the processing time of job i on machine j under scenario s.

Theorem 5 $(\overline{FSS})$ *is NP-hard even for* $m = 2$.

Proof: We reduce the $(\overline{FSS})$ problem to the 2-partition problem.

For a given finite set I and a size $s_i \in Z^+$ for $i \in I$, construct the $(\overline{FSS})$ problem with 2 machines, 2 scenarios, $n + 1$ jobs with job J_0 and jobs $J_i, i \in I = \{1, ..., n\}$. Construct the following processing time vector:

$$P_0^1 = (\frac{1}{2} \sum_{i=1}^{n} s_i, \frac{1}{2} \sum_{i=1}^{n} s_i)$$

$$P_0^2 = (0, \sum_{i=1}^{n} s_i)$$

$$P_i^1 = (0, s_i) \quad i = 1, ..., n$$

$$P_i^2 = (s_i, 0) \quad i = 1, ..., n.$$

We claim that there exists a 2-partition $\sum_{i \in I'} s_i = \sum_{i \in I \setminus I'} s_i$ with $I' \subset I$ if and only if the $(\overline{FSS})$ problem has a solution $z_{\overline{FSS}} = \frac{3}{2} \sum_{i \in I} s_i$.

Assume that a set of $I_0 \subseteq I$ jobs are scheduled before job J_0.

Case 1: $\sum_{i \in I_0} s_i > \frac{1}{2} \sum_{i \in I} s_i > \sum_{i \in I \setminus I_0} s_i$. For scenario $s = 1$, we have $C_{max}^1 = \sum_{i \in I_0} s_i + \frac{1}{2} \sum_{i \in I} s_i + \sum_{i \in I \setminus I_0} s_i = \frac{3}{2} \sum_{i \in I} s_i$. For scenario $s = 2$, we have $C_{max}^2 = \sum_{i \in I_0} s_i + \sum_{i \in I} s_i > \frac{3}{2} \sum_{i \in I} s_i$. Thus $z_{\overline{FSS}} = \max\{C_{max}^1, C_{max}^1\} = C_{max}^2 > \frac{3}{2} \sum_{i \in I} s_i$.

Case 2: $\sum_{i \in I_0} s_i < \frac{1}{2} \sum_{i \in I} s_i < \sum_{i \in I \setminus I_0} s_i$. For scenario $s = 1$, we have $C_{max}^1 = \frac{1}{2} \sum_{i \in I} s_i + \frac{1}{2} \sum_{i \in I} s_i + \sum_{i \in I \setminus I_0} s_i > \frac{3}{2} \sum_{i \in I} s_i$ and $C_{max}^2 = \sum_{i \in I_0} s_i + \sum_{i \in I} s_i < \frac{3}{2} \sum_{i \in I} s_i$. Thus $z_{\overline{FSS}} = C_{max}^1 > \frac{3}{2} \sum_{i \in I} s_i$.

We are left with case 3 where $\sum_{i \in I_0} s_i = \sum_{i \in I \setminus I_0} s_i = \frac{1}{2} \sum_{i \in I} s_i$. This case will give the desired solution $z_{\overline{FSS}} = \frac{3}{2} \sum_{i \in I} s_i$. The Gantt chart in figure 2 illustrates the schedule.

∎

7 Remarks

From the results in previous sections, we see that several classes of polynomially solvable discrete optimization problems are NP-hard in their corresponding min-max versions. We conjecture that all robust discrete optimization problems are NP-hard in general, efficient polynomial algorithms are possible only for some special cases, i.e. special scenario classes. We expect to answer the question through our future work that by placing what type of restrictions to the scenarios, the min-max optimization problems are polynomially solvable. In [14] we will report results on a general algorithmic framework for solving robust discrete optimization problems.

References

[1] Baker, K.R. (1974), *Introduction to Sequencing and Scheduling*, Wiley, New York.

[2] Balinski, M.L. (1985), "Signature Methods for the Assignment Problem", *Operations Research* 33, 527-536.

[3] Dijkstra, E.W. (1959), "A Note on Two Problems in Connection with Graphs", *Numerische Mathematik*, 1, 269-271.

[4] Garey, M.R. and D.S. Johnson (1979), *Computers and Intractability*, W.H. Freeman, San Francisco.

[5] Gupta, S.K. and J. Rosenhead (1972), "Robustness in Sequential Investment Decisions", *Management Science* 15, 2, 18-29.

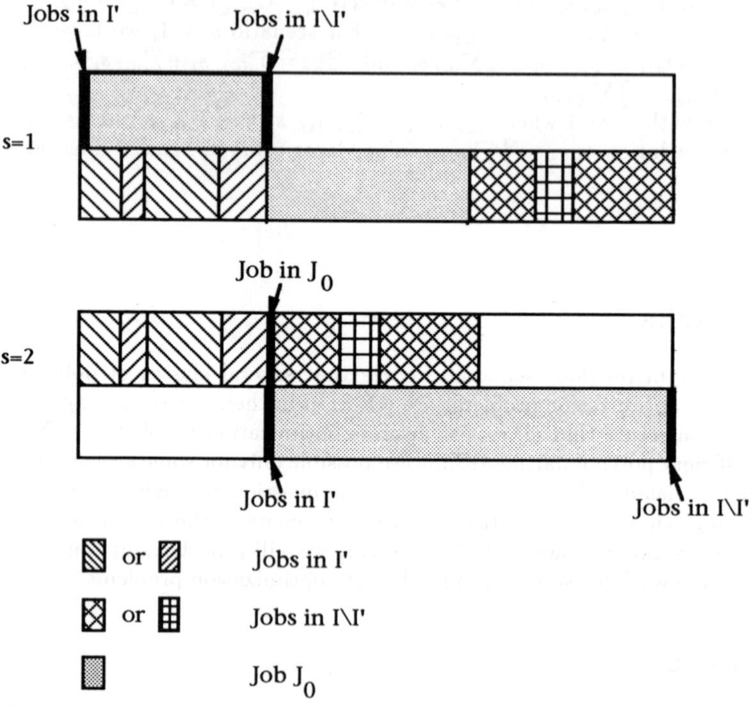

Figure 2: Gantt chart for the NP-hard proof of $(\overline{FSS})$ problem.

[6] Ibaraki, T. and N. Katoh (1988), *Resource Allocation Problems: Algorithmic Approaches*, the MIT Press, Cambridge, Massachusetts.

[7] Johnson, S.M. (1954), "Optimal Two- and Three-stage Production Schedules with Setup Times Included", *Naval Res. Logist. Quart.* 1, 61-68.

[8] Karp, R.M. (1972), "Reducibility among Combinatorial Problems", in R.E. Miller and J.W. Thatcher (eds.), *Complexity of Computer Communications*, Plenum Press, NY, 85-103.

[9] Kouvelis, P., A.A. Karawarwala and G.J. Gutierrez (1992), "Algorithms for Robust Single and Multiple Period Layout Planning for Manufacturing Systems", *European Journal of Operational Research*, 63, 2, 287-303.

[10] Rosenhead, J., M. Elton and S.K. Gupta (1972), "Robustness and Optimality as Criteria for Strategic Decisions", *Operational Research Quarterly* 23, 4, 413-430.

[11] Sengupta, J.K. (1991), "Robust Decisions in Economic Models", *Computers and Operations Research* 18, 2, 221-232.

[12] Toth, P. (1980), "Dynamic Programming Algorithms for the Zero-one Knapsack Problem", *Computing* 25, 29-45.

[13] White, D.J. (1976), *Fundamentals of Decision Theory*, North Holland: Amsterdam.

[14] Kouvelis P. and G. Yu, "Discrete Robust Optimization Models", in preparation.